W9-CLW-984

de Gruyter Expositions in Mathematics 16

de Gruyter Expositions in Mathematics

1 The Analytical and Topological Theory of Semigroups, *K. H. Hofmann, J. D. Lawson, J. S. Pym (Eds.)*

2 Combinatorial Homotopy and 4-Dimensional Complexes, *H. J. Baues*

3 The Stefan Problem, *A. M. Meirmanov*

4 Finite Soluble Groups, *K. Doerk, T. O. Hawkes*

5 The Riemann Zeta-Function, *A. A. Karatsuba, S. M. Voronin*

6 Contact Geometry and Linear Differential Equations, *V. R. Nazaikinskii, V. E. Shatalov, B. Yu. Sternin*

7 Infinite Dimensional Lie Superalgebras, *Yu. A. Bahturin, A. A. Mikhalev, V. M. Petrogradsky, M. V. Zaicev*

8 Nilpotent Groups and their Automorphisms, *E. I. Khukhro*

9 Invariant Distances and Metrics in Complex Analysis, *M. Jarnicki, P. Pflug*

10 The Link Invariants of the Chern-Simons Field Theory, *E. Guadagnini*

11 Global Affine Differential Geometry of Hypersurfaces, *A.-M. Li, U. Simon, G. Zhao*

12 Moduli Spaces of Abelian Surfaces: Compactification, Degenerations, and Theta Functions, *K. Hulek, C. Kahn, S. H. Weintraub*

13 Elliptic Problems in Domains with Piecewise Smooth Boundaries, *S. A. Nazarov, B. A. Plamenevsky*

14 Subgroup Lattices of Groups, *R. Schmidt*

15 Orthogonal Decompositions and Integral Lattices, *A. I. Kostrikin, P. H. Tiep*

The Adjunction Theory of Complex Projective Varieties

by

Mauro C. Beltrametti
Andrew J. Sommese

Walter de Gruyter · Berlin · New York 1995

Authors

Mauro C. Beltrametti
Dipartimento di Matematica
Università di Genova
I-16132 Genova
Italy
e-mail: beltrame@dima.unige.it

Andrew J. Sommese
Department of Mathematics
University of Notre Dame
Notre Dame, IN 46556-0398
U.S.A.
e-mail: sommese.1@nd.edu

1991 Mathematics Subject Classification: 14-02; 14C20, 14J40, 14N99, 14E99

Keywords: Algebraic geometry, projective varieties, adjunction theory

∞ Printed on acid-free paper which falls within the guidelines of the ANSI to ensure permanence and durability.

Library of Congress Cataloging-in-Publication Data

Beltrametti, Mauro, 1948–
The adjunction theory of complex projective varieties / by Mauro Beltrametti, Andrew Sommese.
p. cm. – (De Gruyter expositions in mathematics ; 16)
Includes bibliographical references and index.
ISBN 3-11-014355-0 (acid-free)
1. Adjunction theory. 2. Embeddings (Mathematics). 3. Algebraic varieties. 4. Projective spaces. I. Sommese, Andrew John. II. Title. III. Series.
QA564.B443 1994
516.3′5–dc20 94-27981
CIP

Die Deutsche Bibliothek – Cataloging-in-Publication Data

Beltrametti, Mauro C.:
The adjunction theory of complex projective varieties / by Mauro C. Beltrametti ; Andrew J. Sommese. – Berlin ; New York : de Gruyter, 1995
(De Gruyter expositions in mathematics ; 16)
ISBN 3-11-014355-0
NE: Sommese, Andrew John:; GT

Printed in Germany.
Typeset using the authors' LATEX files: Lewis & Leins, Berlin. Printing: Gerike GmbH, Berlin. Binding: Lüderitz & Bauer GmbH, Berlin. Cover design: Thomas Bonnie, Hamburg.

To Barbara and Rossana Beltrametti
To Rebecca Rooze DeBoer Sommese

Preface

The goal of this book is to cover the active developments of adjunction theory in the last 15 years. In the following we give a more precise definition of adjunction theory, and flesh the definition out with a brief survey of the subject, starting with classical work of the nineteenth century and leading up to the modern work.

Adjunction theory is the study (mainly by means of appropriate adjoint bundles) of the interplay between the intrinsic geometry of a projective variety and the geometry connected with some "embedding" of the variety into a projective space. To make this precise we need some definitions. A line bundle, $\mathcal{L}$, on a projective variety, X, is said to be very ample if there is an embedding, $\phi : X \to \mathbb{P}_{\mathbb{C}}$, of X into a complex projective space, such that $\mathcal{L} \cong \phi^*\mathcal{O}_{\mathbb{P}_{\mathbb{C}}}(1)$, where $\mathcal{O}_{\mathbb{P}_{\mathbb{C}}}(1)$ denotes the hyperplane section bundle on $\mathbb{P}_{\mathbb{C}}$. A line bundle is said to be ample, if some positive power of the line bundle is very ample. The basic objects of study in adjunction theory are pairs, $(X, \mathcal{L})$, where $\mathcal{L}$ is an ample or very ample line bundle on a projective variety, X. For the more precise results X satisfies some restrictions on its singularities. Centering on manifolds exclusively is not practical since the study of manifolds quickly leads to singular varieties. On the other hand results become fragmentary and very technical, and eventually fail to hold as the conditions on the singularities and ampleness are relaxed.

For simplicity we will assume throughout this introduction that, unless otherwise specified, X is a smooth, connected, n-dimensional projective manifold, and $\mathcal{L}$ is a very ample line bundle on X. In this case we can consider $X \subset \mathbb{P}_{\mathbb{C}}$ by using the embedding of X associated to $\mathcal{L}$. There is a rich geometry connected with this embedding, e.g., there are the intersections of X with linear subspaces of $\mathbb{P}_{\mathbb{C}}$, and the discriminant variety, $\mathcal{D} \subset |\mathcal{L}|$, of singular elements in $|\mathcal{L}|$. (Here $|\mathcal{L}|$, the complete linear system of $\mathcal{L}$, is the set of zero sets of not identically zero sections of $\mathcal{L}$. It has the natural structure of a projective space.) The geometrical invariants of these associated objects, which are typically combinations of products of Chern classes of the tangent bundle of X with powers of the 1-st Chern class of $\mathcal{L}$ give a large class of invariants of the pairs, $(X, \mathcal{L})$. These invariants usually only depend on Chern classes, and thus if $\mathcal{L}$ is merely ample, the invariants make sense, even though the associated geometrical objects which motivated their definitions might well not exist. The most precise results of adjunction theory, and indeed the results from adjunction theory most used by researchers in other parts of algebraic geometry, require very ampleness of $\mathcal{L}$. Since every projective variety possesses ample and very ample line bundles, almost by definition of what it means to be projective, the class of objects studied by adjunction theory is as broad as algebraic geometry.

The special role of "adjoint" bundles became apparent early in the history of the theory. Over a century ago Castelnuovo and Enriques were studying projective surfaces by relating the geometry of a surface, $S \subset \mathbb{P}_{\mathbb{C}}$, to the geometry of its curve sections, i.e., the curves obtained by intersecting S with hyperplanes of $\mathbb{P}_{\mathbb{C}}$. This was very natural because most projective varieties come with an explicit embedding into a projective space. The survey [148] by Castelnuovo and Enriques, which first aroused the second author's interest in hyperplane sections of projective varieties, gives a good idea of the methods available and results obtained up to about 1900. Given a smooth curve, C, of genus $g \geq 1$, the canonical bundle, K_C, which for curves is just the holomorphic cotangent bundle, is spanned at all points by global sections. The map associated to the sections of K_C gives a map into $\mathbb{P}^{g-1}$ which is called the canonical mapping. It is an embedding except for the case $g = 1$ when the map is onto a point, and the case of hyperelliptic curves, i.e., double covers of $\mathbb{P}^1$ of genus at least 2. For hyperelliptic curves the canonical map is just the hyperelliptic double cover map onto a smooth rational curve of degree $g - 1$, the minimal possible degree for an irreducible curve in $\mathbb{P}^{g-1}$ not contained in any proper linear subspace of $\mathbb{P}^{g-1}$. Much of the precise theory of curves is in terms of the canonical map. Therefore it was and is very natural to try relating the geometry of a smooth surface to the geometry of the curve, by patching all the canonical maps together into some map of the surface.

Before describing how to do this patching of the maps, it is important to note that many of the concepts and results we use now existed in more primitive forms in the period when Castelnuovo and Enriques were active. For example, though line bundles were not used, the roughly equivalent concept of a linear system was used. Even though sheaf cohomology did not exist, the superabundance, which equals the dimension of the first cohomology group, was a well-known number in this period, and Picard's famous theorem on the "regularity of the adjoint" stated that the superabundance was zero for $|K_S + \mathcal{L}|$ where $\mathcal{L}$ is a very ample line bundle on a surface, S. This is of course the precursor of the modern vanishing theorems, and it was used in the same ways as we use the vanishing theorems. So though we use modern terminology, which is broader in scope and more flexible, we can for the most part recast the description into classical language.

The patching of the maps associated to the canonical maps of the curves, $C \in |\mathcal{L}|$, is done by means of the adjunction bundle, $K_S + \mathcal{L}$. In a surprisingly comprehensive way the geometry of all the curves in $|\mathcal{L}|$ is reflected in the geometry of $|K_S + \mathcal{L}|$. If $S \subset \mathbb{P}_{\mathbb{C}}$ is a projective surface and $\mathcal{L} := \mathcal{O}_{\mathbb{P}_{\mathbb{C}}}(1)_S$, then except for $\mathbb{P}^1$-bundles over curves with the restriction, $\mathcal{L}_f \cong \mathcal{O}_{\mathbb{P}^1}(1)$, to a fiber

$f \cong \mathbb{P}^1$, $K_S + \mathcal{L}$ is the unique line bundle on S whose restriction to every smooth curve, $C \in |\mathcal{L}|$, is isomorphic to K_C, the canonical bundle of C.

From the period surveyed by Castelnuovo and Enriques in [148] date the first results "characterizing" surfaces when their curve sections are of genus ≤ 3 or hyperelliptic [216, 145]. These results were often only birational since it was not known that $K_S + \mathcal{L}$ was spanned at all points by global sections. Nevertheless because of the Riemann-Roch theorem for surfaces, and Picard's regularity of the adjoint theorem, geometers of the late 19-th century could compute the number of sections of adjoint bundles, and work out the behavior of the rational map associated to these sections by restricting to the curves in $|\mathcal{L}|$. The adjunction process, which consisted of replacing a linear system $|\mathcal{L}|$ with its iterates $|K_S + \mathcal{L}|$, $|2K_S + \mathcal{L}|$, etc., came to occupy a central place in the classical theory. If S is of negative Kodaira dimension, it is a basic and nontrivial result that if $\mathcal{L}$ is any line bundle on S, then for all sufficiently large positive integers, t, $tK_S + \mathcal{L}$ has no sections, i.e., "adjunction terminates." When we start the process with a very ample line bundle, $\mathcal{L}$, the last stage of the adjunction process before it terminates gives a rational mapping which lays out much of the structure of the original pair, $(S, \mathcal{L})$.

A very clear picture of the scope of these techniques is given in Roth's excellent 1937 article [526], which stands at the end of the classical period. Among other things Roth obtained a birational classification of surfaces in projective space of degree ≤ 10 or with a hyperplane section of genus ≤ 6. It was by seeing clearly his methods that the second author realized that the birational classical theory could be improved to a more precise biregular theory if $K_S + \mathcal{L}$ was spanned except for the obvious exceptions. This now complete part of adjunction theory on the spannedness and very ampleness of $K_X + (n-1)\mathcal{L}$ for all n is called *classical adjunction theory*. It contains many far reaching, beautiful results, which give a precision to the adjunction process it did not have classically, and allow establishment along classical lines of the biregular versions of the classical classification theorems. Also the results are much stronger, e.g., where classically all smooth curves in a very ample linear system might be assumed to have some property, such as being hyperelliptic, only the assumption that at least one has the property, such as being hyperelliptic, is needed in the modern work.

Classical adjunction theory has proved useful to many mathematicians. For this reason we have given a complete development of it. It is also important as the suggestive model for conjectured results for other adjunction bundles such as $K_X + (n-2)\mathcal{L}$, where the role of the canonical map of a curve is replaced by the more intricate canonical mapping of a surface. Generalizing the spannedness and very ampleness results known for the bundles $K_X + (n-1)\mathcal{L}$ to the bundle, $K_X + (n-2)\mathcal{L}$, is a major problem. We cover the main known results in this direction and discuss the working conjectures.

We have given the history of the theory, which was contributed to by many mathematicians, in Chapters 8, 9, 10, 11, and 12.

In Chapter 8, we develop the background material used in classical adjunction theory.

In Chapter 9 we prove the spannedness of $K_X + (n-1)\mathcal{L}$ for an ample and spanned $\mathcal{L}$ on a normal Gorenstein projective variety, X, with $\mathrm{cod}_X \mathrm{Sing}(X) \geq 3$ and the set of nonrational singularities at most finite. Here we follow papers by Andreatta and Sommese [24], by Lanteri, Palleschi, and Sommese [382], by Sommese [581, 582], and by Sommese and Van de Ven [586].

In Chapters 10 and 11 we thoroughly investigate very ampleness properties of $K_X + (n-1)\mathcal{L}$ for a very ample line bundle, $\mathcal{L}$, on a projective manifold, X. Here we follow closely the second author's paper [571] which started the modern work on adjunction theory and the paper [586] of the second author and A. Van de Ven where optimal results are proven (see also Serrano [554]). It is now known that either $(X, \mathcal{L})$ is one of a short list of very special and well understood examples or $K_X + (n-1)\mathcal{L}$ is spanned by global sections. In this case the map, $\psi : X \to \mathbb{P}_{\mathbb{C}}$, given by sections of $K_X + (n-1)\mathcal{L}$ is called the adjunction mapping and except for a short explicit list of exceptions, ψ expresses X as the blowup at a finite set, B, of a projective manifold, $X' := \psi(X)$. There is a natural ample line bundle, $\mathcal{L}'$, on X', defined as $\mathcal{L}' := (\psi_*\mathcal{L})^{**}$. It turns out that $\mathcal{L}' \cong \mathcal{O}_X(A')$ where A' is the image of a smooth $A \in |\mathcal{L}|$. Indeed there is a one-to-one correspondence between smooth $A \in |\mathcal{L}|$ and smooth $A' \in |\mathcal{L}'|$ that contain B. Also $K_X + (n-1)\mathcal{L} \cong \psi^*(K_{X'} + (n-1)\mathcal{L}')$ with $K_{X'} + (n-1)\mathcal{L}'$ ample. Except for four simple examples with $n = 2$ and degree 7, 8, and 9, $K_{X'} + (n-1)\mathcal{L}'$ is also very ample. Explicitly the positive dimensional fibers of ψ are linear $\mathbb{P}^{n-1}$'s with respect to $\mathcal{L}$. The pair, $(X', \mathcal{L}')$, called the first reduction, or the relative minimal model, of $(X, \mathcal{L})$, is very closely related to $(X, \mathcal{L})$ and has much better properties than $(X, \mathcal{L})$. In general going from $(X, \mathcal{L})$ to $(X', \mathcal{L}')$, the invariants change in such a way that the restrictions imposed by such results, as the Hodge index theorem and Castelnuovo's bound on the genus of a curve section in terms of degree of the curve section, become more stringent.

Chapters 8 and 10 also include a discussion of the recent generalizations for surfaces of these results when $\mathcal{L}$ is k-very ample. A line bundle is k-very ample if given any length $k+1$, 0-dimensional subscheme, $Z \subset S$, it follows that the restriction map, $\Gamma(\mathcal{L}) \to \Gamma(\mathcal{L} \otimes \mathcal{O}_Z)$, is onto. A line bundle is 1-very ample (respectively 0-very ample) if and only if it is very ample (respectively spanned at all points by global sections). We explain the close relations to the study of adjoint bundles of very ample line bundles on the Hilbert scheme of 0-dimensional subschemes of length k on a smooth surface.

In Chapter 11 a number of applications of the theory are given. In §11.4 there is a description of the pairs $(X, \mathcal{L})$ with $\mathcal{L}$ very ample on a projective manifold, X, and such that there is a hyperelliptic curve on X equal to the transversal intersection of $n-1$ elements of $|\mathscr{L}|$. A short discussion of the projective normality of adjoint bundles is given in §11.5. A classification of the projective manifolds with curve sections of genus ≤ 4 is given in §11.6.

In Chapter 12 we work out the structure of the second reduction for an ample and spanned line bundle on a connected projective manifold of dimension 3. These results complement the much simpler structure results for the second reduction in dimension greater than or equal to 4 developed in Chapter 7. They imply very strong restrictions which a general type surface must satisfy if it is a very ample divisor on a smooth projective 3-fold. For example, assume that S is a general type surface which is a very ample divisor on a smooth projective 3-fold, X. If there is a configuration of smooth rational curves of self-intersection -2 on the minimal model of S which is neither A_1, A_2, A_3, nor D_4, then X is a $\mathbb{P}^1$-bundle and S is a meromorphic section.

There is a large number of general results used in adjunction theory on how certain properties of an ample divisor are shared by the ambient manifold, with the relationship getting closer as the dimension increases. Many of these results are included in Chapters 1–4, which cover background material which is usually not part of a first course in algebraic geometry. Much of the material is scattered throughout the literature. Though full proofs would take us too far afield, we usually give proofs reducing to standard material. Some of the material here, especially the theory of the Δ-genus, low degree classification in the ample case, and the theory of Del Pezzo varieties, is covered more thoroughly in the very useful book of Fujita [257].

Here in Chapters 1–4 for the first time we had to make decisions on what to include. When possible we have tried to assume at least terminal singularities, unless the work is not too much more to prove results with less restrictions on the singularities. We have made an exception to this policy for those basic results, e.g., vanishing theorems, which are used everywhere in the theory and will be important in future research. When the irrational singularities of a variety become positive dimensional, simple and very well-behaved examples (see (2.2.10)) show that the vanishing theorems fail and the main structure theorems only hold in a weak birational sense. In §1.7 we have included a number of results on the behavior of singularities and varieties when passing to appropriately general Cartier divisors.

Though our main goal is to cover the mainstream of the subject, we also want to cover the basics of topics where work on the subject is or likely will be developing. For this reason we have paid some attention to covering such topics as k-ampleness and k-very ampleness, and at least pointing out in places what results analogous to the usual results for ample divisors are true.

We could have avoided introducing Rossi's extension theorems in §2.6 and results of Andreotti-Grauert and Griffiths in §2.7 on cohomology of tubular neighborhoods of varieties for the applications in which we used them in Chapter 5.

We include them because we feel that these results give an important insight about ample divisors and are likely to be useful in other future work in adjunction theory.

The Hilbert scheme and deformation theory pose problems for anyone writing an advanced book in algebraic geometry. It is a real gap in the literature that no book exists giving motivation and proofs of the results everyone uses. We have settled for summarizing what we need in the form we need it with references to the literature.

Some applications of these results show that most projective varieties cannot (except in trivial ways) be ample divisors, and the varieties that can be ample divisors practically determine the varieties they are ample divisors on. We develop results of this sort throughout the book as needed, but especially in Chapter 5, where we study the restrictions imposed by the existence of some special geometrical property such as a morphism of an ample divisor. One novelty is that we allow worse singularities than are found in papers in the literature, and we prove many of the results for more general k-ample line bundles, i.e., line bundles $\mathcal{L}$ such that there is some $N > 0$ with $|N\mathcal{L}|$ basepoint free, and with the associated morphism having at most k-dimensional fibers. We sketch in §5.4 the main known results, due for the most part to Bădescu, Fujita, L'vovsky, Wahl, and Zak on when a projective variety can only be an ample divisor on a cone. Though not used elsewhere in the book, these results illuminate some aspects of adjunction theory.

The natural approach to the use of the adjunction process in the higher dimensions is to replace $\mathcal{L}$ by $K_X+\mathcal{L}$ and by iteration with $tK_X+\mathcal{L}$, for positive integers t. Assume for the moment that $n = 3$. Here Fano used this adjunction process, but there are many difficulties with making the obtained results rigorous. Morin's approach was to lift the adjunction process on surfaces by using $(K_X+\mathcal{L})+\mathcal{L}$ and by iteration $t(K_X+\mathcal{L})+\mathcal{L}$. This approach is very fruitful. Notice that on a surface section $S \in |\mathcal{L}|$, $(tK_X+\mathcal{L})_S \cong tK_S-(t-1)\mathcal{L}_S$, while $(t(K_X+\mathcal{L})+\mathcal{L})_S \cong tK_S+\mathcal{L}_S$. In the limit as $t \to \infty$ Morin's approach yields information about $K_X + \mathcal{L}$. The Fano-Morin approach is discussed briefly in §11.7, and more thoroughly in Roth's book [527]. Reading [527] led the second author to carry through the Fano-Morin program for a smooth ample divisor on a 3-fold in [575, 576, 577]. The key point here is that if the Kodaira dimension of $K_X+\mathcal{L}$ is negative then the Morin adjunction process terminates and gives a structure theorem for $(X, \mathcal{L})$ at the last stage before it terminates. One corollary of this approach which gives its flavor is the result that if A is a smooth surface of nonnegative Kodaira dimension which is an ample divisor on a smooth threefold, X, then either X is a $\mathbb{P}^1$-bundle, $\phi : X \to Y$, over a smooth surface, Y, with A a meromorphic section of ϕ or the first reduction $(X', \mathcal{L}')$ of $(X, \mathcal{O}_X(A))$ exists, some positive power of $K_{X'} + \mathcal{L}'$ is spanned at all points by global sections, and in particular the first reduction map restricted to A gives the morphism onto its minimal model A'.

Mori's solution of the Hartshorne conjecture [446] and the subsequent development of the ideas from that solution give a powerful tool to study the fractional adjunction bundles $K_X + t\mathcal{L}$ where $\mathcal{L}$ is ample, t is rational, and X has appropriate singularities. This allows a new approach to these results of Fano-Morin, and a development of new types of results going considerably beyond the previous work. Because of the rapid development in this period, we abandon the historical survey, though we have tried to give the history clearly in the course of covering the material of this book.

In Chapter 6 we introduce unbreakable rational curves. This concept includes Mori's extremal rational curves and the classical notion of a line on a projective variety. It turns out to be technically convenient to develop in this greater generality such basic results as Mori's nonbreaking lemma (see (6.3.1)), and the inequality of Ionescu [335] (see (6.3.5) and also Wiśniewski [619]) for the dimension of the locus of deformations of extremal rational curves. Adjunction theory as a matter of course produces lists of degenerate varieties where some ampleness property of an adjunction bundle fails—these varieties are almost always covered by lines. Our ideas for this chapter developed out of our paper [114] with Wiśniewski, from which we have included a number of results. Smoothness plays a greater role in this chapter than in other chapters.

In Chapter 7 we develop the main central results of adjunction theory for merely ample line bundles on appropriately singular varieties. A major tool is the Kawamata rationality theorem, (1.5.2). We follow mainly our paper [106] for dimensions ≥ 5, and Fujita's paper [261] for dimension 4. These results involve the classification of the exceptions to being able to replace $(X, \mathcal{L})$ with a new closely related pair, $(W, \mathcal{H})$, such that some positive power of $K_W + (\dim W - 3)\mathcal{H}$ is spanned at all points by global sections and the map associated to the sections is birational. Typically these exceptions are low dimensional. We make conjectures about the behavior of natural morphisms (called the nefvalue morphisms) associated to the original pair $(X, \mathcal{L})$, based on the results of this chapter and partially on the results of Chapter 6.

Here the concepts of the first and second reduction are introduced and their structure worked out. Roughly these results say that given a positive integer $k \leq 4$, and an ample line bundle, $\mathcal{L}$, on an appropriately singular projective variety, X, of dimension $n \geq k$, then except for an explicit list of degenerate examples, there is a new pair, $(W, \mathcal{H})$, and a simple birational morphism $p : X \to W$ with $\mathcal{H} = (p_*\mathcal{L})^{**}$, and that some positive power of $K_W + (n - k + 1)\mathcal{H}$ is spanned at all points by global sectionw and the map associated to the sections is birational. For $k \geq 5$ the results show that there is no pair $(W, \mathcal{H})$ with properties as nice as the ones when $k \leq 4$, but there is a conjecture, proven for dimension ≤ 8, that the structure of the pair will be simple if the spectral value of the pair, a basic invariant of the pair defined in §7.1, is $\geq n/2 + 1$. Though conceptually and historically classical adjunction theory came before these results, we prove them before we work out classical adjunction theory, because they clarify many aspects of that theory. To keep the presentation of classical adjunction theory as direct as possible, we have not used the results of Chapter 7 in the development of the central results of classical adjunction theory. This has meant that occasionally in later chapters we give different easier proofs of simpler versions of results proven in Chapter 7.

In Chapter 13 we discuss a number of results related to one of the main open problems in adjunction theory—the conjecture that if $\mathcal{L}$ is a very ample line bundle on a connected projective manifold, X, of nonnegative Kodaira dimension, then $K_{X'} + (n-2)\mathcal{L}'$ is spanned by its global sections on the first reduction, $(X', \mathcal{L}')$, of $(X, \mathcal{L})$. We also develop a number of topics connected with Chern inequalities for varieties with $\kappa(K_X + (n-2)\mathcal{L}) \geq 0$.

In Chapter 14 we study some of the properties of the special varieties that occur in adjunction theory, with special attention to scrolls and quadric fibrations. Here there are a number of strong results, and there are many natural questions and conjectures which are being actively researched. In §14.4 we include a fairly complete discussion of the known results on pairs, (X, L), with L a very ample line bundle on a projective manifold, X, and with (X, L) having positive defect, i.e., with the discriminant variety, $\mathcal{D} \subset |L|$, having codimension at least 2 in $|L|$. In §14.5 we survey some of the research on the relations between curves and their hyperplane sections. Since a nonsingular hyperplane section of a curve is a finite set of points, at first sight one does not expect there to be much more that bounds like Castelnuovo's bound, (1.4.9), for the genus of an irreducible curve in terms of its degree. The way around this is to consider the extra information given by the embedding of the points in a projective space. It turns out that the Hilbert function of the set of points, which is a hyperplane section of a curve, C, contains much information about the curve, C, and its embedding into projective space.

Readers interested in classical adjunction theory, which is covered in Chapter 8 to Chapter 11, should look over the table of contents of the first 5 chapters of the book, and then start reading Chapter 8 to Chapter 11, going back to the first 4 chapters only as needed. The second author has given a successful one semester "Topics in algebraic geometry" course at Notre Dame based on Chapter 8 to Chapter 11 plus the needed material from the first 5 chapters. Chapter 12 can be read after Chapters 8 to Chapter 11 and before Chapters 6 and Chapter 7. Chapter 6 and Chapter 7 are almost completely self-contained, depending on material in the first 4 chapters.

In a book which has absorbed a large part of our energies for 5 years, there are many people and organizations to thank.

First and foremost we would like to thank our families—Rossana and Barbara Beltrametti, and Rebecca, Rachel Catherine, and Ruth Francesca Sommese—for their encouragement and patience.

We would like to thank the University of Notre Dame, the University of Genova, our students and colleagues. We would like to give special thanks to Juan Migliore for discussions about the material in §14.5.

The second author would like to thank Frank Castellino, the Dean of the College of Science at the University of Notre Dame, for his encouragement and his generous support of the author's research.

The first author would like to give special warm thanks to G. Bottaro and E. Carletti for their constant help in solving technical problems connected with the use of computer systems. We would both like to express our gratitude for the kind help that G. Bottaro and E. Carletti gave when illness forced the first author to remain at the Hospital and at home for a long period, just in the final critical stage of the preparation of the book. Only because of their friendly help, were we able to communicate during that period.

We thank M.L. Fania and P. Francia for pointing out several misprints during the proofs correction stage.

We would especially like to thank Professor F. Hirzebruch and the Max-Planck-Institut für Mathematik in Bonn, where over a third of the work on this book was done during the 1992–1993 academic year.

We are very grateful to Dr. M. Karbe and De Gruyter for their professional and courteous manner.

We would also like to thank the CNR and the NSF for support at various stages of this book's development.

Mauro C. Beltrametti
Andrew J. Sommese

Contents

List of Tables

Chapter 1

General background results

In this chapter we collect some general background material.

In §1.1 we fix basic notation and definitions. Then we recall some general facts about projectivizations of vector bundles and generalized cones on a polarized variety. We also prove a very general parity result, Lemma (1.1.11), which will be used in Chapter 13.

In §1.2 we recall the possible configurations of the fundamental cycle for rational double point surface singularities. In §1.3 we describe the rather mild singularities which occur in adjunction theory.

In §1.4 we discuss some general results on curves, including a singular version of Clifford's theorem, (1.4.8), and Castelnuovo's bound for the genus of an irreducible curve in projective space.

The concept of nefvalue of a polarized variety, motivated by the Kawamata rationality theorem, plays a key role in the classification of projective varieties (see, e.g., Chapter 7). In §1.5 we give the definition and a few basic properties of the nefvalue of a polarized pair.

In §1.6 we discuss some general facts about the first jet bundle of a line bundle, L, on a complex manifold and the discriminant locus associated to a vector subspace of $\Gamma(L)$.

In §1.7 we collect some useful versions of Bertini's type theorems, see (1.7.8), (1.7.9), including the generalized Seidenberg theorem, (1.7.1). Some key examples are given in §1.8.

1.1 Some basic definitions

We follow the standard terminology in algebraic geometry, e.g., Hartshorne [318].

We work over the complex field $\mathbb{C}$. Unless said otherwise all dimensions are complex dimensions. A *quasi-projective variety* (respectively *projective variety*, respectively *analytic variety*) is an algebraic set V (respectively projective algebraic set, respectively complex analytic set) which is reduced. We note that in the literature the word, variety, is sometimes used to include the condition of irreducible, e.g., Hartshorne [318], and sometimes not, e.g., Gunning and Rossi [301] and Kempf [348]. If an algebraic or analytic set has all its components of the same dimension we say that the algebraic or analytic set is pure dimensional. By an *n-fold* we mean a reduced algebraic (or analytic) set of pure dimension n. We denote the structure sheaf of an algebraic (respectively analytic) set, V, by $\mathcal{O}_V$. With a few exceptions we only deal with algebraic sets, and therefore, except when we say otherwise, structure sheaves, coherent sheaves, etc., refer to algebraic structure sheaves, algebraic coherent sheaves, etc.

For any sheaf, $\mathcal{F}$, defined over $\mathbb{C}$, on a topological space, X, $h^i(X, \mathcal{F})$ (or $h^i(\mathcal{F})$ when there is no confusion) denotes the complex dimension of $H^i(X, \mathcal{F})$. It is a fundamental theorem of Serre (see [318, Appendix B, §2]) that if V is a possibly nonreduced, projective algebraic set, then the natural transformation from the category of algebraic coherent sheaves on V to the category of analytic coherent sheaves on V with V's underlying complex analytic structure is a natural equivalence of categories. In particular on a projective algebraic set,

1. to be an algebraic section of an algebraic vector bundle, $\mathcal{E}$, is equivalent to being a holomorphic section of the underlying complex analytic vector bundle; and

2. we get the same answer whether we compute the cohomology groups of an algebraic coherent sheaf or the cohomology groups of the associated analytic coherent sheaf on the associated complex analytic space.

These results are often referred to as the GAGA theorems.

1.1.1. The canonical sheaf. An excellent reference for the canonical or dualizing sheaf is the book of Altman and Kleiman [4]. If M is a pure n-dimensional smooth quasi-projective manifold, then the *canonical bundle* is the bundle, $K_M := \det \mathcal{T}_M^* := \wedge^n \mathcal{T}_M^*$, where $\mathcal{T}_M^*$ denotes the dual of the tangent bundle, $\mathcal{T}_M$, of M. A local section of K_M is a *holomorphic n-form*. The associated sheaf of germs of algebraic sections is called the *canonical sheaf*, or *dualizing sheaf*. Following standard practice we usually do not distinguish between K_M and its associated sheaf of germs of algebraic or analytic sections, $\mathcal{O}_M(K_M)$. If X is any n-fold contained in an N-dimensional complex projective space, $\mathbb{P}^N$, then the dualizing sheaf, K_X or ω_X, is defined [4, (1.3)] as $\mathcal{E}xt^{N-n}_{\mathcal{O}_{\mathbb{P}^N}}(\mathcal{O}_X, \omega_{\mathbb{P}^N})$. This is independent of the embedding and local, which lets us define ω_X by restriction for any possibly noncompact and nonreduced, n-dimensional algebraic subset, X. Note that $\omega_{\mathbb{P}^N} \cong \mathcal{O}_{\mathbb{P}^N}(-N-1)$. When we do want to emphasize the sheaf or dualizing nature of K_M, we use the standard notation, ω_M. If V is a normal n-fold, i.e., a pure n-dimensional reduced algebraic or analytic set, then K_V (or ω_V) is a reflexive rank 1 sheaf. From this it follows that for a normal n-fold, V, the *canonical sheaf*, K_V, can equivalently be defined to be $j_*K_{\mathrm{reg}(V)}$ where $j : \mathrm{reg}(V) \hookrightarrow V$ is the inclusion of the smooth points of V and $K_{\mathrm{reg}(V)}$ is the sheaf of holomorphic n-forms.

Let $\mathrm{Div}(V)$ be the group of *Cartier divisors* on an algebraic set, V, and let $\mathrm{Pic}(V)$ denote the group of the line bundles on V. Let $\mathcal{L}$ be a holomorphic line bundle on V. The line bundle $\mathcal{L}$ is said to be *spanned by* W, where W is a vector space of global sections, contained in $\Gamma(\mathcal{L})$, the space of global sections of $\mathcal{L}$, if given any point $v \in V$ there is a section $s \in W$ such that $s(v) \neq 0$. The line bundle $\mathcal{L}$ is said to be *spanned* if $\mathcal{L}$ is spanned by $\Gamma(\mathcal{L})$. Unless said otherwise, by a section we will mean an algebraic section.

Spanned line bundles are of particular interest because of the identification between them and algebraic maps into projective space. Given $(V, \mathcal{L}, W)$ where W spans $\mathcal{L}$ and $\dim W < \infty$, let $\mathbb{P}(W)$ denote the set of one dimensional quotients of W. Sending $v \in V$ to the quotient of W by the vector subspace of $s \in W$ with $s(v) = 0$ gives a holomorphic map $\phi_W : V \to \mathbb{P}(W)$, such that $\mathcal{L} \cong \phi_W^*\mathcal{O}_{\mathbb{P}(W)}(1)$, and $W \cong \phi_W^*\Gamma(\mathcal{O}_{\mathbb{P}(W)}(1))$. The map is given by choosing a basis, $s_0, \ldots, s_{\dim \mathbb{P}(W)}$, of W, and sending $v \in V$ to $[s_0(v), \ldots, s_{\dim \mathbb{P}(W)}(v)]$. In this book we will only be concerned with the case when $\dim W$ is finite.

Among the spanned line bundles a central place is occupied by the *very ample* line bundles—these are the bundles, $\mathcal{L}$, such that the map ϕ_W is an embedding. As we will see many special properties of very ample bundles are shared by *ample* line bundles—these are bundles, $\mathcal{L}$, such that there exists some positive integer, t, such that $t\mathcal{L}$ is very ample. For irreducible projective curves, the ample line bundles are precisely the line bundles of positive degree. Ample bundles do not have to have any sections: see Example (1.8.1).

Among the spanned bundles, ample bundles are characterized by the fact that the map $\phi_{\Gamma(\mathcal{L})}$ has finite fibers.

A class of line bundles that partake of some of the key properties of ample bundles are the *semiample* line bundles. $\mathcal{L}$ is said to be *semiample* if there is a positive integer k such that $k\mathcal{L}$ is spanned. Unlike ample bundles, it does not follow that $k\mathcal{L}$ is spanned for *all* sufficiently large k: see Example (1.8.2.1).

A line bundle, $\mathcal{L}$, is called *nef* if given any effective curve, $C \subset V$, it follows that $\mathcal{L} \cdot C \geq 0$, and *nonnef* otherwise. Every semiample line bundle is nef. The converse is false: see Examples (1.8.2.2) and (1.8.3). If $\mathcal{L}$ is a nef line bundle on a projective variety, V, and $\mathcal{L}^{\dim V} \geq 1$, then Lemma (2.5.7) will show that multiples of $\mathcal{L}$ have many sections, but as the Example (1.8.3) shows, it can easily be the case that no multiple is spanned by global sections.

Let $\varphi : V \to Y$ be a proper morphism of algebraic sets. We say that a line bundle L on V is *φ-ample* (respectively *φ-nef*) if there exists an ample line bundle H on Y such that $L + \varphi^*H$ is ample (respectively nef).

There is a very useful theorem due to Fujita [249, Theorem (1.10)] (based on a classical three dimensional result of Zariski [633]) that guarantees semiampleness. The algebraic version follows from the analytic version by using the GAGA results discussed above. See Lemma (2.6.5) for a typical use of this result.

Theorem 1.1.2. (Fujita-Zariski) *Let L be a line bundle on a compact complex analytic space, X. Assume that for some $t > 0$, the restriction of L to the set of common zeroes of $\Gamma(tL)$ is ample. Then L is semiample.*

Note that Theorem (1.1.2) does not apply if the restriction of L to the set of common zeroes of $\Gamma(tL)$ is merely nef.

The following simple corollary of the Remmert-Stein factorization theorem [318, Chapter III, Corollary 11.5] is very useful.

Lemma 1.1.3. *Let L be a semiample line bundle on a projective variety, V. Let μ be the greatest common divisor of the integers $t \geq 1$ such that tL is spanned, e.g., $\mu = 1$ if tL is spanned for all sufficiently large t. Then there is a morphism $r : V \to Y$, where Y is a projective variety with an ample line bundle, $\mathcal{L}$, on Y, such that the fibers of the map, r, are connected and $\mu L \cong r^*\mathcal{L}$. Moreover $(Y, \mathcal{L})$ has the universal property that given any pair $(Y', \mathcal{L}')$ consisting of a projective variety Y' with $\mathcal{L}'$ an ample line bundle on Y', and a morphism $r' : V \to Y'$ with $eL \cong r'^*\mathcal{L}'$, it follows that there is a morphism $\sigma : Y \to Y'$ such that $\sigma \circ r = r'$ and $c := e/\mu$ is integral with $c\mathcal{L} \cong \sigma^*\mathcal{L}'$. If V is normal then Y is normal.*

Proof. For a positive integer, $a > 0$, such that aL is spanned by global sections, let $\phi_a : V \to \mathbb{P}_{\mathbb{C}}$ be the morphism associated to $\Gamma(aL)$. Let $\phi_a := s_a \circ r_a$ be the Remmert-Stein factorization [318, Chapter III, Corollary 11.5] of ϕ_a. Here $r_a : V \to Y_a$ has connected fibers and $s_a : Y_a \to \mathbb{P}_{\mathbb{C}}$ has finite fibers. Note that $aL \cong r_a^*\mathcal{L}_a$ with $\mathcal{L}_a$ an ample line bundle on Y_a. Let b be another positive integer such that bL is spanned by global sections, and let $\phi_b = s_b \circ r_b$ be the Remmert-Stein factorization of the morphism given by $\Gamma(bL)$. Note that given a connected fiber, F, of r_a the bundle bL is trivial in some neighborhood of F. To see this note that since bL is spanned in a neighborhood of F, there is a section, v, of bL that is not identically zero on F. Since $abL = b(r_a^*\mathcal{L}_a)$ is trivial in a neighborhood of F, we see that $v^{\otimes a}$ is nowhere zero on F. Thus v gives a trivialization of bL in a neighborhood of F. From this we conclude that $bL \cong r_a^*H$ for some line bundle H on Y_a. Thus r_b factors through r_a and reversing the role of a and b we have also that r_a factors through r_b. Thus we conclude there is an isomorphism, $i : Y_a \to Y_b$, of Y_a and Y_b such that $i \circ r_a = r_b$ and such that $i^*(a\mathcal{L}_b) \cong b\mathcal{L}_a$, where $bL \cong r_b^*\mathcal{L}_b$. Let $Y = Y_a$, $r = r_a$, for any a with aL spanned. A combinatorial argument (of exactly the same type, but simpler than the argument used below in Lemma (1.5.6)) shows that $\mu L \cong r^*\mathcal{L}$ for some ample line bundle, $\mathcal{L}$, on Y. From [318, Chapter III, Corollary 11.5] it follows that Y is normal if V is normal.

Finally let us show the universal property. Since $\mathcal{L}'$ is ample, all sufficiently large multiples of $\mathcal{L}'$ are spanned. Thus all sufficiently large multiples of eL are spanned. Therefore $e = c\mu$ for some positive integer c. The argument used in the last paragraph shows that r' factors through r, i.e., $r' = \sigma \circ r$, for some morphism $\sigma : Y \to Y'$. To see that $c\mathcal{L} \cong \sigma^*\mathcal{L}'$, note that $r^*(c\mathcal{L}) \cong eL \cong r'^*\mathcal{L}' \cong r^*\sigma^*\mathcal{L}'$ and hence $r^*(c\mathcal{L} + \sigma^*(-\mathcal{L}')) \cong \mathcal{O}_V$. Since by the proof of the Remmert-Stein factorization [318, Chapter III, Corollary 11.5] we have that $r_*\mathcal{O}_V \cong \mathcal{O}_Y$, it follows that $c\mathcal{L} + \sigma^*(-\mathcal{L}') \cong \mathcal{O}_Y$. □

Given a holomorphic line bundle, $\mathcal{L}$, on a variety, V, and a finite dimensional vector space $W \subset \Gamma(\mathcal{L})$ of global sections, $\phi_W : V \to \mathbb{P}(W)$ denotes the rational mapping associated to W. The *Kodaira dimension*, $\kappa(\mathcal{L})$, of $\mathcal{L}$ (or of $(V, \mathcal{L})$), is defined as $\kappa(\mathcal{L}) = -\infty$ if $h^0(t\mathcal{L}) = 0$ for all $t > 0$, and

$$\kappa(\mathcal{L}) = \max_{t>0}\{\dim \phi_{\Gamma(t\mathcal{L})}(V)\}$$

otherwise. Note that given any positive integer m, $\kappa(\mathcal{L}) = \kappa(m\mathcal{L})$.

A line bundle on an irreducible variety, V, with $\kappa(\mathcal{L}) = \dim V$ is said to be *big*. For a nef line bundle, $\mathcal{L}$, on an irreducible n-dimensional projective variety, V, big is equivalent to $\mathcal{L}^n > 0$: see Lemma (2.5.7) for a proof of this fact.

A line bundle, $\mathcal{L}$, on an irreducible variety, V, is said to be *k-big* if $\kappa(\mathcal{L}) \geq \dim V - k$, i.e., for some N, the dimension of the image of the rational map associated to $\Gamma(N\mathcal{L})$ is $\geq \dim V - k$. Note that this is equivalent to saying that

$$\min_{N>0}\{\dim F \mid F \text{ general fiber of } \phi_{\Gamma(N\mathcal{L})}\} \leq k.$$

Note also that $\mathcal{L}$ being 0-big is equivalent to $\mathcal{L}$ being big. Lemma (2.5.8) shows that for a k-big and nef line bundle, $\mathcal{L}$, on an irreducible n-dimensional projective variety, V, $\mathcal{L}^{n-k} \cdot H^k > 0$ where H is any nef and big line bundle on V. The Example (1.8.2.2) shows that the converse does not hold for $k \neq 0$.

Let $Z_{n-1}(V)$ denote the group of *Weil divisors*, i.e., the free Abelian group generated by prime, i.e., irreducible and reduced, divisors on an algebraic set V. An element of $Z_{n-1}(V) \otimes \mathbb{Q}$ (respectively $\mathrm{Div}(V) \otimes \mathbb{Q}$) is called a *$\mathbb{Q}$-divisor* (respectively a *$\mathbb{Q}$-Cartier divisor*). If γ is a subcycle of V and $D \in Z_{n-1}(V) \otimes \mathbb{Q}$ with $mD \in \mathrm{Div}(V)$ for some integer m, $m \neq 0$, then the intersection symbol $D \cdot \gamma$ stands for $(mD \cdot \gamma)/m$. We also say that a divisor $D \in Z_{n-1}(V)$ is *e-Cartier* if e is a positive integer such that $eD \in \mathrm{Div}(V)$. If K_V is e-Cartier for some positive integer e, then we say that V is *$\mathbb{Q}$-Gorenstein*, or more precisely *e-Gorenstein*. If V is $\mathbb{Q}$-Gorenstein, then the smallest positive integer e such that V is e-Gorenstein is called the *index* (*of the singularities*) of V. An algebraic set, V, is called *Gorenstein* if V is 1-Gorenstein, i.e., K_V is Cartier, and if in addition V has at worst Cohen-Macaulay singularities. We say that an n-dimensional normal variety V is *$\mathbb{Q}$-factorial* if for any divisor $D \in Z_{n-1}(V)$ there exists a positive integer

m (depending on D) such that $mD \in \mathrm{Div}(V)$. We say that V is *m-factorial* if $mD \in \mathrm{Div}(V)$ for all $D \in Z_{n-1}(V)$.

Two elements $D, D' \in Z_{n-1}(V)\otimes\mathbb{Q}$ are said to be $\mathbb{Q}$*-linearly equivalent*, denoted by $D \approx D'$, if there exists a positive integer m such that mD, $mD' \in \mathrm{Div}(V)$ and that mD and mD' are linearly equivalent in the ordinary sense. Two elements $D, D' \in Z_{n-1}(V)\otimes\mathbb{Q}$ are said to be $\mathbb{Q}$*-numerically equivalent*, denoted by $D \sim D'$, if there exists a positive integer m such that mD, $mD' \in \mathrm{Div}(V)$ and that mD and mD' are numerically equivalent in the ordinary sense. For any divisor $D \in Z_{n-1}(V)$ we shall denote by $\mathcal{O}_V(D)$ the associated reflexive sheaf of rank 1. Note that the correspondence

$$Z_{n-1}(V)/\approx\;\to\;\{\text{ reflexive sheaves of rank } 1\}/\cong$$

given by $D \mapsto \mathcal{O}_V(D)$ is a bijection. Recall that, for any $D, D' \in Z_{n-1}(V)$,

$$\mathcal{O}_V(D_1 + D_2) \cong (\mathcal{O}_V(D_1) \otimes \mathcal{O}_V(D_2))^{**}, \text{ the double dual.}$$

We say that a divisor $D \in Z_{n-1}(V)$ is *ample* (respectively *nef*, *nonnef*, *semiample*, or *big*) if mD is an ample (respectively nef, nonnef, semiample, or big) Cartier divisor for some positive integer m. Similarly the *Kodaira dimension*, $\kappa(D)$, of a $\mathbb{Q}$-Cartier divisor, D, on a projective variety, V, is defined to be $\kappa(ND)$ where N is a positive integer such that ND is Cartier. We leave it as an exercise to check that this definition does not depend on the positive N chosen, and therefore that given any positive rational number, m, $\kappa(D) = \kappa(mD)$.

Let V be an irreducible normal variety. Linear equivalence classes of divisors on V and isomorphism classes of reflexive sheaves of rank 1 are used with little (or no) distinction. Hence we shall freely switch from the multiplicative to the additive notation and vice versa. E.g., if $\mathcal{L}$ is a rank 1 reflexive sheaf on V we shall often use the notation $K_V + \mathcal{L}$ with the meaning $K_V + \mathcal{L} = (\mathcal{O}_V(K) \otimes \mathcal{L})^{**}$, where $\mathcal{L} \cong \mathcal{O}_V(D)$ with $D \in Z_{n-1}(V)$ and $K \in Z_{n-1}(V)$ is the canonical divisor of V. Also, for a positive integer m, mK_V, $-mK_V$ are rank one reflexive sheaves defined as

$$mK_V = (\mathcal{O}_V(K)^{\otimes m})^{**} = \mathcal{O}_V(mK) = K_V^{[m]}\,; -mK_V = (\mathcal{O}_V(K)^{\otimes m})^{*}.$$

Note also that $mK_V = j_*(K_{\mathrm{reg}(V)}^{\otimes m})$, where $j : \mathrm{reg}(V) \hookrightarrow V$ is the inclusion of the regular points.

Let V be an irreducible normal projective variety. If $h^0(mK_V) = 0$ for all integers $m > 0$ we define the *Kodaira dimension* or *arithmetic Kodaira dimension*, $\kappa(V)$, of V to be $-\infty$. If $h^0(mK_V) > 0$ for some $m > 0$, we define the *Kodaira dimension* or *arithmetic Kodaira dimension*, $\kappa(V)$, of V to be the smallest nonnegative integer κ such that

$$\limsup_{m\to\infty} \frac{h^0(mK_V)}{m^{\kappa+1}} = 0,$$

i.e., $\kappa(V)$ is the maximum of the dimensions of the images of V under the rational mappings given by $\Gamma(mK_V)$. We define the *geometric Kodaira dimension*, $\overline{\kappa}(V)$, of V to be $\kappa(\overline{V})$ where $\overline{V}$ is a desingularization of V. In general $\kappa(V) \neq \overline{\kappa}(V)$. For example, if $V \subset \mathbb{P}^3$ is a cone over a smooth curve in $\mathbb{P}^2$, then $\overline{\kappa}(V) = -\infty$, but if $\deg(V) \geq 4$, then $\kappa(V) \geq 0$. For most considerations in this book, we have equality. See e.g., §1.7 for the definition of rational singularities.

Lemma 1.1.4. *Let V be an irreducible normal projective variety. Then $\kappa(V) \geq \overline{\kappa}(V)$ with equality if V has at worst rational singularities.*

Proof. Let $p : \overline{V} \to V$ be a desingularization with the restriction $p_{p^{-1}(\mathrm{reg}V)}$ an isomorphism. Thus $p_*(mK_{p^{-1}(\mathrm{reg}(V))}) \cong mK_{\mathrm{reg}(V)}$ for each integer $m \geq 1$. Using this, the fact that $p_*(mK_{\overline{V}})$ is coherent and torsion free, and the fact that $i_*K_{\mathrm{reg}(V)} \cong K_V$ for the inclusion $i : \mathrm{reg}(V) \hookrightarrow V$, we see that there is a natural inclusion $0 \to p_*(mK_{\overline{V}}) \to mK_V$. This gives the inequality, $\kappa(V) \geq \overline{\kappa}(V)$.

If V has at most rational singularities, then by Kempf's theorem, [347], the inclusion $0 \to p_*(mK_{\overline{V}}) \to mK_V$ is an isomorphism for $m = 1$ and hence for every $m \geq 1$. □

1.1.5. Further definitions and notation. If $\mathcal{F}$ is a coherent algebraic sheaf on an n-dimensional projective algebraic set, then $\chi(\mathcal{F}) := \Sigma_{i=0}^{n}(-1)^i h^i(\mathcal{F})$ is the *Euler characteristic* of $\mathcal{F}$.

Let V be an irreducible projective variety. We let $q(V) := h^1(\mathcal{O}_{\overline{V}})$ denote the *irregularity* of V, and $p_g(V) = h^0(K_{\overline{V}})$, the *geometric genus*, of V, where $\overline{V}$ is a desingularization of V. The *arithmetic genus* of V is defined to be $p_a(V) := (-1)^{\dim V}(\chi(\mathcal{O}_V) - 1)$.

For V smooth of dimension n, we denote by $e(V) = c_n(V)$ the *topological Euler characteristic* of V, where $c_n(V)$ is the n-th Chern class of the tangent bundle, $\mathcal{T}_V$, and hence of V. If V is a surface, $e(V) = 12\chi(\mathcal{O}_V) - K_V^2$.

For a prime divisor $D \in Z_{n-1}(V) \otimes \mathbb{Q}$, define

$$\{D\} = \text{ the integral part of } D \text{ and } \lceil D \rceil = -\{-D\}, \text{ the roundup of } D.$$

If $\mathcal{L}$ is a line bundle on an algebraic set, V, then, $|\mathcal{L}|$, *the complete linear system of* $\mathcal{L}$, is defined to be $\mathbb{P}(\Gamma(\mathcal{L})^*)$, the set of 1-dimensional quotients of $\Gamma(\mathcal{L})^*$, i.e., the set of 1-dimensional vector subspaces of $\Gamma(\mathcal{L})$. A *linear system*, $\mathfrak{d}$, on V is a subset of the complete linear system $|\mathcal{L}|$ of the form $\mathbb{P}(W^*)$, where W is a vector subspace of $\Gamma(\mathcal{L})$, $W \subset \Gamma(\mathcal{L})$. If W is finite dimensional, as it always will be for us, then $\mathbb{P}(W^*)$ has the natural structure of a projective space, whose dimension, $\dim W - 1$, is called the *dimension* of the linear system $\mathfrak{d}$.

In line with usually using additive notation for tensor products of line bundles we write $E \otimes (tL)$ for the tensor product of a vector bundle E and the t-th power of a line bundle L.

If Y is a local complete intersection of a nonsingular variety V, we denote by $\mathcal{N}_{Y/V} := (\mathcal{J}_Y/\mathcal{J}_Y^{\otimes 2})^*$ the *normal bundle* of Y in V, where $\mathcal{J}_Y$ denotes the ideal sheaf of Y in V. $\mathcal{N}_{Y/V}$ is a locally free sheaf of rank $r = \mathrm{cod}_V Y$.

The following useful lemma we first learned from the very nice paper of Van de Ven [603].

Lemma 1.1.6. *Let E and F be two vector bundles on a compact algebraic set X. Let V_E and V_F be vector spaces of sections of E and F respectively. Then the image, $\mathcal{W}$, of $V_E \otimes V_F$ in $H^0(X, E \otimes F)$ has dimension at least* $\dim V_E + \dim V_F - 1$.

Proof. Consider the induced map $\mu : \mathbb{P}(V_E) \times \mathbb{P}(V_F) \to \mathbb{P}(H^0(X, E \otimes F))$. Since the map $V_E \overset{\otimes s}{\to} H^0(X, E \otimes F)$ is injective for any nonzero element $s \in V_F$, it follows that μ is injective on $\mathbb{P}(V_E) \times \{s'\}$ for any element $s' \in \mathbb{P}(V_F)$. A similar statement holds with E replaced by F and $s' \in \mathbb{P}(V_F)$ replaced by $t' \in \mathbb{P}(V_E)$. Let $W := \mathbb{P}(\mathcal{W})$. Thus the pullback, $\omega \in H^2(\mathbb{P}(V_E) \times \mathbb{P}(V_F), \mathbb{Z})$, under μ of a generating class of $H^2(W, \mathbb{Z})$ restricts to a generating class of $H^2(\mathbb{P}(V_E) \times \{s'\}, \mathbb{Z})$ and also to a generating class of $H^2(\{t'\} \times \mathbb{P}(V_F), \mathbb{Z})$. From this it is straightforward to check that ω^n is a nonzero element of $H^n(\mathbb{P}(V_E) \times \mathbb{P}(V_F), \mathbb{Z})$ for $n = \dim(\mathbb{P}(V_E) \times \mathbb{P}(V_F))$. This implies

$$\dim \mathcal{W} - 1 = \dim W \geq \dim \mathbb{P}(V_E) + \dim \mathbb{P}(V_F) = \dim V_E - 1 + \dim V_F - 1.$$

□

1.1.7. The projectivization of a vector bundle. The paper [313] and book [318, Chapter II, §7] of Hartshorne are very good references for this topic. Let $\mathcal{E}$ be a rank $d+1$ vector bundle on a variety, Y, and let $\mathcal{E}^{(t)}$ or $S^t(\mathcal{E})$ denote the t-th symmetric power of $\mathcal{E}$ for $t \geq 0$: here we follow the standard convention that $\mathcal{E}^{(0)} := \mathcal{O}_Y$. Let $\mathcal{S} := \bigoplus_{t=0}^{\infty} \mathcal{E}^{(t)}$ be the symmetric algebra of $\mathcal{E}$. Then $\mathbb{P}(\mathcal{E}) := \mathrm{Proj}(\mathcal{S})$ is a $\mathbb{P}^d$-bundle, $\pi : \mathbb{P}(\mathcal{E}) \to Y$, over Y (see also §3.2). Each point $x \in \mathbb{P}(\mathcal{E})$ corresponds to a quotient space of $\mathcal{E}_{\pi(x)}$ of dimension one. Over $\mathbb{C}$, $\mathbb{P}(\mathcal{E})$ is the quotient $(\mathcal{E}^* - Y_0)/\mathbb{C}^*$, where $\mathcal{E}^*$ is the dual bundle of $\mathcal{E}$, Y_0 is the zero section and $\mathbb{C}^* = \mathbb{C} - \{0\}$. Thus there is an invertible line bundle, $\xi_{\mathcal{E}}$ (or ξ for short when no confusion can result), which over $x \in \mathbb{P}(\mathcal{E})$ is the quotient space of $(\pi^*\mathcal{E})_x$ corresponding to the one dimensional quotient of $\mathcal{E}_{\pi(x)}$ defining x. On $\mathbb{P}(\mathcal{E})$, $\xi_F \cong \mathcal{O}_{\mathbb{P}^d}(1)$ for any fiber, F, of π. The sheaf $\xi_{\mathcal{E}}$ is called the *tautological* *line bundle* on $\mathbb{P}(\mathscr{E})$ and it is also denoted by $\mathbb{O}_{\mathbb{P}(\mathscr{E})}(1)$. We have that $\pi_*\xi_{\mathscr{E}} \cong \mathscr{E}$ and further $\pi_*(t\xi_{\mathscr{E}}) = \mathscr{E}^{(t)}$, the t-th symmetric product of $\mathscr{E}$, for any integer $t \geq 0$. Recall the canonical bundle formula

$$K_{\mathbb{P}(\mathscr{E})} \approx \pi^*(K_Y + \det \mathscr{E}) - (d+1)\xi_{\mathscr{E}}. \tag{1.1}$$

Projectivizations of vector bundles occur naturally as a major class of terminal objects in adjunction theory. Moreover many examples can be constructed starting from projectivizations of vector bundles. For example, given any connected projective manifold, A, it is easy to construct a $\mathbb{P}^1$-bundle, X, over A, and a very ample divisor, A', on X which is mapped birationally to A under the product projection. Indeed let $X := A \times \mathbb{P}^1$, let $p : X \to A$ and $q : X \to \mathbb{P}^1$ be the product projections, and let $\mathscr{L} := p^*L + q^*\mathbb{O}_{\mathbb{P}^1}(1)$ where L is a very ample line bundle on A. Then $\mathscr{L}$ is very ample and a smooth $A' \in |\mathscr{L}|$ meets the general fiber of p in one point and is therefore birational to A under the restriction morphism, $p_{A'}$. In particular every curve is a very ample divisor on some smooth surface. We will see later in the book that it is a very stringent condition for a projective manifold to be birational to an ample, smooth divisor, A, on a projective manifold, X, and often can only be true in the trivial case when $X = \mathbb{P}(\mathscr{E})$ for some ample vector bundle $\mathscr{E}$ and $A \in |\xi_{\mathscr{E}}|$.

We say that a vector bundle $\mathscr{E}$ of rank r on a variety, Y, is *ample* (respectively *very ample*) if the tautological bundle, $\xi_{\mathscr{E}}$, on $\mathbb{P}(\mathscr{E})$ is ample (respectively very ample). We leave it to the reader to see that direct sums of very ample line bundles are very ample (a proof of a more general fact is given in Lemma (3.2.3)). Let $\mathbb{O}_Y(\mathscr{E})$ be the sheaf of sections of $\mathscr{E}$. Then $\mathbb{O}_Y(\mathscr{E})$ is a locally free sheaf of $\mathbb{O}_Y$-modules of rank r. If $W \subseteq \Gamma(\mathbb{O}_Y(\mathscr{E}))$ is a subspace of sections of $\mathscr{E}$, there is a canonical homomorphism from the trivial bundle $Y \times W$ to $\mathscr{E}$. We say that $\mathscr{E}$ is spanned by W if this homomorphism is surjective. We say that $\mathscr{E}$ is *spanned* if it is spanned by $\Gamma(\mathbb{O}_Y(\mathscr{E}))$. Note that $W \subseteq \Gamma(\mathbb{O}_Y(\mathscr{E}))$ spans $\mathscr{E}$ if and only if the subspace of $\Gamma(\xi_{\mathscr{E}})$ corresponding to W spans $\xi_{\mathscr{E}}$ on $\mathbb{P}(\mathscr{E})$.

1.1.8. Generalized cones and normal generalized cones. Let L be a very ample line bundle on a projective variety V of pure dimension n. Let $E = \oplus^{N-n}\mathbb{O}_V$ where N is an integer $\geq n$. Let $\mathscr{C} = \mathbb{P}(E \oplus L)$ and let ξ denote the tautological line bundle on $\mathscr{C}$. Let $\varphi : \mathscr{C} \to \mathbb{P}_{\mathbb{C}}$ denote the map associated to $\Gamma(\xi)$. We denote $\varphi(\mathscr{C})$ by $C_N(V, L)$ and we denote the restriction of $\mathbb{O}_{\mathbb{P}_{\mathbb{C}}}(1)$ to $C_N(V, L)$ by ξ_L. We call $(C_N(V, L), \xi_L)$, or $C_N(V, L)$ for short, the *generalized cone of dimension N on* (V, L). We call ξ_L the *tautological bundle* of the generalized cone. If $N = n$ then $C_n(V, L) = V$ and $\xi_L = L$. If $N > n$ we say that the generalized cone $C_N(V, L)$ is *nondegenerate*. We call (V, L) the *base* of the cone and $(\varphi(D), \xi_{L|\varphi(D)})$ the *vertex* of the cone where $D \cong \mathbb{P}(E)$ is the variety on $\mathscr{C}$ determined by the quotient $L \oplus E \to E \to 0$. Note that $\varphi(D) \cong \mathbb{P}^{N-n}$. If $\varphi(D)$ is a single point, i.e., $E \cong \mathbb{O}_V$, then $C_{n+1}(V, L)$ is said to be the *cone* on (V, L). Appendix V-B of Fulton [Fu] is a good reference for basic results on cones.

Unfortunately, cones and generalized cones do not have to be normal when V is normal. Indeed assume V is normal, then normality of the generalized cone is equivalent to (V, L) being projectively normal (recall that (V, L) is said to be *projectively normal* if the natural map $\Gamma(L)^{\otimes t} \to \Gamma(tL)$ is onto for all $t \geq 1$). It is convenient to have a notion of normal generalized cones. Let L be an ample line bundle on a normal projective variety, V, of pure dimension n. Let $E = \oplus^{N-n}\mathcal{O}_V$ where N is an integer $\geq n$. Let $\mathcal{C} = \mathbb{P}(E \oplus L)$ and let ξ denote the tautological line bundle on $\mathcal{C}$. Note that ξ is semiample on $\mathcal{C}$. Let $r : \mathcal{C} \to Y$ with $\xi \cong r^*\mathcal{L}$ be the pair given by Lemma (1.1.3). We denote Y by $\mathfrak{C}_N(V, L)$ and $\mathcal{L}$ by ν_L. We call $(\mathfrak{C}_N(V, L), \nu_L)$, or $\mathfrak{C}_N(V, L)$ for short, the *normal generalized cone on* (V, L). We call ν_L the *tautological bundle* of the normal generalized cone. As in the last paragraph it follows that if $N = n$ then $\mathfrak{C}_n(V, L) = V$ and $\nu_L = L$. If $N > n$ we say that the normal generalized cone $\mathfrak{C}_N(V, L)$ is *nondegenerate*. We call (V, L) the *base* of the normal generalized cone and $(r(D), \nu_{L|r(D)})$ the *vertex* of the cone where $D \cong \mathbb{P}(E)$ is the variety on $\mathcal{C}$ determined by the quotient $L \oplus E \to E \to 0$. Note that $r(D) \cong \mathbb{P}^{N-n}$. If $E \cong \mathcal{O}_V$ then $\mathfrak{C}_{n+1}(V, L)$ is said to be the *normal cone* on (V, L).

The above definition of a normal generalized cone is consistent with the definition of a generalized cone in the sense that given a very ample line bundle, L, on a pure dimensional, normal projective variety, V, the normal generalized cone of dimension N on (V, L) is the normalization of the generalized cone of dimension N on (V, L).

We need in the sequel the following general fact.

Lemma 1.1.9. *Let A be an ample divisor on a smooth manifold, Y. Let W denote the normal generalized cone, $\mathfrak{C}_{\dim Y+1}(Y, A)$, and w denote the vertex of the cone, i.e., let $X := \mathbb{P}(\mathcal{E})$ where $\mathcal{E} := \mathcal{O}_Y \oplus \mathcal{O}_Y(A)$ and let $p : X \to W$ be the contraction of the section, E, of X corresponding to the quotient $\mathcal{O}_Y \oplus \mathcal{O}_Y(A) \to \mathcal{O}_Y$ such that $w = p(E)$. Then the local embedding dimension of W at the vertex, w, is bounded below by $h^0(A)$.*

Proof. It is a straightforward consequence of Nakayama's lemma, that we have heard attributed to Andreotti, that the local embedding dimension at a point w of an algebraic or analytic set is $\dim \mathfrak{m}_w/\mathfrak{m}_w^2$ where $\mathfrak{m}_w$ denotes the maximal ideal of the local ring of functions at the point, w. Since $\xi_{\mathcal{E}}$ is semiample and trivial on E, we have a natural map $\Gamma(\xi_{\mathcal{E}} - E) \to \mathfrak{m}_w/\mathfrak{m}_w^2$. The kernel of this map is zero. Indeed if a section of $\xi_{\mathcal{E}} - E$ went to 0 it would vanish to the second order at $F \cap E$ on a general fiber, F, of $\mathbb{P}(\mathcal{E}) \to Y$. Since $\xi_{\mathcal{E}|F} \cong \mathcal{O}_{\mathbb{P}^1}(1)$ we would conclude that the section is identically zero on the general fiber and hence zero on X. Since

$$h^0(\xi_{\mathscr{E}} - E) \geq h^0(\xi_{\mathscr{E}}) - 1 = h^0(\mathscr{E}) - 1 = h^0(A),$$

we are done. □

1.1.10. The sectional genus formula. Let V be an irreducible normal projective variety of dimension n and $\mathscr{L}$ a line bundle on V. The *sectional genus*, or *curve genus*, $g(\mathscr{L})$, of the pair $(V, \mathscr{L})$ is defined by

$$2g(\mathscr{L}) - 2 = (K_V + (n-1)\mathscr{L}) \cdot \mathscr{L}^{n-1}.$$

If V is Cohen-Macaulay and there are $n-1$ elements, $D_1, \ldots, D_{n-1}$, of $|\mathscr{L}|$ meeting in a curve, C, then as is discussed in the proof of Lemma (1.7.6),

$$K_C \cong (K_V + (n-1)\mathscr{L})_C.$$

Thus if $h^0(\mathcal{O}_C) = 1$, $g(\mathscr{L}) = h^1(\mathcal{O}_C)$, the arithmetic genus of C. Hence in particular $g(\mathscr{L}) \geq 0$ if $\mathscr{L}$ is spanned and big. A result due to Fujita (see (3.1.3)) shows that $g(\mathscr{L}) \geq 0$ if $\mathscr{L}$ is ample. Note also that if $\mathscr{L}$ is nef and big and $n \leq 3$, then $g(\mathscr{L}) \geq 0$. If $\mathscr{L}$ is ample this is an easy consequence of Mori's contraction theorem (see e.g., (4.3.1)) and the proof can be found in the paper [91, §3] of the first author and Palleschi. In the general case the result has been proved in Fujita [255] by using Mori's theory and it is a consequence of the Flip conjecture (see Mori [451]). Fujita conjectured in [255] that $g(V, \mathscr{L}) \geq 0$ for a nef and big line bundle $\mathscr{L}$ on any normal variety V.

We have the following basic parity result.

Lemma 1.1.11. *Let V be an irreducible normal projective variety of dimension n. Let $L_1, \ldots, L_m$ be line bundles on V. If $n = 2m$ is even then*

$$K_V \cdot L_1 \cdot \prod_{i=2}^{m} L_i^2 \equiv \prod_{i=1}^{m} L_i^2 \bmod 2.$$

If $n = 2m - 1$ is odd then

$$K_V \cdot L_1 \cdot L_2 \cdot \prod_{i=3}^{m} L_i^2 \equiv (L_1 + L_2) \cdot L_1 \cdot L_2 \cdot \prod_{i=3}^{m} L_i^2 \bmod 2,$$

and in particular $K_V \cdot \prod_{i=2}^{m} L_i^2 \equiv 0 \bmod 2$.

Proof. We can choose an ample line bundle, H, on V such that $L_i + tH$ is very ample for $t \geq 1$ and all i.

Assume first that $n = 2m$ and let C be the transversal intersection of general $D_1 \in |L_1 + 2H|$ and $D_i, D_{i+m-1} \in |L_i + 2H|$ for $i = 2, \ldots, m$. Then

$$2g(C) - 2 = (K_V + \sum_{j=1}^{2m-1} D_j) \cdot \prod_{j=1}^{2m-1} D_j.$$

Thus modulo 2 we have $(K_V + L_1) \cdot L_1 \cdot \prod_{i=2}^{m} L_i^2 \equiv 0$. The argument is similar in the case when $n = 2m - 1$. □

One simple consequence of the above which we will use without comment is that if $\mathcal{L}$ is a line bundle on an irreducible projective variety, V, then $g(\mathcal{L}) \in \mathbb{Z}$. Indeed

$$2g(\mathcal{L}) - 2 = (K_V + (n-1)\mathcal{L}) \cdot \mathcal{L}^{n-1}.$$

If n is even this is equal to $n\mathcal{L}^n$ modulo 2 and if n is odd this is equal to $(n-1)\mathcal{L}^n$ modulo 2.

1.2 Surface singularities

Surface singularities play an important role in adjunction theory. In what follows x will denote a singular point of a normal surface, S. By possibly passing to a neighborhood of x, or by simply dealing only with the germ of S at x, it can be assumed without loss of generality that x is the only singularity of S. Throughout this section $p : \overline{S} \to S$ will be a desingularization of S at x with $\ell_1, \ldots, \ell_k$ the irreducible and reduced exceptional divisors of $p^{-1}(x)$, and the restriction $p_{\overline{S}-p^{-1}(x)}$ an isomorphism. Moreover the desingularization is *minimal* in the sense that no ℓ_i is a smooth $\mathbb{P}^1$ satisfying $\ell_i^2 = -1$. Note that if one of the ℓ_i was such a curve then it could be smoothly blown down. See Laufer [400, Chapter V] for a careful discussion of the minimal resolution.

Artin [34] showed that given a normal surface singularity, $x \in S$, as above, there exists a unique effective Cartier divisor, $\mathcal{E}$, on $\overline{S}$, whose support is $p^{-1}(x)$, with the properties:

1. $\mathcal{E} \cdot \ell_i \leq 0$ for all $\ell_i \subset p^{-1}(x)$;

2. given any effective divisor, Z, whose support is $p^{-1}(x)$, and such that $Z \cdot \ell_i \leq 0$ for all i, then $Z - \mathcal{E}$ is effective.

This divisor is called the *fundamental cycle* of the singularity x. Artin [35] showed that if x is a rational singularity, then given the maximal ideal sheaf, $\mathfrak{m}_x$, of x,

$p^*\mathfrak{m}_x$ is scheme theoretically $-\mathscr{E}$ (see e.g., §1.7 for the definition of rational singularities).

Lemma 1.2.1. *Let S be a normal $\mathbb{Q}$-Gorenstein surface with rK_S Cartier for a positive integer, r. We have $rK_{\overline{S}} \approx p^*(rK_S) + \sum_{i=1}^k a_i\ell_i$ for integers $a_1, \ldots, a_k$.*

Proof. To see this note that $\mathscr{L} := rK_{\overline{S}} - p^*(rK_S)$ is an invertible sheaf which is trivial on $\overline{S} - p^{-1}(x)$. Thus there exists a meromorphic section, s, of $\mathscr{L}$ which is nowhere vanishing on $\overline{S} - p^{-1}(x)$. Since zero and pole sets are divisors, the zero and pole sets of s form a divisor supported on $p^{-1}(x)$. Since $\mathscr{L}$ is the line bundle associated to the divisor of any section, we are done. □

In the following lemma we will need the basic theorem of Artin [35] that x is rational if and only if $p^{-1}(x)$ forms a tree of rational curves, i.e., each ℓ_i is a smooth rational curve, and there is no sequence of distinct irreducible and reduced curves, $C_1, \ldots, C_m \subset p^{-1}(x)$ among the ℓ_i, such that $C_i \cdot C_{i+1} \geq 1$ for $i = 1, \ldots, m$ with the convention that $C_{m+1} = C_1$.

Lemma 1.2.2. *Let x be a normal Gorenstein singularity of S. Then x is a rational singularity if and only if $K_{\overline{S}} \cong p^*K_S$. If x is not a rational singularity then $K_{\overline{S}} + \Delta \cong p^*K_S$ where Δ is an effective Cartier divisor whose support is $p^{-1}(x)$.*

Proof. By (1.2.1) we have $K_{\overline{S}} \approx p^*K_S + \sum_{i=1}^k a_i\ell_i$. We claim that $a_i \leq 0$ for all i. If not then $\sum_{i=1}^k a_i\ell_i = A - B$ where A, B are effective divisors with no common components and with A nonempty. Thus $K_{\overline{S}} \cdot A = A^2 - A \cdot B \leq A^2$. Since the intersection matrix of the singularity x is negative definite (see, e.g., Mumford [461], [285], or Sakai [531]), it follows that $K_{\overline{S}} \cdot A < 0$. Thus $K_{\overline{S}} \cdot \ell_i < 0$ for at least one of the ℓ_i. Since $\ell_i^2 < 0$ by the negative definiteness of the intersection matrix of the ℓ_i, we conclude that $K_{\overline{S}} \cdot \ell_i = \ell_i^2 = -1$ and thus ℓ_i is a smooth rational curve of self intersection -1. Since the resolution p is minimal, this cannot happen.

If the singularity is rational then all ℓ_i are smooth rational curves, and therefore by minimality of the desingularization, $\ell_i^2 \leq -2$ for all i. Thus $K_{\overline{S}} \cdot \ell_i \geq 0$ for all i. If any a_i is nonzero, $(\sum_{i=1}^k a_i\ell_i)^2 < 0$, by the negative definiteness of the intersection matrix. Therefore

$$0 \leq K_{\overline{S}} \cdot (\sum_{i=1}^{k} a_i\ell_i) = (\sum_{i=1}^{k} a_i\ell_i)^2 < 0.$$

Thus we conclude that all the a_i are zero.

Conversely if all the a_i are zero, then the singularity is rational. To see this note that if all the a_i are zero, then $K_{\overline{S}} \cdot \ell_i = 0$ and hence $\ell_i^2 = -2$, so that ℓ_i are smooth rational curves for all i. If the singularity is not rational, then as pointed out before the statement of the lemma, there exist irreducible curves, $C_1, \ldots, C_m \subset p^{-1}(x)$, such that $C_i \cdot C_{i+1} \geq 1$ for $i = 1, \dots, m$ with the convention that $C_{m+1} = C_1$. Hence

$$(C_1 + \cdots + C_m)^2 = -2m + 2\sum_{i=1}^{m} C_i \cdot C_{i+1} \geq 0$$

which contradicts the negative definiteness of the intersection matrix. Thus we have that if x is not rational, not all the a_i are zero.

Now assume that the singularity is not rational. We have already shown that $a_i \leq 0$ for all indexes i, but not all of the a_i are 0. Thus we can assume without loss of generality that at least one of the a_i is 0. If there are any ℓ_j with $a_j = 0$, then since $p^{-1}(x)$ is connected, we can choose an ℓ_j with $a_j = 0$ that meets an ℓ_i with $a_i < 0$. Thus $\ell_j \cdot (\sum_{i=1}^k a_i \ell_i) < 0$. Hence

$$K_{\overline{S}} \cdot \ell_j = (p^* K_S) \cdot \ell_j + (\sum_{i=1}^{k} a_i \ell_i) \cdot \ell_j < 0.$$

Thus again we get a contradiction to the minimality of the resolution, p. □

The rational Gorenstein singularities are exceedingly nice singularities. We refer to the nice article of Durfee [195] for a discussion of the many characterizations of rational Gorenstein singularities. We will occasionally refer to these singularities as rational double points, a name that refers to the characterization of these singularities as multiplicity 2 hypersurface singularities. We will need to know the fundamental cycles (which are just Dynkin diagrams) of the rational Gorenstein singularities. To be a rational Gorenstein singularity is so restricted that this is an easy computation. The fundamental cycles are enumerated in Table (1.1). For reader's convenience let us sketch the computation of the possible configurations (we leave to the reader the easier exercise of finding the multiplicities of the ℓ_i in the fundamental cycle). We will use the negative definiteness of the intersection matrix for the ℓ_i repeatedly without mentioning it.

Let $x \in S$ be a rational Gorenstein surface singularity. Let $p : \overline{S} \to S$ be a minimal desingularization as at the start of this section, and with the same notation. As pointed out in the proof of Lemma (1.2.2), $\ell_i^2 = -2$ for all i. Note that if ℓ_i and ℓ_j are distinct they are either disjoint or meet transversally in one point. To see this note that $0 > (\ell_i + \ell_j)^2 = -2 - 2 + 2\ell_i \cdot \ell_j$, i.e., $\ell_i \cdot \ell_j \leq 1$.

If no more that two of the ℓ_i meet any given ℓ_i then we have the A_n configuration of the table. Thus without loss of generality we can assume that there is at least one ℓ_i that meets at least three other distinct ℓ_i. Note that no more than 3 distinct ℓ_i can meet any other of the ℓ_i. To see this let X denote one of the curves ℓ_i. Let $C_1, \dots, C_s$ denote distinct ℓ_i all different from X with $s \geq 4$ and such that $C_i \cdot X \geq 1$ for all i. Then we have the contradiction

$$0 > (2X + C_1 + \cdots + C_s)^2 \geq -8 - 2s + 4s = 2(s - 4).$$

Next note that there cannot be two distinct ℓ_i, say X and Y, such that X meets three ℓ_i distinct from itself and Y meets three ℓ_i distinct from itself. If there was such an X and Y then by the connectedness of $p^{-1}(x)$ we can find a chain of ℓ_i, say $A_1, \ldots, A_s$, with $X = A_1$, $Y = A_s$, $A_i \cdot A_{i+1} = 1$ for $i = 1, \ldots, s-1$, and $A_i \cdot A_j = 0$ for $j \geq i+2$. Let X_1, X_2 be the two ℓ_i not included in the chain that meet X and let Y_1, Y_2 be the two ℓ_i not in the chain that meet Y. Note that $(A_1 + \cdots + A_s)^2 = -2$. Since x is a rational singularity and the ℓ_i form a tree, it follows that $X_i \cdot Y_j = 0$ for $1 \leq i, j \leq 2$, and that $Y_1 \cdot Y_2 = 0 = X_1 \cdot X_2$. For the same reason it also follows that $(A_1 + \cdots + A_s) \cdot X_i = 1$ and $(A_1 + \cdots + A_s) \cdot Y_i = 1$ for $i = 1, 2$. Thus we have the contradiction

$$0 > (X_1 + X_2 + 2(A_1 + \cdots + A_s) + Y_1 + Y_2)^2 = -2 - 2 - 8 - 2 - 2 + 16 = 0.$$

Thus we have shown that the configuration of curves looks as follows. Let X denote the curve in the configuration that meets the other distinct curves in the configuration. There are chains $A_1, \ldots, A_s$; $B_1, \ldots, B_t$; and $C_1, \ldots, C_u$ of curves with $s \geq 1$, $t \geq 1$, and $u \geq 1$ and the following intersection properties:

1. $A_1 \cdot X = B_1 \cdot X = C_1 \cdot X = 1$;
2. $A_i \cdot A_{i+1} = 1$ for $i = 1, \ldots, s-1$;
3. $B_i \cdot B_{i+1} = 1$ for $i = 1, \ldots, t-1$;
4. $C_i \cdot C_{i+1} = 1$ for $i = 1, \ldots, u-1$;
5. all intersections of ℓ_i not enumerated above are 0.

We claim that at least one of s, t, u must be 1. Assume otherwise and consider the cycle

$$D := 3X + 2(A_1 + \cdots + A_{s-1}) + 2(B_1 + \cdots + B_{t-1}) + 2(C_1 + \cdots + C_{u-1}) + A_s + B_t + C_u.$$

We have the contradiction

$$0 > D^2 = -18 - 8 - 8 - 8 - 2 - 2 - 2 + 36 + 12 = 0.$$

If two of the s, t, u are 1 then we have the configuration D_n. Thus without loss of generality we can assume that $s \geq t \geq 2$ and $u = 1$. We can further assume without loss of generality that $s \leq 4$. If not then we can find a subconfiguration with $s = 5$ and $t = 2$. Consider the cycle

$$D := A_5 + 2A_4 + 3A_3 + 4A_2 + 5A_1 + 6X + 4B_1 + 2B_2 + 3C_1.$$

Note the contradiction that $0 > D^2 = 0$. We can further assume without loss of generality that $t = 2$. If not then we can find a subconfiguration with $s = 3$ and $t = 3$. Consider the cycle

$$D := A_3 + 2A_2 + 3A_1 + 4X + 3B_1 + 2B_2 + B_3 + 2C_1.$$

Note the contradiction that $0 > D^2 = 0$. If $s \leq 4$, $t = 2$ then we have one of the 3 configurations E_6, E_7, E_8.

We close the section with two simple lemmas we will need for the next section. Let x be a singularity of a normal $\mathbb{Q}$-Gorenstein surface S. Let $p : \overline{S} \to S$ be the minimal desingularization as at the beginning of this section. If in Lemma (1.2.1) all the $a_i > 0$ (respectively $a_i \geq 0$), then x is a *terminal* (respectively a *canonical*) singularity.

Lemma 1.2.3. *If x is a terminal singularity of a surface S then x is a nonsingular point.*

Proof. Since x is a terminal singularity we have

$$rK_{\overline{S}} \approx p^*(rK_S) + \sum_{i=1}^{k} a_i \ell_i$$

for some positive integers $a_1, \ldots, a_k$, r. By the negative definiteness of the intersection matrix of the ℓ_i we have that $(\sum_{i=1}^k a_i \ell_i)^2 < 0$ and thus that $\ell_j \cdot (\sum_{i=1}^k a_i \ell_i) < 0$ for some j. Therefore $rK_{\overline{S}} \cdot \ell_j = (\sum_{i=1}^k a_i \ell_i) \cdot \ell_j < 0$. Thus $K_{\overline{S}} \cdot \ell_j < 0$. Since $\ell_j^2 < 0$ by the negative definiteness of the intersection matrix of the ℓ_i, the genus formula, $2g(\ell_j) - 2 = K_{\overline{S}} \cdot \ell_j + \ell_j^2$, where $g(\ell_j)$ denotes the arithmetic genus of ℓ_j, gives $g(\ell_j) = 0$, $K_{\overline{S}} \cdot \ell_j = \ell_j^2 = -1$. Thus $\ell_j \cong \mathbb{P}^1$ contradicting the fact that the desingularization, $\overline{S}$, was chosen to be minimal. □

We leave the proof of the following lemma, which is a variant of the proof of (1.2.3), as an exercise.

Lemma 1.2.4. *A point $x \in S$ is a canonical singularity for a surface S if and only if x is a rational Gorenstein singularity.*

1.3 On the singularities that arise in adjunction theory

Even though our main interest is in the adjunction theory of smooth varieties, auxiliary varieties are introduced (reductions, divisors with special properties, etc.) which are not smooth. A good part of this section collects useful results that are standard for manifolds, but not so well-known for singular varieties.

For most of our results we need the assumption that varieties are *Cohen-Macaulay*, i.e., that the singularities of the varieties are at worst Cohen-Macaulay. This allows us to use Serre duality and consequently get vanishing for duals of ample line bundles. The canonical sheaf is especially well-behaved for varieties with

Table 1.1. Fundamental cycles for rational double point singularities

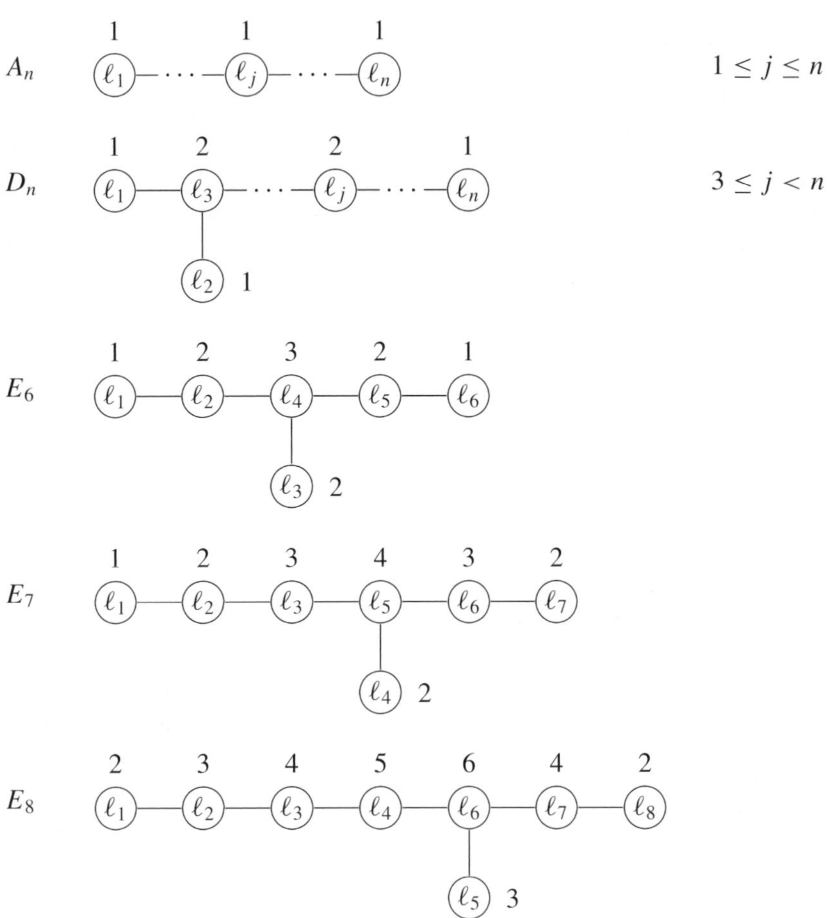

The above graphs describe the possible minimal resolutions, $p : \overline{S} \to S$, of a normal Gorenstein rational surface singularity, $x \in S$. Thus $\overline{S}$ is a smooth surface, $p_{\overline{S}-p^{-1}(x)}$ is an isomorphism, and $K_{\overline{S}} \cdot \ell = 0$ for any irreducible curve, $\ell \subset p^{-1}(x)$. Then the scheme-theoretic fiber $p^*\mathfrak{m}_x$ where $\mathfrak{m}_x$ is the maximal ideal sheaf of $x \in S$ is a divisor, $-\mathfrak{D} := -\sum_{i=1}^{n} a_i\ell_i$ where each ℓ_i is a smooth rational curve of multiplicity $-a_i$ and self intersection -2. For each type of singularity, a_i is given by the integer next to the circled ℓ_i, which represents the curve ℓ_i. Moreover ℓ_i and ℓ_k with $i \neq k$ meet only if there is a solid line segment connecting the circles, and then $\ell_i \cdot \ell_j = 1$. The divisor, $\mathfrak{D}$, equals the fundamental cycle, i.e., $\mathfrak{D} \cdot \ell_i \leq 0$ for all i and any other divisor $\sum_{i=1}^{n} b_i\ell_i$ with this property and with $b_i > 0$ for all i satisfies $b_i \geq a_i$ for all i.

at worst Cohen-Macaulay singularities, and the adjunction formula can be used for Cartier divisors without exception. Note that irreducible curves and normal surfaces are Cohen-Macaulay.

To use vanishing theorems for the adjoint sheaf of a line bundle, we need the assumption that the *irrational locus*, i.e., the locus of nonrational singularities, is finite (see §2.2). Example (2.2.10) presents a very well-behaved 3-dimensional ample divisor, A, on a smooth projective 4-fold such that A has a 1-dimensional irrational locus for which the Kodaira vanishing theorem fails. Many of the basic structure theorems of adjunction theory fail in one way or another for this example.

To use the full strength of the Lefschetz hyperplane section theorem we need an assumption of local complete intersection (see §2.3). The condition that the variety has at most rational singularities (which implies that the variety is Cohen-Macaulay) lets us carry out many arguments. For example, arguments that use the Albanese variety or holomorphic forms work pretty much the same as in the manifold case, see §2.4.

We also prove a general Hodge index theorem for singular varieties and some consequences in §2.5.

Starting with a smooth manifold we are forced to deal with $\mathbb{Q}$-factorial varieties. In particular to use various results from Mori theory we need assumptions on singularities being canonical or terminal (which imply that the singularities are at worst rational, but not conversely). In this section and §4.2, we recall the definitions and basic well-known facts about projective varieties with canonical or terminal singularities from Mori's theory of extremal rays. For full details we refer to Mori [447], to the surveys articles of Mori [450, 452], Kawamata, Matsuda, and Matsuki [346], Kollár [358], and to the book [164] of Clemens, Kollár, and Mori.

We say that a normal variety V has *terminal singularities* (respectively *canonical singularities*) if the following conditions are satisfied:

1. $rK_V \in \operatorname{Pic}(V)$ for some positive $r \in \mathbb{Z}$, (i.e., $K_V \in \operatorname{Pic}(V) \otimes \mathbb{Q}$);

2. for a resolution of singularities $\varphi : \overline{V} \to V$ one can write

$$rK_{\overline{V}} \approx \varphi^*(rK_V) + \sum_i a_i E_i,$$

with $a_i > 0$ (respectively $a_i \geq 0$), where the E_i's vary among *all* exceptional divisors of φ.

Elkik [214] has proved that canonical singularities are rational and therefore Cohen-Macaulay.

As we have seen in Lemma (1.2.3), the terminal singularities occur only for $n \geq 3$. More generally, the locus of terminal singularities has codimension bigger or equal to 3. Hence for $n = 3$ the terminal singularities are isolated. This is in fact a crucial point in the theory which distinguishes the 3-dimensional case from the general case when $n \geq 4$. The 3-dimensional terminal singularities have been classified by Mori [449]. Let us only note that, as has been proved by Reid in [516], the 3-dimensional terminal singularities are either cyclic quotients, i.e., locally V is isomorphic to $\mathbb{A}^3/\mu_r$, where μ_r denotes the cyclic group of the r-th roots of the unity, or are points $p \in V$ such that there exists a suitable section H, $p \in H$, and p is a rational Gorenstein singularity for H, i.e., p is a rational double point for H. For further properties and a detailed study of terminal and canonical singularities we refer to Reid [515, 518].

Lemma 1.3.1. *Let V be a normal projective variety of dimension $n \geq 2$. Let T be the locus of terminal singularities on V. Then, with the convention that the empty set has dimension -1, $\mathrm{cod}_V T \geq 3$.*

Proof. The result is true for $n = 2$, since a terminal singularity of a surface is a smooth point by Lemma (1.2.3). Then, by induction, we can assume that $n \geq 3$ and that the result is true in dimension $n - 1$. Let $\varphi : V' \to V$ be a resolution of the singularities of V. Let H be a general hyperplane section of V. Since H is general we can assume that the inverse image $\varphi^{-1}(H)$ coincides with the proper transform H' of H and that the restriction $\varphi' : \varphi^{-1}(H) \to H$ is a resolution of the singularities of H. We have

$$eK_{V'} \approx \varphi^*(eK_V) + \sum_i a_i E_i,$$

where e is the index of V and the a_i's are positive integers for each index i. Since H is general, we can further assume, without loss of generality, that H contains none of the $\varphi(E_i)$. Since V is normal and Cohen-Macaulay the adjunction formula applies to give

$$\mathcal{O}_{H'}(\varphi^*(eK_V)) \cong \varphi'^*\mathcal{O}_H(eK_V) \cong \varphi'^*(\mathcal{O}_H(eK_H - eH)),$$

and also $eK_{H'} \approx (eK_{V'} + eH')_{H'}$. Since $\varphi'^*\mathcal{O}_H(H) \approx \mathcal{O}_{H'}(H')$ we conclude that

$$eK_{H'} \approx \varphi'^*(eK_H) + \sum_i a_i E'_i,$$

where the restrictions E'_i of E_i to H' vary among all the exceptional divisors of φ'. Therefore H has terminal singularities and thus, by the induction assumption, $\mathrm{cod}_H T_H \geq 3$ where T_H denotes the locus of terminal singularities of H. By the above, we have $T \cap H \subset T_H$. So we get $\dim(T \cap H) \leq \dim T_H \leq n - 4$. This implies $\dim T \leq n - 3$ and we are done. □

It is worthwhile to examine in detail the following classical examples of a 3-dimensional terminal singularity.

Example 1.3.2. Let $S \subset \mathbb{P}^5$ be the Veronese surface, i.e., S equals $\mathbb{P}^2$ embedded in $\mathbb{P}^5$ by $|\mathcal{O}_{\mathbb{P}^2}(2)|$. Let V be the cone over S. Let $\sigma : X \to V$ be a resolution of the singularity and let E be the exceptional divisor. Then $E \cong \mathbb{P}^2$, $\mathcal{O}_E(E) \cong \mathcal{O}_{\mathbb{P}^2}(-2)$. We identify S with a generic hyperplane section of the cone, $V \subset \mathbb{P}^6$, and by abuse of notation, also with the pullback to X of a generic hyperplane section of the cone. Note that under the induced projection, $p : X \to S$, we have $X \cong \mathbb{P}(\mathcal{O}_{\mathbb{P}^2}(2) \oplus \mathcal{O}_{\mathbb{P}^2})$. Thus $K_X \cong -2S + p^*\mathcal{O}_{\mathbb{P}^2}(-1)$. Since $\mathrm{Pic}(X) = \mathbb{Z}[S] \oplus \mathbb{Z}[p^*\mathcal{O}_{\mathbb{P}^2}(1)]$, it is clear that $E + p^*\mathcal{O}_{\mathbb{P}^2}(2) \cong S$. Thus

$$2K_X - E \cong -4S + p^*\mathcal{O}_{\mathbb{P}^2}(-2) - E \cong -5S.$$

Thus letting V' denote the affine neighborhood, $V - S$, of the vertex of the cone, and letting X' denote $X - S$ we have $2K_{X'} \approx \sigma^*(2K_{V'}) + E$. Thus the vertex of the cone is a terminal singularity.

Example 1.3.3. (See Ueno [602, §16]) Let A be a nonsingular Abelian variety of dimension 3. Let us consider the involution $i : A \to A$ which sends an element $a \in A$ to its opposite $-a$. This involution has $2^6 = 64$ fixed points, the points of order two. Therefore the quotient $A/ < i >$ has 64 isolated singular points. Let X be the Kummer 3-fold obtained as canonical resolution, $\varphi : X \to A/ < i >$, of the singularities of $A/ < i >$ and let $E_1, \dots, E_{64}$ be the exceptional divisors. For every $j = 1, \dots, 64$ one has

$$E_j \cong \mathbb{P}^2 \quad ; \quad \mathcal{N}_{E_j/X} \cong \mathcal{O}_{\mathbb{P}^2}(-2). \tag{1.2}$$

Note that for any line $\ell \cong \mathbb{P}^1$ in E_j, we find $K_{E_j} \cdot \ell = K_X \cdot \ell + E_j \cdot \ell$ and therefore $K_X \cdot \ell = -1$. Hence in particular K_X is not numerically effective.

In a neighborhood of any point p of order 2 the involution i can be given analytically by the map $i : \mathbb{C}^3 \to \mathbb{C}^3$ defined by $i(x, y, z) = (-x, -y, -z)$. Moreover $\mathbb{C}^3/ < i >$ can be seen geometrically as the affine cone over the Veronese surface in $\mathbb{P}^5$. Note that the Cartier divisor $2K_A$ is locally defined at p (up to coefficients in $\mathcal{O}_{A,p}$) by the 2-canonical form $(dx \wedge dy \wedge dz)^{\otimes 2}$ which is invariant with respect to the action of i. Thus there is a trivializing section of $2K_A$ that descends to give a trivializing section of $2K_{\mathrm{reg}(A/<i>)}$. Thus $2K_{A/<i>} \cong \mathcal{O}_{A/<i>}$ and therefore $2K_{A/<i>}$ is a Cartier divisor. From relations (1.2) we get, for each index j,

$$K_{X|E_j} \approx K_{E_j} - E_{j|E_j} \cong \mathcal{O}_{\mathbb{P}^2}(-3) \otimes \mathcal{O}_{\mathbb{P}^2}(2) \cong \mathcal{O}_{\mathbb{P}^2}(-1).$$

Since X is smooth and X and $A/ < i >$ are isomorphic outside of the singular points of $A/ < i >$, we have further that

$$2K_X \approx \varphi^*(2K_{A/<i>}) + \sum_j a_j E_j,\ a_j \in \mathbb{Z}.$$

Since $2K_{A/<i>}$ is linearly trivial we obtain, for each index j, $2K_{X|E_j} \approx a_j E_{j|E_j} \approx \mathcal{O}_{\mathbb{P}^2}(-2a_j)$ that is $\mathcal{O}_{\mathbb{P}^2}(-2) \cong \mathcal{O}_{\mathbb{P}^2}(-2a_j)$. Then we conclude that $a_j = 1$ for all indexes j and that $A/ < i >$ has terminal singularities (of index 2).

Example 1.3.4. In $\mathbb{C}^{n+1}$ for $n \geq 2$ let $f_d(x_0, \dots, x_n)$ be a degree d homogeneous equation defining a cone, V_d, with vertex the origin, $O \in \mathbb{C}^{n+1}$. Note that V_d is 1-Gorenstein, i.e., K_{V_d} is a Cartier divisor. Assume that V_d is smooth outside of O. Let $\sigma : X \to V_d$ be the blowing up at O. The exceptional divisor E of σ is the smooth degree d hypersurface in $\mathbb{P}^n$ defined by $f_d = 0$, and $\mathcal{N}_{E/X} \cong \mathcal{O}_E(-1)$. We have

$$\mathcal{O}_E(K_X + E) \cong K_E \cong \mathcal{O}_E(d - n - 1)$$

and hence $\mathcal{O}_E(K_X) \cong \mathcal{O}_E(d - n)$. Furthermore, since K_{V_d} is a Cartier divisor, we have $K_X \cong \sigma^* K_{V_d} + aE$ for some integer a and $\mathcal{O}_E(\sigma^* K_{V_d}) \cong \mathcal{O}_E$. Therefore $\mathcal{O}_E(aE) \cong \mathcal{O}_E(d - n)$ so that $-a = d - n$. It therefore follows that O is a terminal singularity for $d < n$. Note that if $n = 2$ this implies—in agreement with Lemma (1.2.3)—that O is a terminal singularity only if $d = 1$ and $V_d \cong \mathbb{C}^2$.

Example 1.3.5. (See (7.5.6) and (7.5.7)) Let $\mathcal{M}$ be a connected projective manifold of dimension n and $\mathcal{L}$ an ample and spanned line bundle on $\mathcal{M}$. Let $(X, \mathcal{D})$ be the second reduction of $(\mathcal{M}, \mathcal{L})$. Then X has isolated terminal singularities of index 2. Moreover, if $n = 3$, X is 2-Gorenstein (i.e., $2K_X$ is a Cartier divisor) while, for $n \geq 4$, X is 2-factorial and it is Gorenstein in even dimensions.

1.4 Curves

Curves play a key role in many arguments. Given a reduced curve, C, we denote its arithmetic genus, $h^1(\mathcal{O}_C)$, by $g(C)$. The arithmetic genus, $g(\overline{C})$, of the desingularization, $\overline{C}$, of C is called *the geometric genus* of C, and is denoted by $\overline{g}(C)$. A *rational curve* is an irreducible curve whose normalization is biholomorphic to $\mathbb{P}^1$.

The following simple lemma is very useful.

Lemma 1.4.1. *Let C be a reduced, connected, projective curve and let $p : \overline{C} \to C$ denote the normalization mapping. Letting $\overline{C} = C_1 + \cdots + C_m$ denote the decomposition of $\overline{C}$ into connected components, we have $\overline{g}(C) = \sum_{i=1}^m h^1(\mathcal{O}_{C_i})$, and $\overline{g}(C) + s + 1 = m + g(C)$, where s is a sum over the points in the singular set of C of positive integers. Thus $g(C) \geq \overline{g}(C)$, and if C is irreducible and satisfies $g(C) = 0$, then C is biholomorphic to $\mathbb{P}^1$.*

Proof. Since the normalization mapping is finite, the first direct image $p_{(1)}(\mathcal{O}_{\overline{C}})$ is 0, and therefore by the Leray spectral sequence $h^1(p_*(\mathcal{O}_{\overline{C}})) = h^1(\mathcal{O}_{\overline{C}}) = \overline{g}(C)$. Considering the exact sequence

$$0 \to \mathcal{O}_C \to p_*(\mathcal{O}_{\overline{C}}) \to \mathcal{S} \to 0,$$

where $\mathcal{S}$ is a skyscraper sheaf supported on the set of singular points of C, we have $\chi(\mathcal{O}_{\overline{C}}) = \chi(\mathcal{O}_C) + s$ with $s := h^0(\mathcal{S})$. This gives $m - \sum_{i=1}^{m} h^1(\mathcal{O}_{C_i}) = 1 + s - g(C)$, which is the desired equality. Note that $s \geq m - 1$ since $\overline{C}$ has at least $m - 1$ singular points. □

There is one useful simple link between the arithmetic genus of a curve, C, and the number of sections of a line bundle on C.

Lemma 1.4.2. *Let C be an irreducible and reduced curve of arithmetic genus, $g := h^1(\mathcal{O}_C)$, and L a line bundle on C of degree $d \geq 1$. Then $h^0(L) \geq d - g + 1$. Moreover $h^0(L) \leq 1 + d$, with equality only if $(C, L) \cong (\mathbb{P}^1, \mathcal{O}_{\mathbb{P}^1}(d))$.*

Proof. By the Riemann-Roch theorem,

$$h^0(L) = h^1(L) + d - g + 1 \geq d - g + 1$$

which proves the first assertion. From the exact sequence

$$0 \to \mathcal{O}_C \to L \to \mathcal{S} \to 0,$$

where $\mathcal{S}$ is a skyscraper sheaf supported on the zero-section of L, we get $h^0(L) \leq 1 + d$.

Thus we have reduced to the case when $h^0(L) = 1 + d$.

First assume that C is smooth. Note that L is spanned. Indeed, if not, let $x \in C$ be a basepoint of $|L|$. Then $h^0(L - x) = h^0(L) = d + 1$ and by the above we have the contradiction $h^0(L - x) \leq 1 + (d - 1) = d$. Next, note that if $d = 1$, then $\Gamma(L)$ gives a map of degree one, so that $C \cong \mathbb{P}^1$. Thus we can assume $d \geq 2$. By induction, assume that the result is true for line bundles of degree $< d$. Take any point $x \in C$ and consider the exact sequence

$$0 \to L - x \to L \to L_x \to 0.$$

Since L is spanned, $\Gamma(L)$ surjects on $\Gamma(L_x) \cong \mathbb{C}$, so that $h^0(L - x) = h^0(L) - 1 = d$. Since $\deg(L - x) = d - 1$, the induction assumption implies that $(C, L - x) \cong (\mathbb{P}^1, \mathcal{O}_{\mathbb{P}^1}(d - 1))$ and thus $(C, L) \cong (\mathbb{P}^1, \mathcal{O}_{\mathbb{P}^1}(d))$.

Now assume that C is singular and let $p : \overline{C} \to C$ be the normalization mapping. Let $\overline{L} := p^*L$ and note that $\deg \overline{L} = \deg L = d$. Note also that since $1 + d \geq h^0(\overline{L}) \geq h^0(L) = 1 + d$, we conclude by the last paragraph that $(\overline{C}, \overline{L}) \cong (\mathbb{P}^1, \mathcal{O}_{\mathbb{P}^1}(d))$. If p is not a biholomorphism then the composition, π, of p with the map given by $H^0(L)$ cannot be an embedding. Note that π is given by $H^0(\overline{L})$. Since $H^0(\overline{L}) = H^0(\mathcal{O}_{\mathbb{P}^1}(d))$ gives an embedding, we conclude that p is an isomorphism, contradicting the assumption on C. □

Corollary 1.4.3. *Let L be a degree 1, spanned line bundle on an irreducible and reduced projective curve, C. Then $(C, L) \cong (\mathbb{P}^1, \mathcal{O}_{\mathbb{P}^1}(1))$.*

Proof. Since L is spanned and nontrivial, $h^0(L) \geq 2$. Since $\deg L = 1$, the result follows immediately from Lemma (1.4.2). □

Corollary 1.4.4. *Let L be a degree 2 line bundle on an irreducible and reduced projective curve, C. If the morphism to the projective space associated to $|L|$ is generically one-to-one, e.g., if L is very ample, then $(C, L) \cong (\mathbb{P}^1, \mathcal{O}_{\mathbb{P}^1}(2))$.*

Proof. To see this note that since L is spanned and nontrivial, $h^0(L) \geq 2$. If $h^0(L) \geq 3$, then we are done by Lemma (1.4.2). Thus we can assume that $h^0(L) = 2$. In this case the morphism given by $|L|$ is to $\mathbb{P}^1$, and since L is isomorphic to the pullback of $\mathcal{O}_{\mathbb{P}^1}(1)$, we see that $2 = \deg L$ equals the sheet number of the morphism contradicting the generic one-to-one hypothesis. □

Remark 1.4.5. Let C be an irreducible and reduced curve. Without the generic one-to-one hypothesis in Corollary (1.4.4), the result is false. Counterexamples are given by *hyperelliptic curves*, i.e., smooth double covers of $\mathbb{P}^1$ of genus ≥ 1. Note that these important double covers, $p : C \to \mathbb{P}^1$, can be constructed with C having any genus $g \geq 1$. Let $L := p^*\mathcal{O}_{\mathbb{P}^1}(1)$. Note that L is ample and spanned and $\deg L = 2$.

Lemma 1.4.6. *Let C be an irreducible and reduced Gorenstein curve of arithmetic genus g. Let L be a line bundle on C of degree $2g - 2$. Then $h^0(L) \geq g$ if and only if $L \approx K_C$.*

Proof. We have to prove the "only if" part. The Riemann-Roch theorem yields $h^1(L) = h^0(L) - g + 1$. Then $h^1(L) = h^0(K_C - L) > 0$. Since $\deg(K_C - L) = 0$ it thus follows $K_C \approx L$. □

Lemma 1.4.7. (Clifford's theorem) *Let C be a smooth curve of genus g. Let L be a line bundle on C of degree $d \leq 2g$. Then $h^0(L) \leq d/2 + 1$.*

Proof. If $h^0(K_C - L) > 0$ then L is special and the result is the usual Clifford inequality. Note that in this case $d \leq 2g - 2$. If $h^0(K_C - L) = 0$, the Riemann-Roch theorem gives $h^0(L) = d - g + 1 \leq d/2 + 1$, where the inequality follows from the assumption $d \leq 2g$. □

We will need the Clifford inequality in the singular case also (see Beauville [77, (1.3.1)]).

Lemma 1.4.8. (Clifford's theorem; singular case) *Let C be a reduced, irreducible, Gorenstein curve of arithmetic genus g. Let L be a spanned line bundle on C of degree $d \leq 2g$. Then $h^0(L) \leq d/2 + 1$.*

Proof. If $h^0(K_C - L) = 0$ the same argument as in (1.4.7) gives the result.

If $h^0(K_C - L) > 0$, take a not identically zero section $s \in \Gamma(K_C - L)$. Since L is spanned, we can choose a section $t \in \Gamma(L)$ such that $t^{-1}(0) \cap s^{-1}(0) = \emptyset$. It thus follows that the section $\sigma = s \oplus t$ of the vector bundle $\mathscr{E} := (K_C - L) \oplus L$ is nowhere zero. Thus the multiplication, $\mathbb{C}s$, of s by constants gives a rank one vector subbundle of $\mathscr{E}$ (isomorphic to the trivial bundle) and we have a line bundle as a quotient

$$0 \to \mathcal{O}_C \to \mathscr{E} \to \det \mathscr{E} \to 0.$$

This is in fact the Koszul complex of σ. Note that $\det \mathscr{E} \cong (K_C - L) + L \cong K_C$. From the exact sequence above, we get

$$h^0(L) + h^1(L) = h^0(L) + h^0(K_C - L) = h^0(\mathscr{E}) \leq 1 + g.$$

The Riemann-Roch theorem gives $h^0(L) - h^1(L) = d - g + 1$. Thus we conclude that $2h^0(L) \leq d + 2$, or $h^0(L) \leq d/2 + 1$. □

Any finite set of smooth points is the zero set of a section of an ample line bundle on C. From this it is clear that regarded as a variety with no extra structure, an ample divisor on a curve imposes no restrictions on the genus of the curve. As Remark (1.4.5) points out even ample and spanned line bundles have zero sets whose degrees have little relation to the genus of the underlying curve. There is a useful strong link between the genus of an irreducible curve, C, and the number of sections of L, in the case that the morphism associated to $|L|$ is generically one-to-one. A subvariety, $X \subset \mathbb{P}^N$, is said to be a *nondegenerate subvariety of* $\mathbb{P}^N$, if there is no linear $\mathbb{P}^{N-1}$ that contains X.

1.4.9. Castelnuovo's bound. Let C be a reduced, irreducible projective curve. Assume that $\varphi : C \to \mathbb{P}^N$ is a generically one-to-one morphism, and that $\varphi(C)$ does not lie in any hyperplane. Let d denote the degree of $\varphi(C)$ in $\mathbb{P}^N$. Then the arithmetic genus, $g(C)$, of C satisfies the inequality (see, e.g., Hartshorne [307, (3.7)] and [310])

$$g(C) \leq \mathrm{Castel}(d, N) := \left[\frac{d-2}{N-1}\right]\left(d - N - \left(\left[\frac{d-2}{N-1}\right] - 1\right)\frac{N-1}{2}\right), \quad (1.3)$$

where $[x]$ means the greatest integer $\leq x$. Note that $g(C) \leq g(\varphi(C))$ and hence if equality holds in (1.3) then φ is an embedding. Moreover if equality is taken on then $\varphi(C)$ is projectively normal, i.e., for all $t \geq 0$, the restriction mapping

$$H^0(\mathbb{P}^N, \mathcal{O}_{\mathbb{P}^N}(t)) \to H^0(\varphi(C), \mathcal{O}_{\mathbb{P}^N}(t)_{\varphi(C)})$$

is onto.

If $N \geq 3$ and $g(C)$ does not reach the maximum with respect to (1.3), then we have the bound of Gruson-Peskine [299],

$$g(C) \leq d(d-3)/6 + 1. \tag{1.4}$$

1.5 Nefvalue results

A major theorem of Kawamata [346] and Shokurov [563] is the following.

Theorem 1.5.1. (Basepoint free theorem) *Let V be a normal, irreducible projective variety of dimension $n \geq 3$ with at most canonical singularities. Let D be a nef Cartier divisor such that $aD - K_V$ is nef and $(aD - K_V)^n > 0$ for some positive integer a. Then $|mD| \neq \emptyset$ and $|mD|$ has no basepoints for all $m \gg 0$.*

For the proof of the following theorem we refer to Kawamata [343], Kawamata, Matsuda, and Matsuki [346, §4], and Clemens, Kollár, and Mori [164]. This is one of the main tools we use in Chapter 7. We state the result in the case of terminal singularities, which occur in our theory, even though it holds true in the more general case of log-terminal singularities.

Theorem 1.5.2. (Kawamata rationality theorem) *Let V be a normal, irreducible projective variety of dimension n with terminal singularities and let $r = r(V)$ be the index of the singularities of V. Let $\pi : V \to Y$ be a projective morphism onto a variety Y. Let L be a π-ample Cartier divisor of V. If K_V is not π-nef then*

$$\tau = \min\{t \in \mathbb{R} \mid K_V + tL \textit{ is } \pi\textit{-nef}\}$$

is a rational number. Furthermore expressing $r\tau = u/v$ with u, v coprime positive integers, we have $u \leq r(b+1)$ where $b = \max_{y \in Y}\{\dim \pi^{-1}(y)\}$.

Theorem (6.3.15) characterizes the smooth V for which equality occurs in Theorem (1.5.2).

Definition 1.5.3. Let V be a normal variety of dimension n with terminal singularities and let $r = r(V)$ be the index of the singularities of V. Let $\pi : V \to Y$ be a projective morphism onto a variety Y. Let $\mathcal{L}$ be a π-ample Cartier divisor of V. Assume that K_V is not π-nef. Let τ be the positive rational number given by the Kawamata rationality theorem, (1.5.2).

We say that the rational number τ is the *π-nefvalue* of $(V, \mathcal{L})$. If Y is a point, τ is called the *nefvalue* of $(V, \mathcal{L})$. Note also that, if Z is a point, then $K_V + \tau\mathcal{L}$ is nef and hence by the Kawamata rationality theorem, (1.5.2), we have that $r\tau = u/v$ for two coprime positive integers, u and v. Thus by the Kawamata-Shokurov basepoint free theorem, (1.5.1), we know that $|m(vrK_V+u\mathcal{L})|$ is basepoint free for all $m \gg 0$. Therefore for such m, $|m(K_V+\tau\mathcal{L})|$ defines a morphism $f : V \to \mathbb{P}_{\mathbb{C}}$. Let $f = s \circ \phi$ be the Remmert-Stein factorization of f where $\phi : V \to Y$ is a morphism with connected fibers onto a normal projective variety, Y, and $s : Y \to \mathbb{P}_{\mathbb{C}}$ is a finite-to-one surjective morphism. By Lemma (1.1.3) we know that the morphism, ϕ, is the same for any $m > 0$ such that $|m(rvK_V + u\mathcal{L})|$ is basepoint free, and thus only depends on $(V, \mathcal{L})$. Note that, by Lemma (1.1.3), s is an embedding for $m \gg 0$ and therefore $f = \phi$ for $m \gg 0$. We call $\phi : V \to Y$ the *nefvalue morphism* of $(V, \mathcal{L})$. We also know by Lemma (1.1.3) that there is an ample line bundle H on Y such that

$$vrK_V + u\mathcal{L} \cong \phi^* H. \tag{1.5}$$

Let us recall the following simple useful fact.

Remark 1.5.4. Let V be as in (1.5.2) and $\mathcal{L}$ an ample line bundle on V. Let τ be the nefvalue of $(V, \mathcal{L})$ and ϕ the nefvalue morphism of $(V, \mathcal{L})$. Then $\mathcal{L}$ is ϕ-ample and

$$\tau = \min\{t \in \mathbb{R} \mid K_V + t\mathcal{L} \text{ is nef}\} = \min\{t \in \mathbb{R} \mid K_V + t\mathcal{L} \text{ is } \phi\text{-nef}\}.$$

That is τ coincides with the ϕ-nefvalue of $(V, \mathcal{L})$.

Lemma 1.5.5. *Let* $(V, \mathcal{L})$ *be as in* (1.5.2). *A real number* τ *is the nefvalue of* $(V, \mathcal{L})$ *if and only if* $K_V + \tau\mathcal{L}$ *is nef but not ample.*

Proof. To prove the only if part, assume $K_V + \tau\mathcal{L}$ to be ample. Then for any small rational number ε, $0 < \varepsilon \ll 1$, $K_V + (\tau - \varepsilon)\mathcal{L}$ is ample by Kleiman's ampleness criterion (see Theorem (4.2.1), [349, Chapter 4, §2, Proposition 2] and also Kawamata, Matsuda, and Matsuki [346]). Thus $\tau - \varepsilon < \tau$ contradicts the minimality of τ.

To show the converse, assume that $K_V + s\mathcal{L}$ is nef but not ample. Assume that the nefvalue, τ, is $< s$. Then $K_V + s\mathcal{L} = (K_V + \tau\mathcal{L}) + (s - \tau)\mathcal{L}$ is ample. This contradicts the assumption that $K_V + s\mathcal{L}$ is not ample. □

The following lemma occasionally lets us "remove the denominators" arising in Theorem (1.5.2).

Lemma 1.5.6. *Let* $\phi : V \to Y$ *be a morphism between irreducible projective varieties. Assume that* V *is* $\mathbb{Q}$*-Gorenstein of index* r*. Assume that there are positive, coprime integers,* u, v, *and a line bundle,* L, *on* V, *such that* $vrK_V + uL \cong \phi^* H$ *for an ample line bundle* H *on* Y*. Then there exists an ample line bundle,* $\mathcal{L}$, *on* V *and a positive integer* p *such that* $rK_V + u\mathcal{L} \cong p\phi^* H$*. Furthermore if* ϕ *has at least one positive dimensional fiber,* u/r *is the nefvalue of* $(V, \mathcal{L})$.

Proof. Choose positive integers p, q such that $vp - uq = 1$. Let $\mathcal{L} = qrK_V + pL$. To see that $\mathcal{L}$ is ample, note that

$$v\mathcal{L} = v(qrK_V + pL) = q\phi^*H - uqL + pvL = q\phi^*H + L.$$

Furthermore,

$$rK_V + u\mathcal{L} = rK_V + u(qrK_V + pL) = rpvK_V + upL = p\phi^*H.$$

If ϕ has at least one positive dimensional fiber, then $rK_V + u\mathcal{L}$ is nef but not ample, so that u/r is the nefvalue of $(V, \mathcal{L})$ by Lemma (1.5.5). □

We conclude this section by recalling that Kollár [359] recently obtained the following effective bound for the basepoint freeness.

Theorem 1.5.7. *Let V, D, n, a be as in* (1.5.1). *Then $|2(n+2)!(a+n)D|$ is basepoint free.*

1.6 Universal sections and discriminant varieties

The constructions in this section go back to the last century. We refer to Mumford [634, Appendix to Chapter 6] and Harris [311, Lecture 4] for further discussion. Let L be a line bundle on a complex manifold, X, and let $V \subset \Gamma(L)$ be a finite dimensional space of sections of L. Let $|V| = \mathbb{P}(V^*)$, the space of 1-dimensional linear subspaces of V. Let $\pi_1 : |V| \times X \to |V|$ and $\pi_2 : |V| \times X \to X$ be the product projections, and let $\mathcal{L} := \pi_1^*\mathcal{O}_{|V|}(1) \otimes \pi_2^*L$. The *universal section of $\mathcal{L}$ associated to V* is defined to be $s_V := \sum_i \pi_1^* z_i \otimes \pi_2^* s_i$ where $\{s_i\}$ is a basis of V and $\{z_i\}$ are the homogeneous coordinates corresponding to the dual basis of V^*. Note that s_V is independent of the choice of the basis and corresponds to the identity under the functorial isomorphism

$$\pi_1^*V^* \otimes \pi_2^*V \cong \mathrm{Hom}_{\mathbb{C}}(\pi_1^*V^*, \pi_2^*V) \cong \mathrm{Hom}_{\mathbb{C}}(V, V).$$

The *universal zero set, Z_V, associated to V* or the *space of zero sets of sections of L coming from V* is defined as $s_V^{-1}(0)$. Note that Z_V equals the subset of $|V| \times X$ given by $Z_V := \{(v, x) \mid v(x) = 0\}$.

Lemma 1.6.1. *We use the above notation. Z_V is smooth at a point (v, x) if and only if either V spans L at x or v defines a nonsingular divisor on X at x.*

Proof. If there is only one section, the assertion is clear. Therefore we can assume that $r := \dim V \geq 2$. Fix a trivialization of $\mathcal{L}$ on some neighborhood of (v, x).

Choose a hyperplane $H \subset |V|$ not containing v so that we can choose linear coordinates, $u_1, \ldots, u_{r-1}$ over $|V| - H$ such that $\pi_1^* u_i(v) = 0$ for all i, i.e.,

$$s_V = v + \sum_i (\pi_1^* u_i) \otimes (\pi_2^* s_i).$$

By the implicit function theorem, smoothness at (v, x) is equivalent to at least one derivative of s_V being nonzero at (v, x). The differential of s_V at (v, x) is

$$d_X v + \sum_i (\pi_1^* du_i) \otimes (\pi_2^* s_i)$$

with d_X denoting the differential in the directions along X. If V spans L at x, then for some i, $s_i(x) \neq 0$, and the assertion is clear. If all the elements of V are zero at x, then the derivative of the section, s_V, at (v, x) is $d_X v$ which is nonzero if v defines a smooth divisor at x. □

The following lemma is straightforward. It guarantees that Z_V inherits good properties from X, e.g., smoothness, normality, at most rational singularities, at most Cohen-Macaulay singularities, at most local complete intersection singularities, irreducibility, etc.

Lemma 1.6.2. *We use the above notation. Assume that V spans L at all points of X. Then Z_V is isomorphic to the $\mathbb{P}^{\dim V - 1}$-bundle over X, $\mathbb{P}(\mathcal{K}^*)$, where $\mathcal{K}$ is the kernel of the evaluation map $X \times V \to L \to 0$.*

1.6.3. The first jet bundle. Let L be a line bundle on a complex manifold X. By the *first jet bundle* of L, denoted $J_1(X, L)$ or $J_1(L)$ when no confusion will result, we mean the vector bundle of rank $\dim X + 1$, associated to the sheaf $p^*L/(p^*L \otimes \mathcal{J}_\Delta^2)$ where $p : X \times X \to X$ is the projection on the first factor, the tensor product is with respect to $\mathcal{O}_{X \times X}$ and $\mathcal{J}_\Delta$ is the sheaf of ideals of the diagonal, Δ, of $X \times X$. Note that there is a natural map, $j_1 : L \to J_1(X, L)$, defined on the sheaf level and which is not a bundle map, which sends a germ of a section to its 1-jet. Furthermore if L is very ample, the evaluation map $X \times \Gamma(L) \to J_1(X, L)$ induced by j_1 is surjective and hence $J_1(X, L)$ is spanned. Restricting to a section of p through (x, x) it is seen that the fiber of $J_1(L)$ at x corresponds to the constant and linear terms of the Taylor expansion of a section of L at x. A good reference for a detailed discussion is Kumpera and Spencer [363]. Evaluation gives a morphism onto L with kernel isomorphic to

$$p^*L \otimes (\mathcal{J}_\Delta / \mathcal{J}_\Delta^2) \cong p^*L \otimes \mathcal{N}^*_{\Delta / X \times X} \cong L \otimes \mathcal{T}_X^*,$$

i.e., we have the basic exact sequence of vector bundles

$$0 \to \mathcal{T}_X^* \otimes L \to J_1(L) \to L \to 0, \tag{1.6}$$

where $\mathcal{T}_X^*$ is the cotangent bundle of X. We have

$$H^i(X, J_1(L)) \cong H^i(X, L) \oplus H^i(X, \mathcal{T}_X^* \otimes L) \quad \text{for all } i \geq 0$$

since as a sequence of sheaves the sequence (1.6) splits over $\mathbb{C}$, i.e., sends a local section, s, of L to the 1-jet, $j_1(s)$, which at any point is the constant and first order terms of the Taylor series of s at the point. In general, the sequence (1.6), does not split as a bundle sequence, and in fact the Atiyah obstruction to splitting is the image of $c_1(L)$ in $H^1(X, \mathcal{T}_X^*)$. Note that the 1-jet of a section, s, is zero at a point $x \in X$, exactly if the section is zero at the point and all the partial derivatives of the section in some trivialization are zero at the point, i.e., only if $D := s^{-1}(0)$ contains x and is singular at x.

Lemma 1.6.4. *Let $\mathcal{L}$ denote a line bundle on a connected n-dimensional projective manifold, X. Then for $0 \leq i \leq n$*

$$c_i(J_1(\mathcal{L})) = \sum_{t=0}^{i} \binom{n+1-t}{i-t} c_t(\mathcal{T}_X^*) \cdot \mathcal{L}^{i-t}.$$

Proof. A straightforward calculation shows that given any vector bundle, $\mathcal{E}$,

$$c_i(\mathcal{E} \otimes \mathcal{L}) = \sum_{t=0}^{i} \binom{n-t}{i-t} c_t(\mathcal{E}) \cdot \mathcal{L}^{i-t}.$$

This formula with $\mathcal{E} = \mathcal{T}_X^*$ combined with $c(J_1(\mathcal{L})) = c(\mathcal{T}_X^* \otimes \mathcal{L})(1 + \mathcal{L})$, which follows from (1.6), gives the desired formula. □

Let L be a line bundle on X and V a vector subspace of $\Gamma(L)$. The *discriminant variety* of V, $\mathcal{D}(V)$, is the subset of $|V|$ corresponding to zero sets with at least one singular point. We also use the notation $\mathcal{D}(|V|)$, or simply $\mathcal{D}$ when no confusion can occur, for $\mathcal{D}(V)$. If $V = \Gamma(L)$, we denote $\mathcal{D}(V) = \mathcal{D}(L)$. Let $\mathcal{S}(V) \subset Z_V$ denote the set $\{(v, x) \mid j_1(v)(x) = 0\}$, i.e., $\mathcal{S}(V)$ denotes the set of singular points of divisors of V.

Lemma 1.6.5. *We use the above notation. Assume that V spans L, and the map associated to V gives an immersion of the connected manifold, X. Then $\mathcal{S}(V)$ is isomorphic to the $\mathbb{P}^{\dim V - \dim X - 1}$-bundle over X, $\mathbb{P}(\mathcal{K}_1^*)$, where $\mathcal{K}_1$ is the kernel of the evaluation map $X \times V \to J_1(L) \to 0$. In particular the discriminant variety of V is irreducible.*

In the situation of very ample lines bundles, L, where we have a space of sections spanning L and embedding X, some very nice things happen. The following important result is due to Katz [340, Proposition 3.5] (see also Zak [632]).

Theorem 1.6.6. *We use the above notation. Let V be a vector space of global sections of L, a line bundle on a connected projective manifold, X. Assume that V spans L and that $|V|$ gives an embedding of X. Then either $\mathcal{D}(V)$ is of codimension > 1 in $|V|$ or the general $D \in \mathcal{D}(V)$ has a single isolated nondegenerate quadratic singularity. In the last case, the map $\mathcal{S}(V) \to \mathcal{D}(V)$ is a desingularization.*

In the situation of Theorem (1.6.6) note that $\mathcal{D}(V)$ is irreducible since $|V|$ embeds X.

The discriminant variety, $\mathcal{D} \subset |L|$, is usually of codimension 1. Let $\mathbb{P} = \mathbb{P}(V^*)$ and set $k(V) := \mathrm{cod}_{\mathbb{P}}\mathcal{D}(V) - 1$ with the convention that $k(V) = \dim |V|$ if $\mathcal{D}(V)$ is empty. We will also use the notation $k(|V|)$ for $k(V)$. We will see in (1.6.10) below that $k(V)$ is independent of V. Thus we will often denote this basic number, the *defect*, as $k(|L|)$ or $k(X, L)$. When the defect is positive, i.e., when the discriminant variety is of codimension at least 2, we say that the pair (X, L) has *positive defect*. We will later in §14.4 show that, in this case, the pair (X, L) satisfies a very stringent structure theorem.

At this point we need a simple result that will guarantee that the codimension of $\mathcal{D}$ is 1. See §2.3 and §14.4 for similar but much stronger results.

Lemma 1.6.7. *Let L be a very ample line bundle on a smooth connected projective surface, X. If the discriminant variety, $\mathcal{D} \subset |L|$, of (X, L) is of codimension at least 2, then $(X, L) \cong (\mathbb{P}^2, \mathcal{O}_{\mathbb{P}^2}(1))$.*

Proof. If $\mathrm{cod}_{|L|}\mathcal{D} \geq 2$ then we can choose a linear pencil $P := \mathbb{P}^1 \subset |L| - \mathcal{D}$. Let V denote the 2-dimensional subspace of $\Gamma(L)$ corresponding to P. For $x \in P$, let H_x denote the divisor corresponding to x. Let $Z_V \subset P \times X$ be the universal zero set. Let $p : Z_V \to P$ and $q : Z_V \to X$ be the projections induced by the product projections. Let $H = H_x$ and $H' = H_{x'}$ for distinct points $x, x' \in P$. Note that p is of maximal rank by the assumption that $\mathcal{D} \cap P = \emptyset$. Thus p is a differentiable fiber bundle. Since P is simply connected the Leray sheaves $p_{(i)}\mathbb{Z}$ are isomorphic to $P \times H^i(H_x, \mathbb{Z})$. Thus H and H' are transverse, and $H \cap H' \neq \emptyset$ since $L^2 > 0$. Since q blows up X along $H \cap H'$, we see that each point, z, of the nonempty set, $H' \cap H$, gives rise to a section, $q^{-1}(z)$, of p. Thus the group $H^2(P, p_{(0)}\mathbb{Q}) \cong \mathbb{Q}$ survives to the E_∞ term of the spectral sequence. Thus we see that the differentials of the spectral sequence all are zero and we conclude that $H^2(Z_V, \mathbb{Q}) \cong \mathbb{Q}^2$. Since q blows up X along $H \cap H'$ we see that $\dim_{\mathbb{Q}} H^2(X, \mathbb{Q}) \leq 1$. Since $L^2 > 0$ we conclude that $\dim_{\mathbb{Q}} H^2(X, \mathbb{Q}) = 1$. Also since q blows up X along a finite set we see that X and Z_V have the same first homotopy groups, and thus the same first Betti numbers. We also see that $e(Z_V) = 2e(H_x)$ where we denote the Euler characteristic of a simplicial complex, W, by $e(W)$. Since π_2 blows up X along $H \cap H'$ we see that $e(Z_V) = e(X) + e(H \cap H')$. For any simplicial complex W let $b_i(W)$ denote the i-th Betti number. Thus

$$3 - 2b_1(X) = 2 - 2b_1(X) + b_2(X) = e(X) = 2e(H_x) - L^2 = 4 - 2b_1(H_x) - L^2$$

gives $L^2 = 1$. Thus $|L|$ embeds X as a degree 1 subspace of $\mathbb{P}^N$. From this we see that $(X, L) \cong (\mathbb{P}^2, \mathcal{O}_{\mathbb{P}^2}(1))$. □

Corollary 1.6.8. *Let L be a very ample line bundle on a smooth surface, S, with $(S, L) \not\cong (\mathbb{P}^2, \mathcal{O}_{\mathbb{P}^2}(1))$. Let C be a singular hyperplane section corresponding to a general point of $\mathfrak{D}(L)$. Then C has a single quadratic singularity. Moreover C is irreducible and reduced except in the cases:*

1. *$S \cong \mathbb{P}^2$ and $C \in |\mathcal{O}_{\mathbb{P}^2}(2)|$ is a union of two lines;*

2. *when S is a $\mathbb{P}^1$-bundle over a curve, $L_f \cong \mathcal{O}_{\mathbb{P}^1}(1)$ for a fiber f of the bundle, and C is the union of a section of the bundle and a fiber.*

Proof. From Lemma (1.6.7) it follows that $\mathrm{cod}_{|L|}\mathfrak{D} = 1$ where $\mathfrak{D} \subset |L|$ is the discriminant variety, i.e., the set of singular hyperplane sections. Using (1.6.6) we see that C, a singular hyperplane section corresponding to a general point of $\mathfrak{D}$, has a single quadratic singularity. From this it follows that C is reduced. If C is not irreducible, then since C has a single quadratic singularity, we can write $C = C_1 + C_2$, where C_1 and C_2 are smooth and meet only at the singular point. Since the singularity is quadratic, $C_1 \cdot C_2 = 1$. Here we could use the second Lefschetz theorem on hyperplane sections to finish the argument, but we prefer to use a simpler Hodge index theorem argument following Van de Ven [604]. By the Hodge index theorem it follows that $C_1^2 C_2^2 \leq 1$. If $C_i^2 \geq 1$ for $i = 1, 2$ we conclude that $C_i^2 = 1$ for $i = 1, 2$ and thus that $C^2 = C_1^2 + 2C_1 \cdot C_2 + C_2^2 = 4$. From this it follows that $C_1 \sim C_2$ and thus $L \sim 2C_1$. Thus C_1 is an ample divisor. Since $L \cdot C_1 = 1$ it follows from Corollary (1.4.3) that C_1 is a smooth $\mathbb{P}^1$. Let $L_1 := \mathcal{O}_S(C_1)$ and note that $L_1 \cdot C_1 = L_1^2 = L^2/4 = 1$. It is not difficult to show that $(S, L_1) \cong (\mathbb{P}^1, \mathcal{O}_{\mathbb{P}^1}(1))$ giving the first case of the corollary. In the interest of economy we refer the reader to the general results (3.1.3) and (3.1.2) which immediately show that we are in the first case of the corollary. Alternately we could proceed by using the Kodaira vanishing theorem to show that $h^1(-L_1) = 0$, and use this and the exact sequence

$$0 \to -L_1 \to \mathcal{O}_S \to \mathcal{O}_{\mathbb{P}^1} \to 0$$

to conclude that $h^1(\mathcal{O}_S) = 0$. From this and the sequence

$$0 \to \mathcal{O}_S \to L_1 \to \mathcal{O}_{\mathbb{P}^1}(1) \to 0$$

we conclude that L_1 is spanned and $h^0(L_1) = 3$. From this and $L_1^2 = 1$ we see that $|L_1|$ gives a generically one-to-one finite map, $S \to \mathbb{P}^2$. From Zariski's main theorem we conclude that $S \cong \mathbb{P}^2$. From this it follows that we are in the first case of the corollary.

Thus we can assume without loss of generality that one of $C_i^2 \le 0$. After renumbering we can assume that $C_2^2 \le 0$. Since $C_2^2 = C_2 \cdot C - C_2 \cdot C_1 \ge 1 - 1$, we conclude that $C_2^2 = 0$ and $C \cdot C_2 = 1$. From Corollary (1.4.3) we see that $(C_2, L_{C_2}) \cong (\mathbb{P}^1, \mathcal{O}_{\mathbb{P}^1}(1))$. It is straightforward, but tedious, to finish the argument showing f can be taken to be C_2. Again in the interest of economy we refer the reader to the more general Proposition (3.2.1), which is independent of the result here. □

Remark 1.6.9. If $k(V) \ge 1$ it is a classical fact that there is a dense Zariski open subset $\mathcal{D}(V)_0$ contained in $\mathcal{D}(V)$ with the following properties (see also Ein [203] and Kleiman [353, 352]):

i) if $H \in \mathcal{D}(V)_0$, then $\mathrm{Sing}(H)$ is a linear $\mathbb{P}^{k(V)}$;

ii) the union of these linear $\mathbb{P}^{k(V)}$'s is dense in X.

These results are consequences of Theorem (1.6.6). Let L be a very ample line bundle on a smooth connected projective variety, X, of dimension n with V a vector space of sections of L that span L and embed X. Let $\mathcal{D} \subset |V|$ denote the discriminant variety of V. Let $A \in |V| - \mathcal{D}$. Consider the map from $|V| - A$ to $|V|_A$ given by sending $A' \in |V| - A$ to $A \cap A' \in |V|_A$. This map can be identified with the projection map from A in $|V| - A$. Indeed given a linear pencil $P := \mathbb{P}^1 \subset |V|$ with $A \in P$ let V' denote the linear subspace of V corresponding to P. Let B denote the common zeroes of $s \in V'$. Note that $B = A \cap A'$ for any $A' \in P - A$. It can be checked that B is singular precisely when the union of singular points of the divisors $s^{-1}(0)$ for $s \in V' - \{0\}$ meet B. We have by a simple dimension count that singular sets of $H \in \mathcal{D}$ are of dimension at least $k(|V|)$. Thus if $k(|V|) > 0$ we see that $A \cap A'$ is singular for $A' \in |V| - A$ precisely if the pencil generated by A and A' meets $\mathcal{D}$. Thus we see that

$$k(|V|_A) = \max\{k(|V|) - 1, 0\}. \tag{1.7}$$

Moreover it can be checked that given a smooth $A \in |V| - \mathcal{D}$ and a linear pencil, $P := \mathbb{P}^1 \subset |V|$, that contains A and is transverse to $\mathcal{D}$, then the $A' \in P \cap \mathcal{D}$ have precisely a $k(|V|)$-dimensional singular set and $A \cap A'$ has precisely a $k(A, |V|_A)$-dimensional singular set. By induction and (1.6.6) it follows that $L^{k(|V|)} \cdot \mathrm{Sing}(A') = 1$ for a manifold point $A' \in \mathcal{D}$. Thus we see that

$$(\mathrm{Sing}(A'), L_{\mathrm{Sing}(A')}) \cong (\mathbb{P}^{k(|V|)}, \mathcal{O}_{\mathbb{P}^{k(|V|)}}(1)).$$

The fact noted above that the set of singular points, $\mathcal{S}(L)$, of $A' \in \mathcal{D}$ is a bundle over X shows that the union of these linear $\mathbb{P}^{k(|V|)}$'s is dense in X.

One consequence, due to Griffiths and Harris [294], of the fact that X with $k(L) > 0$ is covered by linear spaces is that such an X has negative Kodaira dimension.

Lemma 1.6.10. *Let L be a line bundle on a complex manifold X and let V be a vector subspace of $\Gamma(L)$. Let $J_1(L)$ be the first jet bundle of L. Then, with the notation as above,* $\operatorname{cod}_{\mathbb{P}(V^*)}\mathcal{D}(V) = k(V) + 1$ *if and only if* $c_r(J_1(L)) = 0$ *for* $r \geq n - k(V) + 1$, *and* $c_{n-k(V)}(J_1(L)) \neq 0$.

Proof. Write $k = k(V)$. Note that the 1-jets, $\{j_1(s)|s \in V\}$ span $J_1(L)$. Choose general elements $s_1, \dots, s_v \in V$ where $\dim V = v$. Using the classical representation of a Chern class of a spanned vector bundle (see Fulton [267, 14.3.2]) we have that $c_{n-k+1}(J_1(L))$ is represented by the set

$$\{x \in X | \dim \operatorname{Span}(j_1(s_1)(x), \dots, j_1(s_{k+1})(x)) \leq k\}.$$

Noting that $\operatorname{cod}_{\mathbb{P}(V^*)}\mathcal{D}(V) = k + 1$ implies that a general $\mathbb{P}^k$ in $|V|$ does not meet $\mathcal{D}(V)$ we see that $\sum_{i=1}^{k+1} \lambda_i s_i$ is not singular for any $(\lambda_1, \dots, \lambda_{k+1}) \in \mathbb{P}^k$. This is easily checked to be equivalent to

$$\dim \operatorname{Span}(j_1(s_1)(x), \dots, j_1(s_{k+1})(x)) = k + 1$$

for all $x \in X$. This implies by the above representation that $c_{n-k+1}(J_1(L)) = 0$ and moreover that $c_r(J_1(L)) = 0$ for $r \geq n - k + 2$. Furthermore $c_{n-k}(J_1(L))$ is represented by the set

$$\{x \in X | \dim \operatorname{Span}(j_1(s_1)(x), \dots, j_1(s_{k+2})(x)) \leq k + 1\}.$$

This set is the union of the singular sets of the hyperplane sections $\mathcal{D}(V) \cap \mathbb{P}^{k+1}$ for a general $\mathbb{P}^{k+1}$ contained in $|V|$ and $\mathcal{D}(V) \cap \mathbb{P}^{k+1}$ is contained in $\mathcal{D}(V)_0$. By a standard fact (see properties i), ii) above) these singular sets are linear $\mathbb{P}^k$'s. Therefore $c_{n-k}(J_1(L))$ is represented by a nontrivial sum of k-dimensional varieties. Thus $c_{n-k}(J_1(L)) \neq 0$. This shows the "only if" part of the statement. The converse is an immediate corollary of the argument above. □

Remark 1.6.11. Note that the above argument shows that $k(V) > 0$ if and only if $c_n(J_1(X, L)) = 0$. Moreover it can be shown that

$$\deg \mathcal{D}(V) = c_{n-k}(J_1(L)) \cdot L^k, \quad k = k(V).$$

As mentioned earlier in this section, the above lemma shows that the integer $k(V)$ only depends on (X, L). This integer is called the *defect* of (X, L) and denoted by $\operatorname{def}(X, L)$. We will often write $k(X, L)$, or simply $k(L)$ instead of $\operatorname{def}(X, L)$. Note that $\mathcal{D}(V) = \mathcal{D}(L) \cap |V|$. Assume that $\Gamma(L)$ embeds X in $\mathbb{P}^N$ and let X^* be the *dual variety* of X, i.e., X^* is the closure of the set of the tangent hyperplanes at the regular points of X. Note also that $\mathcal{D} := \mathcal{D}(L) \cong X^*$ and

$$k(L) = N - 1 - \dim X^*. \tag{1.8}$$

Corollary 1.6.12. (Landman) *Let L be a very ample line bundle on a connected projective manifold, X. Then* $\operatorname{def}(X, L) < n$ *except if* $(X, L) \cong (\mathbb{P}^n, \mathcal{O}_{\mathbb{P}^n}(1))$.

Proof. If $\operatorname{def}(X, L) \geq n$, then by (1.6.10) it follows that $\det J_1(L) \cong \mathcal{O}_X$. Thus by Lemma (1.6.4) we conclude that $K_X + (n+1)L \cong \mathcal{O}_X$. Choose $n-1$ general elements of $|L|$. They intersect tranversally to give a smooth curve, C. By the adjunction formula we have $K_C + 2L_C \cong (K_X + (n+1)L)_C \cong \mathcal{O}_C$. Thus $(C, L_C) \cong (\mathbb{P}^1, \mathcal{O}_{\mathbb{P}^1}(1))$. Therefore $L^n = 1$ and $|L|$ embeds X as a degree 1 submanifold of a projective space. □

1.7 Bertini theorems

In this section we collect some theorems of Bertini type asserting that various good properties are preserved when passing from a variety to a divisor satisfying appropriate generality conditions.

Let X be a variety and $p : \overline{X} \to X$ a desingularization. The direct image of the structure sheaf of $\overline{X}$, $p_*(\mathcal{O}_{\overline{X}})$, and its higher derived functors, $p_{(i)}(\mathcal{O}_{\overline{X}}) := R^i p_*(\mathcal{O}_{\overline{X}})$, are supported on the singular set, $\operatorname{Sing}(X)$. It is a theorem of Hironaka [322] that they depend only on X and not on the desingularization. We denote the sheaf $p_{(i)}(\mathcal{O}_{\overline{X}})$ by $\mathcal{S}_i(X)$. Note that $\mathcal{S}_0(X) = \mathcal{O}_X$ if and only if X is normal. If the sheaves $\mathcal{S}_i(X)$ are zero for $i > 0$, then X is said to have at most *rational singularities*. The union of the supports of the sheaves $\mathcal{S}_i(X)$ for $i > 0$ with the reduced structure induced from X is called the *irrational locus* of X and is denoted $\operatorname{Irr}(X)$. Note that $\overline{K}_X := p_* K_{\overline{X}}$, which is called the *Grauert-Riemenschneider canonical sheaf,* is independent of the desingularization (see Grauert and Riemenschneider [286] for the general theory of this sheaf). The following is a generalization of Seidenberg's classical theorem [551] that general hyperplane sections of normal varieties are normal.

Theorem 1.7.1. (Generalized Seidenberg theorem) *Let X be a normal analytic variety and let $\mathcal{L}$ be a line bundle on X spanned by a finite dimensional space V of global sections. There is a Zariski dense open set $U \subset |V|$ of divisors D such that:*

1. *D is normal,* $\operatorname{Sing}(D) \subset \operatorname{Sing}(X)$, *and no irreducible component of* $\operatorname{Sing}(X)$ *belongs to D;*

2. *for $i > 0$ the support of $\mathcal{S}_i(D)$ is a subset of the support of $\mathcal{S}_i(X)$, and no (possibly embedded) irreducible component of the support of $\mathcal{S}_i(X)$ is a subset of D.*

In particular $\mathrm{cod}_X\mathrm{Sing}(X) \geq \mathrm{cod}_D\mathrm{Sing}(D)$ *and* $\mathrm{cod}_X\mathrm{Irr}(X) \geq \mathrm{cod}_D\mathrm{Irr}(D)$. *If* X *has terminal singularities then* D *has terminal singularities. If* X *is compact, or if* X *is an algebraic set and* $\mathcal{L}$ *is an algebraic line bundle on* X*, then* U *can be chosen to be Zariski open.*

Proof. The condition that $D \in |V|$ contains either an irreducible component of $\mathrm{Sing}(X)$ or a (possibly embedded) irreducible component of the support of one of the $\mathcal{S}_i(X)$ for some $i > 0$ is an analytic condition, i.e., the condition defines a proper analytic subset of $|V|$. Therefore there is a dense Zariski open set $U' \subset |V|$ such that if $D \in U'$, then neither irreducible components of $\mathrm{Sing}(X)$ nor irreducible (possibly embedded) components of the support of $\mathcal{S}_i(X)$ belong to D.

Let $p : \overline{X} \to X$ be a desingularization of X. Since $|p^*V|$ spans $p^*\mathcal{L}$ we know by the usual Bertini theorem that elements of a dense Zariski open set of $|p^*V|$ are smooth. Let D belong to the dense set, U, of elements of U' such that $\overline{D} = p^{-1}(D)$ is smooth. Since $p : \overline{X} - p^{-1}(\mathrm{Sing}(X)) \to X - \mathrm{Sing}(X)$ is a biholomorphism it follows that $\mathrm{Sing}(D) \subset \mathrm{Sing}(X)$ for these D.

Let $s \in V$ be such that $s^{-1}(0) = D$ and let $s' = p^*s \in p^*V$. Consider the sequence,

$$0 \to p^*\mathcal{L}^{-1} \to \mathcal{O}_{\overline{X}} \to \mathcal{O}_{\overline{D}} \to 0,$$

and the associated long exact sequence of direct image sheaves under p, i.e.,

$$0 \to \mathcal{L}^{-1} \to \mathcal{O}_X \to p_*\mathcal{O}_{\overline{D}} \to \mathcal{S}_1(X) \otimes (-\mathcal{L}) \to \mathcal{S}_1(X) \to \mathcal{S}_1(D).$$

By the choice of D satisfying (1.7.1.2) it follows that for all $i > 0$ we have an injection

$$0 \to \mathcal{S}_i(X) \otimes (-\mathcal{L}) \to \mathcal{S}_i(X). \tag{1.9}$$

Since we have the exact sequence

$$0 \to -\mathcal{L} \to \mathcal{O}_X \to \mathcal{O}_D \to 0,$$

we see that $p_*\mathcal{O}_{\overline{D}} \cong \mathcal{O}_D$. From this we see that D is normal. The other statements of 1) and 2) of Theorem (1.7.1) follow from injection (1.9).

The fact that D has terminal singularities if X has terminal singularities follows from the proof of Lemma (1.3.1).

To see the last statement that U can be chosen to be Zariski open, note that the condition that $D \in |V|$ contains an irreducible analytic set is an analytic condition, and is therefore satisfied for D in an algebraic subset of $|V|$. Thus the condition that $D \in |V|$ contains an analytic set, Z, with finitely many components (including embedded points), e.g., if Z is algebraic, is satisfied for D in an algebraic subset of $|V|$. Therefore the divisors $D \in |\mathcal{L}|$ that do not contain some such set, Z, form a Zariski open set. □

Remark 1.7.2. If X has at worst Cohen-Macaulay singularities, then we can draw some very useful consequences from the above result. A Cartier divisor on a variety with at worst Cohen-Macaulay singularities has at worst Cohen-Macaulay singularities. A connected variety, W, with at worst Cohen-Macaulay singularities is pure dimensional, and if $\mathrm{cod}_X \mathrm{Sing}(X) \geq 2$, it is normal. In particular let B denote the base locus of a finite dimensional linear subspace, $V \subset H^0(\mathcal{L})$, of sections of a line bundle, $\mathcal{L}$, on a pure dimensional normal variety, X, with at worst Cohen-Macaulay singularities. A general element $A \in |V|$ contains B, and satisfies all the conditions of the above theorem off B. In particular if B is of codimension at least 3 in X then A will be normal. This observation is particularly useful in the situation when V spans $\mathcal{L}$ off a finite set.

There is a Bertini type theorem due to Hironaka ([321, §2]) that is very useful for doing induction arguments.

Theorem 1.7.3. (Hironaka) *Let $\mathcal{L}$ be an ample line bundle on a projective variety, X. Let $f : X \to \mathbb{P}_{\mathbb{C}}$ be a rational mapping that is holomorphic on a Zariski open set X_0. Then there exists a positive integer, N, and a nonempty Zariski open subset, $U \subset |NL|$, such that for all $y \in f_{X_0}(X_0)$, $D \in U$ does not contain any positive dimensional irreducible component of $f_{X_0}^{-1}(y)$.*

Proof. See [562, Theorem (3.39)]. □

The following simple lemma gives a useful criterion for smoothness.

Lemma 1.7.4. *Let L be a line bundle on a manifold, X. Let $s \in \Gamma(L)$ be a section which is zero on a submanifold, $Z \subset X$. Then $Z \subset \mathrm{reg}(s^{-1}(0))$ if and only if the image, $\nu(s)$, of s in $\Gamma(\mathcal{N}^*_{Z/X} \otimes L) \cong \Gamma((\mathcal{J}_Z \otimes L)/(\mathcal{J}_Z^2 \otimes L))$ is nowhere zero, where $\nu : \Gamma(\mathcal{J}_Z \otimes L) \to \Gamma(\mathcal{N}^*_{Z/X} \otimes L)$ is the quotient map, and $\mathcal{J}_Z$ is the ideal sheaf of Z in X.*

Proof. Since we have an injection $0 \to \mathcal{J}_Z/\mathcal{J}_Z^2 \to \mathcal{T}^*_{X|Z}$ it follows that the image of s in $\mathcal{T}^*_{X|Z} \otimes L$ obtained by composition is nowhere zero if and only if $\nu(s)$ is nowhere zero. Choosing a local trivialization of L in a neighborhood, U, of a point $z \in Z$, it is a local check that up to multiplication with a nowhere vanishing function on $U \cap Z$, the image of s in $\mathcal{T}^*_{U|U\cap Z}$ is the restriction of the differential, df, of the function, f, equal to s in the trivialization of L_U. Since $f(Z \cap U) = s(Z \cap U) = 0$, we have $df(v) = 0$ for any tangent vector, v, of $\mathcal{T}_{Z\cap U}$, and thus the image of s in $\mathcal{T}^*_{U|U\cap Z}$ is nowhere zero if and only if $s^{-1}(0) \cap U = f^{-1}(0) \cap U$ is smooth on $U \cap Z$. □

The following simple corollary of this lemma we learned from Hironaka [321].

Corollary 1.7.5. *Let L be a line bundle on a smooth variety, X. Assume that there is a smooth $Z \subset X$, such that $L \otimes \mathcal{J}_Z$ is spanned by global sections, where $\mathcal{J}_Z$ is* *the ideal sheaf of Z in X. If $\mathcal{N}^*_{Z/X} \otimes L$ has a nowhere vanishing section, e.g., if $2\dim Z < \dim X$, then there is a smooth $D \in |L|$ which contains Z. In particular if $Z \subset X$ is any smooth submanifold with $2\dim Z < \dim X$ and L is ample, then there exists a positive integer, t_0, such that for $t \geq t_0$, there is a smooth $D \in |tL|$ which contains Z.*

Proof. Since $\mathcal{N}^*_{Z/X} \otimes L$ is of rank $\dim X - \dim Z$, it has a nowhere vanishing section if $2\dim Z < \dim X$. Since $L \otimes \mathcal{J}_Z$ is spanned by global sections, we know by Bertini's theorem, (1.7.1), that general $D \in |L|$ vanishing on Z are smooth away from Z. Thus by Lemma (1.7.4), choosing a smooth $D \in |L|$ that vanishes on Z is equivalent to the image in $\Gamma(\mathcal{N}^*_{Z/X} \otimes L)$ of a general section of $\Gamma(L \otimes \mathcal{J}_Z)$ being a nowhere vanishing section of $\mathcal{N}^*_{Z/X} \otimes L$.

If L is ample, note that $(tL) \otimes \mathcal{J}_Z$ is spanned for $t \gg 0$ and hence apply the argument above. □

The following result will be used in §2.2 to prove Theorem (2.2.5).

Lemma 1.7.6. *Let $\mathcal{L}$ be a line bundle on a projective n-fold, X, spanned by a finite dimensional space, V, of global sections. Then there is a nonempty Zariski open subset, $U \subset |V|$, such that for $D \in U$, $(\omega_X + \mathcal{L})_D \cong \omega_D$, where ω_X and ω_D are the dualizing sheaves of X and D.*

Proof. If X has at worst Cohen-Macaulay singularities, then this result is true with $U = |V|$, and it is shown by Altman and Kleiman in [4, Propositions (2.3), (2.4)]. We modify their argument for our case. For coherent sheaves $\mathcal{F}$, $\mathcal{G}$ let $\mathcal{E}xt^a_{\mathcal{O}_X}(\mathcal{G}, \mathcal{F})$ denote the a-th sheaf of local Ext groups. By [4, Proposition (2.4)],

$$(\omega_X + D)_D \cong \mathcal{E}xt^1_{\mathcal{O}_X}(\mathcal{O}_D, \omega_X).$$

Thus we will be done if we show that $\omega_D \cong \mathcal{E}xt^1_{\mathcal{O}_X}(\mathcal{O}_D, \omega_X)$. Fix an embedding of X into $\mathbb{P}^N$ for some N. Following [4, Proposition (2.3)], there is a spectral sequence

$$E_2^{p,q} = \mathcal{E}xt^p_{\mathcal{O}_X}(\mathcal{O}_D, \mathcal{E}xt^q_{\mathcal{O}_{\mathbb{P}^N}}(\mathcal{O}_X, \omega_{\mathbb{P}^N})) \Rightarrow \mathcal{E}xt^{p+q}_{\mathcal{O}_{\mathbb{P}^N}}(\mathcal{O}_D, \omega_{\mathbb{P}^N}).$$

Since, by definition, $\omega_D = \mathcal{E}xt^{N-n+1}_{\mathcal{O}_{\mathbb{P}^N}}(\mathcal{O}_D, \omega_{\mathbb{P}^N})$ and since by [4, Lemma (5.1)],

$$\mathcal{E}xt^q_{\mathcal{O}_{\mathbb{P}^N}}(X, \omega_{\mathbb{P}^N}) = 0 \text{ for } q < N - n,$$

we will be done if we show that $\mathcal{E}xt^0_{\mathcal{O}_X}(\mathcal{O}_D, \mathcal{E}xt^q_{\mathcal{O}_{\mathbb{P}^N}}(\mathcal{O}_X, \omega_{\mathbb{P}^N})) = 0$, for $q = N - n + 1, N - n + 2$. Note that, for $q = N - n + 1, N - n + 2$,

$$\mathcal{E}xt^0_{\mathcal{O}_X}(\mathcal{O}_D, \mathcal{E}xt^q_{\mathcal{O}_{\mathbb{P}^N}}(\mathcal{O}_X, \omega_{\mathbb{P}^N})) \cong \mathcal{H}om_{\mathcal{O}_X}(\mathcal{O}_D, \mathcal{E}xt^q_{\mathcal{O}_{\mathbb{P}^N}}(\mathcal{O}_X, \omega_{\mathbb{P}^N})).$$

For divisors $D \in |V|$ not containing any component (including possibly embedded components) of the support of $\mathcal{E}xt^q_{\mathcal{O}_{\mathbb{P}^N}}(\mathcal{O}_X, \omega_{\mathbb{P}^N})$, $q = N-n+1, N-n+2$, the right-hand sheaf in the above isomorphism is zero. By Theorem (1.7.1) such divisors, D, vary in a Zariski dense open set $U \subset |V|$, so we are done. □

The following general Lemma is useful. The second assertion was used by Andreatta and the second author in [24].

Lemma 1.7.7. *Let X be an irreducible projective variety of dimension $n \geq 2$ and let L be a line bundle on X. Let $\mathcal{P}$ be a linear $\mathbb{P}^k \subset \operatorname{reg}(X)$, i.e., $L_{\mathcal{P}} \cong \mathcal{O}_{\mathbb{P}^k}(1)$. Let $p : \overline{X} \to X$ be the blowing up of X along $\mathcal{P}$. Let $E = p^{-1}(\mathcal{P})$.*

1. *If L is very ample, then $p^*L - E$ is spanned and $p^*(2L) - E$ is very ample.*

2. *If L is ample and spanned and $k = 0$, then $p^*L - E$ is nef.*

Proof. Assume L is very ample. Since $|L|$ embeds X into $\mathbb{P}^N$, with $N+1 = h^0(L)$ and with $\mathcal{P}$ going to a linear subspace of $\mathbb{P}^N$, it suffices to prove 1) under the assumption that $(X, L) \cong (\mathbb{P}^N, \mathcal{O}_{\mathbb{P}^N}(1))$. Since $\mathcal{P}$ is defined as the zeroes of sections of homogeneous coordinates, which are sections of L, we see that $L \otimes \mathfrak{J}_{\mathcal{P}}$ is spanned by global sections, where $\mathfrak{J}_{\mathcal{P}}$ is the ideal sheaf of $\mathcal{P}$ (see also Lemma (6.6.1)). Since $p^*L - E \cong p^*(L \otimes \mathfrak{J}_{\mathcal{P}})$, we see that $p^*L - E$ is spanned. Note that choosing homogeneous coordinates, $z_0, \dots, z_N$, in $\mathbb{P}^N$ such that $\mathbb{P}^k$ is defined by $z_{k+1} = \cdots = z_N = 0$, then the projection onto $\mathbb{P}^{N-k-1}$ from $\mathbb{P}^k$ is given on $\mathbb{P}^N - \mathbb{P}^k$ by sending $[z_0, \dots, z_N] \mapsto [z_{k+1}, \dots, z_N]$. Since $L \otimes \mathfrak{J}_{\mathcal{P}}$ is spanned by $z_{k+1}, \dots, z_N$, it follows that the morphism given by $|p^*L - E|$ coincides on $\overline{X} - E$ with the composition of p with the projection $\mathbb{P}^N \to \mathbb{P}^{N-k-1}$ from $\mathbb{P}^k$, and hence can be interpreted as the projection from $\mathcal{P}$. A straightforward check using this shows that the morphism $q : \overline{X} \to \mathbb{P}^{N-k-1}$ given by $|p^*L - E|$ is a $\mathbb{P}^{k+1}$-bundle.

Let $q' : \mathbb{P}^{N-k-1} \times \mathbb{P}^N \to \mathbb{P}^{N-k-1}$, $p' : \mathbb{P}^{N-k-1} \times \mathbb{P}^N \to \mathbb{P}^N$ be the product projections. Let $\mathcal{L} := q'^*\mathcal{O}_{\mathbb{P}^{N-k-1}}(1) \otimes p'^*\mathcal{O}_{\mathbb{P}^N}(1)$. Then we see that $p^*(2L) - E$ is just the line bundle induced on $\overline{X}$ from $\mathcal{L}$ by pullback under the product morphism $\overline{X} \overset{(q,p)}{\to} \mathbb{P}^{N-k-1} \times \mathbb{P}^N$, i.e., $p^*(2L) - E \approx (q, p)^*\mathcal{L}$. Since $\mathcal{L}$ is easily seen to be very ample, it suffices to check that (q, p) is an embedding.

First note that it is one-to-one. Let $a, b \in \overline{X}$ be two distinct points that go to the same point under (q, p). Then since $q(a) = q(b)$ we see that both a and b lie on the same fiber of q. Using the interpretation of q in terms of projection from $\mathcal{P}$, each fiber of q goes isomorphically under p onto a linear $\mathbb{P}^{k+1}$ which gives a contradiction.

Now assume that b is a tangent vector of $\mathcal{T}_{\overline{X},a}$ at a point $a \in \overline{X}$ which goes to zero under the differential of (q, p). Since q is a $\mathbb{P}^{k+1}$-bundle, we conclude that this implies that b is tangent to the fiber of q that contains a. We get a contradiction since as noted in the last paragraph, p maps each fiber of q isomorphically onto a $\mathbb{P}^{k+1}$.

To show 2), let C be a curve on $\overline{X}$ such that $C \subset E$. Then the normal bundle, $\mathcal{N}_{E/\overline{X}}$, has negative degree so that $C \cdot E < 0$ and hence $(p^*L - E) \cdot C = -E \cdot C > 0$.

Thus we can assume that $C \cap E$ is a (possibly empty) finite set. Then $p(C)$ is a curve on X. Note that a general $D \in |p^*L - E|$ does not contain C. Otherwise all $D' \in |L \otimes \mathfrak{m}_x|$ would contain $p(C)$, where $\mathfrak{m}_x$ denotes the ideal sheaf of x in X, and hence $p(C)$ would be contained in the base locus $B := \mathrm{Bs}|L \otimes \mathfrak{m}_x|$ of $|L \otimes \mathfrak{m}_x|$. This contradicts the fact that B is finite. To see this note that a set theoretic computation yields $B = \varphi^{-1}\varphi(x)$, where φ is the morphism given by $\Gamma(L)$. Since L is ample, $\varphi^{-1}(\varphi(x))$ is finite. □

We need some useful versions of the Bertini theorem from the second author's papers [579, (0.6.2)] and [571, (0.10.2)]. The first result follows immediately from the above results.

Theorem 1.7.8. *Let X be an irreducible n-dimensional algebraic subvariety of $\mathbb{P}^N$ and let $|L - x|$ denote the linear system of hyperplane sections H of X such that $x \in H$. Then there exists a dense Zariski open set $U \subset |L - x|$ such that if $H \in U$ then $H \cap \mathrm{reg}(X)$ is smooth.*

Theorem 1.7.9. *Let X be a connected n-dimensional submanifold of $\mathbb{P}^N$. Given two distinct points $\{x, y\} \subset X$ (respectively a point $x \in X$ and a tangent direction $y \in \mathcal{T}_{X,x}$), let $|L - x - y|$ denote the linear system of hyperplane sections H of X such that $\{x, y\} \subset H$ (respectively $x \in H$ and y tangent to H). If $n \geq 3$, then there exists a dense Zariski open set $U \subset |L - x - y|$ such that if $H \in U$ then $H \cap \mathrm{reg}(X)$ is smooth. If $n = 2$, then either:*

1. *there exists a dense Zariski open set $U \subset |L - x - y|$ such that $H \in U$ is smooth; or*

2. *the line $\ell \subset \mathbb{P}^N$ with $\{x, y\} \subset \ell$ (respectively with $x \in \ell$ and y tangent to ℓ) belongs to X, $L - \ell$ is spanned by global sections, and there is a dense Zariski open set $U \subset |L - x - y|$ such that $H \in U$ is of the form $\ell + \mathcal{R}$, where $\mathcal{R} \in |L - \ell|$ is a smooth curve that meets ℓ transversally.*

Proof. We will prove the theorem in the case of two distinct points $\{x, y\} \subset X$. The proof in the case of one point, as in (1.7.8), is easier and in the case of one point and a tangent vector is almost identical.

Theorem (1.7.1) implies that there is a dense Zariski open set $U \subset |L - x - y|$ of hyperplane sections that are smooth outside of the base locus of $|L - x - y|$. Therefore it suffices to study the smoothness of $H \in |L - x - y|$ at the basepoints of $|L - x - y|$. Let $\ell \subset \mathbb{P}^N$ denote the line with $\{x, y\} \subset \ell$. Since the intersection of the hyperplanes in $\mathbb{P}^N$ that contain ℓ is scheme-theoretically equal to ℓ, it follows that the base locus of $|L - x - y|$ is precisely the scheme-theoretic intersection in $\mathbb{P}^N$ of ℓ and X.

At each point z of $\ell \cap X$, the hyperplanes that contain ℓ and are tangent to X at z form a proper algebraic subset of the hyperplanes that contain ℓ. Thus at each point z of $\ell \cap X$, there is a Zariski open subset $U_z \subset |L - x - y|$ such that $H \in U_z$ is smooth at z and on $X - X \cap \ell$. In particular the theorem is true if $X \cap \ell$ is finite.

Therefore without loss of generality we can assume that $\ell \subset X$. The sections of $L \otimes \mathcal{J}_\ell$, where $\mathcal{J}_\ell$ denotes the ideal sheaf of ℓ in X, span $\mathcal{N}^*_{\ell/X} \otimes L$. If $n \geq 3$, the theorem follows from Corollary (1.7.5).

If $n = 2$ then it follows from Lemma (1.7.7) that $L - \ell$ is spanned by global sections and hence the generic element of $|L - \ell|$ is smooth. This implies that a dense Zariski open set $U \subset |L - x - y| \cong |L - \ell|$ is of the form $\ell + \mathcal{R}$, where $\mathcal{R}$ is a smooth curve that meets ℓ transversally. □

See the paper [385] of Lanteri, Palleschi, and Sommese for improvements of the above results, including the following [385, Theorem (0.3)].

Theorem 1.7.10. *Let $P_1, \ldots, P_r$ be r totally disjoint linear subspaces of $\mathbb{P}^n$, i.e., the linear space that they generate has dimension $a_1 + \cdots + a_r + r - 1$, where $a_i = \dim P_i$. Let $\pi : X \to \mathbb{P}^n$ be the blowing up of the union of the P_i's, and let $E_i := \pi^{-1}(P_i)$. If $t \geq r \geq 3$, then the line bundle, $L := \pi^*\mathcal{O}_{\mathbb{P}^n}(t) - \sum_i E_i$, is very ample.*

1.8 Some examples

The following examples show that certain natural properties of line bundles are not equivalent. Example (1.8.1) shows that an ample line bundle does not have to have any sections, even though all sufficiently high positive multiples of the line bundle are very ample. Example (1.8.2.1) shows that a line bundle can be spanned for infinitely many positive multiples, but still have no sections for infinitely many different positive multiples. This emphasizes the strength of the conclusion of the basepoint free theorem, (1.5.1). Example (1.8.2.2) shows that there exist nef line bundles which are k-big for $k > 0$, but no positive multiples have any sections, even though high positive multiples of 0-big, i.e., big, line bundles have many sections. Example (1.8.3) shows that there are big and nef line bundles which have no positive multiple spanned by global sections.

Example 1.8.1. Let C be a smooth projective curve of genus $g \geq 2$. The ample line bundles on C coincide with the set of positive degree line bundles, and the degree d line bundles on C are parameterized by the Jacobian of C which is of dimension g. Given a line bundle, L, of degree 1 on C with a not identically zero section, s, we have that s vanishes at a single point x and $L \cong \mathcal{O}_C(x)$. Thus the degree one line bundles with nontrivial sections are parameterized by the points of the one dimensional curve C. Since $g > 1$, there exist ample line bundles of degree 1 which have no sections.

Example 1.8.2. Let $V = E \times M$ where E and M are positive dimensional, connected, projective manifolds. Let $p : V \to E$ and $q : V \to M$ denote the product projections. Let H be a very ample line bundle on M and let μ denote a numerically trivial line bundle on E. Let $\mathcal{L} := p^*\mu \otimes q^*H$. Assume that $H^1(E, \mathbb{C}) \neq (0)$, e.g., E an elliptic curve.

1. For any integer $d \geq 1$, a numerically trivial μ can be chosen so that $t\mu \cong \mathcal{O}_E$ if and only if t is a multiple of d. In this case $t\mathcal{L}$ with $t \geq 1$ is spanned if and only if t is a positive multiple of d.

2. Numerically trivial line bundles μ can be chosen such that no positive multiple of μ is isomorphic to $\mathcal{O}_E$. In this case $\mathcal{L}$ is nef, but no positive multiple has any sections. Note that given any ample line bundle $\mathcal{H}$ on V, $\mathcal{L}^{\dim M+1} = 0$, and that $\mathcal{L}^{\dim M} \cdot \mathcal{H}^{\dim E} \neq 0$.

Example 1.8.3. Let $\mathcal{E} := H \oplus \mu$ where H is a very ample line bundle on a projective manifold E, and μ is any numerical trivial line bundle, such that $t\mu$ is not isomorphic to $\mathcal{O}_E$ for any positive integer, t. Such μ exist on any projective manifold E with $H^1(E, \mathbb{C}) \neq (0)$. Let $\xi_{\mathcal{E}}$ denote the tautological line bundle on $\mathbb{P}(\mathcal{E})$. Let σ denote the section of $X := \mathbb{P}(\mathcal{E}) \to E$ corresponding to the quotient mapping $\mathcal{E} \to \mu$. Then $\xi_{\mathcal{E}|\sigma} \cong \mu$ from which it follows that $t\xi_{\mathcal{E}}$ cannot be spanned for any positive t. Indeed otherwise $t\xi_{\mathcal{E}|\sigma} \cong t\mu$ would be spanned and therefore, since $\deg \mu = 0$, we would have the contradiction $t\mu \cong \mathcal{O}_E$. Note that global sections of $\xi_{\mathcal{E}}$ span $\xi_{\mathcal{E}|(X-\sigma)}$ and that $\xi_{\mathcal{E}}^{\dim X} = H^{\dim E} \geq 1$.

Chapter 2

Consequences of positivity

In this chapter we introduce some positivity conditions generalizing ampleness, e.g., k-ampleness and k-bigness. We also discuss an assortment of global results involving positivity conditions.

In §2.1 we prove some general facts about k-ampleness and k-bigness.

In the main result of §2.2, Theorem (2.2.4), we prove a useful extension of the usual relative vanishing theorem, (2.2.1), for a nef and big line bundle. We call the reader's attention to Example (2.2.10) which shows one of the boundaries beyond which the most precise results of adjunction theory are no longer true.

In §2.3 we state Hamm's Lefschetz theorem and we prove a number of useful consequences of it. We also discuss a generalized version of the Lefschetz theorem, (2.3.11). Furthermore we prove a vanishing theorem, (2.3.8), for the fundamental group of a positive defect pair (X, L).

In §2.4 we recall some basic facts about the Albanese variety and the Albanese mapping in presence of rational singularities.

In §2.5, following an argument of Fujita [246], we prove a very general version of the Hodge index theorem and a number of useful consequences of it. We also characterize bigness for $\mathbb{Q}$-Cartier divisors and nef line bundles on an irreducible projective variety.

In §2.6 we study pairs (X, L) where X is either an analytic space or a normal projective variety, L is a very ample line bundle on X, and $A \in |L|$ is a Cartier divisor such that the normal bundle $\mathcal{N}_{A/X}$ is ample. We use in a critical way an important extension theorem, (2.6.1), due to Rossi. Rossi's papers [524, 525] are the main sources in this section.

Finally in §2.7 we discuss some very useful results of Andreotti-Grauert and Griffiths about the cohomology of analytic coherent sheaves.

2.1 k-ampleness and k-bigness

In [569], the second author defined a line bundle, L, on a projective variety, X, to be *k-ample* for an integer $k \geq 0$, if NL is spanned for some $N > 0$, and the morphism $X \to \mathbb{P}_{\mathbb{C}}$ defined by $\Gamma(NL)$ for such an N has all fibers of dimension $\leq k$. A vector bundle, E, on X is *k-ample* if the tautological bundle ξ_E on $\mathbb{P}(E)$ is k-ample. For $k = 0$, the definition is one of the basic characterizations of ampleness.

The concept of k-ampleness comes up very naturally in the study of submanifolds of homogeneous complex manifolds, (see [572, 578, 585]), and provides a basic connection between algebraic geometric properties of a normal bundle and convexity properties of complements of submanifolds. In that theory the key bundle was the normal bundle, $\mathcal{N}_{X/M}$, of the projective submanifold, X, of the not necessarily compact homogeneous complex manifold, M. Homogeneity implies that $\mathcal{T}_M$ and hence that $\mathcal{N}_{X/M}$ is spanned. Since the quotient of a k-ample bundle is k-ample, an upper bound for k can be obtained by computing the minimum k for which $\mathcal{T}_M$ is k-ample, which was done by Goldstein [278] (see also Snow [565] for the case of a general homogeneous bundle). For example, submanifolds of the Grassmannian, $\mathrm{Grass}(a, b)$, of a-dimensional quotient vector spaces of $\mathbb{C}^b$ have $(\dim \mathrm{Grass}(a, b) - b + 1)$-ample normal bundles where $\dim \mathrm{Grass}(a, b) = a(b-a)$, and submanifolds of smooth quadrics have 1-ample normal bundles (see Goldstein [279, 280] for more information in the case of the 4-dimensional quadric). Usually the actual k, such that the normal bundle of every proper submanifold is k-ample, is better than this bound, for example if M is an Abelian variety, then k can be taken as the dimension of the maximum proper Abelian subvariety of M; see [569] for a proof of this result following the proof in the ample case due to Hartshorne [316, Proposition 4.1]. The computation of the minimum k is properly part of algebraic geometry. The link with finiteness theorems for cohomology and theorems of Lefschetz type comes from a result of Sommese [570, 580] (see also Okonek [490]) which implies that given a k-ample spanned normal bundle, $\mathcal{N}_{X/M}$, then $M - X$ possesses a $(k + \mathrm{cod}_M X)$-convex exhaustion function in the sense of Andreotti-Grauert. See §2.6 for some definitions and further discussion of convexity and exhaustion functions.

For a submanifold X of $\mathbb{P}^N$, $\mathcal{N}_{X/\mathbb{P}^N}(-1)$, the normal bundle of X tensored with $\mathcal{O}_{\mathbb{P}^N}(-1)$, is spanned by global sections, and the global sections of $\mathcal{T}_{\mathbb{P}^N}(-1)$ which span $\mathcal{N}_{X/\mathbb{P}^N}(-1)$ map $\mathbb{P}(\mathcal{N}_{X/\mathbb{P}^N}(-1))$ onto the discriminant variety of $X \subset \mathbb{P}^N$ (cf., §1.6). The relation between the k-ampleness of this bundle and the geometry of $X \subset \mathbb{P}^N$ is therefore quite interesting. Peternell, Le Potier, and Schneider [502, Corollary 1.4] show that $\mathcal{N}_{X/\mathbb{P}^N}(-1)$ is $(N - \dim X - 1)$-ample as a consequence of a theorem of Zak [628].

Though our main interest is in ample divisors, we give k-ample versions of results, especially in Chapter 5, when the extra work is minor. The reader interested only in ample divisors should just let $k = 0$ in any result about k-ampleness.

A k-ample line bundle, L, on a projective variety, X, is nef since NL is spanned for some $N \geq 1$. One simple fact that follows from noting that the image of the morphism associated to $|NL|$ is at least $(\dim X - k)$-dimensional is that k-ampleness implies k-bigness and in particular, given any ample line bundle, H on X (compare with §1.1 and Lemma (2.5.8)),

$$H^t \cdot L^{\dim X - t} > 0 \text{ if } t \geq k. \tag{2.1}$$

From this it follows that vanishing theorems of Kodaira-Kawamata-Viehweg type hold for k-ample bundles (see Theorem (2.2.5)).

Theorem 2.1.1. *Let E be a k-ample vector bundle on a projective variety, X. Then given any subvariety $V \subset X$, with $\dim V \geq k+1$, there is no surjective quotient map $E \to \mathcal{O}_V$. If E is spanned this property implies that E is k-ample.*

Proof. Assume that there is a surjective quotient map $E \to \mathcal{O}_V \to 0$. By the defining universal property of $\mathbb{P}(E)$, we associate to $V \subset X$ with $E \to \mathcal{O}_V \to 0$ a subvariety $V' \subset \mathbb{P}(E)$ such that V' maps isomorphically to V under the bundle projection, p, and such that there exists an exact sequence, $\xi_E \to \mathcal{O}_{V'} \to 0$, going to $E \to \mathcal{O}_V \to 0$ under the direct image under p. Thus $\xi_{E|V'} \cong \mathcal{O}_{V'}$. Hence any power of ξ_E is trivial on V', and the morphism associated to $|N\xi_E|$, where $N > 0$ is chosen so that $N\xi_E$ is spanned, has a variety, V', of dimension $> k$ as fiber. This contradicts the k-ampleness. If E is spanned, we can consider the morphism associated to ξ_E and the characterization is straightforward; see for example [569, Theorem 1.7]. □

The following corollary improves the result [255, (1.4)] of Fujita.

Corollary 2.1.2. *Let X be an irreducible projective variety of dimension n and let $\varphi : X \to Y$ be a proper surjective morphism. Let L be a k-big line bundle on X, for an integer $k \geq 0$. Let $U \subset Y$ be the Zariski open subset such that the restriction $\varphi_{\mathcal{U}} : \mathcal{U} \to U$, with $\mathcal{U} := \varphi^{-1}(U)$, is flat. Let F_u be a fiber of φ with $u \in U$. Then the restriction L_B is k-big for some irreducible component, B, of F_u. In particular, if F is an irreducible fiber contained in $\varphi^{-1}(U)$ then the restriction, L_F, of L to F is k-big.*

Proof. Since L is k-big there exists some multiple NL of L such that the morphism, $\psi : X_0 \to \mathbb{P}_{\mathbb{C}}$, associated to $|NL|_{X_0}$ with $X_0 := X - \mathrm{Bs}|NL|$, has an image of dimension $\geq n - k$. Let $F \subset \mathcal{U}$ be a fiber of $\varphi_{\mathcal{U}}$. Since $\varphi_{\mathcal{U}}$ is flat, we conclude that F is pure dimensional of dimension $f := n - \dim Y$. The restriction, L_B, of

L to some irreducible component, B, of F is k-big. Otherwise the restriction L_F would not be k-big. Therefore for any integer $t > 0$, the dimension of the image of F under the rational map given by $\Gamma(tL)$ would be $< f - k$. This implies that for any real constant $c > 0$, $h^0(tL_F) < ct^{f-k}$ for t large. Thus

$$\lim_{t\to\infty} \frac{h^0(tL_F)}{t^{f-k}} = 0.$$

By flatness of $\varphi_{\mathcal{U}}$ and the semi-continuity theorem for dimensions of spaces of sections on fibers of a flat morphism [318, Chapter III, Theorem 12.8], we conclude

$$\lim_{t\to\infty} \frac{h^0(tL_{F_y})}{t^{f-k}} \le \lim_{t\to\infty} \frac{h^0(tL_F)}{t^{f-k}} = 0$$

for $y \in V \subset Y$ where V is some Zariski neighborhood of $\varphi(F)$ and where $F_y = \varphi^{-1}(y)$. In particular we conclude that for any positive integer t the dimension of the image of the rational map given by $\Gamma(tL_{F_y})$ is less than $f - k$. It thus follows that $\dim \psi(F_y \cap X_0) \le \dim F_y - k - 1$. By replacing V by a smaller nonempty Zariski open set, V', not necessarily containing $\varphi(F)$, it can be assumed that $F_y \not\subset \mathrm{Bs}|NL|$ for $y \in V'$. Thus, since $\dim \psi(X_0 \cap \varphi^{-1}(V')) \le \dim \psi(F_y \cap X_0) + \dim Y$, we have

$$\dim \psi(X_0 \cap \varphi^{-1}(V')) \le \dim Y + \dim F_y - k - 1 = n - k - 1.$$

This contradicts $\dim \psi(X_0) \ge n - k$. □

2.2 Vanishing theorems

Vanishing theorems play a central role in algebraic geometry. Good references are the book [217] by Esnault and Viehweg and the book [562] by Shiffman and Sommese. Recall that given a continuous mapping, $\varphi : X \to Y$, between topological spaces, and a sheaf, $\mathcal{F}$, on X, we let $\varphi_{(i)}(\mathcal{F})$ be the i-th higher derived functor of the direct image, $\varphi_*(\mathcal{F})$.

The following theorem [346, (1.2.5), (1.2.6)] is a consequence of the Kawamata-Viehweg vanishing theorem [341, 607].

Theorem 2.2.1. (Relative vanishing theorem) *Let X be a normal variety of dimension n with only terminal singularities. Let $\varphi : X \to Y$ be a proper morphism onto a variety, Y, and let $D \in Z_{n-1}(X) \cap (\mathrm{Div}(X) \otimes \mathbb{Q})$, i.e., let D be a $\mathbb{Q}$-Cartier integral Weil divisor. If $D - K_X$ is φ-nef and φ-big then $\varphi_{(i)}\mathcal{O}_X(D) = 0$ for $i > 0$.*

We also need the following simple variant of the Grauert-Riemenschneider vanishing theorem [286] for the i-th direct image of the canonical sheaf.

Theorem 2.2.2. (Grauert-Riemenschneider vanishing theorem) *Let X and Y be normal projective varieties, and let $\varphi : X \to Y$ be a birational morphism. If X has rational singularities then $\varphi_{(i)}(K_X) = 0$ for $i > 0$.*

Proof. We reduce to the usual Grauert-Riemenschneider vanishing theorem. Let $p : \overline{X} \to X$ be a desingularization of X, and let $\beta : \overline{X} \to Y$ denote the composition $\varphi \circ p$. Since X has rational singularities we have $p_*K_{\overline{X}} \cong K_X$. By the usual Grauert-Riemenschneider vanishing theorem we also have $p_{(i)}(K_{\overline{X}}) = 0$ for $i > 0$. Therefore, by using the Grothendieck spectral sequence for the higher direct images of the composition $\beta = \varphi \circ p$, we have $\varphi_{(i)}(K_X) \cong \beta_{(i)}(K_{\overline{X}})$ for all $i \geq 0$. Thus we are done since $\beta_{(i)}(K_{\overline{X}}) = 0$ for $i > 0$ again by the usual Grauert-Riemenschneider vanishing theorem. □

Lemma 2.2.3. *Let X be a normal irreducible projective variety with terminal singularities. Let L be an ample line bundle on X. Assume that K_X is nonnef. Let τ be the nefvalue of (X, L) and let $\phi : X \to Y$ be the nefvalue morphism of (X, L). Assume that ϕ is birational. Then Y has rational singularities.*

Proof. We have to show that the higher direct images sheaves $p_{(i)}(\mathcal{O}_{\overline{X}})$ are zero for all $i > 0$ for some desingularization $p : \overline{X} \to X$. By a similar argument to that in (2.2.2), reducing to the usual Grauert-Riemenschneider vanishing theorem, it suffices to show that $\phi_{(i)}(\mathcal{O}_X)$ are zero for all $i > 0$. Let m be an integer such that $m\tau = b$ is an integer. Then $m(K_X + \tau L) = mK_X + bL$ and by the projection formula $\phi_{(i)}(mK_X + bL) \cong H \otimes \phi_{(i)}(\mathcal{O}_X)$, where $mK_X + bL \cong \phi^*H$ for some ample line bundle H on Y. Write $mK_X + bL = K_X + ((m-1)K_X + bL)$ and note that $(m-1)K_X + bL$ is ample since $b/(m-1) > \tau$. Therefore, by the vanishing theorem, (2.2.1), $\phi_{(i)}(mK_X + bL) = 0$ for $i > 0$ and hence $\phi_{(i)}(\mathcal{O}_X) = 0$ for $i > 0$. □

The following extension of the usual vanishing theorems to take into account the irrational locus, $\mathrm{Irr}(X)$, of nonrational singularities of a variety, X, is very useful [581, 583].

Theorem 2.2.4. *Let $\varphi : X \to Y$ be a morphism from an irreducible normal projective variety, X, to a projective variety, Y. Let $\mathcal{L}$ be a nef line bundle on X such that $\mathcal{L}^{\dim X - k} \cdot H^k \neq 0$ for some ample Cartier divisor, H, on X and some integer $k \geq 0$, e.g., $\mathcal{L}$ is nef and k-big. Then $\varphi_{(i)}(K_X + \mathcal{L}) = 0$ for*

$$i \geq \max\{k, \max_{y \in \varphi(\mathrm{Irr}(X))} \dim(\varphi^{-1}(y) \cap \mathrm{Irr}(X))\} + 1.$$

Proof. First we prove the version for $k = 0$, i.e., $\mathcal{L}$ big and nef.

Lemma 2.2.5. *Let $\varphi : X \to Y$ be a morphism from an irreducible normal projective variety, X, to a projective variety, Y. Let $\mathcal{L}$ be a nef and big line bundle on*

X. Then $\varphi_{(i)}(K_X + \mathcal{L}) = 0$ *for*

$$i \geq \max_{y \in \varphi(\mathrm{Irr}(X))} \dim(\varphi^{-1}(y) \cap \mathrm{Irr}(X)) + 1.$$

Proof. We prove the absolute version when Y is a point. The general case is the same, except that we use the relative form, (2.2.1), of the Kawamata-Viehweg vanishing theorem, [341, 607].

Let $p : \overline{X} \to X$ be a desingularization of X. Consider the exact sequence

$$0 \to p_* K_{\overline{X}} \to K_X \to \mathcal{S} \to 0. \tag{2.2}$$

By Kempf's theorem [347] the sheaf $\mathcal{S}$ is supported on the set $\mathrm{Irr}(X)$. Tensoring the above sequence with $\mathcal{L}$, we are reduced to proving that $h^i(X, p_* K_{\overline{X}} + \mathcal{L}) = 0$ for $i > \dim \mathrm{Irr}(X)$. Since $h^i(\overline{X}, K_{\overline{X}} + p^*\mathcal{L}) = 0$ for $i > 0$ by the Kawamata-Viehweg vanishing theorem, using the Leray spectral sequence for p and $K_{\overline{X}} + p^*\mathcal{L}$, the desired result will follow if $p_{(j)}(K_{\overline{X}}) + \mathcal{L} = 0$ for $j > 0$. This is just the usual Grauert-Riemenschneider vanishing theorem. □

We turn now to the proof of Theorem (2.2.4). We let $\mathrm{VT}(n, I)$ denote the theorem with a given $n := \dim X$ and

$$I := \max\{k, \max_{y \in \varphi(\mathrm{Irr}(X))} \dim(\varphi^{-1}(y) \cap \mathrm{Irr}(X))\}.$$

By Theorem (1.7.3) there is a Zariski open set of divisors $A \in |NH|$ for some $N > 0$ such that A meets every positive dimensional irreducible component of any fiber of φ and of any fiber of the restriction $\varphi_{\mathrm{Irr}(X)}$ in a lower dimensional algebraic set. By Theorem (1.7.1) it can be assumed further that A is normal, and $\mathrm{Irr}(A) \subset \mathrm{Irr}(X) \cap A$. Let φ_A be the restriction of φ to A. In particular

$$\max_{y \in \varphi(\mathrm{Irr}(X))} \dim(\varphi^{-1}(y) \cap \mathrm{Irr}(X)) \geq 1 + \max_{y \in \varphi(\mathrm{Irr}(A))} \dim(\varphi_A^{-1}(y) \cap \mathrm{Irr}(A)).$$

By Lemma (1.7.6) it can further be assumed that we have an exact sequence

$$0 \to K_X + \mathcal{L} \to K_X + \mathcal{L} + A \to K_A + \mathcal{L}_A \to 0. \tag{2.3}$$

Note that if $k \geq 1$ then

$$\mathcal{L}_A^{\dim A - (k-1)} \cdot H^{k-1} = \mathcal{L}^{\dim X - k} \cdot A \cdot H^{k-1} = N\mathcal{L}^{\dim X - k} \cdot H^k \neq 0.$$

Therefore by using the long exact sequence of sheaves associated to the direct image φ_* applied to the short exact sequence, (2.3), we see that the statement $\mathrm{VT}(n, I)$ will follow from Lemma (2.2.5) applied to the ample bundle $\mathcal{L} + A$ and $\mathrm{VT}(n-1, I-1)$ for the bundle $\mathcal{L}_A$. By induction this reduces to the statement $\mathrm{VT}(n - I, 0)$ which follows from Lemma (2.2.5), since $I = 0$ implies $k = 0$. □

The following is simply a restatement of Theorem (3.42) of [562].

Corollary 2.2.6. *Let L be a line bundle on an irreducible, n-dimensional projective variety X. Assume that $|L|$ has a base locus of codimension ≥ 2 with the image under the rational map associated to $|L|$ having dimension ≥ 2. Then there exists a nonempty Zariski open set, $U \subset |L|$, such that $D \in U$ is irreducible.*

Proof. By Hartshorne [318, Example 7.17.3] we can find a birational morphism, $p : V \to X$, from a projective variety V onto X, and an effective Cartier divisor Z on V such that $p^*L - Z$ is spanned, $p(Z) \subset \mathrm{Bs}|L|$, and the natural map $\Gamma(p^*L - Z) \to \Gamma(L)$ is an isomorphism. By passing to a desingularization, we see that V can be chosen smooth. Since $|p^*L - Z|$ gives rise to a map with an image of dimension at least 2, we see that $\mathcal{L} := p^*L - Z$ is $(n-2)$-big and therefore there is a nonempty Zariski open set, $U' \subset |p^*L - Z|$, with $D' \in U'$ being both smooth and connected. The smoothness assertion follows from Theorem (1.7.1). To see that D' is connected, note that Theorem (2.2.4), applied in the absolute case when Y, from the theorem, is a point, gives $h^{n-1}(K_V + \mathcal{L}) = h^1(-\mathcal{L}) = 0$. This implies $h^0(\mathcal{O}_{D'}) = 1$ and therefore D' is connected. Letting $U \subset |L|$ be the set corresponding to U' under the natural isomorphism $|p^*L - Z| \cong |L|$ mentioned above, we have that $D \in U$ is of the form $p(D')$ for $D' \in U'$, and hence irreducible. □

Finally we note that the usual slicing trick used above gives a slight variant of Mumford's vanishing theorem [464]. Note that by Lemma (2.5.8), the hypothesized inequality, $L^{n-k} \cdot H^k > 0$, in the next Lemma is satisfied if L is nef and k-big.

Lemma 2.2.7. *Let L be a nef line bundle on an n-dimensional irreducible normal projective variety, X. Assume that $L^{n-k} \cdot H^k > 0$ for some nef and big Cartier divisor, H, on X and some integer $k \geq 0$. Then if $n \geq k+2$, $h^1(-L) = 0$.*

Proof. We follow Mumford's argument. Let $p : \overline{X} \to X$ be a desingularization of X. From the Leray spectral sequence for p we have an injection $H^1(-L) \hookrightarrow H^1(-p^*L)$. Therefore it suffices to show that $h^{n-1}(K_{\overline{X}} + p^*L) = 0$. Note that for some ample divisor, $\mathcal{H}$, on $\overline{X}$, the divisor $A := p^*H + \mathcal{H}$ is ample and

$$(p^*L)^{n-k} \cdot A^k \geq (p^*L)^{n-k} \cdot (p^*H)^k = L^{n-k} \cdot H^k > 0.$$

Thus Theorem (2.2.4), applied in the absolute case when Y, from the theorem, is a point, gives $h^{n-1}(K_{\overline{X}} + p^*L) = 0$ as soon as $n - 1 \geq k + 1$, or $n \geq k + 2$. □

We also need the following fact (see [583, (0.2.2)]).

Corollary 2.2.8. *Let X be a normal projective variety and let the set,* $\mathrm{Irr}(X)$, *of irrational singularities be finite and nonempty. Then given any nef and big line* *bundle, L, on X one has $h^0(K_X + L) \geq \#(\mathrm{Irr}(X)) > 0$, where $\#(\mathrm{Irr}(X))$ is the number of points in* Irr(X). *In particular if $A \in |L|$ lies in the set of Cohen-Macaulay points of X, then $h^0(K_X) + h^0(K_A) \geq \#(\mathrm{Irr}(X)) > 0$.*

Furthermore if L is spanned, then A can be chosen so that $h^0(K_A) \geq 1$, and if K_X is invertible then global sections of $K_X + L$ span $K_X + L$ on Irr(X).

Proof. Let $p : \overline{X} \to X$ be a desingularization of X. Consider the exact sequence (2.2) as in the proof of Lemma (2.2.5). By tensoring with L we obtain the exact sequence

$$0 \to p_*K_{\overline{X}} + L \to K_X + L \to \mathcal{S} \otimes L \to 0. \tag{2.4}$$

Note that $h^1(p_*K_{\overline{X}} + L) = 0$ by the Leray spectral sequence for p and the Kawamata-Viehweg vanishing theorem. Therefore we get

$$h^0(K_X + L) \geq h^0(\mathcal{S} \otimes L) = h^0(\mathcal{S}) \geq \#(\mathrm{Irr}(X)).$$

Assume now that $A \in |L|$ is in the set of Cohen-Macaulay points of X. By the adjunction formula, we have the exact sequence

$$0 \to K_X \to K_X + L \to K_A \to 0,$$

which yields $h^0(K_X) + h^0(K_A) \geq h^0(K_X + L) \geq \#(\mathrm{Irr}(X))$. Finally, assume that L is spanned. By the above, we have $h^0(K_X + L) \geq 1$. Take a section $s \in \Gamma(K_X + L)$. Since L is spanned we can choose $A \in |L|$ such that A is not contained in the zero locus $s^{-1}(0)$. Then the restriction s_A is not identically zero on A. This implies that $h^0(K_A) \geq 1$. If K_X is invertible, then, by the sequence (2.4), global sections of $K_X + L$ span $K_X + L$ on Irr(X). □

Remark 2.2.9. Very often we will need vanishing of $H^1(X, K_X + \mathcal{L})$ where $\mathcal{L}$ is an ample line bundle on a normal variety X. This is the main technical reason that we often require the locus of irrational singularities, Irr(X), to be finite. This is more than just a restriction of technique. The following examples of the second author from [583] of normal divisors, X, on a smooth four dimensional manifold are very well-behaved, except that Irr(X) is 1-dimensional. The Kodaira vanishing theorem fails badly for these very ample line bundles. We will come back to this example later because many of the structure theorems of this book fail for it (though with appropriate rephrasing they remain true on $X -$ Irr(X)).

Example 2.2.10. (Failure of vanishing theorems) Let $P := \mathbb{P}(E)$ where $E := \mathcal{O}_{\mathbb{P}^1} \oplus \mathcal{O}_{\mathbb{P}^1}(1) \oplus \mathcal{O}_{\mathbb{P}^1}(1) \oplus \mathcal{O}_{\mathbb{P}^1}(1)$. Let $\pi : P \to \mathbb{P}^1$ denote the canonical projection with F a fiber of π. Letting ξ_E be the tautological line bundle, we have $\pi_*\xi_E = E$. Let X_d be a general element of $|d(\xi_E - F)|$. Note that for $d \geq 1$, the restriction $\pi_{X_d} : X_d \to \mathbb{P}^1$ is a fibration whose general fiber is biholomorphic to a cone in $\mathbb{P}^3$ on a smooth curve of degree d. Note also that

$$K_P \cong \pi^*(K_{\mathbb{P}^1} + \det E) - 4\xi_E \cong F - 4\xi_E.$$

Therefore by the adjunction formula, $K_{X_d} \cong ((d-4)\xi_E + (1-d)F)_{X_d}$. Note that $L = \xi_E + F$ is very ample. A straightforward calculation shows that for $d \geq 3$, $h^1(K_{X_d} + L_{X_d}) = \binom{d}{4}$. This shows that the Kodaira vanishing theorem fails to hold in this case. Note also that $(2d-5)\xi_E \cong K_{X_d} + (d-1)L_{X_d}$ is spanned by global sections, since E, and hence ξ_E, are spanned and that the morphism, ϕ, associated to $\Gamma((2d-5)\xi_E)$ has a 3-dimensional image with its only positive dimensional fiber being $\mathrm{Irr}(X_d)$, the curve, C, associated to the quotient $E \to \mathcal{O}_{\mathbb{P}^1} \to 0$. To see this note that, by construction, $C = \mathrm{Sing}(X_d)$. Since the Kodaira vanishing theorem fails, we conclude that $\dim \mathrm{Irr}(X_d) \geq 1$ and hence $C = \mathrm{Sing}(X_d) = \mathrm{Irr}(X_d)$.

Note that, for $d = 3$, the morphism, ϕ, associated to $\Gamma(\xi_E)$ is the first reduction map of the pair (X_3, L_{X_3}) (we refer to Chapter 7, §7.3, for definitions and more details). By the above, the only positive dimensional fiber of ϕ is the curve C. Hence in particular ϕ does not contract any linear $\mathbb{P}^2$'s, which are the only possible positive dimensional fibers for the first reduction map studied in Chapter 7, §7.3, under stronger hypotheses on the singularities allowed. Thus this example shows that the structure theorem, (9.2.2), fails to be true if the locus of irrational singularities has positive dimension.

Note also that, if $d > 3$, then $K_{X_d} + 2L_{X_d}$ is spanned on the complement of C but not on C.

2.3 The Lefschetz hyperplane section theorem

Theorems of Lefschetz type form an elaborate subject. The following very useful general result is due to Hamm [305] in the local complete intersection case and to Andreotti [29] and Bott [136] in the case when the complement of the divisor is smooth. We refer to Fulton-Lazarsfeld [270] and Goresky-McPherson [282] for a detailed discussion.

Theorem 2.3.1. (Hamm's Lefschetz theorem) *Let L be an ample line bundle on an irreducible projective variety, X. Let $D \in |L|$ be a divisor such that $X - D$ is a local complete intersection. Then given any point $x \in D$ it follows that the j-th relative homotopy group, $\pi_j(X, D, x)$, vanishes for $j \leq \dim X - 1$. In particular it follows that under the restriction mapping, $H^j(X, \mathbb{Z}) \cong H^j(D, \mathbb{Z})$ for $j \leq \dim X - 2$, and $H^j(X, \mathbb{Z}) \to H^j(D, \mathbb{Z})$ is injective with torsion free cokernel for $j = \dim X - 1$.*

In practice the Lefschetz theorem is used for homology or cohomology. The key fact here is the following standard immediate consequences of the Hurewicz isomorphism theorem, see e.g., Spanier [589, Chapter 7, §5, Proposition 1], and the universal-coefficient theorem for cohomology, see, e.g., [589, Chapter 5, §5, Corollary 4].

Lemma 2.3.2. *If D is a connected subvariety of a connected projective variety, X, and if for some $x \in D$ and some $k \geq 1$, $\pi_j(X, D, x)$ vanishes for $j \leq k$, then*

1. *the inclusion mapping, $H_j(D, \mathbb{Z}) \to H_j(X, \mathbb{Z})$, is an isomorphism for $j \leq k-1$, and a surjection for $j = k$;*
2. *the restriction mapping, $H^j(X, \mathbb{Z}) \to H^j(D, \mathbb{Z})$, is an isomorphism for $j \leq k-1$, and an injection with torsion free cokernel for $j = k$.*

Corollary 2.3.3. *Let L be a k-ample line bundle on an irreducible projective variety, X. Let $D \in |L|$ be a divisor such that $X - D$ is a local complete intersection. Then given any point $x \in D$ it follows that under the restriction mapping, $H^j(X, \mathbb{Z}) \cong H^j(D, \mathbb{Z})$ for $j \leq \dim X - 2 - k$, and $H^j(X, \mathbb{Z}) \to H^j(D, \mathbb{Z})$ is injective with torsion free cokernel for $j = \dim X - 1 - k$.*

Proof. In Sommese [569, Proposition (1.16)] this is proved for smooth D and X by a simple slicing argument that reduces the result to the manifold version of (2.3.1). Using (2.3.1) in that argument gives this result with no change of wording. □

Corollary 2.3.4. *Let L be a k-ample line bundle on a normal irreducible projective variety, X, with at worst Cohen-Macaulay singularities. Assume that $\dim \operatorname{Irr}(X) \leq 0$ and that $D \in |L|$ is a divisor such that $X - D$ is a local complete intersection. Then under the restriction mapping, $\operatorname{Pic}(X) \cong \operatorname{Pic}(D)$ if $\dim X \geq 4 + k$, and $\operatorname{Pic}(X) \to \operatorname{Pic}(D)$ is injective with torsion free cokernel if $\dim X = 3 + k$.*

Proof. In this proof the structure sheaves are the sheaves of germs of holomorphic functions. Associated to the diagram

$$\begin{array}{ccccccccc} 0 & \to & \mathbb{Z} & \to & \mathcal{O}_X & \xrightarrow{\exp(2\pi\sqrt{-1}\bullet)} & \mathcal{O}_X^* & \to & 0 \\ & & \downarrow & & \downarrow & & \downarrow & & \\ 0 & \to & \mathbb{Z} & \to & \mathcal{O}_D & \xrightarrow{\exp(2\pi\sqrt{-1}\bullet)} & \mathcal{O}_D^* & \to & 0 \end{array}$$

with the vertical arrows being restrictions, and $\exp(2\pi\sqrt{-1}\bullet)$ being the exponential function applied to elements of the structure sheaf, we have the diagram of long exact cohomology sequences

$$\begin{array}{ccccccccc} \cdots & \to & H^j(X, \mathbb{Z}) & \to & H^j(\mathcal{O}_X) & \xrightarrow{\exp(2\pi\sqrt{-1}\bullet)} & H^j(\mathcal{O}_X^*) & \to & \cdots \\ & & \downarrow & & \downarrow & & \downarrow & & \\ \cdots & \to & H^j(D, \mathbb{Z}) & \to & H^j(\mathcal{O}_D) & \xrightarrow{\exp(2\pi\sqrt{-1}\bullet)} & H^j(\mathcal{O}_D^*) & \to & \cdots \end{array}$$

Note that $\operatorname{Pic}(X) \cong H^1(X, \mathcal{O}_X^*)$ since X is projective. Using the vanishing theorem, (2.2.4), we have that the restriction mapping, $H^j(\mathcal{O}_X) \to H^j(\mathcal{O}_D)$ is an isomorphism for $j \leq \dim X - 2$ and an injection for $j = \dim X - 1$. Using this, the result follows from (2.3.3) and a diagram chase. □

For surfaces there is a very useful variant (2.3.6) of the above result on Picard groups due to Weil [615] for non-ruled surfaces and Sommese [571, (0.9)] in general. We need the following general fact to prove (2.3.6).

Lemma 2.3.5. *Let $p : X \to C$ be a proper surjection from an irreducible reduced variety, X, of dimension n, onto a smooth, not necessarily compact, curve C. Assume that all fibers of p are irreducible and that there is a section, $C' \subset X$, of p. Let $\mathcal{H}$ be a line bundle on X such that $\mathcal{H}_{C'} \approx \mathcal{O}_{C'}$ and $\mathcal{H}_F \approx \mathcal{O}_F$ for fibers $F = p^{-1}(y)$ with y in some open set, U, of C. Then $\mathcal{H} \approx \mathcal{O}_X$.*

Proof. Note that $p_*\mathcal{H}$ is a torsion free coherent sheaf on C, and that, by the assumptions made, it satisfies the condition $p_*\mathcal{H}_{|U} \approx \mathcal{O}_U$. Thus $p_*\mathcal{H}$ is a rank one torsion free sheaf on C. Since C is a smooth curve it follows that $\mathcal{E} := p_*\mathcal{H}$ is invertible on C.

Consider the line bundle $\mathcal{H} - p^*\mathcal{E}$ on X. Note that $\mathcal{H} - p^*\mathcal{E} \approx \mathcal{O}_X$ implies $\mathcal{H} \approx \mathcal{O}_X$. Indeed, restricting to the section C', we have $\mathcal{O}_{C'} \approx \mathcal{H}_{C'} \approx (p^*\mathcal{E})_{C'}$, which gives $p_*\mathcal{H} \approx \mathcal{O}_C$ and hence $\mathcal{H} \approx \mathcal{O}_X$.

Thus we are reduced to show that $\mathcal{H} - p^*\mathcal{E} \approx \mathcal{O}_X$. We claim that $\mathcal{H} - p^*\mathcal{E}$ has a section which is not identically zero on any fiber of p. To see this, let F be a fiber of p and consider the exact sequence

$$0 \to \mathcal{H} - p^*\mathcal{E} - F \to \mathcal{H} - p^*\mathcal{E} \to (\mathcal{H} - p^*\mathcal{E})_F \to 0.$$

Applying p_*, since $p_*(\mathcal{H} - p^*\mathcal{E}) \approx \mathcal{O}_C$ and all fibers of p are reduced, we obtain the long exact sequence

$$0 \to \mathfrak{m}_y \to \mathcal{O}_C \to \mathbb{C}_y \to p_{(1)}(\mathcal{H} - p^*\mathcal{E} - F) \to \cdots,$$

where $\mathfrak{m}_y$ is the ideal sheaf of $y = p(F)$ in C. Thus we conclude that the section $s \in \Gamma(\mathcal{H} - p^*\mathcal{E})$, corresponding to $1 \in \Gamma(\mathcal{O}_C)$, is not vanishing identically on F, and hence on any fiber of p.

Let $D := s^{-1}(0)$. If D is not empty, then D is a divisor. By the above we know that $D \cap F \neq F$ for any fiber F of p. Therefore by counting constants we see that the restriction map $p_D : D \to C$ is onto. Then $\mathcal{O}_F(D \cap F) \not\approx \mathcal{O}_F$. Since $\mathcal{O}_F(D \cap F) \approx (\mathcal{H} - p^*\mathcal{E})_F$ we contradict the fact that, by the assumptions on $\mathcal{H}$, $(\mathcal{H} - p^*\mathcal{E})_F \approx \mathcal{O}_F$ for a general fiber F of p.

Thus we conclude that D is the empty set. Therefore the section s is nowhere vanishing. This implies that $\mathcal{H} - p^*\mathcal{E} \approx \mathcal{O}_X$. □

Proposition 2.3.6. *Let L be a very ample line bundle on a smooth connected projective surface, S. Let H be a line bundle on S such that the restriction, H_C, of H to C is trivial for an open set U of curves $C \in |L|$. If S is not a $\mathbb{P}^1$-bundle, $\pi : S \to B$, over a curve, B, with $L_f \cong \mathcal{O}_{\mathbb{P}^1}(1)$ for a fiber f of π, then H is trivial.*

Proof. Let X denote $S \times \mathbb{P}^1$ with the two product projections $p : X \to \mathbb{P}^1$ and $q : X \to S$. Let $\mathcal{L} := p^*\mathcal{O}_{\mathbb{P}^1}(1) + q^*L$. Choose two general sections s_0, s_1 of L such that the zero sets of s_0, s_1 are contained in U, and two distinct sections z_0, z_1 of $\mathcal{O}_{\mathbb{P}^1}(1)$. Let $\overline{S}$ be the zero set of the section $p^*z_0 \otimes q^*s_0 + p^*z_1 \otimes q^*s_1$ of $\mathcal{L}$. Note that the restriction map $q_{\overline{S}} : \overline{S} \to S$ is just the blowup of S at the finite set of common zeroes of s_0 and s_1. Thus $q^*_{\overline{S}}$ gives an isomorphism of $\Gamma(H)$ and $\Gamma(\mathcal{H})$ where $\mathcal{H} := q^*_{\overline{S}}H$. Therefore it suffices to prove that $\mathcal{H} \cong \mathcal{O}_{\overline{S}}$.

Note that $\mathcal{H}_F$ is trivial for fibers, F, of the restriction $p_{\overline{S}}$, over an open set V of $\mathbb{P}^1$, since such fibers, F, are the proper transforms under $q_{\overline{S}}$ of the curves $C \in U$ such that the restriction H_C is trivial. Note also that, by construction, any exceptional divisor, C', of the blowup $q_{\overline{S}} : \overline{S} \to S$ is a section of $p_{\overline{S}}$ and $\mathcal{H}_{C'} \approx \mathcal{O}_{C'}$. By the assumption on S and by using Corollary (1.6.8) we know that all fibers of $p_{\overline{S}}$ are irreducible and reduced. Therefore we can apply Lemma (2.3.5) to conclude that $\mathcal{H} \approx \mathcal{O}_{\overline{S}}$. □

Remark 2.3.7. Notation as in Proposition (2.3.6). In (8.7.1), the condition $q(S) = g(L) > 0$ is shown to be equivalent to the condition that S is a $\mathbb{P}^1$-bundle, $\pi : S \to B$, over a smooth curve, B, with $L_f \cong \mathcal{O}_{\mathbb{P}^1}(1)$ for fibers, f, of π.

If S is a $\mathbb{P}^1$-bundle, $\pi : S \to B$, over a curve, B, with $L_f \cong \mathcal{O}_{\mathbb{P}^1}(1)$ for a fiber f of π, then the line bundle $H := K_S + L - \pi^*K_B$ is trivial on all smooth curves $C \in |L|$. Indeed $(K_S + L)_C \cong K_C$ and, since π is an isomorphism on such curves, C, one has $-\pi^*K_{B|C} \cong -K_C$. But $H \not\cong \mathcal{O}_S$ since $H_f \cong -L_f \cong \mathcal{O}_{\mathbb{P}^1}(-1)$.

Thus for a fixed very ample line bundle, L, on a smooth connected projective surface, S, $K_S + L$ is the unique line bundle with the property that $(K_S + L)_C \approx K_C$ for all smooth $C \in |L|$, unless S is a $\mathbb{P}^1$-bundle, $\pi : S \to B$, onto a smooth curve, B, and $L_f \approx \mathcal{O}_f(1)$ for fibers, f, of π.

Let L be a very ample line bundle on a connected n-dimensional projective manifold, X. Let $\mathcal{D} \subset |L|$ denote the discriminant variety of (X, L), e.g., see §1.6. If $\mathrm{cod}_{|L|}\mathcal{D} = k+1 \geq 2$ then we can choose a linear pencil $P := \mathbb{P}^1 \subset |L| - \mathcal{D}$. Let V denote the 2-dimensional subspace of $\Gamma(L)$ corresponding to P. Let $Z_V \subset P \times X$ be the universal zero set. Let $p : Z_V \to P$ and $q : Z_V \to X$ be the projections induced by the product projections. For $x \in P$, let $H := H_x$ be the divisor corresponding to x. Let $H' = H_{x'}$, $x' \in P$. Note that $U := X - H$ is isomorphic to $Z_V - q^{-1}(H)$. Moreover p_U is a differentiable fiber bundle map over $\mathbb{C} = \mathbb{P}^1 - \{x\}$ with fiber diffeomorphic to $H - H \cap H'$. Thus $X - H$ is homotopic to $H - H \cap H'$. If $k(V) > 1$ then the analogous topological constraints get very severe. These restrictions are explored in the very interesting papers by Lanteri and Struppa [390, 391, 392, 393, 394]. Here we only give a sharpened Lefschetz theorem in this direction. We suppress basepoints of homotopy groups since our spaces are pathwise connected.

Theorem 2.3.8. *Let L be a very ample line bundle on a connected n-dimensional projective manifold, X. Assume that the codimension of the discriminant variety*

$\mathfrak{D} \subset |L|$ is $k+1$. Then if $(X, L) \not\cong (\mathbb{P}^n, \mathcal{O}_{\mathbb{P}^n}(1))$, $\pi_i(X, A) = 0$ for $i \leq n+k-1$ and $A \in |L|$. Moreover $k \leq n$.

Proof. If $k = 0$ then this is simply the usual Lefschetz theorem, (2.3.1). Thus we can assume that $k > 0$ without loss of generality. By Corollary (1.6.12) we can assume that $k \leq n-1$ without loss of generality. Choose a general linear $\mathbb{P}^k \subset |L| - \mathfrak{D}$. Denote it by P and let V denote the linear subspace of $\Gamma(L)$ corresponding to P. Let $Z_V \subset P \times X$ be the universal zero set. Let $p : Z_V \to P$ and $q : Z_V \to X$ be the projections induced by the product projections. Let $H = H_x$ for an $x \in P$. As seen in §1.6, Z_V is an ample divisor on $P \times X$. Thus the usual first Lefschetz theorem, (2.3.1), gives that $\pi_j(P \times X, Z_V) = 0$ for $j \leq \dim Z_V = n+k-1$.

Since $k \leq n-1$ we see that $|V|$ has basepoints and thus p has sections, e.g., $q^{-1}(z)$ for any basepoint z of $|V|$. Thus the long exact homotopy sequence of the differential fiber bundle, p, splits to give $\pi_i(Z_V) = \pi_i(H_x) \oplus \pi_i(\mathbb{P}^1)$ for all $i \geq 0$. Since the section of p is a section of the product projection $P \times X \to P$ we can compare the two homotopy sequences and conclude from $\pi_j(P \times X, Z_V) = 0$ for $j \leq \dim Z_V = n+k-1$ that $\pi_i(X, A) = 0$ for $i \leq n+k-1$ and $A \in |L|$. □

Remark 2.3.9. We refer to the papers of Lanteri and Struppa for a converse to the above. To see why there should be a converse consider the special case when $n = 3$. Using the second Lefschetz theorem, e.g., Mumford's appendix to [634, Chapter 6], we see that the kernel of $H_2(A, \mathbb{Q}) \to H_2(X, \mathbb{Q})$ for $A \in |L|$ is generated by the so-called vanishing cycles. Since $\dim A$ is even it follows from the intersection formulae for the vanishing cycles that they have nonzero self-intersection numbers. Thus if $H_2(A, \mathbb{Q}) \cong H_2(X, \mathbb{Q})$ there are no vanishing cycles. The vanishing cycles are in one-to-one correspondence with the points of intersection of $\mathfrak{D} \subset |L|$ with a general $\mathbb{P}^1 \subset |L|$, and thus it follows that $\mathrm{cod}_{|L|}\mathfrak{D} \geq 2$. This argument works if n is odd. For n even, the self intersections of the vanishing cycles are zero in $H^{n-1}(A, \mathbb{Q})$. In this case $k(X, L) > 0$ is implied by $H_j(A, \mathbb{Q}) \cong H_j(X, \mathbb{Q})$ for $j \leq n$. The vanishing cycles are still in one-to-one correspondence with the points of intersection of $\mathfrak{D} \subset |L|$ with a general $\mathbb{P}^1 \subset |L|$, but in this case a vanishing cycle can be used to construct a nonzero element of $H^n(X, \mathbb{Q})$ that restricts to 0 in $H^n(A, \mathbb{Q})$. That this extra isomorphism is needed is not surprising since as we will see in §14.4 there is a parity result of Landman that if $k(X, L) > 0$ it follows that $k(X, L) \equiv n \bmod 2$.

2.3.10. Generalizations of Lefschetz's theorem. Barth [68] generalized the first Lefschetz theorem for cohomology with complex coefficients to the case of submanifolds of $\mathbb{P}_{\mathbb{C}}$. Barth and Larsen [71, 397, 70] improved the result to show that given a connected submanifold, X, of $\mathbb{P}^N$, and any point $x \in X$, the relative homotopy groups $\pi_j(\mathbb{P}^N, X, x)$ are zero for $j \leq 2n - N + 1$. Sommese's extension

[568, 569, 570, 572, 578, 580] of the Barth-Larsen theorem to pairs of submanifolds in arbitrary homogeneous complex manifolds included the introduction of k-ampleness and the construction of the necessary machinery to convert the algebraic geometric calculation of k-ampleness of normal bundles to the appropriate exhaustion functions to do Morse theory. See Sommese and Van de Ven [585] for a version of the Barth-Larsen theorems covering morphisms to homogeneous spaces, and Okonek [491, 490] for the full singular versions of the results. See Fulton and Lazarsfeld [270] and Goresky and McPherson [282] for background and further references on Lefschetz-type results. See also Schneider and Zintl [548] for a reduction of the original Barth-Lefschetz theorem to the Le Potier vanishing theorem.

Theorem 2.3.11. (Generalized Lefschetz theorem) *Let B be a compact connected complex submanifold of a homogeneous complex manifold, M, and let A be a connected, not necessarily compact, complex submanifold of M. Let $N = \dim M$, $n = \dim B$. Assume that the normal bundle of B in M is k-ample. Then for any $x \in A \cap B$, the relative homotopy groups $\pi_j(A, A \cap B, x) = 0$ for $j \leq \min\{\dim A, \dim B + 1\} - \mathrm{cod}_M B - k$. In particular taking $M = A$ we conclude that $\pi_1(M, x) \cong \pi_1(B, x)$ for any $x \in B$ if $2n - N - k \geq 1$. Taking $A = M$ and assuming M to be projective, we conclude:*

1. *if $2n - N - k = 1$, the restriction map $r : \mathrm{Pic}(M) \to \mathrm{Pic}(B)$ is injective with torsion free cokernel; and*

2. *if $2n - N - k \geq 2$, then $\mathrm{Pic}(M) \cong \mathrm{Pic}(B)$ via r.*

Remark 2.3.12. As was noted in §2.1, the number k in Theorem (2.3.11) has been computed in a wide variety of cases, e.g., it can be taken to be 1 if M is a quadric, 0 if M is $\mathbb{P}_{\mathbb{C}}$ or an Abelian variety with no proper Abelian subvarieties, and $\dim \mathrm{Grass}(a, b) - b + 1 = a(b - a) - b + 1$, where $\mathrm{Grass}(a, b)$ is the Grassmannian of a-dimensional quotient vector spaces of $\mathbb{C}^b$.

There are also some Lefschetz type results for the part of the middle dimensional homology of a general hyperplane section that is generated by algebraic cycles. For these results which go under the name of Noether-Lefschetz theorems we refer to [161, 174, 202, 287, 288, 295, 588, 608, 609, 610], and particularly to the nice survey in Spandaw [587].

2.4 The Albanese mapping in the presence of rational singularities

We follow the presentation in [583]. Ueno [602] is a good reference for general properties of the Albanese mapping.

Let $\varphi : \overline{X} \to X$ be a resolution of singularities of a normal, irreducible projective variety, X. Let $\mathrm{Alb}(\overline{X})$ be the Albanese variety of $\overline{X}$ and $\mathrm{alb}_{\overline{X}}$ the Albanese mapping for $\overline{X}$. We suppress basepoints. We say that the *Albanese mapping is defined* for X if there exists a holomorphic mapping $\beta : X \to \mathrm{Alb}(\overline{X})$, which makes the following diagram commute.

$$\begin{array}{ccc} \overline{X} & \xrightarrow{\mathrm{alb}_{\overline{X}}} & \mathrm{Alb}(\overline{X}) \\ \varphi \downarrow & \nearrow \beta & \\ X & & \end{array}$$

In this case β and $\mathrm{Alb}(\overline{X})$ are independent of the resolution. By the *Albanese variety* of X, $\mathrm{Alb}(X)$, we mean $\mathrm{Alb}(\overline{X})$ with the map β, which we refer to as the Albanese mapping of X and denote alb or alb_X when there is possible confusion. Let $\mathcal{S}_i(X) = \varphi_{(i)}(\mathcal{O}_{\overline{X}})$ be the higher derived functors of $\varphi_*(\mathcal{O}_{\overline{X}})$.

Lemma 2.4.1. *Notation as above. Let X be a normal projective variety. Then the Albanese mapping of X is defined if $h^0(\mathcal{S}_1(X)) = 0$, e.g., if X has at most rational singularities.*

Proof. Let $\varphi : \overline{X} \to X$ be a resolution of singularities and let $\mathrm{alb}_{\overline{X}} : \overline{X} \to \mathrm{Alb}(\overline{X})$ be the Albanese mapping. By the normality of X the map $\mathrm{alb}_{\overline{X}}$ will descend to a holomorphic mapping $\beta : X \to \mathrm{Alb}(\overline{X})$ on X if and only if for each $x \in \mathrm{Sing}(X)$ it follows that $\mathrm{alb}_{\overline{X}}(\varphi^{-1}(x))$ is a point of $\mathrm{Alb}(\overline{X})$. If this failed we could pull back an element of $H^1(\mathcal{O}_{\mathrm{Alb}(\overline{X})})$ which would restrict to a nontrivial element of $H^1(\mathcal{O}_{\varphi^{-1}(x)})$, which would give a nontrivial element of $\Gamma(\mathcal{S}_1(X))$. □

Remark 2.4.2. The above lemma holds more generally if the differential from $\Gamma(\mathcal{S}_1(X))$ to $H^2(\mathcal{O}_X)$ in the Leray spectral sequence for φ and $\mathcal{O}_{\overline{X}}$ is injective.

Remark 2.4.3. (Forms on varieties with rational singularities) Let X be a normal variety with rational singularities and let $\varphi : \overline{X} \to X$ be a desingularization. We define a holomorphic q-form to be a section of the sheaf $\wedge^q \mathcal{T}_X^* := \varphi_*(\wedge^q \mathcal{T}_{\overline{X}}^*)$. We leave it to the reader to check that such forms, and the operations we now define are independent of the desingularization chosen. We define $a \wedge b$ for a holomorphic i-form, a, and a holomorphic j-form, b, to be the holomorphic $(i+j)$-form, $\varphi_*(\overline{a} \wedge \overline{b})$, where $\overline{a}$ (respectively $\overline{b}$) is the holomorphic i-form (respectively holomorphic j-form) on $\overline{X}$ such that $\varphi_*(\overline{a}) = a$ (respectively $\varphi_*(\overline{b}) = b$). Note that since $H^i(\mathcal{O}_X) \cong H^i(\mathcal{O}_{\overline{X}})$, $i \geq 0$, and since $H^0(X, \wedge^q \mathcal{T}_X^*) \cong H^0(\overline{X}, \wedge^q \mathcal{T}_{\overline{X}}^*)$ we can use the conjugation from Hodge theory to give an anti-linear isomorphism, for $q \geq 0$

$$H^q(\mathcal{O}_X) \cong H^0(X, \wedge^q \mathcal{T}_X^*). \tag{2.5}$$

The reader can simply treat these concepts as a shorthand for working on a desingularization.

Proposition 2.4.4. *Let A be a normal connected nef and big Cartier divisor on a normal irreducible projective variety X. Assume that A and X have at worst rational singularities. Then the map* $\mathrm{Alb}(A) \to \mathrm{Alb}(X)$ *induced by inclusion is surjective with connected fibers if* $\dim X \geq 2$ *and an isomorphism when* $\dim X \geq 3$.

Proof. If A is nef and big, then it follows from the Kawamata-Viehweg vanishing theorem, (2.2.1), that the restriction map $H^1(\mathcal{O}_X) \to H^1(\mathcal{O}_A)$ is injective if $\dim X \geq 2$ and an isomorphism if $\dim X \geq 3$. By using the anti-linear isomorphism (2.5) we get that the map $\mathrm{Alb}(A) \to \mathrm{Alb}(X)$ is surjective if $\dim X \geq 2$ and finite-to-one if $\dim X \geq 3$.

Assume that $\dim X \geq 2$ and the map $\mathrm{Alb}(A) \to \mathrm{Alb}(X)$ does not have connected fibers. Then we see that the image, Γ, of $H_1(A, \mathbb{Z})$ in $H_1(X, \mathbb{Z})$ is of finite positive index. Passing to the cover, $p : \overline{X} \to X$, of X corresponding to this finite index subgroup, Γ, we get a normal irreducible projective variety $\overline{X}$ which contains a disconnected nef and big divisor, $\overline{A} := p^*A$. This contradicts $h^1(-\overline{A}) = 0$. □

In the special case that the image of the Albanese mapping is one dimensional, the Albanese mapping has connected fibers and a smooth image.

Lemma 2.4.5. *Let X be an irreducible reduced normal projective variety with at worst rational singularities. Assume that the Albanese mapping,* $\mathrm{alb} : X \to \mathrm{Alb}(X)$*, has a 1-dimensional image, Y. Then Y is a smooth curve and* alb *has connected fibers.*

Proof. Let $Y := \mathrm{alb}(X)$. Let $s \circ r$ denote the Remmert-Stein factorization of $\mathrm{alb} : X \to Y$ with $r : X \to Y'$ being a morphism with a normal, and hence smooth curve, Y', as its image and with s a finite map. Since Y' maps nontrivially into an Abelian variety it has positive genus, and thus the map into its Albanese variety, $\mathrm{Alb}(Y')$, which is its Jacobian variety, is an embedding. Since the map $X \to Y'$ induces a map $X \to \mathrm{Alb}(Y')$, which factors through the map $\mathrm{alb} : X \to \mathrm{Alb}(X)$, we conclude that $Y' \cong Y$ and that alb has connected fibers. □

2.5 The Hodge index theorem and the Kodaira lemma

The following result is a basic tool.

Proposition 2.5.1. (Hodge index theorem) *Let D_i for $0 \leq i \leq k$ be $\mathbb{Q}$-Cartier divisors on an irreducible n-dimensional projective variety, X. Assume that $n \geq 2$ and that for $i \geq 1$, D_i is nef (we are not assuming that D_0 is nef). Then if $n_1 + \cdots + n_k = n - 1$ and $n_1 \geq 1$ we have*

$$(D_0 \cdot D_1^{n_1} \cdots D_k^{n_k})^2 \geq (D_0^2 \cdot D_1^{n_1-1} \cdots D_k^{n_k})(D_1^{n_1+1} \cdots D_k^{n_k}).$$

Proof. We follow the argument of Fujita [246, Proposition (1.2)].

The truth or falsity of the inequality we are trying to prove is unaffected by passing to the analogous inequality for the desingularization of X and the pullbacks of the D_i to the desingularization. Therefore we may assume without loss of generality that X is smooth.

It suffices to prove the result when the D_i are ample for $i = 1, \ldots, k$. To see this, let H be an ample Cartier divisor on X. Note that for $i = 1, \ldots, k$, $D_i + tH$ is ample for any positive $t \in \mathbb{Q}$. The inequality with $D_i + tH$ in place of D_i for $i = 1, \ldots, k$ has the desired inequality for the D_i in the limit as t goes to 0. Thus without loss of generality we can assume that the D_i are ample for $i = 1, \ldots, k$.

Note that the truth or falsity of the statement is unaffected by scaling of the divisors by positive rational numbers, and in particular we can assume without loss of generality that the D_i are Cartier divisors, and are very ample for $i = 1, \ldots, k$.

By using not necessarily distinct divisors, we see that the inequality to be proven would follow from the special inequality with $k = n - 1$ and each $n_i = 1$. Let S be the intersection of general elements $A_i \in |D_i|$ for $i \geq 2$. Then S is a smooth surface, and the special inequality is equivalent to $(D_{0|S} \cdot D_{1|S})^2 \geq D_{0|S}^2 D_{1|S}^2$, the usual Hodge inequality for surfaces. □

We need a simple lemma.

Lemma 2.5.2. *Let A and B be nef $\mathbb{Q}$-divisors on an irreducible n-dimensional, projective variety, X, with $n \geq 2$. Then $(A^{n-1} \cdot B)(A \cdot B^{n-1}) \geq A^n B^n$.*

Proof. As above we can assume that X is smooth, and A and B are smooth ample divisors. If $n = 2$ then this is just (2.5.1). Therefore we can assume $n \geq 3$ and as an induction hypothesis that the result is true for dimensions $\leq n - 1$. Then using the fact from (2.5.1) that $(A^{n-1} \cdot B)^2 \geq (A^n)(A^{n-2} \cdot B^2)$, we get

$$\begin{aligned}(A^{n-1} \cdot B)^2(A \cdot B^{n-1}) &\geq (A^n)(A^{n-2} \cdot B^2)(A \cdot B^{n-1}) \\ &\geq (A^n)[(A_B^{\dim B-1} \cdot B_B)(A_B \cdot B_B^{\dim B-1})].\end{aligned}$$

By the induction hypothesis we have on B,

$$(A_B^{\dim B-1} \cdot B_B)(A_B \cdot B_B^{\dim B-1}) \geq A_B^{\dim B} B_B^{\dim B}.$$

Therefore we get $(A^{n-1} \cdot B)^2(A \cdot B^{n-1}) \geq A^n(A^{n-1} \cdot B)B^n$, which gives the desired inequality after dividing by the positive number $A^{n-1} \cdot B$. □

The following is useful for obtaining lower bounds for intersections of divisors.

Corollary 2.5.3. *Let $D_1, \ldots, D_k$ be nef $\mathbb{Q}$-Cartier divisors on an irreducible n-dimensional projective variety, X, with $n \geq 2$. Then if $n_1 + \cdots + n_k = n$ we have*

$$(D_1^{n_1} \cdots D_k^{n_k})^n \geq (D_1^n)^{n_1} \cdots (D_k^n)^{n_k}.$$

Proof. As in Proposition (2.5.1) it suffices to prove this result in the case of n nef $\mathbb{Q}$-divisors, $D_1, \ldots, D_n$, with the $n_i = 1$ for all i. Moreover as in (2.5.1) it can be assumed that X is smooth, each D_i is a smooth, very ample, Cartier divisor.

For each choice of integers a and b satisfying $1 \leq a < b \leq n$ we have from (2.5.1) that

$$(\Pi_{i=1}^n D_i)^2 \geq (D_a^2 \cdot \Pi_{i \neq a,b} D_i)(D_b^2 \cdot \Pi_{i \neq a,b} D_i). \tag{2.6}$$

When $n = 2$, the result to be proven is just (2.5.1). Thus we can assume that $n \geq 3$ and as an induction hypothesis that the result is true for dimensions $\leq n-1$. Therefore

$$\begin{aligned}(D_a^2 \cdot \Pi_{i \neq a,b} D_i)^{\dim D_a} &= (D_{a|D_a} \cdot \Pi_{i \neq a,b} D_{i|D_a})^{\dim D_a} \geq \Pi_{i \neq b}(D_{i|D_a}^{\dim D_a}) \\ &= D_a^n \Pi_{i \neq b}(D_a \cdot D_i^{n-1}).\end{aligned}$$

Thus the inequality (2.6) raised to the $(n-1)$-st power becomes

$$(\Pi_{i=1}^n D_i)^{2(n-1)} \geq [D_a^n \Pi_{i \neq b}(D_a \cdot D_i^{n-1})][D_b^n \Pi_{i \neq a}(D_b \cdot D_i^{n-1})]. \tag{2.7}$$

Taking the product of the inequalities (2.7) over all a and b satisfying $1 \leq a < b \leq n$ we get, since the number of such a, b is $n(n-1)/2$,

$$(\Pi_{i=1}^n D_i)^{n(n-1)(n-1)} \geq (\Pi_{i=1}^n D_i^n)^{n-1}(\Pi_{i<j}[(D_i \cdot D_j^{n-1})(D_j \cdot D_i^{n-1})]^{n-2}).$$

Using Lemma (2.5.2) we get:

$$\begin{aligned}(\Pi_{i=1}^n D_i)^{n(n-1)(n-1)} &\geq (\Pi_{i=1}^n D_i^n)^{n-1}(\Pi_{i<j} D_i^n D_j^n)^{(n-2)} \\ &= (\Pi_{i=1}^n D_i^n)^{n-1}(\Pi_{i=1}^n D_i^n)^{(n-1)(n-2)} \\ &= (\Pi_{i=1}^n D_i^n)^{(n-1)(n-1)}.\end{aligned}$$

Both sides of the inequality are positive, and taking the $(n-1)^2$ root of both sides gives the desired result. □

Corollary 2.5.4. *Let X be an irreducible normal projective variety of dimension $n \geq 2$ with at worst Cohen-Macaulay singularities. Assume that* $\dim \mathrm{Irr}(X) \leq 0$, *with the convention that* $\dim \emptyset = -1$, *and assume that* $X - F$ *is a local complete intersection for some finite, possibly empty, set* $F \subset X - \mathrm{Irr}(X)$. *Let* $D_1, \dots, D_{n-2}$ *be* $\mathbb{Q}$*-Cartier, ample divisors,* A *a* $\mathbb{Q}$*-Cartier divisor satisfying* $A^2 \cdot \Pi_{i=1}^{n-2} D_i > 0$, *and* B *an arbitrary* $\mathbb{Q}$*-Cartier divisor on* X. *If*

$$(A \cdot B \cdot \Pi_{i=1}^{n-2} D_i)^2 = (A^2 \cdot \Pi_{i=1}^{n-2} D_i)(B^2 \cdot \Pi_{i=1}^{n-2} D_i),$$

then there is a rational number λ *such that* A *is numerically equivalent to* λB.

Proof. Since the hypotheses and conclusions of the theorem are unchanged by scaling divisors, it can be assumed that all divisors are Cartier, and that the D_i are very ample, for $i = 1, \dots, n-2$, contain $\mathrm{Irr}(X)$, are normal and have at worst rational singularities outside of the finite set $F \cup \mathrm{Irr}(X)$. Moreover by induction we can assume that $S := \cap_{i=1}^{n-2} D_i$ is an irreducible normal Cohen-Macaulay surface with $\dim \mathrm{Irr}(S) \le 0$. Passing to the desingularization of S and using the usual Hodge index theorem on the desingularization, we see that there is a $\lambda \in \mathbb{Q}$ such that $(A - \lambda B)_S$ is numerically equivalent to zero. Thus we can assume without loss of generality that $n \ge 3$. Since $\dim \mathrm{Irr}(X) \le 0$ we can also assume that choosing the D_i sufficiently general, S has only rational singularities. Thus the line bundles on S that are numerically equivalent to zero are in the image of the map $H^1(\mathcal{O}_S) \to H^1(\mathcal{O}_S^*)$ induced from the exponential sequence. Consider the restriction map $r : \mathrm{Pic}(X) \to \mathrm{Pic}(S)$. By Corollary (2.3.4) the map r is injective. Since $H^1(\mathcal{O}_X) \cong H^1(\mathcal{O}_S)$ by the Kodaira vanishing theorem, we conclude that there is a line bundle, L, on X with $L_S \approx (A - \lambda B)_S$. Note that $L \sim \mathcal{O}_X$ on X since L is in the image of $H^1(\mathcal{O}_X) \to \mathrm{Pic}(X)$. Therefore, since r is injective, we conclude that $A - \lambda B \approx L$ is numerically trivial. □

Lemma 2.5.5. *Let* $\mathcal{D}$ *be a* $\mathbb{Q}$*-Cartier divisor on an irreducible projective variety,* X, *of dimension* n. *Then the following are equivalent:*

1. *given any ample Cartier divisor* H, *there exist an effective Cartier divisor* D, *and a positive integer* N, *such that* $N\mathcal{D} \approx D + H$;
2. $\kappa(\mathcal{D}) = n$.

Proof. The implication from 1) to 2) is immediate.

Assume 2) is true. Since $\kappa(\lambda\mathcal{D}) = \kappa(\mathcal{D})$ for any positive rational number λ, we can assume without loss of generality that $\mathcal{D}$ is Cartier. Let $L := \mathcal{O}_X(\mathcal{D})$. Let H be an irreducible ample Cartier divisor on X. By replacing H by a positive multiple, $(m+1)H$, such that mH and $(m+1)H$ are very ample, we do not change the conclusion of the lemma. Indeed, if $NL \approx D' + (m+1)H$ with D' effective, then noting that there is an effective $D \in |D' + mH|$ since mH is very ample, we have $NL \approx D + H$. Therefore we can assume without loss of generality that H is very ample. Consider the sequence

$$0 \to tL - H \to tL \to tL_H \to 0.$$

First note that $h^0(tL_H) \le ct^{n-1}$ for some positive constant c independent of t. To see this, note that by taking a desingularization we may assume H to be smooth. Let A be any ample Cartier divisor on H and recall that by the Serre vanishing theorem we have $h^i(N(L_H + A)) = 0$ for $i > 0$, $N \gg 0$. Therefore

$$\begin{aligned} h^0(tL_H) &\le h^0(t(L_H + A)) = \chi(t(L_H + A)) \\ &= \frac{(L_H + A)^{n-1}}{(n-1)!} t^{n-1} + \text{(lower degree terms)}. \end{aligned}$$

Then take $c : (L_H + A)^{n-1}/(n-1)! + \epsilon$ for any real $\epsilon > 0$. Since $\kappa(L) = n$, we have that there is no positive constant, C, such that $h^0(tL) \le Ct^{n-1}$ for all sufficiently large integers t. Thus for a sufficiently large t, which we denoted N, we can find, by using the exact sequence above, a not identically zero section of $NL - H$, i.e., an effective divisor D such that $D := NL - H$. This gives 1). □

Corollary 2.5.6. *Let L be a line bundle on an irreducible projective variety, X. Let D be an effective irreducible Cartier ample divisor on X. Let L_D be the restriction of L to D. If $\kappa(L) < \dim X$ then $h^0(tL_D) \ge h^0(tL)$ for all $t > 0$, and in particular $\kappa(L_D) \ge \kappa(L)$.*

Proof. If $\kappa(L) > \kappa(L_D)$ then $h^0(tL) > h^0(tL_D)$ for some $t \ge 1$. From the short exact sequence,

$$0 \to tL - D \to tL \to tL_D \to 0,$$

we conclude that $h^0(tL - D) > 0$. Thus by Lemma (2.5.5), $\kappa(L) = \dim X$. □

The following useful result is often referred to as Kodaira's Lemma.

Lemma 2.5.7. *Let L be a nef line bundle on an irreducible projective variety, X, of dimension n. Then $\kappa(L) = n$ if and only if $L^n > 0$.*

Proof. Assume $\kappa(L) = n$. By Lemma (2.5.5), we have $NL \approx D + H$ for some positive integer N, where D is an effective divisor and H is an ample Cartier divisor. We see that

$$L^n \ge \frac{1}{N} L^{n-1} \cdot H \ge \cdots \ge \frac{1}{N^n} H^n > 0.$$

Now assume $L^n > 0$. Let $p : \overline{X} \to X$ be a desingularization of X. Let $\overline{L} := p^* L$. Note that $\kappa(L) = \kappa(\overline{L})$ and $\overline{L}^n = L^n > 0$. Thus without loss of generality we can assume that X is smooth. Using the Hirzebruch-Riemann-Roch theorem and the Kawamata-Viehweg vanishing theorem it is clear that for all sufficiently large t, $h^0(K_X + tL) \geq Ct^n$ for some positive constant C, e.g., $C := (1-\epsilon)L^n/n!$, for any real $\epsilon \in (0, 1)$. An argument such as that in the proof of Lemma (2.5.5) shows that given any ample Cartier divisor, A, it follows that for all sufficiently large t, $K_X + tL - A$ is effective. We can assume that A is of the form $K_X + H$ for some ample divisor H. Thus $tL \approx H + D$ where H is ample and D is effective. This shows $\kappa(L) = n$. □

The natural analogue of Lemma (2.5.7) does not hold for L nef with $\kappa(L) < n$. As the next lemma shows, it is true that if a line bundle, L, on an n-dimensional irreducible projective variety, X, is nef, and if $\kappa(L) = n - k$, then $L^{n-k} \cdot H^k > 0$ for any ample line bundle, H, on X. Example (1.8.2.2) shows that the converse fails.

Lemma 2.5.8. *Let L be a nef, k-big line bundle on an irreducible n-dimensional projective variety X. Then for any $t \geq k$ and any nef and big line bundles, H_i, $1 \leq i \leq t$, on X it follows that $L^{n-t} \cdot \prod_{i=1}^{t} H_i > 0$.*

Proof. Let H denote an ample line bundle on X. By (2.5.5), each H_i can be written as $H_i \approx \lambda_i H + D_i$ where $\lambda_i \in \mathbb{Q}$, $\lambda_i > 0$, and D_i is $\mathbb{Q}$-effective. Since $L^{n-t} \cdot \prod_{i=1}^{t} H_i \geq (\prod_{i=1}^{t} \lambda_i) L^{n-t} \cdot H^t$, we will be done if we prove that $L^{n-t} \cdot H^t > 0$.

Since H is ample we have that $H \approx \alpha L + B$ where $\alpha \in \mathbb{Q}$, $\alpha > 0$, and B is $\mathbb{Q}$-effective. Thus $L^{n-t} \cdot H^t \geq \alpha^{t-k} L^{n-k} \cdot H^k$. Thus we will be done if we show that $L^{n-k} \cdot H^k > 0$.

If $k = 0$ then there is nothing to prove by (2.5.7). Therefore we can assume that $k > 0$ and assume as an induction hypothesis that the lemma is true for $k' < k$. In particular we can assume without loss of generality that $\kappa(L) = n - k$, and in particular that there is an integer $N \gg 0$ and a rational map, $\phi := \phi_{\Gamma(NL)} : X \to \mathbb{P}_{\mathbb{C}}$, with image of dimension $n - k$.

By replacing H by mH for some integer $m \gg 0$ we can assume without loss of generality that H is very ample. A general $D \in |H|$ can be assumed irreducible by (2.2.6) and moreover meets a general fiber of ϕ in a nonempty set not contained in the indeterminacy locus of ϕ. Thus the restriction ϕ_D has the same image as ϕ. Since $n - k = \dim D - (k-1)$ and since ϕ_D factors through $\phi_{\Gamma(NL_D)}$ we see that L_D is $(k-1)$-big. Since H_D is ample we conclude by the induction hypothesis that $L_D^{n-k} \cdot H_D^{k-1} > 0$. We are finished upon noting that $L^{n-k} \cdot H^k = L_D^{n-k} \cdot H_D^{k-1}$. □

Lemma 2.5.9. *Let L be an $(n-1)$-big line bundle on an irreducible n-dimensional projective variety, X, i.e., L is a line bundle such that $h^0(NL) \geq 2$ for some $N \geq 1$. Then for any $n-1$ nef and big line bundles, H_i, $1 \leq i \leq n-1$, on X it follows that $L \cdot \prod_{i=1}^{n-1} H_i > 0$.*

Proof. We can without loss of generality replace L by a multiple tL with an at least 2-dimensional space of sections. By pulling back all the line bundles to a desingularization of X we can assume without loss of generality that X is smooth.

It suffice to prove the result in the case that the base locus of L is of codimension at least 2. To see this note that $L \approx A+B$ where B has a base locus of codimension at least 2 and A is an effective Cartier divisor. Thus $L \cdot \prod_{i=1}^{n-1} H_i \geq B \cdot \prod_{i=1}^{n-1} H_i$.

Let $p : \overline{X} \to X$ be a birational morphism such that $p^*L \approx D + \mathcal{L}$ where $\mathcal{L}$ is spanned and D is a divisor such that $p(D)$ is of codimension at least 2 on X: such a morphism, p, exists by using Hartshorne [318, Example 7.17.3] followed by a desingularization if necessary. Since $L \cdot \prod_{i=1}^{n-1} H_i \geq \mathcal{L} \cdot \prod_{i=1}^{n-1} p^*H_i$, the result follows from Lemma (2.5.8). □

2.6 Rossi's extension theorems

Ampleness of a divisor implies the "pseudoconcavity" of appropriate neighborhoods of the ample divisor. In this section and §2.7 some basic consequences of this "pseudoconcavity" are given. The discussion of Hartshorne [315, Chapter 6] is helpful.

Let $\rho : M \to \mathbb{R}$ be a C^∞ function from a complex manifold to the real numbers. The function ρ is said to be *q-convex* (respectively *q-concave*) *in the sense of Andreotti-Grauert* [32] if the *Levi form* associated to ρ

$$\mathcal{L}(\rho) := (dz_1, \ldots, dz_n) \begin{pmatrix} \frac{\partial^2 \rho}{\partial z_1 \partial \bar{z}_1} & \cdots & \frac{\partial^2 \rho}{\partial z_1 \partial \bar{z}_n} \\ \vdots & \vdots & \vdots \\ \frac{\partial^2}{\partial z_n \partial \bar{z}_1} & \cdots & \frac{\partial^2 \rho}{\partial z_n \partial \bar{z}_n} \end{pmatrix} \begin{pmatrix} d\bar{z}_1 \\ \vdots \\ d\bar{z}_n \end{pmatrix}$$

has at most $q-1$ nonpositive eigenvalues (respectively $q-1$ nonnegative eigenvalues) on $\mathcal{T}_{M,x}$ for each $x \in M-K$, for some compact subset $K \subset M$. The extension of the definition to analytic spaces (i.e., not necessarily reduced, complex analytic spaces) is straightforward. One notes that locally any analytic space, X, can be embedded in $\mathbb{C}^N$ for some $N \geq \dim X$. Next one defines a C^∞ function, g, on an analytic space X to be a function on X such that for any point $x \in X$ there is some embedding of a neighborhood of x into $\mathbb{C}^N$ with g the restriction of a C^∞ function on an open set of $\mathbb{C}^N$. It is easily checked that this definition is independent of the local embeddings. A C^∞ function $\rho : X \to \mathbb{R}$ of an analytic space is *q-convex* if given any $x \in X$ there is an embedding of an open neighborhood, U, of x into $\mathbb{C}^N$ for some N such that ρ_U is the restriction on U of a C^∞ function $\rho' : \mathcal{U} \to \mathbb{R}$, for some open set $\mathcal{U}$ of $\mathbb{C}^N$, such that the Levi form $\mathcal{L}(\rho')$ has at most $q-1$ nonpositive eigenvalues.

A C^∞ function, $\rho : X \to [0, \infty) \subset \mathbb{R}$, on an analytic space X is said to be a *q-convex exhaustion in the sense of Andreotti-Grauert* if

1. ρ is proper, e.g., $\rho^{-1}([0, a])$ is compact for all real numbers, $a \geq 0$;

2. there is a compact set $K \subset X$ such that ρ is q-convex on $X - K$.

If K can be taken to be empty in the above definition then ρ is said to be *q-complete*.

The prototype 1-convex exhaustion function is the function $|z_1|^2 + \cdots + |z_n|^2$ on $\mathbb{C}^n$ with Levi form $|dz_1|^2 + \cdots + |dz_n|^2$. More generally given any compact ample Cartier divisor A on an analytic space X there is an open neighborhood, $\mathcal{U}$, of $A \subset X$, and a proper C^∞ function $\rho : \mathcal{U} \to (0, \infty)$ which is 1-convex. By a *strongly 1-pseudoconcave neighborhood*, U, of A in X, we mean any open neighborhood U of A of the form $A \cup \rho^{-1}((a, \infty))$ with $a > 0$ for such a function ρ. If X is compact then there exists a 1-convex exhaustion function ρ on $X - A$. See Fritzsche [238], Sommese [570, 580], and Okonek [490] for further results and references on the construction of convexity functions.

The key fact about an irreducible noncompact analytic space, X, possessing 1-convex exhaustion functions, and such that there is an integer N_0 with each point $x \in X$ having a neighborhood that can be locally embedded into $\mathbb{C}^{N_0}$, is that there exists a generically one-to-one, proper holomorphic surjection from X onto a closed analytic subspace $\mathbb{C}^N$ for some N.

In [524], Rossi proves a wonderful result on when "holes" in a complex space can be filled in. The main corollary of this for us is Theorem 3 of Rossi's paper.

Theorem 2.6.1. (Rossi) *Let A be an effective Cartier divisor on an analytic space, X, of pure dimension $n \geq 3$. Assume that A is compact and the normal bundle $\mathcal{N}_{A/X}$ is ample. Given any strongly 1-pseudoconcave neighborhood, U, of A in X, there exists a projective variety, V, of pure dimension n, and an injection $g : U \to V$ such that $g(A)$ is an ample Cartier divisor on V.*

Remark 2.6.2. One immediate corollary of Rossi's construction, which uses the spectrum of the ring of holomorphic functions on $U - A$, is that any holomorphic map from $U - A$ to a Stein space, Z, extends from $U - A$ to a holomorphic map from $V - g(A)$ to Z.

Corollary 2.6.3. *Let A be an ample Cartier divisor on a pure dimensional projective variety, X, of dimension $n \geq 3$. Assume that there is a finite proper map $\pi_{U'} : U' \to U$ of a pure dimensional complex space, U', onto U, a strongly 1-pseudoconcave neighborhood of the projective variety A. Assume that the restriction map $\pi_{U'-A'}$ is unramified over $U - A$ with A' the inverse image of A. Then, after possibly shrinking U and U', $\pi_{U'}$ extends to a finite branched cover, $\pi : X' \to X$, of a projective variety X' to X.*

Proof. It can be shown that there are strongly 1-pseudoconcave neighborhoods of $A' \subset U'$. Since A is ample on X, $\mathcal{N}_{A/X}$ is ample. Since $\pi_{U'}$ is finite, $\pi_{U'}^* \mathcal{N}_{A/X} \cong \mathcal{N}_{A'/U'}$ and $\mathcal{N}_{A'/U'}$ is ample. Therefore we can apply Rossi's theorem, (2.6.1), to A' on U'. Then, after possibly shrinking U' and U, we find a neighborhood, $\mathcal{U}$, of A' in U' and a projective variety, X', of pure dimension n, and an injection $g : \mathcal{U} \to X'$ such that $g(A')$ is an ample Cartier divisor on X'. By Remark (2.6.2) the holomorphic map from $\mathcal{U} - A'$ to $X - A$ extends to a holomorphic map $\pi : X' - g(A') \to X - A$. Defining π on $g(A')$ as $\pi_{U'} \circ g^{-1}$ we get the requested finite cover $\pi : X' \to X$. □

Remark 2.6.4. Notation as in (2.6.3). Since A is compact, and A and X can be triangulated, there is a neighborhood U of A in X, with A a deformation retract of U. This means that there is a homotopy $\theta_t : U \to U$ defined by $\theta_t(u) = \theta(u, t)$, where $\theta : U \times [0, 1] \to U$ is a continuous map and $[0, 1]$ is the unit interval, with the following properties: $\theta_0 = \mathrm{id}_U$, $\theta_1(U) \subset A$ and $\theta_{1|A}$ is the inclusion $i : A \hookrightarrow U$. Thus θ_1 defines a retraction $r : U \to A$ with $i \circ r = \theta_1$ and $r \circ i = \mathrm{id}_A$ (see, e.g., Spanier's book [589]).

Thus corresponding to an unramified cover of A there is an unramified cover, $\pi_{U'} : U' \to U$, with the same sheet number as the restriction $\pi_{U'|A'}$. Note that if A and X are local complete intersections, then by the first Lefschetz theorem, (2.3.1), we have an isomorphism of fundamental groups under the assumption that $\dim X \geq 3$. Thus in this case the above result is a corollary of the Lefschetz theorem and π is in fact an unbranched cover.

Later in this book we give an application, Theorem (5.4.10), of the above result to the classification of the pairs, (X, A), where A is an ample Cartier divisor on a normal variety, X, and $A \cong \mathbb{P}^n$ with $\dim X \geq 3$. That result shows that the map π is not always a covering on all of X.

The classification theorem for $\mathbb{P}^1$ as an ample divisor gives some examples showing that the assumption, $\dim X \geq 3$, of Rossi's extension theorem is needed. Indeed let A denote the diagonal of the product $X := \mathbb{P}^1 \times \mathbb{P}^1$. If Corollary (2.6.3) held in this case we could find a normal, projective variety X' and a two-to-one branched cover $\pi : X' \to X$, with an ample divisor A' on X' going isomorphically onto A. We would have that X' is smooth in a neighborhood of A' and $A' \cdot A' = 1$. By the results of §3.1 we see that $X' \cong \mathbb{P}^2$. This is absurd since X', as a finite cover of $X := \mathbb{P}^1 \times \mathbb{P}^1$, would have its second rational Betti number at least 2. See Rossi [524] for further examples showing the necessity of $\dim X \geq 3$.

The following well-known fact is useful.

Lemma 2.6.5. *Let A be an effective Cartier divisor on an irreducible normal projective variety, X. If the normal bundle $\mathcal{N}_{A/X}$ is ample, then $\mathcal{O}_X(A)$ is semi-ample. Furthermore there exist a normal variety X' and a birational morphism*

$\varphi : X \to X'$, which is an isomorphism in a neighborhood of A and such that A is ample on X'.

Proof. The fact that $\mathcal{O}_X(A)$ is semiample follows immediately from Theorem (1.1.2). Then there exists some $t > 0$ such that $\mathcal{O}_X(tA)$ is spanned. Thus for some large $N > 0$ the map associated to $\Gamma(NA)$ gives a morphism with connected fibers, $\varphi : X \to X'$, onto a normal variety X', such that φ is an isomorphism on A (see also (1.1.3)).

We claim that for a complex neighborhood, U, of A in X' the restriction $\varphi_U : \varphi^{-1}(U) \to U$ is an isomorphism. Indeed, if not, for an arbitrarily small neighborhood U of A in X' the morphism φ_U would have positive dimensional fibers by the Zariski main theorem. Let $Y \subset X'$ be the set in X' defined by the property that $y \in Y$ if and only if $\dim \varphi^{-1}(y) > 0$. We may assume that Y meets A, since otherwise we can find a neighborhood U of A in X' as we want. Therefore there exists a curve, C, in X such that C meets A and $\varphi(C)$ is a point in Y. Hence in particular $A \cdot C > 0$. This contradicts the fact that $\dim \varphi(C) = 0$. Thus we conclude that φ is an isomorphism on U and hence in particular φ is birational.

To see that A is ample on X', note that $\mathcal{O}_X(A) \approx \varphi^*\mathcal{O}_{X'}(A)$ since $A = \varphi^{-1}(A)$. On the other hand, $NA \approx \varphi^*H$ for some ample Cartier divisor H on X'. Thus $NA \approx H$ on X', and hence A is ample on X'. □

Remark 2.6.6. Let A, X, X', $\varphi : X \to X'$ be as in Lemma (2.6.5). Let U be a neighborhood of $\varphi(A)$ in X' such that $\varphi^{-1}(U) \to U$ is an isomorphism. If $n := \dim X \geq 3$ and $U \cong \varphi^{-1}(U)$ is strongly 1-pseudoconcave (note that this is always the case since $\mathcal{N}_{A/X}$ is ample), Rossi's extension theorem, (2.6.1), gives an extension $g : U \to V$, where V is a projective variety of pure dimension n. We claim that $V \cong X'$. Indeed from Remark (2.6.2) we see that the embedding $g : U \to V$ extends to a meromorphic map $\theta : X' \to V$ and similarly the embedding $\varphi : U \to X'$ extends to a meromorphic map $\theta' : V \to X'$, such that $\theta \circ \theta' = \theta' \circ \theta$ is the identity map on U. Since $g(A)$, $\varphi(A)$ are ample divisors on V, X' respectively, the sets $V - g(A)$, $X' - \varphi(A)$ are affine. By using this we conclude that θ, θ' are indeed holomorphic maps and hence $V \cong X'$. For example, to any given point $y \in X'$ where θ is not defined it corresponds some compact subset $B \subset V$. Since θ is holomorphic on $\varphi(A)$ it must be $B \subset V - g(A)$. This contradicts the fact that $V - g(A)$ is affine.

A second extension theorem of Rossi [525, Theorem 4.3] deals with extensions of meromorphic functions from neighborhoods of intersections of ample divisors on a projective variety to the whole projective variety. This result yields a number of extension theorems for analytic objects from neighborhoods of intersections of ample divisors on a projective variety to the whole projective variety. For us the case of holomorphic functions to projective varieties, which is an immediate consequence of [525, Theorem 4.3], suffices.

Theorem 2.6.7. (Rossi) *Let L be an ample line bundle on an irreducible n-dimensional projective variety, X. Let $\{A_1, \dots, A_r\} \subset |L|$ be ample divisors such that $r < n$ and $Y = \cap_{i=1}^r A_i$ is of pure dimension $n - r$. Given a holomorphic mapping $f : U \to Z$ where U is a connected open neighborhood of Y and Z is a projective variety, then there is a meromorphic mapping, $\overline{f} : X \to Z$ which agrees with f on U.*

Let us briefly discuss here the following question which is related to Rossi's theorem, (2.6.1). To this purpose, recall that a projective variety, X, of dimension n, is called *unirational* if there exists a generically surjective rational map $\mathbb{P}^n \to X$. Recall also that a *weighted projective space* $\mathbb{P}(q_0, \dots, q_n)$, $q_0, \dots, q_n$ positive integers, is the quotient of $\mathbb{C}^{n+1} - \{0\}$ by $\mathbb{C}^* := \mathbb{C} - \{0\}$ which acts by sending $(\lambda, z_0, \dots, z_n)$ to $(\lambda^{q_0} z_0, \lambda^{q_1} z_1, \dots, \lambda^{q_n} z_n)$ (see, e.g., the surveys of Dolgachev [192], and the first author and Robbiano [92] for the definition and general properties of weighted projective spaces).

Ebihara posed in [196] the following question. If a projective complex threefold X contains a smooth rational surface S with an ample normal bundle $\mathcal{N}_{S/X}$, is X unirational? In [196] a positive answer is given in the case of toric surfaces (see Oda [480, 481] for definitions and properties of toric varieties). For the case of very ample divisors $S \subset X$ with X smooth this was already stated by Fano [230] and the first rigorous proof was given by Martynov [426]. For a smooth ample divisor on a smooth 3-fold this was shown by Sommese [576]. By using Mori's theory, Campana and Flenner [143] have proved the following result, and also results when $\kappa(S) \geq 0$. We refer to [143] for the proof and more details.

Theorem 2.6.8. (Campana-Flenner) *Let S be a smooth surface with negative Kodaira dimension, $\kappa(S)$, and assume that $S \subset X$ is a Cartier divisor in a 3-dimensional irreducible projective variety, X, such that the normal bundle $\mathcal{N}_{S/X}$ is ample. Then X is birational to one of the following:*

1. *$S \times \mathbb{P}^1$;*
2. *a sextic in $\mathbb{P}(1, 1, 1, 2, 3)$ or $\mathbb{P}(1, 1, 2, 2, 3)$ with at most terminal singularities;*
3. *a quartic in $\mathbb{P}(1, 1, 1, 1, 2)$ with at most terminal singularities (which is a two-to-one covering of $\mathbb{P}^3$ ramified along a quartic surface);*
4. *a cubic threefold in $\mathbb{P}^4$.*

Furthermore the varieties in 2), 3), 4) are unirational with the only possible exception of a sextic in $\mathbb{P}(1, 1, 1, 2, 3)$.

2.7 Theorems of Andreotti-Grauert and Griffiths

All coherent sheaves in this section are analytic coherent sheaves.

The following is a very special case of the results of Andreotti-Grauert [32] (see the discussion in [572, pg. 529–530] and [315, Chapter 6]).

Theorem 2.7.1. (Andreotti-Grauert) *Let A be a compact smooth divisor on a complex manifold, M, such that the normal bundle $\mathcal{N}_{A/M}$ is ample. Then there exist arbitrarily small neighborhoods $U \subset M$ of A such that the following holds.*

1. *Given a locally free analytic sheaf $\mathcal{E}$ on U, the restriction map $H^j(U, \mathcal{E}) \to H^j(A, \mathcal{E}_{|A})$, with $\mathcal{E}_{|A}$ the set-theoretic restriction of sheaves, is a bijection for $j \le \dim M - 2$ and an injection for $j = \dim M - 1$.*

2. *The inclusion map $A \hookrightarrow U$ is a homotopy equivalence, and in particular $H^j(U, \mathbb{Z}) \cong H^j(A, \mathbb{Z})$ for all $j \ge 0$.*

3. *The sets U satisfy the conclusions in* (2.6.1) *when* $\dim M \ge 3$.

The above open sets, U, are tubular neighborhoods of A constructed using an exhaustion function in a neighborhood of A with appropriate pseudoconvexity properties. The assertion 3) of Theorem (2.7.1) applies since the sets U of Theorem (2.6.1) are any connected open sets chosen to have the exact same pseudoconvexity properties.

Let A be a compact smooth divisor on a complex manifold, M. Given a locally free analytic sheaf $\mathcal{E}$ on some neighborhood, U, of A, and the ideal sheaf $\mathcal{J}_A$ of A in M, we have the *restriction, $\mathcal{E}_\mu$, of $\mathcal{E}$ to the μ*-th *infinitesimal neighborhood of* A,

$$\mathcal{E}_\mu := \mathcal{E}/(\mathcal{E} \otimes \mathcal{J}_A^{\mu+1}).$$

Note that there is a natural map $\mathcal{E}_{|A} \to \mathcal{E}_\mu$. Indeed $\mathcal{E}_{|A} = i^{-1}\mathcal{E}$, where i is the inclusion $A \hookrightarrow U$. Fix $a \in A$. Then each $e \in (\mathcal{E}_{|A})_a$ corresponds to an element in $\mathcal{E}_{U_a}$, where U_a is a neighborhood of a in the complex topology, and therefore to an element in $\mathcal{E}_{U_a}/(\mathcal{E}_{U_a} \otimes \mathcal{J}_{A\cap U_a}^{\mu+1})$.

The following result of Griffiths [291, Theorem III] gives a method which sometimes lets us compute the cohomology group, $H^j(A, \mathcal{E}_{|A})$, in Theorem (2.7.1). We only need the case of a smooth divisor in this book, and refer to Kosarew's paper [362] for the singular case.

Theorem 2.7.2. (Griffiths) *Let A be a compact smooth divisor with ample normal bundle on an n-dimensional connected complex manifold, M. Then there is a positive integer μ_0 such that for any $\mu \ge \mu_0$, the cohomology map, $H^j(\mathcal{E}_{|A}) \to H^j(\mathcal{E}_\mu)$, induced by the natural map $\mathcal{E}_{|A} \to \mathcal{E}_\mu$ is a bijection for $j \le \dim M - 3$, and an injection for $j \le \dim M - 2$.*

Let $\mathcal{J} = \mathcal{J}_A$ be the ideal sheaf of A in M. By taking $\mathcal{E} = \mathcal{O}_M$ in the above definition, let $\mathcal{O}_{A,\mu} := \mathcal{O}_M/\mathcal{J}^{\mu+1}$ be the restriction of $\mathcal{O}_M$ to the μ-th infinitesimal neighborhood of A. Let $L := \mathcal{O}_M(A)$. Note that by Kodaira vanishing, $h^1(-tL_A) = 0$ for $t \geq 1$ if $\dim A \geq 2$.

We claim that $h^1(\mathcal{J}/\mathcal{J}^{\mu+1}) = 0$. To see this look at the exact sequence

$$0 \to \mathcal{J}^{i+1}/\mathcal{J}^{\mu+1} \to \mathcal{J}^i/\mathcal{J}^{\mu+1} \to \mathcal{J}^i/\mathcal{J}^{i+1} \to 0,$$

for $i = 1, \dots, \mu$. Note that $\mathcal{J}^i/\mathcal{J}^{i+1} \cong -iL_A \cong \mathcal{N}^{*\,\otimes i}_{A/M}$. Then $h^1(\mathcal{J}^i/\mathcal{J}^{i+1}) = 0$, $i = 1, \dots, \mu$. Therefore we get

$$h^1(\mathcal{J}/\mathcal{J}^{\mu+1}) \leq h^1(\mathcal{J}^2/\mathcal{J}^{\mu+1}) \leq \dots \leq h^1(\mathcal{J}^\mu/\mathcal{J}^{\mu+1}) = 0.$$

This proves the claimed assertion. From the exact sequence

$$0 \to \mathcal{J}/\mathcal{J}^{\mu+1} \to \mathcal{O}_M/\mathcal{J}^{\mu+1} \to \mathcal{O}_M/\mathcal{J} \to 0$$

we see that $H^1(\mathcal{O}_{A,\mu}) \to H^1(\mathcal{O}_A)$ is injective. By Theorem (2.7.2) we have an injection $H^1(\mathcal{O}_{M|A}) \hookrightarrow H^1(\mathcal{O}_{A,\mu})$. By Theorem (2.7.1.1) we have $H^1(\mathcal{O}_{M|A}) \cong H^1(U, \mathcal{O}_U)$. Therefore we conclude that we get an injection

$$0 \to H^1(\mathcal{O}_U) \to H^1(\mathcal{O}_A). \tag{2.8}$$

Corollary 2.7.3. *Let A be a compact smooth divisor on an n-dimensional connected complex manifold, X, such that $\mathcal{N}_{A/X}$ is ample. Let $U \subset X$ be a neighborhood of A of the type that exists by* (2.7.1). *Then the restriction map* $\operatorname{Pic}(U) \to \operatorname{Pic}(A)$ *is an isomorphism if* $\dim X \geq 3$.

Proof. Using Theorem (2.7.1) and Theorem (2.6.1) we can assume, after taking normalization, that we have a normal projective variety, X, containing A as an ample Cartier divisor, and that the open set $U \subset X$ containing A satisfies the first two conclusions of Theorem (2.7.1). Let $p : \overline{X} \to X$ be a projective desingularization of X which is a biholomorphism from $\overline{X} - p^{-1}(\operatorname{Sing}(X))$ to $X - \operatorname{Sing}(X)$.

Since $L := \mathcal{O}_X(A)$ is ample when restricted to A, we conclude that L is nef and big. To see this note that $L^n = (A_A)^{n-1} > 0$. Furthermore let $C \subset X$ be an irreducible curve. If C is contained in A, then $L \cdot C = A \cdot C = A_A \cdot C > 0$. If C is not contained in A, then $L \cdot C \geq 0$. Therefore $h^i(-L) = 0$ for $i = 1, 2$. Thus the restriction $H^1(\mathcal{O}_{\overline{X}}) \to H^1(\mathcal{O}_A)$ is an isomorphism. Note that since X is a manifold, the open set U above containing A is in fact smooth. Then $A \cap \operatorname{Sing}(X) = \emptyset$ and hence A can be considered as a subvariety of $\overline{X}$. Since we also have the restriction maps, $H^1(\mathcal{O}_{\overline{X}}) \to H^1(\mathcal{O}_U) \to H^1(\mathcal{O}_A)$, with the composition $H^1(\mathcal{O}_{\overline{X}}) \to H^1(\mathcal{O}_A)$ being an isomorphism and with $0 \to H^1(\mathcal{O}_U) \to H^1(\mathcal{O}_A)$ by the injectivity (2.8), we conclude that $H^1(\mathcal{O}_U) \cong H^1(\mathcal{O}_A)$. Moreover, by (2.7.1.2), we have $H^2(U, \mathbb{Z}) \cong H^2(A, \mathbb{Z})$ and therefore by the exponential sequence we obtain the desired conclusion. □

Chapter 3

The basic varieties of adjunction theory

In this chapter we discuss some properties of the basic varieties which occur in the adjunction theoretic classification of the projective varieties.

In §3.1 we recall the classification, due to Fujita, of the polarized pairs, (X, L), of Δ-genus zero, and we prove a generalized version of the ubiquitous Kobayashi-Ochiai characterization of projective spaces and quadrics.

In §3.2 we give a method for recognizing linear $\mathbb{P}^d$-bundles, Proposition (3.2.1), and we prove and recall a number of basic results for $\mathbb{P}^d$-bundles on curves. E.g., we derive in (3.2.10) the well-known classification of normal polarized surfaces, (X, L), with sectional genus $g(L) = 0$. We also state the Grauert and Nakano contractibility criteria in the form we need later in the book.

Finally in §3.3 we recall the precise definition of the special varieties which occur in adjunction theory, and we prove a basic fact, Lemma (3.3.2), about the index of a Fano variety.

3.1 Recognizing projective spaces and quadrics

Adjunction theory makes use of the many ways to recognize projective spaces and quadrics.

Let X be an n-dimensional (irreducible) projective variety and L an ample line bundle on X. Let $\Delta(X, L)$ be the *Δ-genus* of (X, L), defined by the formula

$$\Delta(X, L) := n + L^n - h^0(X, L). \tag{3.1}$$

Let us recall some useful general properties of the Δ-genus from Fujita's book [257].

Theorem 3.1.1. ([257, I, (4.2), (4.12)]) *Let X be an n-dimensional irreducible projective variety and L an ample line bundle on X. Then*

1. $\Delta(X, L) > \dim \mathrm{Bs}|L|$, *assuming by convention that* $\dim \emptyset = -1$. *In particular* $\Delta(X, L) \geq 0$;

2. *L is very ample if* $\Delta(X, L) = 0$.

Proposition 3.1.2. ([257, Chap. I, (5.10), (5.15)]) *Let X be an n-dimensional irreducible projective variety,* $n \geq 2$, *and L an ample line bundle on X. Assume that* $\Delta(X, L) = 0$. *Then either:*

1. $(X, L) \cong (\mathbb{P}^n, \mathcal{O}_{\mathbb{P}^n}(1))$ *if* $L^n = 1$;

2. $(X, L) \cong (Q, \mathcal{O}_Q(1))$, *Q hyperquadric in* $\mathbb{P}^{n+1}$, *if* $L^n = 2$;

3. (X, L) *is a* $\mathbb{P}^{n-1}$*-bundle over* $\mathbb{P}^1$, $X \cong \mathbb{P}(\mathcal{E})$ *for a vector bundle,* $\mathcal{E}$, *on* $\mathbb{P}^1$ *which is a direct sum of lines bundles of positive degrees;*

4. $(X, L) \cong (\mathbb{P}^2, \mathcal{O}_{\mathbb{P}^2}(2))$; *or*

5. (X, L) *is a generalized cone over a smooth submanifold* $V \subset X$ *with* $\Delta(V, L_V) = 0$, *where* L_V *denotes the restriction of L to V.*

Theorem 3.1.3. ([257, II, (12.1)]) *Let L be an ample line bundle on X, an irreducible projective variety. Then* $g(L) \geq 0$ *with equality if and only if* $\Delta(X, L) = 0$.

The following result goes back to Goren [281] and Kobayashi-Ochiai [355]. See also Fujita [257, Chap. I, (1.2)] where the Cohen-Macaulay assumption is removed.

Theorem 3.1.4. ([257, Chap. I, (1.1)]) *Let L be an ample line bundle on X, an n-dimensional irreducible projective variety. If* $L^n = 1$ *and* $h^0(X, L) \geq n+1$, *then* $(X, L) \cong (\mathbb{P}^n, \mathcal{O}_{\mathbb{P}^n}(1))$.

Remark 3.1.5. Note that if L is spanned as it often is in our applications, $h^0(L) \geq n + 1$, and we can choose an $(n + 1)$-dimensional subspace, $V \subset \Gamma(L)$, of the sections of L that span L. In this case the condition that $L^n = 1$ implies that the map associated to V is a finite-to-one and generically one-to-one map from X to $\mathbb{P}^n$. By Zariski's main theorem we see that $(X, L) \cong (\mathbb{P}^n, \mathcal{O}_{\mathbb{P}^n}(1))$.

The following result is a consequence of Fujita [255, (2.2), (2.3)].

Theorem 3.1.6. (Generalized Kobayashi-Ochiai theorem) *Let X be an n-dimensional normal connected projective variety. Let L be an ample line bundle on X. Then we have*

1. $(X, L) \cong (\mathbb{P}^n, \mathcal{O}_{\mathbb{P}^n}(1))$ *if* $K_X + (n + 1)L \sim \mathcal{O}_X$;

2. $(X, L) \cong (Q, \mathcal{O}_Q(1))$, *Q a hyperquadric in* $\mathbb{P}^{n+1}$, *if* $K_X + nL \sim \mathcal{O}_X$.

Proof. We only prove the assertion 2) since the argument for 1) is similar but easier.

Let $\pi : M \to X$ be a desingularization of X. We have a short exact sequence

$$0 \to \pi_* K_M \to K_X \to \mathcal{S} \to 0,$$

where the sheaf $\mathcal{S}$ is supported on the singular set of X. Since $h^0(K_X + tL) = 0$ for $0 < t < n$ we conclude from the above sequence that $h^n(-t\pi^* L) = h^0(K_M + t\pi^* L) = 0$ for $0 < t < n$. Moreover we conclude in the same way from $h^0(K_X + nL) \leq 1$ that $h^n(-n\pi^* L) \leq 1$.

If $h^n(-n\pi^* L) = 0$, then [255, Theorem (2.2)] implies that (X, L) is isomorphic to $(\mathbb{P}^n, \mathcal{O}_{\mathbb{P}^n}(1))$. This is absurd since $K_X + nL \sim \mathcal{O}_X$.

Thus $h^n(-n\pi^* L) = 1$. This implies by [255, Theorem (2.3)], that either $(\pi^* L)^n = 1 = g(M, \pi^* L)$, or (X, L) is $(Q, \mathcal{O}_Q(1))$, Q a hyperquadric in $\mathbb{P}^{n+1}$, as desired. In the former case note that since X is normal one has $g(M, \pi^* L) = g(X, L)$ by [255, Lemma (1.8)]. Thus we conclude that $0 = (K_X + (n-1)L) \cdot L^{n-1} = -L^n < 0$. This absurdity proves the result. □

There is the following result of Lazarsfeld settling a question of Remmert and Van de Ven [402].

Theorem 3.1.7. *Let $\varphi : X \to Y$ be a finite-to-one mapping of a connected projective manifold, X, onto a projective manifold, Y. If $X \cong \mathbb{P}^n$, then $Y \cong \mathbb{P}^n$. If X is isomorphic to a smooth quadric, then either $Y \cong \mathbb{P}^n$ or Y is isomorphic to a smooth quadric. In the latter case when Y is a quadric of dimension at least 3, the map φ is biholomorphic.*

Finally recall the following standard useful fact.

Lemma 3.1.8. *Let $\varphi : X \to Y$ be a surjective morphism of irreducible projective varieties. Let* $\dim X = n$ *and assume* $\mathrm{Pic}(X) \cong \mathbb{Z}[L]$, *for some line bundle L on X. Then either* $\dim \varphi(X) = 0$ *or n. Furthermore φ is finite-to-one if* $\dim \varphi(X) = n$.

Proof. Let $\dim Y = s$, $1 \leq s < n$. Let H be a very ample line bundle on Y and let $\mathcal{L} := \varphi^* H$. Since H is very ample, the line bundle $\mathcal{L}$ is nontrivial. Therefore $\mathcal{L} \approx \alpha L$ for some nonzero integer α. Thus $\alpha L \approx \varphi^* H$ is trivial on the fibers of φ, implying there are no ample line bundles on X. This contradicts the projectivity of X. The same argument shows that φ is finite-to-one if $\dim \varphi(X) = n$. □

For others characterizations of projective spaces and quadric hypersurfaces we refer to Fujita [255, §2], Biancofiore and the authors of the book [85, §2], and to Cho and Miyaoka [156] (see Theorem (6.3.14)).

3.2 $\mathbb{P}^d$-bundles

In this section we collect a few facts about $\mathbb{P}^d$-bundles which will be used throughout the book. By a *$\mathbb{P}^d$-bundle* we mean an algebraic morphism, $p : X \to Y$, from a variety X to a variety Y which is a $\mathbb{P}^d$-bundle in the complex topology. Such bundles do not have to be locally trivial in the Zariski topology (the deviation from triviality in the Zariski topology is measured following Grothendieck [298] by the Brauer-Grothendieck group of Y) but the ones that are of most importance so far in adjunction theory are locally trivial in the Zariski topology.

We say that (X, L) is a *linear $\mathbb{P}^d$-bundle* over a normal variety Y of dimension m if there exists a surjective morphism, $p : X \to Y$, such that all the fibers F of p are $\mathbb{P}^d$, $d = n - m$, and $L_F \cong \mathcal{O}_{\mathbb{P}^d}(1)$. This is equivalent to say that $(X, L) \cong (\mathbb{P}(\mathcal{E}), \mathcal{O}_{\mathbb{P}(\mathcal{E})}(1))$ for the rank $d+1$ ample vector bundle $\mathcal{E} := p_*L$ on Y.

The following method of recognizing linear $\mathbb{P}^d$-bundles is a variation, in the singular case, of a theorem of Fujita [252, (2.12)].

Proposition 3.2.1. *Let X be an n-dimensional normal connected projective variety and let $p : X \to Y$ be a holomorphic surjection from X onto a normal variety, Y. Let L be an ample line bundle on X. Assume that $(F, L_F) \cong (\mathbb{P}^d, \mathcal{O}_{\mathbb{P}^d}(1))$ for a general fiber, F, of p and that all fibers of p are d-dimensional. Further assume that X is Cohen-Macaulay and that* $\operatorname{Sing}(X)$ *contains no fiber of the map, e.g.,* $\operatorname{cod}_X \operatorname{Sing}(X) > \dim Y$. *Then $p : X \to Y$ gives to (X, L) the structure of a linear $\mathbb{P}^d$-bundle with $X \cong \mathbb{P}(p_*L)$. In particular X is smooth if and only if Y is smooth.*

Proof. First, let us prove the case when Y is nonsingular. In this case, since X is Cohen-Macaulay and p has equidimensional fibers, it follows that p is flat (see Theorem (4.1.2) or Fischer [234, pg. 145 and 158]). Note also that every fiber is defined by exactly $n-d$ coordinate functions, so that each fiber is Cohen-Macaulay. Let Z be any fiber of p. Since $L_F^d = 1$ we have by flatness $L_Z^d = 1$, so that Z is irreducible and generically reduced. Since Z is Cohen-Macaulay, Z is in fact reduced. Therefore all fibers of p are irreducible, reduced, and Cohen-Macaulay. By the semicontinuity theorem for dimensions of spaces of sections on fibers of a flat morphism [318, Chapter III, Theorem 12.8], $h^0(L_Z) \geq h^0(L_F) = d + 1$. Then by (3.1.4) we conclude that $(F, L_F) \cong (\mathbb{P}^d, \mathcal{O}_{\mathbb{P}^d}(1))$ for every fiber F of p. Since p is flat it follows by the same semicontinuity theorem for dimensions of spaces of sections on fibers of a flat morphism as above that a basis of the sections of L_F extends to sections, $s_0, \ldots, s_{d+1}$, of L_U on a neighborhood, U of F, in the Zariski topology. Since the sections, $s_0, \ldots, s_{d+1}$, restricted to F span L_F, it can be assumed by possibly shrinking U that the sections, $s_0, \ldots, s_{d+1}$, have no common zeroes on U. Choose a Zariski open set $V \subset Y$ with $F \subset p^{-1}(V) \subset U$. The map to $V \times \mathbb{P}^d$ given by sending a point $x \in p^{-1}(V)$ to $(p(x), [s_0(x), \ldots, s_{d+1}(x)])$ is an isomorphism, showing that $p : X \to Y$ gives to (X, L) the structure of a linear $\mathbb{P}^d$-bundle.

In the general case when Y is singular, by taking suitable hyperplane sections we can assume that Y has only isolated singularities. Let F be any d-dimensional irreducible component of any fiber $p^{-1}(y)$ of p. Then the assumption $\mathrm{cod}_X\mathrm{Sing}(X) > \dim Y$ gives $\dim \mathrm{Sing}(X) < d$, so that F is not contained in $\mathrm{Sing}(X)$. Take a large integer m such that mL is very ample and let $D_1, \ldots, D_d$ be d general elements of $|mL|$. Let $\mathcal{D} = D_1 \cap \cdots \cap D_d$. By applying Bertini's theorem to the restriction of $|mL|$ to each F we see that $\mathcal{D} \cap F$ is a nonsingular subscheme consisting of a finite number of points contained in $\mathrm{reg}(F)$. Now exactly the same argument as in Fujita [252, (2.12)] applies to give the result. □

Remark 3.2.2. Note that the condition $\mathrm{cod}_X\mathrm{Sing}(X) > \dim Y$ is always satisfied if either $\dim Y = 1$ (since X is normal) or $\dim Y = 2$ and X has terminal singularities (since in this case $\mathrm{cod}_X\mathrm{Sing}(X) \geq 3$ by (1.3.1)).

Lemma 3.2.3. *Let E and F be very ample vector bundles on a projective variety X. Then $E \oplus F$ is very ample.*

Proof. The very ampleness of E (respectively F) is equivalent to the map $\varphi_E : E^* \to \mathbb{C}^{h^0(E)}$ (respectively $\varphi_F : F^* \to \mathbb{C}^{h^0(F)}$) given by sections of E (respectively of F) being an embedding from $E^* - X$ to $\mathbb{C}^{h^0(E)} - \{0\}$ (respectively $F^* - X$ to $\mathbb{C}^{h^0(F)} - \{0\}$). Since the restriction maps $\varphi_{E|E^*-X}$ and $\varphi_{F|F^*-X}$ are embeddings, the map $\varphi_{E\oplus F} := (\varphi_E, \varphi_F)$ is an embedding on $E^* \oplus F^* - X$. □

The following lemma is very useful and gives many interesting examples of very ample divisors (see Hartshorne [318, p.380]).

Lemma 3.2.4. *Let $\mathcal{E}$ be a rank $d+1$ vector bundle over $\mathbb{P}^1$ of the form*

$$\mathcal{E} = \mathcal{O}_{\mathbb{P}^1} \oplus \mathcal{O}_{\mathbb{P}^1}(a_1) \oplus \cdots \oplus \mathcal{O}_{\mathbb{P}^1}(a_d)$$

where $a_i \geq 0$ for each $i = 1, \ldots, d$. Let $X = \mathbb{P}(\mathcal{E})$ and let $\varphi : X \to \mathbb{P}^1$ be the $\mathbb{P}^d$-bundle projection. Let ξ be the tautological line bundle on X and F a fiber of φ. Then $a\xi + bF$ is ample if and only if it is very ample. Moreover

1. *$a\xi + bF$ is spanned if and only if $a \geq 0, b \geq 0$; and*
2. *$a\xi + bF$ is ample if and only if it is very ample and if and only if $a > 0$, $b > 0$.*

Proof. Assume $\mathcal{E}$ is spanned at all points by global sections. From the assumption, $\mathcal{E}$ is spanned, it follows that ξ is spanned. Moreover $\mathcal{O}_X(F) = \varphi^*\mathcal{O}_{\mathbb{P}^1}(1)$ is spanned. Then $a \geq 0, b \geq 0$ implies that $a\xi + bF$ is spanned. To see the converse assume $a\xi + bF$ is spanned. Therefore $(a\xi + bF)_F \cong a\xi_F \cong \mathcal{O}_{\mathbb{P}^d}(a)$ is spanned, so that $a \geq 0$. Let σ be the section of φ corresponding to the quotient $\mathcal{E} \to \mathcal{O}_{\mathbb{P}^1} \to 0$.

Then $\xi_\sigma \cong \mathcal{O}_{\mathbb{P}^1}$ and $(a\xi + bF)_\sigma \cong bF_\sigma \cong \mathcal{O}_{\mathbb{P}^1}(b)$ is spanned, so that $b \geq 0$ and (3.2.4.1) is proved.

Note that exactly the same argument as above shows that $a > 0$, $b > 0$ if $a\xi + bF$ is ample. Thus to prove 2) we have first to show that $a\xi + bF$ is ample, assuming $a > 0$, $b > 0$. By (3.2.4.1) we know that $a\xi + bF$ is spanned. Let $\psi : X \to \mathbb{P}_{\mathbb{C}}$ be the morphism defined by $\Gamma(a\xi + bF)$. If $a\xi + bF$ is not ample, there exists a curve, C, such that $(a\xi + bF) \cdot C = 0$. Since ξ, F are both spanned it thus follows that $\xi \cdot C = F \cdot C = 0$ and hence C is contained in a fiber of φ, say F. Therefore $\xi \cdot C = \xi_F \cdot C = \mathcal{O}_{\mathbb{P}^d}(1) \cdot C = 0$ contradicts the ampleness of $\mathcal{O}_{\mathbb{P}^d}(1)$. This proves the ampleness assertion of (3.2.4.2).

It only remains to show that $a\xi + bF$ is very ample if $a > 0$, $b > 0$. Since $a\xi + bF$ is spanned, we have a morphism $\psi : X \to \mathbb{P}_{\mathbb{C}}$, and if we show that ψ is an embedding, then we will be done. Since ξ is spanned and has global sections whose restrictions span $\mathcal{O}_F(1)$ for any fiber F, it follows that the global sections of $a\xi + bF$ span $\mathcal{O}_F(a)$ for any fiber F, and thus the restriction ψ_F is an embedding for any fiber F of φ. Let x, y be two distinct points of X. Since $a\xi + (b-1)F$ is spanned by global sections and since $\mathcal{O}_X(F)$ is spanned and gives the fiber bundle morphism φ, it follows that $\psi(x) = \psi(y)$ implies that $\varphi(x) = \varphi(y)$, and thus that x, y lie on the same fiber of φ. This contradicts the fact that ψ_F is an embedding. Therefore ψ separates distinct points.

Now let $x \in X$ and let $y \in \mathcal{T}_{X,x}$ be a tangent vector such that $d\psi_x(y) = 0$, where $d\psi_x$ denotes the differential map of ψ at x. Arguing as in the last paragraph, we conclude that y is tangent to some fiber F of φ, which is impossible since ψ_F is an embedding. □

Let us recall two further results on $\mathbb{P}^d$-bundles over elliptic curves, due to Ionescu [337].

Lemma 3.2.5. *Let $\mathcal{E}$ be a rank r ample vector bundle of degree d on a smooth elliptic curve, C. Then $h^1(\mathcal{E}) = 0$ and $h^0(\mathcal{E}) = d$.*

Proof. If $r = 1$ this is clear. Assume by induction that we know this for bundles of rank $< r$. Since $\mathcal{E}$ is ample, $d > 0$, and therefore there exists a nonvanishing section of $\mathcal{E}$. So we have an exact sequence

$$0 \to \mathcal{O}_C \to \mathcal{E} \to Q \to 0.$$

Note that $\operatorname{rank} Q = r - 1$ and $\deg Q = \deg \mathcal{E} = d$. Note also that this sequence does not split since if it did, $\mathcal{E}$ would not be ample. Thus the connecting map $\alpha : H^0(\mathcal{O}_C)(\cong \mathbb{C}) \to H^1(Q^*)$ of the dual sequence is non trivial. Hence α is an injection. Therefore $H^0(Q^*) \cong H^0(\mathcal{E}^*)$ and $h^1(\mathcal{E}^*) = h^1(Q^*) - 1 + h^1(\mathcal{O}_C) = h^1(Q^*)$. By Serre duality and the induction hypothesis we have $h^0(\mathcal{E}) = d$ and hence, by the Riemann-Roch theorem, $h^1(\mathcal{E}) = 0$. □

Corollary 3.2.6. *Let $\mathcal{E}$ be a rank r, very ample vector bundle (in the sense that the tautological line bundle, ξ, on $\mathbb{P}(\mathcal{E})$ is very ample) on a smooth elliptic curve C. Let d be the degree of $\mathcal{E}$. Then $d \geq 2r+1$.*

Proof. By the lemma above, $h^0(\mathcal{E}) = d$. Then ξ embeds $\mathbb{P}(\mathcal{E})$ in $\mathbb{P}^{d-1}$. Assume $2r \geq d$. Then the generalized Lefschetz theorem, (2.3.11), gives $h^1(\mathcal{O}_{\mathbb{P}(\mathcal{E})}) = 0$. This contradicts the assumption that $h^1(\mathcal{O}_{\mathbb{P}(\mathcal{E})}) = h^1(\mathcal{O}_C) = 1$. Thus $2r \leq d-1$. □

For reader's convenience we recall the well-known contractibility criterion of Grauert [285] and of Nakano [475] (see also Cornalba [170]), that we use later on in this book. Note that the complex variety, Y, in these results is not necessarily projective, even if X is projective.

Theorem 3.2.7. (Grauert contractibility criterion) *Let X be a normal, complex analytic variety, and let Z be a compact local complete intersection in X such that the conormal bundle, $\mathcal{N}^*_{Z/X}$, of Z in X is ample. Then there exists a holomorphic map $p : X \to Y$ onto a normal, complex analytic variety, Y, such that $p(Z)$ is a point, y, and p induces a biholomorphism $X - Z \cong Y - \{y\}$.*

Theorem 3.2.8. (Nakano contractibility criterion) *Let X be a complex manifold and let $Z \subset X$ be a compact complex submanifold of X which is a $\mathbb{P}^k$-bundle $p' : Z \to Z'$ over a complex manifold, Z'. If $\mathcal{N}_{Z/X|F} \cong \mathcal{O}_{\mathbb{P}^k}(a)$ with $a < 0$ for a fiber F of p', there exists a holomorphic map, $p : X \to Y$, onto a complex analytic variety, Y, and a commutative diagram*

$$\begin{array}{ccc} Z & \hookrightarrow & X \\ p' \downarrow & & \downarrow p \\ Z' & \hookrightarrow & Y. \end{array}$$

The restriction map $p_{X-Z} : X - Z \to Y - Z'$ is a biholomorphism, and if $a = 1$, then Y is smooth, and $p : X \to Y$ is the blowup of Y along Z'.

$\mathbb{P}^1$-bundles over curves. Let us consider now the case of $\mathbb{P}^1$-bundles over curves. Let $\mathcal{E}$ denote a rank 2 vector bundle on a smooth connected curve Y of genus q. The Hartshorne-Nagata invariant, e, of $\mathbb{P}(\mathcal{E})$ is the integer such that $-e$ is the minimum self-intersection of any section of $p : \mathbb{P}(\mathcal{E}) \to Y$. Let E denote such a section for which the minimum is taken on, and let f denote a fiber of p. The direct image of the exact sequence

$$0 \to \mathcal{O}_{\mathbb{P}(\mathcal{E})} \to \mathcal{O}_{\mathbb{P}(\mathcal{E})}(E) \to \mathcal{O}_E(E) \to 0$$

under p gives an exact sequence

$$0 \to \mathcal{O}_Y \to \mathcal{F} \to p_*\mathcal{O}_E(E) \to 0.$$

Here $\mathcal{F} \cong \mathcal{E} \otimes \mathcal{L}$ for some line bundle $\mathcal{L}$ on Y, and therefore $\mathbb{P}(\mathcal{F}) \cong \mathbb{P}(\mathcal{E})$. Thus as far as this discussion goes, we can assume that $\mathcal{E}$ was chosen in the first place so that $\mathcal{E} := p_*\mathcal{O}_{\mathbb{P}(\mathcal{E})}(E)$. In particular $\deg \mathcal{E} = -e$. The canonical bundle formula (1.1) is in this special case

$$K_{\mathbb{P}(\mathcal{E})} \sim -2E + (2q - 2 - e)f. \tag{3.2}$$

3.2.9. The Hirzebruch surfaces. $\mathbb{P}^1$-bundles over $\mathbb{P}^1$, which are usually called the Hirzebruch surfaces, are very classical and come up in many contexts throughout adjunction theory. We collect here some of the standard facts about these surfaces.

By $\mathbb{F}_r$, with $r \geq 0$, we denote the r-th Hirzebruch surface. $\mathbb{F}_r$ is the unique holomorphic $\mathbb{P}^1$-bundle over $\mathbb{P}^1$ with a section E satisfying $E^2 = -r$. Let $p : \mathbb{F}_r \to \mathbb{P}^1$ denote the bundle projection. In the case $r = 0$, $\mathbb{F}_0$ is simply $\mathbb{P}^1 \times \mathbb{P}^1$. In the case $r \geq 1$, E is the unique irreducible curve on $\mathbb{F}_r$ with negative self-intersection. By $\widetilde{\mathbb{F}}_r$, with $r \geq 1$, we denote the normal surface obtained from $\mathbb{F}_r$ by contracting E. In the case $r = 1$, $\widetilde{\mathbb{F}}_1$ is $\mathbb{P}^2$. We shall denote by E, f a basis for the second integral cohomology $H^2(\mathbb{F}_r, \mathbb{Z})$ of $\mathbb{F}_r$, f a fiber of p. Since $\mathbb{F}_r = \mathbb{P}(\mathcal{O}_{\mathbb{P}^1} \oplus \mathcal{O}_{\mathbb{P}^1}(r))$, by using Lemma (3.2.4), it follows that a line bundle $aE + bf$ is ample if and only if it is very ample and it is very ample if and only if $a > 0$ and $b > ar$. Furthermore $aE+bf$ is spanned by global sections if and only if $a \geq 0$ and $b \geq ar$.

In the following E, f denote a section of self-intersection $E^2 = -r$ and a fiber of an Hirzebruch surface $\mathbb{F}_r$.

Corollary 3.2.10. *Let X be a normal surface and L an ample line bundle on X. Assume that $g(L) = 0$. Then either*

1. *$(X, L) \cong (\mathbb{P}^2, \mathcal{O}_{\mathbb{P}^2}(a))$, $a = 1, 2$; or*
2. *$(X, L) \cong (Q, \mathcal{O}_Q(1))$, Q quadric in $\mathbb{P}^3$; or*
3. *$X \cong \mathbb{F}_r$, $r \geq 0$, $L \approx E + bf$, $b \geq r + 1$; or*
4. *$X \cong \widetilde{\mathbb{F}}_r$ and $p^*L \approx E + rf$, where $p : \mathbb{F}_r \to \widetilde{\mathbb{F}}_r$ is the natural map that blows down E, $r \geq 2$.*

Proof. From Theorem (3.1.3) we have $\Delta(X, L) = 0$. Then the result follows from Proposition (3.1.2). If X is smooth, then either we are in case 1) or 2), or $X \cong \mathbb{F}_r$. In this case $L \approx aE + bf$ for some integers a, b such that $a > 0$, $b \geq ar + 1$ by Lemma (3.2.4). A direct computation, by using the genus formula, shows that $a = 1$. Thus we are in case 3).

If X is singular, by (3.1.2) we conclude that X is a cone over a curve, $C \subset X$, with $\Delta(C, L_C) = 0$. That is $C \cong \mathbb{P}^1$ and $X \cong \widetilde{\mathbb{F}}_r$, $r \geq 2$. Let $p : \mathbb{F}_r \to \widetilde{\mathbb{F}}_r$ be the natural map which contracts E. Then $p^*L \approx aE + bf$ and $p^*L \cdot E = 0$ imply $b = ar$. Since $g(L) = g(p^*L) = 0$ we see again that $a = 1$. □

We refer the reader to Nagata [469], Hartshorne [318, Chapter V, §2], and our paper [102, §3] for a discussion of these bundles and in particular for the proof of the result below. We also refer to (8.5.6), (8.5.7), (8.5.8).

Theorem 3.2.11. *Let $\mathcal{E}$ be a rank 2 vector bundle on a smooth curve, Y, of genus $q \geq 1$. Let E be a section of minimal self-intersection on $\mathbb{P}(\mathcal{E})$, and let $e := -E^2$. Let f denote a fiber of the bundle projection $\mathbb{P}(\mathcal{E}) \to Y$. Let $D \sim aE + bf$ be a divisor in $\mathbb{P}(\mathcal{E})$, for integers a, b. Assume that D is neither E nor a fiber f. Then $e \geq -q$ and if $\mathcal{E}$ is not a direct sum of line bundles $e \leq 2q - 2$. Furthermore the following hold.*

1. *Assume $e \geq 0$:*
   (a) *if D is an irreducible curve, then $a > 0$, $b \geq ae$;*
   (b) *D is ample if and only if $a > 0$, $b > ae$;*
   (c) *D is very ample if $a \geq 1$ and $b \geq ae + 2q - 2 + \max\{3, e\}$.*
2. *Assume $e < 0$:*
   (a) *if D is an irreducible curve, then either $a = 1$, $b \geq 0$, or $a \geq 2$, $b \geq ae/2$;*
   (b) *D is ample if and only if $a > 0$, $b > ae/2$.*
   (c) *D is very ample if $a \geq 1$ and $b \geq ae/2 + 2q + 1$.*

The following lemma of Hartshorne [314, §1] is very useful.

Lemma 3.2.12. *Let $\mathcal{E}$ be a rank 2 vector bundle on a smooth curve, Y, of genus q. Let E be a section of minimal self-intersection on $\mathbb{P}(\mathcal{E})$, and let f denote a fiber of the bundle projection $\mathbb{P}(\mathcal{E}) \to Y$. Let C be an irreducible curve on $\mathbb{P}(\mathcal{E})$ with $C \sim xE + yf$, and let $g(C)$ be the arithmetic genus of C. Then*

$$2g(C) - 2 = x(2q - 2) + \frac{x-1}{x}C^2.$$

Proof. Using the canonical bundle formula, (3.2), to write

$$2g(C) - 2 = (x - 1)(2y - xe) + x(2q - 2),$$

and $C^2 = x(2y - xe)$, we get $2g(C) - 2 = \frac{x-1}{x}C^2 + x(2q - 2)$. □

The following lemma will be used in the proof of the nonbreaking lemma, (6.3.1).

Lemma 3.2.13. *Let $\mathscr{E}$ be a rank 2 vector bundle on a smooth curve, Y, of genus q. Let E be a section of minimal self-intersection on $\mathbb{P}(\mathscr{E})$, and let $e := -E^2$. Let f denote a fiber of the bundle projection $\mathbb{P}(\mathscr{E}) \to Y$. Let C be an irreducible curve on $\mathbb{P}(\mathscr{E})$ such that $C^2 < 0$. Then $C = E$ and $e > 0$.*

Proof. Let $C \sim xE + yf$ for some integers x, y. We can assume that $C \neq E$ since otherwise the lemma is proven. Thus $C \cdot E \geq 0$, giving $y \geq xe$.

Since C is irreducible we conclude that $x \geq 1$. Indeed, otherwise, since $x = C \cdot f \geq 0$, we would have $x = 0$ and hence $C^2 = 0$. From this and $C^2 = x(2y - xe) < 0$, we conclude that $2y < xe$. Thus, since $y \geq xe$, we have $xe > 2xe$ implying that $e < 0$ and hence $2y < xe$ implies $y < 0$.

Let $\overline{C} \to C$ be the normalization map. Composing this map with the restriction, p_C, of $p : \mathbb{P}(\mathscr{E}) \to Y$ to C, we get an x-sheeted covering map of Y. By the usual Hurwitz formula we conclude that $2g(\overline{C}) - 2 \geq x(2q - 2)$. Since $g(\overline{C}) \leq g(C)$ we conclude from Lemma (3.2.12) that $0 \leq \frac{x-1}{x}C^2$. Since $x \geq 1$ and $C^2 < 0$ we conclude that $x = 1$. Hence $C = E + yf$ is a section. Thus by the definition of e we have $-e + 2y = C^2 \geq E^2 = -e$, or $y \geq 0$. This contradicts $y < 0$. □

Lemma 3.2.14. *Let $\mathscr{E}$ be a rank 2 vector bundle on a smooth curve, Y, of genus q. Let E be a section of minimal self-intersection on $\mathbb{P}(\mathscr{E})$, and let $e := -E^2$. Let f denote a fiber of the bundle projection $p : \mathbb{P}(\mathscr{E}) \to Y$. Assume $e > 0$. Then there exists a holomorphic surjection $\pi : \mathbb{P}(\mathscr{E}) \to W$ with W a normal analytic space, $\pi(E)$ a point, and the restriction map $\pi_{\mathbb{P}(\mathscr{E})-E} : \mathbb{P}(\mathscr{E}) - E \to W - \pi(E)$ a biholomorphism. Furthermore the variety W is projective if and only if $\mathscr{E}$ splits into a direct sum of line bundles.*

Proof. If $e > 0$, then by Theorem (3.2.7) there exists a holomorphic surjection $\pi : \mathbb{P}(\mathscr{E}) \to W$ with W a normal analytic space, $\pi(E)$ a point, and the restriction map $\pi_{\mathbb{P}(\mathscr{E})-E} : \mathbb{P}(\mathscr{E}) - E \to W - \pi(E)$ is a biholomorphism.

To show the second part of the statement, assume first that $\mathscr{E} \cong \mathcal{O}_Y \oplus p_*\mathcal{O}_E(E)$. Then W is the normal cone of dimension $\dim Y + 1$ over $(Y, p_*\mathcal{O}_E(-E))$ in the sense of §1.1.8, and hence W is projective.

Conversely if W is projective, then we can find an irreducible divisor, $D \subset W$, which does not meet $\pi(E)$. Let D' denote $\pi^{-1}(D)$. Note that $D' \cong D$, and the restriction map $p_{D'}$ is finite-to-one. Since p is a $\mathbb{P}^1$-bundle, we have for any fiber F of p that $F \cong \mathbb{P}^1$. If the intersection $F \cap D'$ consists of smooth points, $z_1, \ldots, z_n$, we can average them, i.e., get a new point $(z_1 + \cdots + z_n)/n$, in $\mathbb{C} \cong F - F \cap E$. The operation of averaging is well-defined. To see this note that since p is a $\mathbb{P}^1$-bundle in the Zariski topology, it is enough to note that averaging is independent of the linear coordinate we choose on $\mathbb{C} \cong F - F \cap E$. Thus we get a section, μ, over $Y - Z$ where Z is the image under p of the branch locus of $p_{D'}$. Choose a point, $z \in Z$, and $U \subset Y$, a complex open neighborhood of z in Y. Since p is a

$\mathbb{P}^1$-bundle, we can choose U small enough so that the bundle is a product $U \times \mathbb{P}^1$ over U. The section, μ, composed with $U \times \mathbb{P}^1 - E \to \mathbb{C}$ is a complex function. Since $D' \cap E = \emptyset$, we see that this function is bounded and extends to a function on all of U. This procedure gives a well-defined section, E', of p that does not meet E. Thus we see that $\mathbb{P}(p_*\mathcal{O}_E(E) \oplus \mathcal{O}_Y) \cong \mathbb{P}(\mathcal{E})$ and thus that there exists a line bundle, $\mathcal{L}$, on Y, such that $\mathcal{E} \cong (p_*\mathcal{O}_E(E) + \mathcal{L}) \oplus \mathcal{L}$. □

Corollary 3.2.15. *Let $\mathcal{E}$ be a rank 2 vector bundle on a smooth curve, Y, of genus q. Let E be a section of minimal self-intersection on $\mathbb{P}(\mathcal{E})$, and let $e := -E^2$. Let f denote a fiber of the bundle projection $\mathbb{P}(\mathcal{E}) \to Y$. Assume $e > 0$, and let $\pi : \mathbb{P}(\mathcal{E}) \to W$ be the holomorphic surjection with W a normal 2-dimensional analytic space, $\pi(E)$ a point, and the restriction map $\pi_{\mathbb{P}(\mathcal{E})-E} : \mathbb{P}(\mathcal{E}) - E \to W - \pi(E)$ a biholomorphism. If $\phi : \mathbb{P}(\mathcal{E}) \to Z$ is a holomorphic map onto a complex 2-dimensional analytic space, Z, then the map ϕ factors as $\phi = s \circ \pi$ for some finite-to-one holomorphic map $s : W \to Z$.*

Proof. Remmert-Stein factorize the map ϕ as $\phi = s \circ r$ where $r : \mathbb{P}(\mathcal{E}) \to S$ is a holomorphic map with connected fibers onto a normal surface, S, and $s : S \to Z$ is a finite-to-one map.

The intersection matrix of the curves making up fibers of a bimeromorphic morphism from a smooth surface onto a normal surface is negative definite. This is proven in the algebraic case by Mumford [461, §1], and the general case follows from the proof in [461, §1] by using the fact that any germ of an analytic space with an isolated singularity is biholomorphic to a germ of a projective variety (see Artin [36]). Using this and (3.2.13) we conclude that $r : \mathbb{P}(\mathcal{E}) \to S$ is the map $\pi : \mathbb{P}(\mathcal{E}) \to W$. Indeed, if r is not biholomorphic there exists a curve, C, such that $r(C)$ is a point. Then $C^2 < 0$ and therefore, by (3.2.13), $C = E$ and $e > 0$. □

3.3 Special varieties arising in adjunction theory

We recall in this section the definitions of some special varieties which occur as "building blocks" in the classification of projective varieties via adjunction theory.

3.3.1. Special varieties. Let V be a normal r-Gorenstein variety of dimension n, and let L be an *ample* line bundle on V. We say that V is an *r-Gorenstein-Fano variety* (or simply that V is *r-Fano* or *$\mathbb{Q}$-Fano*) if $-rK_V$ is ample. We say that (V, L) is a *Del Pezzo variety* (respectively a *Mukai variety*) if $rK_V \approx -(n-1)rL$ (respectively $rK_V \approx -(n-2)rL$).

We also say that (V, L) is a *scroll* (respectively a *quadric fibration*; respectively a *Del Pezzo fibration*; respectively a *Mukai fibration*) over a normal variety Y of dimension m if there exists a surjective morphism with connected fibers $p : V \to Y$, such that $r(K_V + (n - m + 1)L) \approx p^*\mathcal{L}$ (respectively $r(K_V + (n - m)L) \approx p^*\mathcal{L}$; respectively $r(K_V+(n-m-1)L) \approx p^*\mathcal{L}$; respectively $r(K_V+(n-m-2)L) \approx p^*\mathcal{L}$) for some *ample* line bundle $\mathcal{L}$ on Y.

We say that a smooth n-dimensional variety V is a *Fano variety of index i*, if i is the largest positive integer such that $K_V \approx -iH$ for some ample line bundle H on V. Note that $i \leq n + 1$ (see Lemma (3.3.2) below) and $n - i + 1$ is referred to as the *co-index* of V. We say that an n-dimensional smooth polarized variety, (V, L), is a *Fano fibration of co-index* $n - m + 1 - t$ if there exists a surjective morphism with connected fibers, $p : V \to Y$, onto an m-dimensional variety Y such that $K_V + tL \approx p^*\mathcal{L}$ for some ample line bundle $\mathcal{L}$ on Y and positive integer t. Thus a scroll, (V, L), is a Fano fibration of co-index 0, a quadric fibration is a Fano fibration of co-index 1 and so on. Let L be a *very ample* line bundle on a Fano n-fold V. If $\operatorname{def}(V, L) = k > 0$ and $K_V + ((n + k)/2 + 1)L \approx \mathcal{O}_V$, we say that V is a *Fano n-fold of positive defect.* From the results of our paper with Fania [87] (see in particular Corollary (6.6.8), our paper with Wiśniewski [114, (2.4)] and §14.4) we see that the classification of positive defect manifolds reduces to the classification of such Fano n-folds.

We refer to Fujita [257, 258] and Reid [521] for classification results on Del Pezzo varieties. Note that Del Pezzo manifolds are completely described by Fujita [257, I, §8]. We refer to Mukai [458, 459, 460] for results on Mukai varieties. We refer to D'Souza [193], D'Souza and Fania [194], and Corti [171] for results on the structure of Del Pezzo fibrations and to Lanteri, Palleschi, and Sommese [383, 384, 385] for the classification of n-folds with Del Pezzo surfaces as hyperplane surfaces sections. We refer to Ishkovskih and Shokurov [339], Ishkovskih [338], Mori and Mukai [453], Alzati and Bertolini [5], and to the nice survey paper of Murre [467] for general results and classical references on classification of Fano 3-folds. We refer to our paper [105] for relations between scrolls, (V, L), in the adjunction theoretic sense and classical scrolls, i.e., linear $\mathbb{P}^d$-bundles (V, L). We also refer to Chapter 14 for more results on scrolls and quadric fibrations.

Let us also recall the following general fact we use over and over in the sequel.

Lemma 3.3.2. *Let $\mathcal{L}$ be a nef and big line bundle on a normal projective variety, V, of dimension n with only terminal singularities. Then if $t(aK_V + b\mathcal{L}) \approx \mathcal{O}_V$ for some integers $a > 0, b > 0$, $t > 0$, and if $aK_V + b\mathcal{L}$ is invertible, one has $aK_V + b\mathcal{L} \approx \mathcal{O}_V$, and $b/a \leq n + 1$. In particular the index r of V divides a, say $a = ra'$. If a', b are coprime, there exists a nef and big line bundle M on V such that $rK_V \approx -bM, \mathcal{L} \approx a'M$. If $\mathcal{L}$ is ample, then so is M.*

Proof. It runs parallel to that of Lemma (4.3.1) in the paper [227] by Fania and Sommese. Note that V has rational singularities since terminal singularities are rational. Choose the smallest integer $t > 0$ such that $t(aK_V + b\mathcal{L}) \approx \mathcal{O}_V$. Let $q : V' \to V$ be the unramified cover associated to the t-th root of the constant function. By choice of t, V' is irreducible and $aK_{V'} + b\mathcal{L}' \approx \mathcal{O}_{V'}$ where $\mathcal{L}' = q^*\mathcal{L}$. Since $atK_V \approx -bt\mathcal{L}$, $aK_{V'} \approx -b\mathcal{L}'$ we see that $-K_V$, $-K_{V'}$ are nef and big. Then by (2.2.1) we have $h^i(\mathcal{O}_V) = h^i(\mathcal{O}_{V'}) = 0$ for $i > 0$. Thus $\chi(\mathcal{O}_V) = \chi(\mathcal{O}_{V'}) = 1$. But since q is an unramified cover, $\chi(\mathcal{O}_{V'}) = t\chi(\mathcal{O}_V)$ (see Fujita [267, Example 18.3.9]). This implies $t = 1$.

To prove that $b/a \le n+1$, note that $\chi(K_V + s\mathcal{L})$ is a polynomial $p(s)$ of degree $\le n$ in s since $\mathcal{L}$ is a line bundle and $\chi(K_V + s\mathcal{L}) = h^0(K_V + s\mathcal{L})$, for $s > 0$, again by (2.2.1). Moreover $h^0(K_V + s\mathcal{L}) = 0$ whenever $s < b/a$ since $K_V \approx -b\mathcal{L}/a$. Thus we conclude that $p(s)$ would have at least $n+1$ roots if $b/a > n+1$, which is a contradiction. □

Chapter 4

The Hilbert scheme and extremal rays

In the first section of this chapter we recall definitions and some basic properties of the Hilbert scheme and limited families of subvarieties of a projective variety, in the form we will need in this book. We also prove a useful and general vanishing result, Lemma (4.1.5), and a well-known rigidity lemma as well.

In §4.2 and §4.3 we collect what we need from Mori's theory of extremal rays, for completeness and for the convenience of readers not expert in this theory. We refer to the wide original literature for the details. We state the cone theorem, (4.2.5), giving some basic examples, and we prove the contraction theorem, (4.3.1), in the smooth case.

4.1 Flatness, the Hilbert scheme, and limited families

Deformation theory and the Hilbert scheme are very important tools for which there is no single reference. We have found the sources by Grothendieck [296], Mumford [462, 463], Sernesi [553], and Harris [311] useful. Right before this book was sent off to the publisher we received Kollár's paper [360], which is a very useful reference for constructions associated with the Hilbert scheme. In this section we go over basic theory stating the theorems that we will use.

Flatness captures the geometric notion of continuous variation of geometric objects depending on parameters (see Mumford [462, Lecture 7, §2] and [463, Chapter 3, §3]). A *flat map*, $\phi : X \to Y$, is a morphism between varieties such that given any $y \in Y$, the local ring, $\mathcal{O}_{X,x}$, is flat over the local ring, $\mathcal{O}_{Y,y}$, for all $x \in \phi^{-1}(y)$. In this definition the local rings can be taken to be either analytic or algebraic without any difference, i.e., the notions are equivalent for complex schemes in the sense that given an algebraic morphism, $\phi : X \to Y$, the algebraic local ring, $\mathcal{O}_{X,x}$, is flat over the algebraic local ring, $\mathcal{O}_{Y,y}$, if and only if the analytic local ring, $\mathcal{O}_{X_{an},x}$, is flat over the analytic local ring, $\mathcal{O}_{Y_{an},y}$.

Remark 4.1.1. Let $\phi : X \to Y$ be a proper surjective morphism between connected analytic spaces. Some key points:

1. If ϕ is flat, then all fibers of ϕ have the same dimension, d, and represent the same homology class in $H_d(X, \mathbb{Z})$.
2. If ϕ is flat, then the Euler characteristic, $\chi(\mathcal{O}_{\phi^{-1}(y)})$, is locally constant. Thus if ϕ is flat, it follows that for any line bundle, L, on X which is relatively ample with respect to ϕ, the Hilbert polynomial, $p_y(t) := \chi(tL_{\phi^{-1}(y)})$, is locally constant.
3. It is a theorem of Hartshorne [318, III, §9] that if ϕ is a projective morphism between algebraic varieties with L a relatively ample line bundle, and if Y is irreducible, then the converse is true, i.e., the Hilbert polynomial $p_y(t) := \chi(tL_{\phi^{-1}(y)})$ is constant if and only if ϕ is flat.

We have the following useful theorem (see Fischer [234, pg. 145 and 158] or Hartshorne [318, Chapter III, Exercise 10.9]).

Theorem 4.1.2. *Let $\phi : X \to Y$ be a surjective holomorphic map from a connected Cohen-Macaulay complex analytic variety, X, to a complex manifold, Y. Then ϕ is flat if and only if all fibers are equal dimensional. In particular any surjective morphism from a normal surface onto a smooth curve is flat.*

The following corollary of (4.1.2) is the basis of Mori's nonbreaking lemma (see (6.3.1)).

Corollary 4.1.3. *Let $\phi : S \to C$ be a proper morphism from a connected normal complex analytic surface, S, onto a smooth curve, C. If there is a fiber of ϕ that is isomorphic to $\mathbb{P}^1$, and if all fibers of ϕ are irreducible and generically reduced, then ϕ is a $\mathbb{P}^1$-bundle.*

Proof. By Theorem (4.1.2) we know that ϕ is flat. Let $y \in C$ be a point such that $f := \phi^{-1}(y)$ is biholomorphic to $\mathbb{P}^1$. Since the map, ϕ, is flat and the fiber, f, is smooth, it follows that ϕ is smooth in a neighborhood of f and hence a fiber bundle in a neighborhood of f. We will thus be done if we show that every fiber is biholomorphic to $\mathbb{P}^1$. By (4.1.1.2) we conclude that given any fiber, f, satisfies $\chi(\mathcal{O}_f) = 1$. Since C is smooth, f is a Cartier divisor. Note that S is Cohen-Macaulay since any normal 2-dimensional singularity is Cohen-Macaulay. Then f is Cohen-Macaulay. Therefore f is reduced since it is generically reduced. From this we conclude that $h^0(\mathcal{O}_f) = 1$, so $h^1(\mathcal{O}_f) = 0$ and therefore f is biholomorphic to $\mathbb{P}^1$ by Lemma (1.4.1). □

We also recall the following result from [318, III, (9.7)], which generalizes the second part of the statement of Theorem (4.1.2).

Theorem 4.1.4. *Let $\varphi : X \to Y$ be a morphism of a variety to a smooth curve, Y. Assume that every irreducible component of X maps surjectively onto Y. Then φ is flat.*

The Hilbert scheme. Let X be a projective variety with a very ample line bundle, L. The *Hilbert scheme*, Hilb(X), is a countable union of projective algebraic sets that represent the functor which assigns to an analytic set, $\mathcal{H}$, the set of flat families, $\mathcal{X} \subset \mathcal{H} \times X$ of not necessarily reduced algebraic subsets of X. This means that the restriction $\mathcal{X} \to \mathcal{H}$ of the product projection $\pi : \mathcal{H} \times X \to \mathcal{H}$ is a flat morphism, i.e., for every point $p \in \mathcal{X}$, the local ring $\mathcal{O}_{\mathcal{X},p}$ is a flat $\mathcal{O}_{\mathcal{H},\pi(p)}$-module. Saying this in another way, Hilb(X) is the set of not necessarily reduced projective algebraic subsets of X. It possesses the structure of a countable union of projective algebraic subsets in such a way that there is an analytic subset $Z \subset \text{Hilb}(X) \times X$, with the universal property that for every flat family $\mathcal{X} \subset \mathcal{H} \times X$ the classifying map $f : \mathcal{H} \to \text{Hilb}(X)$ is holomorphic, and $\mathcal{X} \subset \mathcal{H} \times X$ is equal to the pullback, $\mathcal{Z}$, of $Z \subset \text{Hilb}(X) \times X$ under f. Note that f is defined by sending each point $b \in \mathcal{H}$ to the point of Hilb(X) corresponding to the fiber $\pi^{-1}(b) = \{(Y, x), x \in Y\}$ where Y is the algebraic subset of X given by b. Moreover $\mathcal{Z}$ is obtained from the product projection $Z \to \text{Hilb}(X)$ making the base extension $f : \mathcal{H} \to \text{Hilb}(X)$. The set, $\text{Hilb}_{p(t)}(X)$, of not necessarily reduced projective algebraic subsets, Y, of X with a given Hilbert polynomial, $p(t) := \chi(Y, tL_Y)$, has the structure of a projective algebraic subset. Thus also $Z_{p(t)}$, the pullback of Z under the inclusion $i : \text{Hilb}_{p(t)}(X) \hookrightarrow \text{Hilb}(X)$, is projective.

To illustrate the use of the Hilbert scheme, we will prove a basic finiteness theorem for the varieties with hyperplane sections specified up to a finite amount of data. The notion of a limited family is convenient. A set, $\mathcal{F}$, of pairs (X, L) each consisting of a very ample line bundle, L, on a not necessarily reduced or irreducible projective algebraic set, X, is called a *limited family* of algebraic subsets of $\mathbb{P}^N$ for some fixed N if there is a flat surjective morphism between projective algebraic sets, $\varphi : \mathcal{F}' \to \mathcal{H}$, such that:

1. $\mathcal{F}' \subset \mathcal{H} \times \mathbb{P}^N$ for some N with the morphism φ given by restricting the product projection $\mathcal{H} \times \mathbb{P}^N \to \mathcal{H}$;

2. each $X \in \mathcal{F}$ is isomorphic to a fiber of φ;

3. the pullback under the induced map $\mathcal{F}' \to \mathbb{P}^N$ of $\mathcal{O}_{\mathbb{P}^N}(1)$ to the fiber of φ corresponding to X is isomorphic to L.

By the existence theorem for the Hilbert scheme, a family is a limited family if and only if the set of Hilbert polynomials for elements of the family form a finite set.

We need the following useful lemma which follows immediately from the original proof of the second author [567, Lemma I-B]; see also Fujita [241, Lemma 2.1] and Bădescu [42, §2].

Lemma 4.1.5. *Let X be an irreducible projective variety and let $\mathcal{E}$ be a locally free sheaf on X. Let L be an ample line bundle on X and let A be an effective divisor in $|L|$. Let $\mathcal{E}_A$, L_A be the restrictions of $\mathcal{E}$, L to A respectively. Assume that there exists some positive integer n_0 such that $H^i(\mathcal{E} \otimes (tL)) \to H^i(\mathcal{E}_A \otimes (tL_A))$ is onto (e.g., $h^i(\mathcal{E}_A \otimes (tL_A)) = 0$) for all $t \geq n_0$ and some fixed i. Then $h^{i+1}(\mathcal{E} \otimes (tL)) = 0$ for $t \geq n_0 - 1$.*

Proof. Consider the exact sequence

$$0 \to \mathcal{E} \otimes (tL) \to \mathcal{E} \otimes ((t+1)L) \to \mathcal{E}_A \otimes ((t+1)L_A) \to 0.$$

From the long exact cohomology sequence associated to it and our present assumption, we have an injection of $H^{i+1}(\mathcal{E} \otimes (tL))$ into $H^{i+1}(\mathcal{E} \otimes ((t+1)L))$, for $t+1 \geq n_0$. By Serre's vanishing theorem $h^{i+1}(\mathcal{E} \otimes (tL)) = 0$ for $t \gg 0$. Therefore we conclude that $h^{i+1}(\mathcal{E} \otimes (tL)) = 0$ for $t \geq n_0 - 1$. □

Corollary 4.1.6. *Let X be an irreducible projective variety with at most Cohen-Macaulay singularities and let $\mathcal{E}$ be a locally free sheaf on X. Let L be an ample line bundle on X and let A be an effective divisor in $|L|$. Let $\mathcal{E}_A$, L_A be the restrictions of $\mathcal{E}$, L to A respectively. Assume that $h^i(\mathcal{E}_A \otimes (-tL_A)) = 0$ for all $t \geq 0$ and some fixed i. Then $h^i(\mathcal{E} \otimes (-tL)) = 0$ for all $t \geq 0$.*

Proof. Since X has at worst Cohen-Macaulay singularities we have, by Serre duality, $h^i(\mathcal{E}_A \otimes (-tL_A)) = h^{n-1-i}(K_A \otimes \mathcal{E}_A^* \otimes (tL_A)) = 0$, for $t \geq 0$. By the same argument as in the proof of Lemma (4.1.5), applied with the exact sequence

$$0 \to K_X \otimes \mathcal{E}^* \otimes (tL) \to K_X \otimes \mathcal{E}^* \otimes ((t+1)L) \to K_A \otimes \mathcal{E}_A^* \otimes (tL_A) \to 0,$$

we get $h^{n-i}(K_X \otimes \mathcal{E}^* \otimes (tL)) = h^i(\mathcal{E} \otimes (-tL)) = 0$, for all $t \geq 0$. □

Theorem 4.1.7. *Let $\mathcal{F}$ be a limited family of projective subvarieties of $\mathbb{P}^N$ for some fixed N. The subvarieties of $\mathbb{P}^{N+1}$ with a hyperplane section in $\mathcal{F}$ form a limited family.*

Proof. The statement that $\mathcal{F}$ is a limited family is equivalent to the set of Hilbert polynomials of elements of $\mathcal{F}$ being finite. Fix one of these polynomials, $p(t)$. Let X be a subvariety of $\mathbb{P}^{N+1}$ and let $L := \mathcal{O}_{\mathbb{P}^{N+1}}(1)_{|X}$. We will be done if we show that a pair (X, L) with a hyperplane section (A, L_A), $A \in |L|$, having Hilbert polynomial $p(t) = \chi(tL_A)$ has at most a finite number of possible Hilbert polynomials, with these determined only by $p(t)$.

Let $Z_{p(t)}$ be as above. Standard cohomology results, e.g., Mumford [462, Lect. 7, §2], applied to the family $Z_{p(t)} \to \mathrm{Hilb}_{p(t)}(X)$ show that there exists a positive integer n_0 depending only on the family and hence only on $p(t)$ such that for $t \geq n_0$ and (A, L_A) a fiber of $Z_{p(t)} \to \mathrm{Hilb}_{p(t)}(X)$, $h^i(A, tL_A) = 0$ for $i > 0$, and $\alpha : \Gamma(\mathcal{O}_{\mathbb{P}^N}(t)) \to \Gamma(tL_A)$ is onto. Since $\beta : \Gamma(\mathcal{O}_{\mathbb{P}^{N+1}}(t)) \to \Gamma(\mathcal{O}_{\mathbb{P}^N}(t))$ is onto and $\alpha \circ \beta$ is the composition of $\varphi : \Gamma(tL) \to \Gamma(tL_A)$ with $\Gamma(\mathcal{O}_{\mathbb{P}^{N+1}}(t)) \to \Gamma(tL)$, we see that φ is onto. Hence in particular the assumption of Lemma (4.1.5) is satisfied for $i > 0$. Thus we conclude that for $t \geq n_0$, $h^i(X, tL) = 0$ for $i > 0$.

Let $p_{(X,L)}(t) := \chi(tL)$ and $n = \dim X$. Thus for $t = n_0, \dots, n_0 + n$,

$$\begin{aligned} p_{(X,L)}(t) = h^0(X, tL) &= h^0(\mathcal{O}_X) + \sum_{j=1}^{t} (h^0(X, jL) - h^0(X, (j-1)L)) \\ &\leq 1 + \sum_{j=1}^{t} (h^0(A, jL_A)) \\ &\leq 1 + th^0(A, tL_A) \leq 1 + (n_0 + n)p(n_0 + n). \end{aligned}$$

To see the last inequality note that, for $t \geq n_0$, $p(t) = h^0(tL_A)$ and $p(n+n_0) \geq p(t)$. Thus we see that there are at most finitely many values (which depend only on $p(t)$) for $p_{(X,L)}(t)$ for the values $t = n_0, \dots, n_0 + n$. Since $p_{(X,L)}(t)$ is determined by $n + 1$ distinct values of t, we see that there are at most finitely many $p_{(X,L)}(t)$ for a given $p(t)$. □

Corollary 4.1.8. *The pairs, (X, L), with L a very ample line bundle on an n-dimensional, irreducible projective variety, X, satisfying $L^n \leq d$ for a fixed integer d form a limited family.*

Proof. Let $X \to \mathbb{P}_{\mathbb{C}}$ be the embedding given by $\Gamma(L)$. Since $\deg X \geq \mathrm{cod}_{\mathbb{P}_{\mathbb{C}}} X + 1$ we have $h^0(L) \leq d + n$. Then we can use induction and the above theorem to reduce to the case of sets of $\leq d$ points in a projective space of dimension bounded by d. These clearly form a limited family. □

We need some facts about the differential calculus of the Hilbert scheme. In what follows Y can be replaced by a local complete intersection subvariety and B by any not necessarily reduced subvariety of Y. For our purposes the results in Namba [477, Chapter 0], and Mori [446, Prop. 3] are more than adequate. See also Ran [512].

Theorem 4.1.9. *Let X be a connected projective manifold. Let $B \subset Y \subset X$ be smooth submanifolds of X, with B possibly empty. Let $\mathcal{J}_B$ be the ideal sheaf of B in X. Let* Hilb(X, B) *denote the subspace of* Hilb(X) *consisting of algebraic subsets containing B. The Zariski tangent space of* Hilb(X, B) *at Y is isomorphic* *to* $H^0(Y, \mathcal{N}_{Y/X} \otimes \mathcal{J}_B)$ *under the natural morphism, e.g., see the proof of* (4.1.11). *There is an holomorphic map of a neighborhood of the origin in* $H^0(Y, \mathcal{N}_{Y/X} \otimes \mathcal{J}_B)$ *to a neighborhood of the origin in* $H^1(Y, \mathcal{N}_{Y/X} \otimes \mathcal{J}_B)$ *such that the fiber over the origin is biholomorphic to a neighborhood of* Y *in* $\mathrm{Hilb}(X, B)$. *In particular the dimension of* $\mathrm{Hilb}(X, B)$ *is bounded below by* $h^0(Y, \mathcal{N}_{Y/X} \otimes \mathcal{J}_B) - h^1(Y, \mathcal{N}_{Y/X} \otimes \mathcal{J}_B)$, *and thus if* $h^1(Y, \mathcal{N}_{Y/X} \otimes \mathcal{J}_B) = 0$ *then* $\mathrm{Hilb}(X, B)$ *is smooth at* Y.

Lemma 4.1.10. *Let* S *be an irreducible normal projective surface. Assume that there is a nef and big line bundle* L *on* S. *Assume that there is an irreducible curve,* $C \subset \mathrm{reg}(S)$, *with* $C \cong \mathbb{P}^1$, $C^2 = 0$, *and* $L \cdot C = 1$. *Then there is a morphism with connected fibers,* $p : S \to Y$, *of* S *onto a smooth curve* Y *of genus* $q := h^1(\mathcal{O}_S)$, *with* C *a fiber of* p. *If there is an ample line bundle* L *on* S *with* $L \cdot C = 1$, $p : S \to Y$ *is a* $\mathbb{P}^1$*-bundle.*

Proof. If $q = 0$, then using $\mathcal{O}_C(C) \cong \mathcal{O}_{\mathbb{P}^1}$ we see from the exact sequence

$$0 \to \mathcal{O}_S \to \mathcal{O}_S(C) \to \mathcal{O}_C(C) \to 0,$$

that $\mathcal{O}_S(C)$ is spanned by two global sections. The map $p : S \to \mathbb{P}^1$ given by the global sections has C as a fiber. Thus all the fibers are connected.

If $q \neq 0$, then note that $(K_S + L) \cdot C < 0$. Note that C moves by Theorem (4.1.9), and thus $h^0(K_S + L) = 0$. Hence by (2.2.8) it follows that S has at worst rational singularities. Therefore we have the Albanese map, $\mathrm{alb} : S \to \mathrm{Alb}(S)$, by (2.4.1). Since C moves and since holomorphic 1-forms are zero when restricted to rational curves, we see that the image of S in $\mathrm{Alb}(S)$ is at most 1-dimensional. Thus since $q > 0$, $\mathrm{alb}(S)$ is 1-dimensional. By (2.4.5) we see that $Y := \mathrm{alb}(S)$ is smooth and $p := \mathrm{alb}$ has connected fibers. By construction q equals the genus of Y.

Assume now that L is ample. If $L \cdot C = 1$ it follows that $L \cdot f = 1$ for all fibers of p. Thus all fibers are irreducible and generically reduced. This implies that p is a $\mathbb{P}^1$-bundle by Lemma (4.1.3). □

Lemma 4.1.11. *Notation as above. Let* $\mathcal{F} \subset \mathcal{H} \times X$ *be an irreducible flat family of connected smooth submanifolds of a projective manifold,* X. *Assume that the image of the map* $q : \mathcal{F} \to X$ *induced by the product projection contains an open set of* X. *Then given a general submanifold,* Y, *in the family,* $H^0(Y, \mathcal{N}_{Y/X})$ *spans* $\mathcal{N}_{Y/X}$ *at some point of* Y.

Proof. Let $p : \mathcal{F} \to \mathcal{H}$ be the restriction of the product projection. At a general point of $z \in \mathcal{F}$ the differential map $dq : \mathcal{T}_{\mathcal{F}|z} \to \mathcal{T}_{X|q(z)}$ is onto. Since we have an exact sequence

$$0 \to \mathcal{T}_{p^{-1}(p(z))} \to \mathcal{T}_{\mathcal{F}|p^{-1}(p(z))} \to p^*\mathcal{T}_{\mathcal{H}|p(z)} \to 0,$$

we conclude that the composition of $dq : p^*\mathcal{T}_{\mathcal{H}|p(z)} \to \mathcal{N}_{q(p^{-1}(p(z)))/X}$ with the restriction $\mathcal{N}_{q(p^{-1}(p(z)))/X} \to \mathcal{N}_{q(p^{-1}(p(z)))/X|q(z)}$ is onto. Note that $p^*\mathcal{T}_{\mathcal{H}|p(z)}$ is trivial. □

Remark 4.1.12. On $\mathbb{P}^1$ every vector bundle is a direct sum of line bundles. From this it follows that a vector bundle on $\mathbb{P}^1$ which is spanned at one point by global sections is spanned at all points. This fact is often used in conjunction with the above lemma.

We will need the following result.

Lemma 4.1.13. (Rigidity lemma) *Let $p : X \to Y$ be a proper holomorphic map with connected fibers from a normal irreducible variety X onto a normal irreducible variety Y. Let f be a holomorphic map from X to an analytic space, Z, such that $f(F)$ is a point for some fiber F of p. Then there exists an open set U of Y with $F \subseteq p^{-1}(U)$ and a holomorphic map $h : U \to Z$ such that $f_{p^{-1}(U)} = h \circ p_{p^{-1}(U)}$.*

If p has equidimensional fibers, e.g., p is flat, and f is proper, then U can be taken to be Y.

Proof. Choose a neighborhood V of $z := f(F)$ and an embedding $g : V \to \mathbb{C}^N$ with $g(z) = 0$. Let $w_1, \ldots, w_N$ be the coordinates in $\mathbb{C}^N$. Then the composition $w_i \circ g \circ f$ is a holomorphic function on some neighborhood $\mathcal{U}$ of F, for $i = 1, \ldots, N$.

Since $p_*\mathcal{O}_X \approx \mathcal{O}_Y$ by normality of X, Y and the connectedness of the fibers of p, it follows that there is an open set U of Y containing $p(F)$ and holomorphic functions $y_1, \ldots, y_N$ on U such that $y_i \circ p = w_i \circ g \circ f$, $i = 1, \ldots, N$. Letting $h : U \to V \subseteq \mathbb{C}^N$ be the map given by $(y_1, \ldots, y_N)$ we have proved the first part of the lemma.

Now assume that p has equal dimensional fibers and f is proper. Let $\psi := (p, f) : X \to Y \times Z$. Let $\nu : W \to \psi(X)$ be the normalization map and let $\phi : X \to W$ be the map induced by the surjection $\psi : X \to \psi(X)$. Note that p factors as $p = q \circ \phi$, where $q : W \to Y$ is the composition of ν and the product projection $Y \times Z \to Y$. If we show that q is an isomorphism then we are done. Assume that q is not an isomorphism. Then by the Zariski main theorem there is a point $y \in Y$ such that $q^{-1}(y)$ is positive dimensional. By the first part of the lemma, q is birational. Let $n := \dim X$. Thus for all $w \in W$, one has $\dim \phi^{-1}(w) \geq n - \dim W = n - \dim Y$. Then we find the absurdity $n - \dim Y = \dim p^{-1}(y) \geq \dim q^{-1}(y) + n - \dim Y$. □

Remark 4.1.14. In the above lemma, we can replace the condition that the fibers of p have the same dimension with a "maximality condition" on p. For example if X, Y, Z are projective varieties and p is a fiber type Mori contraction (see §4.2), we have $f = h \circ p$.

4.2 Extremal rays and the cone theorem

Let X be a connected normal projective scheme of dimension n. First, let us recall some definitions.

$Z_{n-1}(X)$, the group of Weil divisors, i.e., the free Abelian group generated by irreducible and reduced divisors on X (cf., §1.1);

$N_1(X) = (\{1\text{-cycles}\}/\sim) \otimes \mathbb{R}$;

$NE(X)$, the convex cone in $N_1(X)$ generated by the effective 1-cycles;

$\overline{NE}(X)$, the closure of $NE(X)$ in $N_1(X)$ with respect to the Euclidean topology;

$\rho(X) = \dim_{\mathbb{R}} N_1(X)$, the *Picard number* of X;

$\overline{NE}_D(X) = \{Z \in \overline{NE}(X) \mid Z \cdot D \geq 0\}, D \in \mathrm{Pic}(X) \otimes \mathbb{Q}$.

If γ is a 1-dimensional cycle in X we denote by $\mathbb{R}_+[\gamma]$ or $[\gamma]$ its class in $\overline{NE}(X)$. Note that the vector spaces $N_{n-1}(X) = (\mathrm{Div}(X)/\sim) \otimes \mathbb{R}$ and $N_1(X)$ are dual each other via the usual intersection of cycles "$\cdot$".

The following result of Kleiman is important.

Theorem 4.2.1. (Kleiman's ampleness criterion [349]) *Let X be a nonsingular projective variety. A divisor $D \in \mathrm{Div}(X)$ is ample if and only if $D \cdot Z > 0$ for any $Z \in \overline{NE}(X) - \{0\}$.*

4.2.2. Assume that X has at most canonical singularities. Following Mori [447] we say that a half line $R = \mathbb{R}_+[Z]$, where $\mathbb{R}_+ = \{x \in \mathbb{R}, x \geq 0\}$, in $\overline{NE}(X)$ is an *extremal ray* if $K_X \cdot Z < 0$ and $Z_1, Z_2 \in R$ for every $Z_1, Z_2 \in \overline{NE}(X)$ such that $Z_1 + Z_2 \in R$.

An extremal ray $R = \mathbb{R}_+[Z]$ is *nef* if $D \cdot Z \geq 0$ for every Cartier divisor D on X. An extremal ray which is not nef is said to be *nonnef*.

Let $H \in \mathrm{Pic}(X) \otimes \mathbb{Q}$ be a nef $\mathbb{Q}$-Cartier divisor. Let

$$F_H := H^{\perp} \cap (\overline{NE}(X) - \{0\})$$

where "$\perp$" means the orthogonal complement of H in $N_1(X)$. Then F_H is called an *extremal face* of $\overline{NE}(X)$ and H is the *supporting function* of F_H if F_H is entirely contained in the set $\{Z \in N_1(X), K_X \cdot Z < 0\}$ (see also Kawamata [343]). An extremal ray is a 1-dimensional extremal face. Indeed, for any extremal ray R there exists a nef $H \in \mathrm{Pic}(X) \otimes \mathbb{Q}$ such that $R = H^{\perp} \cap (\overline{NE}(X) - \{0\})$ (see, e.g., Mori [448, (3.1)]).

Let us recall the following standard facts.

Lemma 4.2.3. *Let X be a projective variety with at most canonical singularities. Let $H \in \mathrm{Pic}(X) \otimes \mathbb{Q}$ be a nef $\mathbb{Q}$-Cartier divisor and $F_H = H^{\perp} \cap (\overline{NE}(X) - \{0\})$ be an extremal face. Then*

1. *$mH - K_X$ is ample for $m \gg 0$;*

2. *Assume that F_H is 1-dimensional, i.e., $F_H = R$, R an extremal ray on X. Let $D \in \mathrm{Pic}(X)$ be a Cartier divisor such that $D \cdot R > 0$. Then $mH + D$ is ample for $m \gg 0$.*

Proof. Let $B = \{Z \in N_1(X) \mid \|Z\| = 1\} \cap \overline{NE}(X)$ and $\mathcal{U}_m = \{Z \in B \mid (mH - K_X) \cdot Z > 0\}$. We have $B = \cup_m \mathcal{U}_m$. Indeed the inclusion " $\supset$ " is clear and to show the converse it is enough to note that $-K_X \cdot Z > 0$ if $H \cdot Z = 0$. Since B is a compact set in $\mathbb{R}^\rho$, $\rho = \rho(X)$, one has that B is a finite union, $B = \cup_i \mathcal{U}_{m_i}$. Let $m' = \max_i\{m_i\}$. Then $(m'H - K_X) \cdot Z > 0$ for every $Z \in B$ and hence $m'H - K_X$ is ample by (4.2.1).

To prove 2), use the same argument with D in the place of $-K_X$. □

Lemma 4.2.4. *Let X be a projective variety with at most canonical singularities of dimension n. Let $R = H^\perp \cap (\overline{NE}(X) - \{0\})$ be an extremal ray. Then R is nonnef if and only if $H^n > 0$.*

Proof. It is simply an extension of the proof given by Mori [448, (3.3)] in the case $n = 3$. If R is not numerically effective, then there exists an irreducible and reduced divisor D such that $D \cdot R < 0$. Then $aH - D$ is ample for $a \gg 0$ by (4.2.3). Thus $h^0(maH - mD)$ is a polynomial of degree n in m if $m \gg 0$. Since by (4.2.3.1)

$$h^0(maH) = \chi(maH) = \frac{a^n m^n H^n}{n!} + \text{(lower degree terms)}$$

and $h^0(maH) \geq h^0(maH - mD)$ we conclude that $H^n > 0$.

Conversely if $H^n > 0$, then $h^0(mH) = \chi(mH)$ is a polynomial of degree n in m by the Riemann-Roch formula. This means that H has Kodaira dimension n (see Iitaka [329]). Let L be an ample divisor on X. Since H has Kodaira dimension n, there is a positive number a and an effective divisor D such that $aH \sim L + D$ by Lemma (2.5.5). Then $D \cdot R < 0$. □

The following theorem is essentially due to Mori, which proved it in the non-singular case. The extension to the singular case is due to Kawamata and Kollár. For the proof of the theorem we refer to [447, 343, 356, 164]. Note that the proof of the cone theorem given in [164] is new and it is based on the following logical order of steps: basepoint free theorem $\Rightarrow$ rationality theorem $\Rightarrow$ cone theorem. The original proof given by Mori in [447] is different and it is based on a reduction to characteristic p argument.

Theorem 4.2.5. (Cone theorem (Mori, Kawamata, Kollár)) *Let X be a projective variety of dimension n with at most canonical singularities. Then there exists* *a countable set of curves $C_i, i \in I$, with $K_X \cdot C_i < 0$, such that one has the decomposition*

$$\overline{NE}(X) = \sum_{i \in I} \mathbb{R}_+[C_i] + \overline{NE}_{K_X}(X).$$

The decomposition has the properties:

i) *the set of the curves C_i is minimal, no smaller set is sufficient to generate the cone;*

ii) *given any $\mathbb{R}_+$ -invariant neighborhood U of $\overline{NE}_{K_X}(X)$, only finitely many $\mathbb{R}_+[C_i]$ do not belong to U.*

The $\mathbb{R}_+[C_i]$ which, together with $\overline{NE}_{K_X}(X)$, form a minimal generating set for $\overline{NE}(X)$ are the extremal rays of X. If X is nonsingular, then the curves C_i's are possibly singular reduced irreducible rational curves which satisfy the condition $1 \leq -K_X \cdot C_i \leq n+1$. Such curves are called extremal rational curves.

Remark 4.2.6. Since $\overline{NE}(X) - U$ is a closed convex cone, the property ii) above can be rephrased as follows. For any real number $\varepsilon > 0$ and an ample line bundle L, any cycle Z, such that $(K_X + \varepsilon L) \cdot Z \leq 0$, can be written in the form

$$Z = \sum_{i \in I} \lambda_i C_i + \gamma_\varepsilon$$

where I is a finite set and γ_ε is a 1-cycle depending on ε and satisfying $(K_X + \varepsilon L) \cdot \gamma_\varepsilon = 0$.

Let us illustrate the above with some examples.

Example 4.2.7. (Existence of countably many extremal rays) A well-known example due to Nagata (see Hartshorne [318, p. 409] and also Franchetta [235]) shows that there exists a nonsingular surface S with countably many exceptional curves ℓ of the first kind, i.e., $\ell \cong \mathbb{P}^1$, $\ell^2 = \ell \cdot K_S = -1$. The surface S is obtained by blowing up $\mathbb{P}^2$ in 9 points in general position. Indeed any such a curve ℓ is an extremal ray (see Mori [447]).

Example 4.2.8. (See Mori [452]) Let $\mathbb{F}_r$, $r \geq 0$, be the r-th Hirzebruch surface as in §3.2. Let $p : \mathbb{F}_r \to \mathbb{P}^1$ be the bundle projection. We denote by E, f a basis for the second integral cohomology of $\mathbb{F}_r$, f a fiber of p. Here $\rho(X) = 2$ and $[E]$, $[f]$ span $NE(X)$. Thus $NE(X)$ is closed and $NE(X) = \mathbb{R}_+[f] + \mathbb{R}_+[E]$.

Example 4.2.9. ($NE(X)$ may be not closed) Mumford gave an example of a complete nonsingular surface, S, containing a divisor D such that $D \cdot C > 0$ for every curve C of S and $D \cdot D = 0$ (see Hartshorne [315, p. 56]). In this case $NE(S) \neq \overline{NE}(S)$. Indeed, otherwise, we would have $D \cdot Z > 0$ for every

$Z \in \overline{NE}(S) - \{0\}$ and hence D would be ample by the Kleiman ampleness criterion (4.2.1). Here $S = \mathbb{P}(\mathscr{E})$ where $\mathscr{E}$ is a "generic" stable vector bundle of rank 2 over a nonsingular projective curve of genus $g \geq 2$.

By using the previous example, C.P. Ramanujam constructed a complete nonsingular threefold X containing an effective divisor D' such that $D' \cdot C > 0$ for any curve C of X and $D'^3 > 0$ but not ample (see Hartshorne [315, p. 57]). Hence, again by (4.2.1), $NE(X) \neq \overline{NE}(X)$. Here X denotes $\mathbb{P}(\mathcal{O}_S(D - H) \oplus \mathcal{O}_S)$ where S, D are as in Mumford's example and H is an ample effective divisor on S.

See also our paper [113, Ex. 3.14] for a new example of a quadric fibration X with $NE(X)$ not closed.

Example 4.2.10. ($\overline{NE}(X)$ can be polyhedral) If X is a Fano variety, i.e., $-K_X$ is ample, then $\overline{NE}(X)$ is a rational polyhedral cone. Indeed, since $-K_X$ is ample, $\overline{NE}_{K_X}(X) = 0$ and there exists a finite number of rational curves $\ell_1, \dots, \ell_t$ such that $(-K_X \cdot \ell_i) \leq n+1$ for every $i = 1, \dots, t$, $NE(X)$ is closed and $NE(X) = \mathbb{R}_+[\ell_1] + \cdots + \mathbb{R}_+[\ell_t]$ (see Mori [448, (1.2)]). E.g., a cubic surface X in $\mathbb{P}^3$ is a Fano 2-fold with $\mathcal{O}_X(-K_X) \cong \mathcal{O}_X(1)$. It is a classical result that $\rho(X) = 27$ and X contains 27 lines $\ell_1, \dots, \ell_{27}$ (see Nagata [468]). In this case $NE(X) = \mathbb{R}_+[\ell_1] + \cdots + \mathbb{R}_+[\ell_{27}]$.

In the case of a nonsingular surface, the extremal rays can be easily and completely described.

Theorem 4.2.11. *Let X be a nonsingular projective surface and let C be a curve on X such that $R = \mathbb{R}_+[C]$ is an extremal ray. Then either*

1. *C is an exceptional curve of the first kind (i.e., $C \cong \mathbb{P}^1, C^2 = K_X \cdot C = -1$) and R is nonnef;*
2. *X is a $\mathbb{P}^1$-bundle over a nonsingular curve B, $p : X \to B$, $C \cong \mathbb{P}^1$ is a fiber of p and R is nef; or*
3. *$(X, \mathcal{O}_X(C)) \cong (\mathbb{P}^2, \mathcal{O}_{\mathbb{P}^2}(1))$ and R is nef.*

Remark 4.2.12. (Existence of minimal models for surfaces) Let us recall the following classical result (Castelnuovo, Enriques, Zariski). *Let X be a projective nonsingular surface such that X is not ruled. Then there exists a nonsingular model X_0, such that the canonical divisor K_{X_0} is numerically effective (hence in particular X_0 is a relatively minimal model). Moreover X_0 is unique up to isomorphisms and it is the minimal model of X.*

By using Theorem (4.2.11) we can rephrase the classical definition above of minimal surface as follows. We say that a surface X is *minimal* if either K_X is nef or X belongs to one of the classes 2), 3) of (4.2.11) (compare with (4.3.3)).

Let $H \in \mathrm{Pic}(X) \otimes \mathbb{Q}$ be a nef $\mathbb{Q}$-Cartier divisor and let $F_H = H^{\perp} \cap (\overline{NE}(X) - \{0\})$ be an extremal face. Then $aH - K_X$ is ample for $a \gg 0$ by Lemma (4.2.3).

Therefore the linear system $|mH|$ is basepoint free by Theorem (1.5.1), for $m \gg 0$, so that it defines a morphism, say $\sigma : X \to W$. Let $\varphi : X \to Y$ be the morphism with connected fibers and normal image obtained by the Remmert-Stein factorization of σ. If C is an irreducible curve on X, then $[C] \in F_H$ if and only if $H \cdot C = 0$ and if and only if $\dim \varphi(C) = 0$, i.e., φ contracts the extremal face F_H. Note that $\varphi_* \mathcal{O}_X \cong \mathcal{O}_Y$, the pair (Y, φ) is unique up to isomorphism and $H \in \varphi^*$ $\mathrm{Pic}(Y)$. We will call such a contraction, φ, as *contraction of* F_H. If $F_H = R$, R an extremal ray, we will denote $\varphi = \mathrm{cont}_R : X \to Y$ the contraction of the extremal ray R.

The following lemma expresses the contractions of extremal faces in terms of nefvalues morphisms and vice versa. It is an easy consequence of the results and definitions above.

Lemma 4.2.13. *Let X be an irreducible normal projective variety with at most canonical singularities. Let e be the index of X.*

1. *Let L be an ample line bundle on X, let $\tau := \tau(X, L)$ be the nefvalue and let ϕ be the nefvalue morphism of (X, L). Then ϕ is the contraction of an extremal face of $\overline{NE}(X)$ whose supporting function is $H := e(K_X + \tau L)$.*

2. *Let $F := H^\perp \cap (\overline{NE}(X) - \{0\})$ be an extremal face of $\overline{NE}(X)$, with $H \in$ $\mathrm{Pic}(X) \otimes \mathbb{Q}$ the supporting function of F. Let φ be the contraction of F. Assume that φ is not an isomorphism. Then there exists an ample line bundle $\mathcal{L}$ on X such that φ coincides with the nefvalue morphism of $(X, \mathcal{L})$ and the nefvalue, $\tau(X, \mathcal{L})$, of $(X, \mathcal{L})$ is $\tau(X, \mathcal{L}) = 1/e$.*

Proof. To prove 1), recall that the nefvalue morphism ϕ of (X, L) is associated to $|m(K_X + \tau L)|$ for some integer m such that $m\tau$ is integral and $e|m$. Then $H := e(K_X + \tau L)$ is nef and clearly $F := H^\perp \cap (\overline{NE}(X) - \{0\})$ is contained in $\{Z \in N_1(X), K_X \cdot Z < 0\}$. Therefore F is an extremal face and ϕ coincides with the contraction of F.

To show 2), note that $\mathcal{L} := eaH - eK_X$ is an ample line bundle by Lemma (4.2.3) for some integer $a > 0$. Since φ is given by $|mH|$ for $m \gg 0$ we have that $eaH = eK_X + \mathcal{L}$ is nef and not ample. Therefore by Lemma (1.5.5), $\tau(X, \mathcal{L}) = 1/e$. Moreover φ coincides with the nefvalue morphism of $(X, \mathcal{L})$. □

Let $\varphi = \mathrm{cont}_R : X \to Y$ be a contraction of an extremal ray R. Let

$$E := \{x \in X \mid \varphi \text{ is not isomorphism at } x\}.$$

Note that E is the locus of curves whose numerical classes are in R. We will refer to E simply as the *locus* of R. If $a = \dim E$, $b = \dim \varphi(E)$ we will say that R is an *extremal ray of type* (a, b). Clearly $n \geq a > b \geq 0$. In particular φ is birational if and only if $a \leq n - 1$. If $a = n - 1$, R is called *divisorial*. If $a \leq n - 2$, the contraction φ is also called a *small contraction*. If $a = n$, then $\dim Y < \dim X$ and φ is said to be *a contraction of fiber type*.

If X is smooth we define the *length of an extremal ray*,

$$\operatorname{length}(R) = \min\{-K_X \cdot C \mid C \text{ rational curve, } [C] \in R\}.$$

Note that the cone theorem yields the bound $0 < \operatorname{length}(R) \leq n+1$ (see Proposition (6.3.12) and Remark (6.3.13) for a description of the cases when $\operatorname{length}(R) = n+1,\ n$). We will also use the notation $\operatorname{length}(R) = \ell(R)$.

The following easy consequence of the cone theorem is useful.

Lemma 4.2.14. *Let V be a projective variety with at most canonical singularities. Let L be an ample line bundle on V and let t be some positive rational number such that K_V+tL is nef. Let C be an effective curve in $NE(V)$ such that $(K_V+tL)\cdot C = 0$. Then C can be written in $NE(V)$ as finite sum $C = \sum_i \lambda_i C_i$ where $\lambda_i \in \mathbb{R}_+$, and $\mathbb{R}_+[C_i]$ are extremal rays such that $(K_V + tL)\cdot C_i = 0$ for all i. In particular if V is nonsingular the curves, C_i, can be chosen as extremal rational curves.*

Proof. By the cone theorem we can write C in $NE(V)$ as a finite sum $C = \sum_i \lambda_i C_i + \gamma$ where $\lambda_i \in \mathbb{R}_+$, $\mathbb{R}_+[C_i]$ are extremal rays and γ satisfies the condition $K_V \cdot \gamma \geq 0$. We have

$$0 = (K_V + tL)\cdot C = \sum_i \lambda_i (K_V + tL)\cdot C_i + (K_V + tL)\cdot \gamma.$$

Since $K_V + tL$ is nef, $K_V \cdot \gamma \geq 0$ and L is ample we get a contradiction unless $\gamma = 0$ and $(K_V + tL)\cdot C_i = 0$ for all indices i. □

Remark 4.2.15. Let V be an n-dimensional smooth projective variety and let $R = \mathbb{R}_+[C]$ be an extremal ray on V, with C an extremal rational curve. Let L be an ample line bundle on V and let $t \in \mathbb{Q}$ be such that $(K_V + tL)\cdot C = 0$. Thus we can choose an extremal rational curve, ℓ, with $R = \mathbb{R}_+[\ell]$, $(K_V + tL)\cdot \ell = 0$ and $-K_V \cdot \ell = \operatorname{length}(R)$. Indeed, if $L \cdot C = 1$, then $\ell = C$. Note that in any case there exists a rational curve, ℓ, $[\ell] \in R$, with $\operatorname{length}(R) = -K_V \cdot \ell$. Then we have only to show that ℓ is extremal, i.e., $1 \leq -K_V \cdot \ell \leq n+1$. The inequality $1 \leq -K_V \cdot \ell$ is obvious since $[\ell] \in R$. If $-K_V \cdot \ell > n+1$ we would have $-K_V \cdot \ell > -K_V \cdot C$ since $-K_V \cdot C \leq n+1$. This contradicts the minimality of ℓ.

Let us also recall the following lemma of Fujita [252, (2.5)], which is weaker than (6.3.5), but works in the singular case too. Note that rational Gorenstein singularities imply canonical singularities, so Mori's theory applies on the variety X as in the two statements below.

Lemma 4.2.16. (Fujita) *Let X be an n-dimensional projective variety with only rational normal Gorenstein singularities. Let R be a nonnef extremal ray on X*

and let $\varphi : X \to Y$ be the contraction (in particular birational) of R. Suppose that $d = \dim \varphi^{-1}(y) > 0$ for some point y on Y. Then $(K_X + dA) \cdot R \geq 0$ for any ample line bundle A on X.

By using the results above we can prove the following general facts. The first is a consequence of (4.2.16).

Lemma 4.2.17. *Let X be an n-dimensional projective variety with only rational normal Gorenstein singularities. Let L be an ample line bundle on X. Let R_1, R_2 be two distinct extremal rays on X and let E_1, E_2 be the loci of R_1, R_2 respectively. Assume that $(K_X + tL) \cdot R_i = 0$ for some rational number t and that R_i is nonnef, $i = 1,\ 2$. Let $\lceil t \rceil$ be the smallest integer $\geq t$. If $\lceil t \rceil \geq (n+1)/2$ and if $E_1 \cap E_2 \subset \operatorname{reg}(X)$, then E_1 and E_2 are disjoint.*

Proof. Assume $E_1 \cap E_2 \neq \emptyset$ and take a point $x \in E_1 \cap E_2$. Let ρ_i be the contractions associated to R_i, $i = 1, 2$. Let F_i be a fiber of the restriction $\rho_i : E_i \to \rho_i(E_i)$ passing through x and let $d_i = \dim F_i$, $i = 1, 2$. From (4.2.16) we know that $(K_X + d_i L) \cdot R_i \geq 0$. Since $(K_X + tL) \cdot R_i = 0$ it thus follows that $d_i \geq t$, i.e., $d_i \geq \lceil t \rceil$, $i = 1, 2$. Since $\emptyset \neq F_1 \cap F_2 \subset \operatorname{reg}(X)$ and $\lceil t \rceil \geq (n+1)/2$ we have

$$\dim(F_1 \cap F_2) \geq 2\lceil t \rceil - n > 0.$$

Then there exists a curve, C, contained in $F_1 \cap F_2$ which contracts to a point under ρ_1, ρ_2. Therefore $[C] \in R_1$, $[C] \in R_2$, this leading to the contradiction $R_1 = R_2$. □

Proposition 4.2.18. *Let X be an n-dimensional projective variety with only rational normal Gorenstein $\mathbb{Q}$-factorial singularities. Let R be an extremal ray on X and let $\varphi = \operatorname{cont}_R : X \to Y$ be the contraction of R. Let $\mathcal{L}$ be a line bundle on X. Assume that $\mathcal{L}$ is φ-ample, $K_X + (n-1)\mathcal{L}$ is nef and big, and that $K_X + (n-1)\mathcal{L} \cong \varphi^* H$ for some line bundle H on Y. Let E be the locus of R. Then $(E, \mathcal{L}_E) \cong (\mathbb{P}^{n-1}, \mathcal{O}_{\mathbb{P}^{n-1}}(1))$.*

Proof. Note that φ is birational since $K_X + (n-1)\mathcal{L}$ is nef and big. Assume that $d := \dim \varphi^{-1}(y) \leq n-2$ for some $y \in Y$. Then by (4.2.16), $(K_X + (n-2)\mathcal{L}) \cdot R \geq (K_X + d\mathcal{L}) \cdot R \geq 0$, this contradicting $(K_X + (n-1)\mathcal{L}) \cdot R = 0$. Therefore the positive dimensional fibers of φ are divisorial, i.e., R is of divisorial type. Take an irreducible component, D, of a divisorial fiber $\varphi^{-1}(y)$, with its reduced structure. Then $D \cdot R < 0$. Hence $D \cdot C < 0$ for any curve C in X, $[C] \in R$, so that D contains all curves on X such that $[C] \in R$. It thus follows that $\varphi^{-1}(y)$ is in fact an irreducible divisor and coincides with E. Since X is Gorenstein we have the exact sequence

$$0 \to K_X + t\mathcal{L} - E \to K_X + t\mathcal{L} \to (K_X + t\mathcal{L})_E \to 0,$$

where $-E := (\mathcal{J}_E)^{**}$ and $\mathcal{J}_E$ is the ideal sheaf of E in X. Note that, since E is a prime divisor and X is normal, it is easy to see that $-E = \mathcal{J}_E$. Since $-E$, $\mathcal{L}$ are φ-ample, the relative Kodaira vanishing theorem (see (2.2.1)) gives $\varphi_{(i)}(K_X + t\mathcal{L}) = 0$ for $i > 0$, $t > 0$ as well as $\varphi_{(i)}(K_X - E + t\mathcal{L}) = 0$ for $i > 0$, $t \geq 0$, where $\varphi_{(i)}(\mathcal{F}) = R^i\varphi_*\mathcal{F}$ is the i-th direct image of a sheaf $\mathcal{F}$. Therefore we have

$$\varphi_{(i)}((K_X + t\mathcal{L})_E) \cong H^i((K_X + t\mathcal{L})_E) = 0$$

for $i > 0, t > 0$. Let $q(t) := \chi((K_X + t\mathcal{L})_E)$. Thus $q(t) = h^0((K_X + t\mathcal{L})_E)$ for $t \geq 1$. Since $(K_X + (n-1)\mathcal{L})_E \approx \mathcal{O}_E$ we have that $q(t) = 0$ for $t = 1, \dots, n-2$, and $q(n-1) = 1$.

We claim that $q(0) = \chi(K_{X|E}) = 0$. To see this consider the exact sequence

$$0 \to K_X - E \to K_X \to K_{X|E} \to 0.$$

Note that $h^0(K_{X|E}) = h^0(-(n-1)\mathcal{L}_E) = 0$. Furthermore $\varphi_{(i)}(K_X - E) = 0$ for $i > 0$ by (2.2.1) since $-E$ is φ-ample and $\varphi_{(i)}(K_X) = 0$ for $i > 0$ by (2.2.2). Therefore $\varphi_{(i)}(K_{X|E}) \cong H^i(K_{X|E}) = 0$ for $i > 0$, so $q(0) = \chi(K_{X|E}) = 0$.

Thus we conclude that $q(t) = 0$ for $t = 0, \dots, n-2$. Since $q(n-1) = 1$ we see that

$$q(t) = t(t-1)\cdots(t-(n-2))/(n-1)!.$$

Then the coefficient $\mathcal{L}_E^{n-1}/(n-1)!$ of the leading term is $1/(n-1)!$, that is $\mathcal{L}_E^{n-1} = 1$. Consider now the exact sequence

$$0 \to t\mathcal{L} - E \to t\mathcal{L} \to t\mathcal{L}_E \to 0.$$

Recalling that $(K_X + (n-1)\mathcal{L})_E \approx \mathcal{O}_E$ and $\mathcal{L}$, E are φ-ample, we see that $t\mathcal{L} - K_X$ is φ-ample for $t > 1-n$ as well as $t\mathcal{L} - K_X - E$ is φ-ample for $t \geq 1-n$. Therefore $\varphi_{(i)}(t\mathcal{L}) = \varphi_{(i)}(t\mathcal{L} - E) = 0$ for $i > 0$, $t > 1-n$ and hence $\varphi_{(i)}(t\mathcal{L}_E) \cong H^i(t\mathcal{L}_E) = 0$ for $i > 0$, $t > 1-n$. Let $p(t) := \chi(t\mathcal{L}_E)$. Thus $p(t) = h^0(t\mathcal{L}_E)$ for $t > 1-n$, so that $p(t) = 0$ for $t = -1, \dots, -(n-2)$ and $p(0) = h^0(\mathcal{O}_E) = 1$. Let $-c$ be the remaining zero of $p(t)$. By the above we have

$$p(t) = \frac{(t+1)\cdots(t+n-2)(t+c)}{(n-2)!c}.$$

Then $\mathcal{L}_E^{n-1}/(n-1)! = 1/(c(n-2)!)$. Recalling that $\mathcal{L}_E^{n-1} = 1$ we find $c = n-1$ and

$$p(t) = (t+1)\cdots(t+n-1)/(n-1)!.$$

Hence in particular $h^0(\mathcal{L}_E) = p(1) = n$. It follows that $\Delta(E, \mathcal{L}_E) = n - 1 + \mathcal{L}_E^{n-1} - h^0(\mathcal{L}_E) = 0$. Thus by (3.1.2) we conclude that $(E, \mathcal{L}_E) \cong (\mathbb{P}^{n-1}, \mathcal{O}_{\mathbb{P}^{n-1}}(1))$. □

4.3 Varieties with nonnef canonical bundle

Let X be a connected projective variety of dimension $n \geq 3$ with at most terminal singularities, $\mathbb{Q}$-factorial and with K_X nonnef. Then in particular $\overline{NE}(X) \neq \overline{NE}_{K_X}(X)$, so by the cone theorem there exists an extremal ray, R. In the following $R = H^{\perp} \cap (\overline{NE}(X) - \{0\})$ is an extremal ray of type (a, b), $H \in \text{Pic}(X) \otimes \mathbb{Q}$ is nef and E is the locus of R, i.e., the locus of X where the contraction $\varphi = \text{cont}_R$ is not an isomorphism.

Theorem 4.3.1. (Contraction theorem) *Let X be a connected projective variety of dimension $n \geq 3$ with at most terminal, $\mathbb{Q}$-factorial singularities and such that K_X is nonnef. Let $R = R_+[\ell]$, ℓ extremal rational curve, an extremal ray on X and let E be the locus of R. Then*

1. *There exist a normal projective variety Y and a morphism $\varphi = \text{cont}_R : X \to Y$ with connected fibers such that the sequence $0 \to \text{Pic}(Y) \xrightarrow{\varphi^*} \text{Pic}(X) \xrightarrow{\cdot \ell} \mathbb{Z}$ is exact and $\rho(X) = \rho(Y) + 1$. Moreover $-K_X$ is φ-ample.*

2. *If R is nonnef, then $a < n$ and φ is a birational morphism. In the case $a = n - 1$, one has*

   (a) *E is an irreducible and reduced divisor such that $E \cdot R < 0$, φ is an isomorphism on $X - E$, $\dim \varphi(E) \leq n - 2$;*

   (b) *Y is $\mathbb{Q}$-factorial with terminal singularities. Moreover if X is factorial, then Y is $(-E \cdot R)$-factorial.*

3. *If R is nef, then $a = n$, the general fiber of φ is a $\mathbb{Q}$-Fano variety with terminal singularities and Y is $\mathbb{Q}$-factorial.*

Proof. We give here the proof in the nonsingular case. For the general case we refer to Kollár [358]. Since $\varphi_* \mathcal{O}_X \cong \mathcal{O}_Y$ we have $\varphi_* \varphi^* = \text{id}_X$ and hence φ^* is injective. To show the exactness in $\text{Pic}(X)$, let $D \in \text{Pic}(X)$, $D \cdot \ell = 0$. For $\lambda \gg 0$, $D + \lambda H$ is a Cartier divisor which satisfies the conditions that $D + \lambda H$ is nef, and that $D + \lambda H - K_X$ is ample. To show the first condition note that H is nef and if $H \cdot Z = 0$ for some $Z \in \overline{NE}(X)$ then necessarily $Z \in R$ so that $D \cdot Z = 0$. The second condition follows from Lemma (4.2.3).

Thus Theorem (1.5.1) applies to say that the linear system $|m(D + \lambda H)|$ is basepoint free for $m \gg 0$. Note that for any curve C in X, $(D + \lambda H) \cdot C = 0$ if $H \cdot C = 0$. Therefore the Remmert-Stein factorization of the morphism defined by $|m(D + \lambda H)|$ coincides, up to isomorphisms, with the contraction $\varphi = \text{cont}_R$ associated to R. In fact, by looking over the proof of Theorem (3-1-1) in Kawamata, Matsuda, and Matsuki [346] one sees that $\text{Bs}|p^t(D + \lambda H)| = \text{Bs}|q^s(D + \lambda H)| = \emptyset$ for coprime integers p, q and positive integers s, t. Then $p^t(D +$

$\lambda H), q^s(D + \lambda H) \in \varphi^*\mathrm{Pic}(Y)$ and hence in particular $D + \lambda H \in \varphi^*\mathrm{Pic}(Y)$. It follows that $D \in \varphi^*\mathrm{Pic}(Y)$ and thus $\mathrm{Ker}(\cdot\ell) \subset \varphi^*\mathrm{Pic}(Y)$. The converse is clear since for any $M \in \mathrm{Pic}(Y)$, $\varphi^*M \cdot \ell = M \cdot \varphi_*\ell = 0$.

The fact that $-K_X$ is φ-ample follows immediately from Lemma (4.2.3). Indeed, for $m \gg 0$, $mH - K_X$ is ample and $(mH - K_X)_F \sim -K_{X|F}$ for any fiber F of φ. This proves (4.3.1.1).

Suppose now R nonnef. By Lemmas (4.2.3), (4.2.4) it thus follows that, for $m \gg 0, \chi(mH) = h^0(mH)$ goes to the infinity as m^n. This means that the morphism with connected fibers φ is birational. Since R is nonnef there exists an irreducible and reduced divisor D such that $D \cdot R < 0$. Hence $D \cdot C < 0$ for any curve C in X, $[C] \in R$ and therefore D contains the curve C. If $a = n - 1$, this implies that D is unique and coincides with E. By Lemma (4.2.3) we know that $mH - E$ is ample for $m \gg 0$ and then φ is an isomorphism on $X - E$. Moreover the equality $\rho(X) = \rho(Y) + 1$ implies that $\dim \varphi(E) \leq n - 2$.

To show that Y has terminal singularities, let r be a positive integer such that $rK_Y \in \mathrm{Pic}(Y)$. Since φ is an isomorphism on $X - E$, we have for some integer a that $rK_X \approx \varphi^*(rK_Y) + aE$. The conditions $(K_X \cdot R) < 0$, $(E \cdot R) < 0$, $(\varphi^*K_Y \cdot R) = 0$ give $a > 0$.

To conclude the proof of (4.3.1.2), let D be any irreducible and reduced divisor on Y and let D' be the proper transform of D under φ. Put $-\alpha = (E \cdot \ell)$, $\beta = (D' \cdot \ell)$. Then $(\alpha D' + \beta E) \cdot R = 0$ so that, by (4.3.1.1), $\alpha D' + \beta E \approx \varphi^*\mathcal{F}$, $\mathcal{F} \in \mathrm{Pic}(Y)$. Therefore we have a section, s, of $\mathcal{F}$ over $Y - \varphi(E)$ which defines αD on $Y - \varphi(E)$. Since Y is normal and $\varphi(E)$ is of codimension ≥ 2, we conclude by Levi's extension theorem that s extends to a global section of $\mathcal{F}$. This means that αD is a Cartier divisor and hence Y is $\alpha = (-E \cdot R)$-factorial. This shows (4.3.1.2).

Finally, let us assume that R is nef. Then $H^n = 0$ and $h^i(mH) = 0$, $m \gg 0$, by (4.2.3), (4.2.4). Therefore the Riemann-Roch theorem yields

$$h^0(mH) = k(-K_X \cdot H^{n-1})m^{n-1} + (\text{terms of lower degree}),$$

k positive integer. This means that $\dim Y \leq n - 1$, i.e., $a = n$. Since $-K_X$ is φ-ample, the general fiber of φ is a nonsingular Fano variety. The proof of the $\mathbb{Q}$-factoriality is similar to the previous one (see Kawamata [343, (5.2)]). □

Remark 4.3.2. It is worthwhile to compare the theorem above with the corresponding Theorem (4.2.11) in the 2-dimensional case. The morphism $\varphi = \mathrm{cont}_R$ is, for $n = 2$, either a contraction of an exceptional curve of the first kind, a $\mathbb{P}^1$-bundle over a nonsingular curve, or $\varphi : X \to \mathrm{Spec}(\mathbb{C})$, respectively. The case (4.3.1.2) can be viewed as a generalization of (4.2.11.1) while the classes (4.2.11.2), (4.2.11.3) are particular cases of (4.3.1.3).

We refer to the surveys papers quoted at the beginning and especially to the book of Clemens, Kollár, and Mori [164] for any further background material and more details about Mori's theory of extremal rays and Mori's program on minimal models. Here we only recall that Mori [451] has solved the minimal model problem in the 3-dimensional case.

To summarize Mori's result, recall that a variety V of dimension n is called *ruled* if it is birational to $\mathbb{P}^1 \times Y$ for some variety Y of dimension $n-1$ and it is called *uniruled* if there exists an $(n-1)$-dimensional variety Y and a rational map $\varphi : Y \times \mathbb{P}^1 \to V$ which is generically surjective. This is equivalent to ruledness if $n \leq 2$, but not in higher dimensions.

Theorem 4.3.3. (Mori [451]) *Let X be a projective $\mathbb{Q}$-factorial variety with at most terminal singularities of dimension $n = 3$. Then, if X is not uniruled, there exists a minimal model Y, birational to X, with K_Y nef.*

We give here some comments on the contraction theorem, (4.3.1), which point out substantial differences between the case $n = 3$ and the general one. The notation are as in (4.3.1).

First consider the case when R is nonnef, i.e., $\varphi = \mathrm{cont}_R$ is a birational morphism.

4.3.4. Let X be a smooth projective variety of dimension $n = 3$. Assume that K_X is nonnef. Let R be a nonnef extremal ray and let E be the locus of R. Then E is completely described (see Mori [447, 448]). Precisely, the following cases are possible.

$E \cong \mathbb{P}^2$, $\mathcal{O}_E(-E) \cong \mathcal{O}_{\mathbb{P}^2}(m)$, $m = 1, 2$;

$E \cong \mathbb{P}^1 \times \mathbb{P}^1$, $\mathcal{O}_E(-E) \cong \mathcal{O}_E(1, 1)$;

$E \cong Q \subset \mathbb{P}^3$, Q quadric cone, $\mathcal{O}_E(-E) \cong \mathcal{O}_Q(1)$;

E is isomorphic to a $\mathbb{P}^1$-bundle over $\varphi(E)$, $\varphi(E)$ nonsingular curve, and $\mathcal{O}_f(E) \cong \mathcal{O}_{\mathbb{P}^1}(-1)$ for any fiber f of the restriction of φ to E.

Moreover φ is the blowing up of Y along $\varphi(E)$. Note that this is no longer true if $n \geq 4$ (see Ando [7]).

If X is singular there exist examples of extremal rays of type $(1, 0)$ (see Francia [236] and Reid [520]). To this purpose we have the following result due to Benveniste [115]: *if $n = 3$, X has at most canonical singularities and $R = \mathbb{R}_+[C]$ is an extremal ray of type* $(1, 0)$, *then $C \cong \mathbb{P}^1$ and $-1 < K_X \cdot C < 0$. In particular X is not Gorenstein since $K_X \cdot C \notin \mathbb{Z}$* (compare also with Example (2.2.10)).

If X is smooth of dimension $n \geq 3$, and $a = n - 1$, then E is uniruled (see Kawamata, Matsuda, and Matsuki [346]). More precise results are obtained for $n = 4, 5$ (see Ando [7] and Beltrametti [82]). E.g., if $n = 4$ and E contracts to a point then E is a Fano variety (i.e., K_E is a Cartier divisor and $-K_E$ is ample) possibly singular and not normal of degree $(-K_E)^3 \leq 72$ and E can be completely described (see Beltrametti [83] and Fujita [258]).

If R is of type $(n-1, n-2)$ with $n=3$, the restriction $\varphi_E : E \to \varphi(E)$ is a $\mathbb{P}^1$-bundle as above. This is no longer true if $n \geq 4$, as the following example shows.

Example 4.3.5. (See Beltrametti [82, (2.6)]) Let Y be an ordinary double point of a variety of dimension $n=4$, given in $\mathbb{A}^5$ by the equation $xy - zt + w^2 = 0$ and let $\varphi : X \to Y$ be the blowing up of the plane $x = y = w = 0$. Then X is nonsingular, E is a $\mathbb{P}^1$-bundle outside of the origin, O, but $\varphi^{-1}(O)$ is 2-dimensional.

If $n \geq 4$, even if X is nonsingular, we can find extremal rays of type (a, b) with $a < n-1$.

Example 4.3.6. (See Reid [520, (3.9)]) Let Y be the cone over the Segre embedding of $\mathbb{P}^1 \times \mathbb{P}^2$ in $\mathbb{P}^5$ and let $f : X' \to Y$ be a resolution of the singularities. The locus, E, where f is not an isomorphism is $E = \mathbb{P}^1 \times \mathbb{P}^2$. We have a commutative diagram

$$\begin{array}{ccccc} & & X' & & \\ & {\scriptstyle\sigma_1}\swarrow & & \searrow{\scriptstyle\sigma_2} & \\ \sigma_1(E) = \mathbb{P}^2 \subset X & & & & X_1 \supset \mathbb{P}^1 = \sigma_2(E) \\ & {\scriptstyle\varphi}\searrow & & \swarrow & \\ & & Y & & \end{array}$$

where σ_1 is the contraction of $\mathbb{P}^1 \subset E$ and σ_2 is the contraction of $\mathbb{P}^2 \subset E$. The varieties X, X_1 are nonsingular and birationally equivalent. Moreover the class $[\ell]$ in $\overline{NE}(X)$ of any line ℓ in $\sigma_1(E) = \mathbb{P}^2$ is an extremal ray $R = \mathbb{R}_+[\ell]$ of type $(2, 0)$ and $\varphi = \operatorname{cont}_R$.

Let us show that $K_X \cdot \ell < 0$. In fact we have $\mathcal{N}_{\mathbb{P}^2/E} \cong \mathcal{O}_{\mathbb{P}^2}$, $\mathcal{N}_{E/X'} \cong \mathcal{O}_E(-1)$ and an exact sequence

$$0 \to \mathcal{O}_{\mathbb{P}^2} \to \mathcal{N}_{\mathbb{P}^2/X'} \to \mathcal{O}_{\mathbb{P}^2}(-1) \to 0.$$

It thus follows that $\det \mathcal{N}_{\mathbb{P}^2/X'} \cong \mathcal{O}_{\mathbb{P}^2}(-1)$. The adjunction formula gives

$$\mathcal{O}_{\mathbb{P}^2}(K_{\mathbb{P}^2}) \cong \mathcal{O}_{\mathbb{P}^2}(K_{X'}) \otimes \det \mathcal{N}_{\mathbb{P}^2/X'}$$

and hence $\mathcal{O}_{\mathbb{P}^2}(K_{X'}) \cong \mathcal{O}_{\mathbb{P}^2}(-2)$. On the other hand

$$\mathcal{O}_{\mathbb{P}^2}(K_{X'}) \cong \sigma_1^*\mathcal{O}_X(K_X) \otimes \mathcal{O}_{\mathbb{P}^2}(E) \cong \sigma_1^*\mathcal{O}_X(K_X) \otimes \mathcal{O}_{\mathbb{P}^2}(-1).$$

So we conclude that $\sigma_1^*\mathcal{O}_X(K_X) \otimes \mathcal{O}_{\mathbb{P}^2} \cong \mathcal{O}_{\mathbb{P}^2}(-1)$. Thus, on $\mathbb{P}^2 = \sigma_1(E)$, $\mathcal{O}_X(K_X) \otimes \mathcal{O}_{\mathbb{P}^2} \cong \mathcal{O}_{\mathbb{P}^2}(-1)$ and then $K_X \cdot \ell < 0$.

Furthermore the singularity of Y at the vertex is not $\mathbb{Q}$-Gorenstein (see (4.3.7) below). Note that by the exceptional locus inequality (6.3.5) the example above is the "worst possible" for $n = 4$.

4.3.7. Notation as in contraction theorem, (4.3.1). If $a < n - 1$ the variety Y in (4.3.1.2) has rational singularities (see Kawamata [343, §5, (B)]) but is not $\mathbb{Q}$-Gorenstein. Otherwise we would have $K_X \approx \varphi^* K_Y$ (note that for any $D \in \operatorname{Pic}(Y) \otimes \mathbb{Q}$ such that $mD \in \operatorname{Pic}(Y)$ we define $\varphi^* D = \varphi^*(mD)/m \in \operatorname{Pic}(X) \otimes \mathbb{Q}$) which gives the contradiction $0 > K_X \cdot R = \varphi^* K_Y \cdot R = 0$.

Suppose now that R is nef, i.e., $\varphi = \operatorname{cont}_R$ is a fibration in $\mathbb{Q}$-Fano varieties as in (4.3.1.3).

If $n = 3$ and X is nonsingular, the morphism φ is flat and either $\dim Y = 2$ and φ is a conic bundle (see Beauville [76]), $\dim Y = 1$ and φ is a Del Pezzo fibering (i.e., every fiber is an irreducible reduced surface F with $-\omega_F$ ample) or $\dim Y = 0$ and X is a Fano variety. We also refer to Miyanishi [439] for a detailed analysis of nef extremal rays, especially in the case of conic bundles.

If $n \geq 4$, φ can be *not equidimensional* as the following example shows.

Example 4.3.8. (See Beltrametti [82, (3.6)]) Let L_1, L_2, L_3 be general hyperplanes in $\mathbb{P}^N$, $N \geq 3$, and let $\mathbb{P}^2_{(\lambda_1,\lambda_2,\lambda_3)}$ be the base of the net of hyperplanes $\sum_{i=1}^3 \lambda_i L_i$. Let $X \subset \mathbb{P}^2_{(\lambda_1,\lambda_2,\lambda_3)} \times \mathbb{P}^N$ be the incidence correspondence divisor, i.e., $X = \{(L, p), p \in L\}$. Then the projection $\varphi : X \to Y = \mathbb{P}^N$ is a contraction, is a $\mathbb{P}^1$-bundle outside of the axis $L_1 = L_2 = L_3 = 0$ but has a $\mathbb{P}^2$ as fiber over any point of the axis.

If $n = 4$ and X is nonsingular the classification of the nonnef extremal rays is almost complete. In fact, if R is not divisorial, then R is either of type $(2, 1)$ or $(2, 0)$. The former case is excluded. Indeed in this case the general fiber of the restriction $\varphi_E : E \to \varphi(E)$ is 1-dimensional so we contradict the Ionescu inequality (6.3.5). The case (2,0) has been described by Kawamata [344]. If R is divisorial see [7], [82], [83] and [258] quoted above.

Theorem 4.3.9. (Kawamata) *Let X be a 4-dimensional connected projective manifold, and let φ be the contraction of an extremal ray R of type* $(2, 0)$. *Then the locus where φ is not an isomorphism is the disjoint union of irreducible components E_j such that $E_j \cong \mathbb{P}^2$ and $\mathcal{N}_{E_j/X} \cong \mathcal{O}_{\mathbb{P}^2}(-1) \oplus \mathcal{O}_{\mathbb{P}^2}(-1)$.*

To this purpose let us recall an interesting generalization of Theorem (4.3.9) for polarized pairs due to Andreatta, Ballico, and Wiśniewski [20, Theorem A]. The morphism φ in the theorem below is given by the basepoint free theorem, (1.5.1).

Theorem 4.3.10. *Let X be a connected projective manifold of dimension $n \geq 4$. Let L be an ample line bundle on X. Assume that $K_X + (n-3)L$ is nef and big. Let $\varphi : X \to Y$ be the birational morphism with connected fibers and normal image associated to $m(K_X + (n-3)L)$, $m \gg 0$. Assume that φ is the contraction of an extremal ray, R, and let E be the locus of R. Assume that* $\dim E \leq n - 2$. *Then* $\dim E = n - 2$ *and E is a disjoint union of its irreducible components, E_i, such that $E_i \cong \mathbb{P}^{n-2}$ and $\mathcal{N}_{E_i/X} \cong \mathcal{O}_{\mathbb{P}^{n-2}}(1)^{\oplus 2}$, $i = 1, \ldots, s$.*

Chapter 5

Restrictions imposed by ample divisors

In this chapter we prove a number of results showing the strong restrictions imposed on a variety X by possession of a special ample or very ample divisor. In many cases the restrictions are so severe that we obtain nonexistence results, i.e., many varieties cannot be ample divisors except in trivial ways.

Theorem (5.2.1), proved originally in the manifold case by Sommese [567], is a general result, central to the chapter. It gives mild conditions under which morphisms defined on ample divisors extend to the global variety. We also give a basic result, Proposition (5.1.5), due to Sommese [567, 583], on when retracts onto ample divisors exist, and Fujita's beautiful generalization, (see §5.4), on when local retracts onto smooth ample divisors exist.

5.1 On the behavior of k-big and ample divisors under maps

In this section we prove a few useful results on the behavior of k-big and ample divisors under holomorphic maps. We will use without explicit comment the fact that if A is an effective divisor on an irreducible variety, X, and if $f : X \to Y$ is a surjective map, which satisfies $f(A) = Y$, then $\dim X > \dim Y$. To see this simply note that $\dim X = \dim A + 1 \geq \dim Y + 1$.

Proposition 5.1.1. *Let L be a k-big line bundle on an irreducible projective variety X, and let $A \in |L|$ be an effective divisor. Let $f : X \to Y$ be a proper surjective holomorphic map. If* $\dim X > \dim Y + k$ *then* $f(A) = Y$.

Proof. If $f(A) \neq Y$, then since A does not meet a general fiber, F, of f, $L_F \cong \mathcal{O}_F$. Since $\dim F = \dim X - \dim Y > k$, this contradicts Corollary (2.1.2). □

Proposition 5.1.2. *Let L be a k-big line bundle on an irreducible projective variety, X, and let $A \in |L|$ be an effective divisor. Let $f : X \to Y$ be a proper surjective holomorphic map. If* $\dim X \leq \dim Y + k$ *then* $\dim A \leq \dim f(A) + k$.

Proof. Assume to the contrary that $\dim A \geq \dim f(A) + k + 1$. Note that we have $k + \dim Y \geq \dim X = \dim A + 1 \geq \dim f(A) + k + 2$. Thus

$$\mathrm{cod}_Y f(A) \geq 2. \tag{5.1}$$

We proceed by induction on $\dim X - \dim Y$. Assume first that $\dim X = \dim Y$.

Note that $h^0(tL) > 0$, for all integers $t > 0$, by assumption. Since $h^0(X, tL) \leq h^0(\overline{X}, t\nu^* L)$ where $\nu : \overline{X} \to X$ is the normalization map, it can be assumed without loss of generality that X is normal. Using the Remmert-Stein factorization of f, we see that we can without loss of generality assume that Y is normal and f has connected fibers. By (5.1), $f(A)$ is of codimension at least 2 in Y. Let $U := Y - f(A)$, let $i : U \hookrightarrow Y$, and denote $\mathcal{R} := i_*(f_{f^{-1}(U)})_*(tL)$. Since $\mathcal{R}$ and $(f_*(tL))^{**}$ are reflexive sheaves which agree outside of a set, $f(A)$, of codimension ≥ 2 we have that $\mathcal{R} \cong (f_*(tL))^{**}$. Since $f^{-1}(U) = X - A$ we have $(f_{f^{-1}(U)})_*(tL) \cong \mathcal{O}_U$, and hence $\mathcal{R} \cong \mathcal{O}_Y$. Thus $(f_*(tL))^{**} \cong \mathcal{O}_Y$. Now $h^0(tL) = h^0(f_*(tL)) \leq h^0((f_*(tL))^{**}) = 1$, so that $h^0(tL) = 1$. This implies that $k \geq \dim X$, which contradicts $\dim A \geq \dim f(A) + k + 1$.

Now assume that we have proven the proposition for $\dim X - \dim Y < N$ where $N > 0$. Assume that $\dim X - \dim Y = N$. Note that $k > 0$ since otherwise the hypothesis, $\dim X \leq \dim Y + k$, would imply the contradiction, $N = 0$.

On X, choose a very ample line bundle, $\mathcal{H}$, and a very ample divisor $D \in |\mathcal{H}|$, such that D is irreducible. This can be done by Corollary (2.2.6). It can also be assumed without loss of generality that D contains no component of A. Note that $\mathcal{H}$ is 0-big. Applying Proposition (5.1.1) to f and $D \in |\mathcal{H}|$, we see that $f(D) = Y$. Since $\mathrm{cod}_Y f(A) \geq 2$ by (5.1), we conclude that for every irreducible component, A' of A, $\dim A' - \dim f(A') \geq 1$. Thus we can apply Proposition (5.1.1) to conclude that

$$f(A \cap D) = f(A). \tag{5.2}$$

Since $k \geq 1$ we conclude from Corollary (2.5.6) that L_D is $(k-1)$-big. As noted above we have $f(D) = Y$. We have $\dim D = \dim X - 1 \leq \dim Y + k - 1$. Note that $\dim D - \dim Y = \dim X - \dim Y - 1$. Thus by induction we conclude that $\dim(A \cap D) \leq \dim f(A \cap D) + k - 1$. By (5.2), we conclude that $\dim A = \dim(A \cap D) + 1 \leq \dim f(A \cap D) + k - 1 + 1 = \dim f(A) + k$. □

Proposition 5.1.3. *Let A be an effective and k-big Cartier divisor on an irreducible projective variety X. Let $f : X \to Y$ be a proper surjective holomorphic map with $\dim A > \dim f(A) + k$. Then $f(A) = Y$.*

Proof. Let $L := \mathcal{O}_X(A)$. If $f(A) \neq Y$, then by Proposition (5.1.1) it follows that $\dim X \leq \dim Y + k$. Thus by Proposition (5.1.2) it follows that $\dim A \leq \dim f(A) + k$ which contradicts the hypothesis that $\dim A > \dim f(A) + k$. □

Lemma 5.1.4. *Let A be an effective nef and k-big Cartier divisor on a normal irreducible projective variety, X, of dimension n. Let $f : X \to Y$ be a morphism with connected fibers from X onto a normal projective variety, Y. If* $\dim A > \dim f(A) + k$ *then the general fiber of the restriction f_A is connected, and if A is irreducible then all fibers of f_A are connected.*

Proof. By using Theorem (1.7.1), we see that a general fiber, F, of f is normal, and since the fibers of f are connected, F is irreducible. By Proposition (5.1.3) we have $f(A) = Y$. By dimension considerations, $\dim F = n - \dim Y = \dim A + 1 - \dim Y \geq k + 2$. Let $D = F \cap A$. By (2.1.2), D is nef and k-big. Therefore $h^1(-D) = 0$ by Lemma (2.2.7). Thus $h^0(\mathcal{O}_D) = 1$, so that D is connected. This shows that the general fibers of f_A are connected. If A is irreducible, then if we replace A by its normalization, $\nu : A' \to A$, we see that all the fibers of $f_A \circ \nu$ are connected, and hence all fibers of f_A are connected. $\square$

The following result, due to Sommese [567, 583], shows that the condition that an ample divisor is a section of a map has extremely strong consequences.

Proposition 5.1.5. *Let A be an ample Cartier divisor on an irreducible projective variety, X, such that $X - A$ is a local complete intersection. If there exists a continuous map $f : X \to A$, such that the restriction f_A induces an isomorphism $f_A^* : H^2(A, \mathbb{Q}) \to H^2(A, \mathbb{Q})$, then* $\dim X \leq 2$.

Proof. We can assume without loss of generality that $n := \dim X \geq 3$. Let i be the inclusion of A in X. By Theorem (2.3.1) the restriction mapping $i^* : H^2(X, \mathbb{Q}) \to H^2(A, \mathbb{Q})$ is injective. Since $f_A^* = i^* \circ f^*$ is an isomorphism we conclude that both i^* and $f^* : H^2(A, \mathbb{Q}) \to H^2(X, \mathbb{Q})$ are isomorphisms. Let $\omega \in H^2(X, \mathbb{Q})$ be the cohomology class, i.e., the first Chern class, of an ample divisor on X. We conclude after multiplying ω by a positive integer that $\omega = f^*\eta$ for some element $\eta \in H^2(A, \mathbb{Q})$. Since $\eta^{\dim A+1} = \eta^n = 0$ we conclude the absurdity that $\omega^n = 0$. $\square$

The following result is a generalization of Sommese's result [574, (1.3)]. A CW complex X is said to be a $K(\Pi, 1)$ if X is an Eilenberg-MacLane space of type $(\Pi, 1)$ (see Whitehead [617, 2.10]), i.e., X is connected, Π is a group, and $\pi_1(X, x) \cong \Pi$ for the first homotopy group with a basepoint $x \in X$, and the higher homotopy groups of X vanish.

Corollary 5.1.6. *Let A be an ample Cartier divisor on an irreducible projective variety X such that $X - A$ is a local complete intersection. If A is a $K(\Pi, 1)$ for some group Π, then* $\dim X \leq 2$.

Proof. By the Lefschetz theorem, (2.3.1), it follows that if $\dim X \geq 3$ the inclusion map $i : A \hookrightarrow X$ with a basepoint of the fundamental group of A going to a basepoint of the fundamental group of X gives $\pi_1(X, A) = \pi_2(X, A) = 0$ and hence an isomorphism of fundamental groups $\pi_1(A) \cong \pi_1(X)$. Since A is a $K(\Pi, 1)$ and $\pi_1(A) \cong \pi_1(X)$ under i_*, by the universal property of a $K(\Pi, 1)$, there is a continuous map $r : X \to A$ giving an isomorphism of fundamental groups. Since the induced map $r \circ i : A \to A$ gives an isomorphism of $\pi_1(A)$ with itself, it follows from the universal property of $K(\Pi, 1)$'s that the induced map from A to A is a homotopy equivalence. Hence in particular $r \circ i$ induces an isomorphism $(r \circ i)^* : H^2(A, \mathbb{Q}) \to H^2(A, \mathbb{Q})$. Therefore Proposition (5.1.5) gives the contradiction $\dim X \leq 2$. □

Remark 5.1.7. To appreciate this result note that any product of Abelian varieties, curves of positive genus, and quotients of bounded symmetric domains by discrete groups acting freely is a $K(\Pi, 1)$. In particular it follows from results of Miyaoka and Yau (see Miyaoka [441]) that nonrational surfaces satisfying $c_1^2 = 3c_2$ are $K(\Pi, 1)$'s, and thus such surfaces cannot be ample divisors on any smooth projective 3-folds.

The following result is based on Sommese's result [567, Proposition I].

Proposition 5.1.8. *Let A be a smooth ample Cartier divisor on a normal irreducible local complete intersection projective variety, X, having at worst rational singularities. Assume A is isomorphic to a submanifold of an Abelian variety. Then the cotangent bundle, $\mathcal{T}_A^*$, of A does not split into a direct sum of two nontrivial holomorphic sub-bundles.*

Proof. We follow the conventions on wedging forms given in Remark (2.4.3).

Assume that $\mathcal{T}_A^* \cong E \oplus F$ for two nontrivial holomorphic sub-bundles, E and F. Since $\dim A = \operatorname{rank} E + \operatorname{rank} F \geq 2$, it follows from Proposition (2.4.4) that $\operatorname{Alb}(A) \cong \operatorname{Alb}(X)$. Since A is a submanifold of an Abelian variety, alb_A is an injection. Therefore the restriction of $\operatorname{alb}_X : X \to \operatorname{Alb}(X)$ to A is an isomorphism onto its image. Hence $\dim \operatorname{alb}_X(X) \geq n - 1$ and if $\dim \operatorname{alb}_X(X) = n - 1$, then $\operatorname{alb}_X(X) \cong \operatorname{alb}_A(A)$. In this case we have a continuous map $f := \operatorname{alb}_A^{-1} \circ \operatorname{alb}_X : X \to A$ such that $f_A = \operatorname{id}_A$. Thus Proposition (5.1.5) applies to give the contradiction $\dim X \leq 2$. It thus follows that $\operatorname{alb}_X(X) \ncong \operatorname{alb}_A(A)$. Therefore $\dim \operatorname{alb}_X(X) = n := \dim X$, and there exist n holomorphic one forms $\omega_1, \ldots, \omega_n$ on X such that $\omega_1 \wedge \ldots \wedge \omega_n$ is a nontrivial section of K_X.

Choose a basis $\{e_i, f_j | i \in I, j \in J\}$ of $H^0(\mathcal{T}_X^*)$ with $e_{i|A} \in H^0(A, E)$ and $f_{j|A} \in H^0(A, F)$. Writing the one forms $\{\omega_1, \ldots, \omega_n\}$ in terms of the e_i and f_j we see that there is a nontrivial section λ of K_X of the form $\lambda = \wedge_{i=1}^a e_i \wedge_{j=a+1}^n f_j$ for some possibly reordered and renumbered set of the e_i and f_j, and some integer $1 \leq a \leq n - 1$.

To see that this is impossible, note that either $a > \operatorname{rank} E$ or $n - a > \operatorname{rank} F$. Indeed otherwise we would have the contradiction $n \leq \operatorname{rank} E + \operatorname{rank} F = \operatorname{rank} \mathcal{T}_A^* =$

$n-1$. Therefore without loss of generality it can be assumed that one can choose an s with $s \leq a$ and $\operatorname{rank} E < s \leq a \leq n-1$ such that $(e_1 \wedge \ldots \wedge e_s)_A = 0$. But since by Proposition (2.4.4) we have an injection $0 \to H^0(\wedge^s \mathcal{T}_X^*) \to H^0(\wedge^s \mathcal{T}_A^*)$ for $s \leq n-1$ we have $e_1 \wedge \ldots \wedge e_s = 0$ on X and hence the contradiction that $\lambda = 0$. □

Corollary 5.1.9. *Let A be a smooth ample Cartier divisor on a normal irreducible local complete intersection projective variety, X, with at worst rational singularities. If* $\dim A \geq 2$ *and A is isomorphic to a submanifold of an Abelian variety, then* K_A *is ample.*

Proof. The main theorem of Ran [511] shows that if K_A is not ample then there is a nontrivial holomorphic vector field χ on A, i.e., there exists a global section, χ, of $\mathcal{T}_A$. Choosing a holomorphic one form ω on A with $\omega(\chi) \neq 0$ we have the splitting $\mathcal{T}_A^* \cong \mathbb{C}[\omega] \oplus \mathcal{K}$ where $\mathcal{K}$ is the kernel of the bundle map $\mathcal{T}_A^* \to A \times \mathbb{C}$ given by evaluation with χ. This contradicts Proposition (5.1.8). □

5.2 Extending morphisms of ample divisors

The extension theorem, (5.2.1), is due to Sommese [567] in the smooth case, with the observation that it holds for a smooth A on a local complete intersection due to Fujita [241].

Theorem 5.2.1. *Let A be a normal k-ample Cartier divisor on an irreducible local complete intersection projective algebraic set, X. Assume that* $\dim \operatorname{Irr}(X) \leq 0$ *and let* $p : A \to Y$ *be a holomorphic surjection onto a projective variety, Y. If*

$$\dim A - \dim Y \geq 2 + \max\{\max_{y \in Y}\{\dim p^{-1}(y) \cap \operatorname{Irr}(A)\}, k\},$$

then p extends to a holomorphic surjection $\overline{p} : X \to Y$.

Proof. Without loss of generality it can be assumed that $\dim Y \geq 1$. Note that X is normal. Indeed X is Cohen-Macaulay since it is a local complete intersection. Therefore if X was not normal, then $\operatorname{Sing}(X)$ would be a divisor on X and therefore $\operatorname{Sing}(X) \cap A$ would be a divisor on A. This contradicts the normality of A.

Since $\dim Y \geq 1$ we have that $\dim X \geq 4+k$ and therefore by Corollary (2.3.4) we see that the restriction map gives an isomorphism, $\operatorname{Pic}(X) \cong \operatorname{Pic}(A)$. Moreover by Remmert-Stein factorizing p it can be assumed without loss of generality that Y is normal and p has connected fibers.

Let $\mathcal{L}$ be a very ample line bundle on Y. Since $\operatorname{Pic}(X) \cong \operatorname{Pic}(A)$ there exists an $H \in \operatorname{Pic}(X)$, whose restriction, H_A, to A is isomorphic to $p^*\mathcal{L}$.

We claim that the restriction to A gives a surjection

$$\Gamma(H) \to \Gamma(H_A) \to 0. \tag{5.3}$$

This will follow if $h^1(H - A) = 0$. Let $L := \mathcal{O}_X(A)$. By Lemma (4.1.6) applied to $\mathcal{E} := H - A$, the claim (5.3) will follow if we show that $h^1(A, (H - tL)_A) = 0$ for $t \geq 1$. By Serre duality we are reduced to showing $h^{\dim A - 1}(A, \omega_A + (tL - H)_A) = 0$ for $t \geq 1$ where ω_A is the dualizing sheaf of A. Using the Leray spectral sequence for $\omega_A + (tL - H)_A$ and p, the vanishing, $h^{\dim A - 1}(A, \omega_A + (tL - H)_A) = 0$, will follow from $h^{\dim A - 1 - j}(Y, p_{(j)}(\omega_A + (tL - H)_A)) = 0$ for $0 \leq j \leq \dim A - 1$. Since $h^{\dim A - 1 - j}(Y, \mathcal{F}) = 0$ for any coherent sheaf $\mathcal{F}$ when $\dim A - 1 - j \geq \dim Y + 1$, we can assume

$$\dim A - 1 - j \leq \dim Y. \tag{5.4}$$

From Theorem (2.2.5) we see that

$$p_{(j)}(\omega_A + (tL - H)_A) = p_{(j)}(\omega_A + tL_A) \otimes (-\mathcal{L}) = 0$$

for $j \geq \max\{\max_{y \in Y}\{\dim p^{-1}(y) \cap \operatorname{Irr}(A)\}, k\} + 1$ and thus we can assume that

$$j \leq \max\{\max_{y \in Y}\{\dim p^{-1}(y) \cap \operatorname{Irr}(A)\}, k\}. \tag{5.5}$$

Inequality (5.4) and inequality (5.5) together imply

$$\dim A - 1 - \dim Y \leq \max\{\max_{y \in Y}\{\dim p^{-1}(y) \cap \operatorname{Irr}(A)\}, k\}$$

which contradicts a hypothesis of the theorem. Thus the restriction map $\Gamma(H) \to \Gamma(H_A)$ is onto.

Since global sections of H span H_A, the base locus of $|H|$ is disjoint from A. Since $H_A \cong p^*\mathcal{L}$ with $\mathcal{L}$ very ample on Y, there exist $\dim Y + 1$ divisors $D_1, \ldots, D_{\dim Y + 1}$ in $|H|$ such that $\operatorname{Bs}|H| \subset \cap_{i=1}^{\dim Y + 1} D_i$ and $D_1 \cap \ldots \cap D_{\dim Y + 1} \cap A = \emptyset$. Since A is k-ample it thus follows from Theorem (2.1.1) that $\dim \cap_{i=1}^{\dim Y + 1} D_i \leq k$. We claim that H is spanned by its global sections. If $\cap_{i=1}^{\dim Y + 1} D_i = \emptyset$, then $\operatorname{Bs}|H| = \emptyset$. If $\cap_{i=1}^{\dim Y + 1} D_i \neq \emptyset$, then

$$\dim(D_1 \cap \ldots \cap D_{\dim Y + 1}) \geq \dim X - \dim Y - 1 \geq 2 + k.$$

This contradicts the above inequality.

Let $\overline{p} : X \to \mathbb{P}_{\mathbb{C}}$ be the map associated to $\Gamma(H)$. Note that, since $\dim X > \dim \overline{p}(X)$, by Proposition (5.1.1) we conclude that $\overline{p}(X) = p(A)$. Noting that, by definition of H, the map p is given by $\Gamma(H_A)$ and that $\overline{p}_A = p$ we are done. □

Remark 5.2.2. As Fujita [241] points out we could drop the local complete intersection condition on X if we knew that there existed some line bundle H on X whose restriction to A is the pullback of an ample line bundle on Y.

By (2.7.3) such a line bundle exists on a neighborhood of A when X is a complex manifold and A is smooth. Using the results of §2.6, the map p extends to a holomorphic map from some neighborhood of A to Y. Using Theorem (2.6.7), this map extends to a rational mapping, $\overline{p} : X \to Y$. Consider a desingularization of X on which the extension of p is holomorphic. Since $\dim A > \dim \overline{p}(A)$, Proposition (5.1.3) gives $\overline{p}(A) = Y$ and hence the image of X under $\overline{p}$ is Y.

Morphisms of ample divisors with general fiber being 1-dimensional do not extend in general, e.g., $\mathbb{P}^1 \times \mathbb{P}^1$ is isomorphic to the smooth quadric $\mathcal{Q} \in |\mathcal{O}_{\mathbb{P}^3}(2)|$. Nonetheless such morphisms often do extend, though the complete story is not yet known. A general meromorphic extension theorem for the case when the general fiber of the morphism $A \to Y$ is $\mathbb{P}^1$ is given by Fania and Sommese [228]. The case when $\dim A = \dim Y = 1$ is particularly interesting. Serrano [555] has made the basic investigation in this case. See also Fania [220] and Andreatta and Ballico [17] which are based on the results of our paper with Francia [88] and our paper [103], (see also (8.5.1)). See Paoletti [498] for some interesting variants of the extension theorems.

The following result, due to Sommese [567] in the smooth case, gives some other conditions under which there is an extension theorem.

Theorem 5.2.3. *Let A be an ample divisor on an irreducible projective variety, X. Assume that X and A are both normal and have at most rational singularities. Let $p : A \to Y$ be a holomorphic surjection onto a projective variety, Y. If $\dim A - \dim Y \geq 1$, and if there exists a finite-to-one morphism from Y into an Abelian variety, then p extends to a holomorphic surjection $\overline{p} : X \to Y$.*

Proof. We can assume without loss of generality that $\dim Y \geq 1$ since otherwise the theorem is trivial. Let $s : Y \to W$ be the finite morphism from Y into an Abelian variety, W. Let $\mathrm{alb}_X : X \to \mathrm{Alb}(X)$ be the Albanese mapping of X which exists by Lemma (2.4.1). Since $\dim X = 1 + \dim A \geq 2 + \dim Y \geq 3$ we conclude that $\mathrm{Alb}(A) \cong \mathrm{Alb}(X)$ by Proposition (2.4.4). Using the map from $\mathrm{Alb}(A) \to W$ that exists by the universal property of the Albanese mapping, we get an extension of the composite map $\sigma := s \circ p : A \to W$ to a map $\overline{\sigma} : X \to W$. We claim that $\overline{\sigma}(X) = \sigma(A)$. Since $\dim A > \dim \sigma(A)$, this follows from Proposition (5.1.3).

Let $\overline{\sigma} = \beta \circ \alpha : X \to Z \to \overline{\sigma}(X)$ be the Remmert-Stein factorization of $\overline{\sigma}$. Since $\alpha : X \to Z$ has connected fibers of dimension ≥ 2, we conclude by Lemma (5.1.4) that the fibers of the restriction $\alpha_A : A \to Z$ are connected. This implies that $\beta_A \circ \alpha_A$ is the Remmert-Stein factorization of the restriction $\overline{\sigma}_A : A \to \overline{\sigma}(X) = \sigma(A)$. Note that $\overline{\sigma}_A$ factors through $A \to Y \to \overline{\sigma}(X)$, where $Y \to \overline{\sigma}(X)$ is a finite map by assumption. Therefore, by the universal property of the Remmert-Stein factorization, there exists a morphism $\psi : Z \to Y$, which makes everything commute. Then take $\overline{p} := \psi \circ \alpha : X \to Y$. □

Corollary 5.2.4. *Let A be a normal ample Cartier divisor on an irreducible local complete intersection projective variety, X. Assume that* $\dim \operatorname{Irr}(X) \leq 0$ *and that A has at most rational singularities. If $A = A_1 \times \cdots \times A_r$ where* $\dim A_i \geq 1$ *for all i and with $r \geq 2$, then $r = 2$ and one of the A_i is a curve. If moreover X has at most rational singularities, then not all of the A_i can map finite-to-one to their Albanese varieties. In particular if* $\dim A = 2$*, then one of the factors is $\mathbb{P}^1$.*

Proof. Let $p_j : A \to A_j$ denote the product projection. By (5.2.1), p_j extends to a morphism $\overline{p}_j : X \to A_j$ if $\sum_{i \neq j} \dim A_i \geq 2$. If each p_j extends then we have a morphism $X \to \prod_i A_i$ which is the identity on A. This contradicts (5.1.5), and thus $\sum_{i \neq j} \dim A_i < 2$ for some j, i.e., $r \leq 2$ and $\dim A_i = 1$ for some i. The last statement follows by the same argument using (5.2.3) in place of (5.2.1). □

Corollary (5.4.4) gives a generalization of Corollary (5.2.4) where X is merely normal, but A is smooth.

The following result (see Sommese [567, Proposition V]) gives a basic restriction for a fiber bundle to be an ample divisor.

Proposition 5.2.5. *Let A be a smooth connected ample divisor on a projective manifold, X. Let $Z \subset Y$ denote the set over which a surjective morphism, $p : A \to Y$, with connected fibers and Y normal is not of maximal rank. If p has an extension to X, $\overline{p} : X \to Y$ (e.g., by* (5.2.1)*, if* $\dim A \geq \dim Y + 2$*), then* $\dim X \geq 2 \dim Y - \dim Z - 2$ *where* $\dim Z := -1$ *if Z is empty.*

Proof. We can choose a submanifold, $W \subset Y$, with $W \cap Z = \emptyset$ and $\dim W = \dim Y - \dim Z - 1$. Let $X_W := \overline{p}^{-1}(W)$ and note that

$$\dim X_W = \dim W + \dim X - \dim Y = \dim X - \dim Z - 1.$$

Since $2 \dim W + \dim Z + 1 = 2 \dim Y - \dim Z - 1$, we see that $\dim X \geq 2 \dim Y - \dim Z - 2$ is equivalent to

$$\dim X_W \geq 2(\dim Y - \dim Z - 1) - 1 = 2 \dim W - 1.$$

Since $W \cap Z = \emptyset$, the restriction of $\overline{p}$ to W is of maximal rank. Thus without loss of generality it can be assumed that p is of maximal rank, i.e., p is a differentiable fiber bundle, and that

$$\dim X \leq 2 \dim Y - 2 \quad \text{or equivalently} \quad \dim F \leq \dim Y - 2 \tag{5.6}$$

where F is a general fiber of $\overline{p}$. Let f be a general fiber of p. We have the following commutative diagram with $j \geq 1$. The rows are the long exact homotopy sequences (as usual we suppress basepoints).

$$\begin{array}{ccccccccccc}
\cdots & \to & \pi_j(f) & \to & \pi_j(A) & \to & \pi_j(A, f) & \to & \pi_{j-1}(f) & \to & \cdots \\
 & & \downarrow & & \downarrow & & \downarrow & & \downarrow & & \\
\cdots & \to & \pi_j(F) & \to & \pi_j(X) & \to & \pi_j(X, F) & \to & \pi_{j-1}(F) & \to & \cdots
\end{array}$$

Since p is of maximal rank, the set where $\overline{p}$ is not of maximal rank is an analytic subset, $\mathcal{S}$, that does not meet A. Since A is ample, it follows that $\mathcal{S}$ is finite. Thus the natural map $\pi_j(X, F) \to \pi_j(Y)$ is an isomorphism for $j \leq 2\dim Y - 2$. Since the natural map $\pi_j(A, f) \to \pi_j(Y)$ is an isomorphism for all j, it follows that the map $\pi_j(A, f) \to \pi_j(X, F)$ is an isomorphism for $j \leq 2\dim Y - 2$. Since $\pi_j(X, A) = 0$ for $j \leq \dim X - 1$ by (2.3.1), and since $\dim X - 1 \leq 2\dim Y - 3$ we see that $\pi_j(F, f) = 0$ for $j \leq \dim X - 2$. Since $h^{2\dim F}(F, f, \mathbb{Q}) \neq 0$, we see $\dim X - 2 \leq 2\dim F - 1$, i.e., $\dim Y \leq \dim F + 1$. This contradicts (5.6). □

Remark 5.2.6. A more careful argument shows that $\dim X \geq 2\dim Y - \dim Z - 1$. In this case equality implies that the general fiber, F, of $\overline{p}$ is isomorphic to $(\mathbb{P}^{\dim F}, \mathcal{O}_{\mathbb{P}^{\dim F}}(1))$. The key point in the study of the equality is to use (2.3.9).

5.3 Ample divisors with trivial pluricanonical systems

The following result shows that it is very restrictive for a variety with a trivial dualizing sheaf to be an ample Cartier divisor.

Theorem 5.3.1. *Let A be an ample Cartier divisor on a normal irreducible local complete intersection projective variety X. Assume that* $\mathrm{Irr}(X)$ *is finite and $tK_A \approx \mathcal{O}_A$ for some $t > 0$. Let $L = \mathcal{O}_X(A)$. Then if $\dim X \geq 3$, $-K_X \approx L$ and $K_A \approx \mathcal{O}_A$. In particular, $h^i(\mathcal{O}_X) = 0$ for $i \geq 1$ and $h^i(\mathcal{O}_A) = 0$ for $0 < i < \dim A$.*

Proof. By Corollary (2.3.4) we have an injection $r : \mathrm{Pic}(X) \hookrightarrow \mathrm{Pic}(A)$ with torsion free cokernel. Thus $t(K_X + L) \approx \mathcal{O}_X$. Therefore $-K_X$ is ample. From this and Kodaira's vanishing theorem we conclude that $h^i(\mathcal{O}_X) = h^i(K_X + A) = 0$ for $i > 0$. Using the exact sequence

$$0 \to -A \to \mathcal{O}_X \to \mathcal{O}_A \to 0$$

we get the vanishing statement for $h^i(\mathcal{O}_A)$. We have an unramified t-to-one cover $q : W \to X$ of X. Since $-K_X$ is ample, $-K_W$ is ample and thus $t = t\chi(\mathcal{O}_X) = \chi(\mathcal{O}_W) = 1$. Therefore $K_X + L \approx \mathcal{O}_X$. □

Remark 5.3.2. Let S be an Enriques surface, i.e., a smooth connected projective surface with fundamental group, $\mathbb{Z}_2$, and $2K_S \cong \mathcal{O}_S$, but $K_S \not\cong \mathcal{O}_S$. The above result implies that for S to be a hyperplane section on X, the variety X cannot have only local complete intersection singularities. There is a detailed and interesting theory based on the special geometry of these varieties. See, e.g., Bayle [75], Conte and Murre [166, 167] and Picco Botta and Verra [505].

5.4 Varieties that can be ample divisors only on cones

Let A be a normal irreducible projective nondegenerate subvariety of $\mathbb{P}^r$ of positive dimension. An *extension* of A in $\mathbb{P}^{r+1}$ is by definition an irreducible projective subvariety X of $\mathbb{P}^{r+1}$ of dimension $\dim X = \dim A + 1$, such that there is a hyperplane $H := \mathbb{P}^r \subset \mathbb{P}^{r+1}$ with $A = X \cap H$ scheme-theoretically. Of course, given $A \subset \mathbb{P}^r$ one can easily construct an extension of A in $\mathbb{P}^{r+1}$. Take a point $x \in \mathbb{P}^{r+1} - \mathbb{P}^r$ and let X be the cone over A with vertex x. The extensions of this type will be called *trivial extensions* of A in $\mathbb{P}^{r+1}$. One very interesting and hard problem in projective geometry, which goes back classically to Scorza, is to understand if there are nontrivial extensions, i.e., extensions that are not cones, of a given subvariety A of $\mathbb{P}^r$ and to classify all such possible extensions. In this section we present some results due to Fujita [245], Bădescu [46], Zak [630], and L'vovsky [418] on projective manifolds, A, such that the only normal varieties that they can be ample Cartier divisors on are cones. We have found the papers of Bădescu [40], where Theorem (5.4.2) appears in the special case when the ample divisor is a Grassmannian, and [46] helpful. Since we do not use these results in the main development of the book we only give detailed sketches of proofs.

Let us recall that there are two canonical exact sequences associated to the embedding $A \subset \mathbb{P}^r$, the normal sequence of $A \subset \mathbb{P}^r$ and the Euler sequence of $\mathbb{P}^r$ restricted to A respectively,

$$0 \to \mathcal{T}_A \to \mathcal{T}_{\mathbb{P}^r|A} \xrightarrow{\alpha} \mathcal{N}_{A/\mathbb{P}^r},$$

$$0 \to \mathcal{O}_A \to \mathcal{O}_A(1)^{\oplus r+1} \xrightarrow{\beta} \mathcal{T}_{\mathbb{P}^r|A} \to 0.$$

Note that the composition $\alpha \circ \beta : \mathcal{O}_A(1)^{\oplus r+1} \to \mathcal{N}_{A/\mathbb{P}^r}$ gives rise to a map on A,

$$\gamma : H^0(\mathcal{O}_A^{\oplus r+1}) \to H^0(\mathcal{N}_{A/\mathbb{P}^r}(-1)), \tag{5.7}$$

which we use in the sequel.

Let L be an ample line bundle on an irreducible, normal projective variety, A, and let $Z := \mathbb{P}(\mathcal{O}_A \oplus L)$. Note that the bundle projection $\pi : Z \to A$ has a section σ corresponding to the surjection $\mathcal{O}_A \oplus L \to \mathcal{O}_A$ and a section σ' corresponding to the surjection $\mathcal{O}_A \oplus L \to L$. Let $s \in H^0(\mathcal{O}_A \oplus L)$ be a section corresponding to the inclusion $\mathbb{C} = H^0(\mathcal{O}_A) \subset H^0(\mathcal{O}_A \oplus L)$, i.e., corresponding to $\mathcal{O}_A \xrightarrow{\otimes s} \mathcal{O}_A \oplus L$. Note that s corresponds to a section, $\tilde{s}$, of the tautological bundle ξ of Z. Since s is nowhere zero, we see that the image, $\sigma'(X)$, under the section, σ', corresponding to the quotient $\mathcal{O}_A \oplus L \to L$, coincides with the zero set of $\tilde{s}$ on Z. We simply say in this case that σ' is the zero set of s. We do not distinguish between sections and their images.

Given any fiber, F, of $\pi : Z \to A$, F is isomorphic to $\mathbb{P}^1$, and the restriction of the tautological bundle ξ of $\mathbb{P}(\mathcal{O}_A \oplus L)$ to F is isomorphic to $\mathcal{O}_{\mathbb{P}^1}(1)$. Let z_0

be a homogeneous coordinate on $F \cong \mathbb{P}^1$ which is 0 on $F \cap \sigma$ and let $z_1 = s_F$ be a homogeneous coordinate on $F \cong \mathbb{P}^1$ which is 0 on $F \cap \sigma'$. Note that, for any positive integer N, $\Gamma(N\xi)$ is isomorphic to $\oplus_{t=0}^N \Gamma(tL)$ with any given section e of $\Gamma(tL) \subset \Gamma(N\xi)$ corresponding under restriction to F to $\lambda z_0^t z_1^{N-t}$, where λ is a constant that depends on the given e. Since L is ample, NL is spanned for all $N \gg 0$ and by the preceding correspondence we see that $N\xi$ is spanned if and only if NL is. We can Remmert-Stein factorize the morphism, $\phi : Z \to \mathbb{P}_{\mathbb{C}}$, associated to $|N\xi|$ as $\phi = \alpha \circ r$, with $r : Z \to \mathcal{C}$ a map with connected fibers onto a normal variety, $\mathcal{C}$. Note that r is a biholomorphism on $Z - \sigma$ and $r(\sigma)$ is a point, since, by the above construction, $\xi_{\sigma(A)} \cong \mathcal{O}_A$. We call $\mathcal{C}$ *the cone on the pair,* (A, L), and denote it $\mathcal{C}(A, L)$ or $\mathcal{C}$ for short when there is no confusion. The line bundle ξ on Z is the pullback of an ample line bundle, $\mathcal{L}$ on $\mathcal{C}$. The line bundle $\mathcal{L}$ is called the *tautological bundle of the cone,* $\mathcal{C}(A, L)$.

Let $\mathbb{A} := \oplus_{t=0}^{\infty} \Gamma(A, tL)$. Let $\tilde{s}$ be the section on Z which corresponds to the section s as above. Then $\tilde{s} = r^*c$ for a section, c, of $\mathcal{L}$ on $\mathcal{C}$. Note that $\oplus_{t=0}^{\infty} \Gamma(t\mathcal{L})$ is isomorphic to the polynomial ring $\mathbb{A}[x]$, where the isomorphism is defined by $x \mapsto c$. Indeed $\Gamma(N\mathcal{L}) \cong \Gamma(N\xi) \cong \oplus_{t=0}^N \Gamma(tL)$ with a section of $\Gamma(tL) \subset \Gamma(N\xi)$ corresponding under restriction to F to $\lambda z_0^t \tilde{s}_F^{N-t}$ where λ is a constant that depends on the given section. Thus we have $\mathcal{C}(A, L) \cong \mathbb{P}(\mathbb{A}[x])$.

For example, note that, by the above construction, $\mathcal{C}(\mathbb{P}^n, \mathcal{O}_{\mathbb{P}^n}(1)) \cong \mathbb{P}^{n+1}$.

From the above discussion we have a simple formal criterion of Bădescu [40] for a variety to be a cone.

Lemma 5.4.1. *Let $\mathcal{L}$ be an ample line bundle on a normal projective variety X. Assume that A is a normal Cartier divisor on X and $s \in \Gamma(\mathcal{L})$ is a section defining A. Then to show that $(X, \mathcal{L})$ is the cone on $(A, \mathcal{L}_A)$ it suffices to show for all $N \geq 0$ that $\Gamma(N\mathcal{L}) \cong \oplus_{t=0}^N \Gamma(t\mathcal{L}_A)s^{N-t}$ with the isomorphism compatible with the maps $\Gamma(a\mathcal{L}) \otimes \Gamma(b\mathcal{L}) \to \Gamma((a+b)\mathcal{L})$ for $a \geq 0$ and $b \geq 0$.*

Theorem 5.4.2. *Let X be a normal irreducible projective variety. Let A be a smooth ample Cartier divisor on X. Let $\mathcal{L} := \mathcal{O}_X(A)$. Assume that* $\dim X \geq 3$. *If there exists a holomorphic map, $r : U \to A$, of a neighborhood, $U \subset X$, of A such that the restriction r_A is the identity map, then r has a meromorphic extension to X and $(X, \mathcal{L})$ is the cone on $(A, \mathcal{L}_A)$.*

Proof. Note that by Theorem (2.6.7), the map r extends as a rational map $\bar{r} : X \to A$.

Choose a neighborhood $V \subset U$ of A such that every section of $\mathcal{L}_V$ extends to $\mathcal{L}$ and the restriction map $\mathrm{Pic}(V) \to \mathrm{Pic}(A)$ is injective, and in fact an isomorphism. Note that this is possible in view of Corollary (2.7.3). Note also that $(r^*\mathcal{L}_A)_A \approx \mathcal{L}_A$ and thus $r_V^*\mathcal{L}_A \approx \mathcal{L}_V$, where r_V denotes the restriction of r to V. Therefore we have a natural injective map, $\Gamma(N\mathcal{L}_A) \overset{\alpha_N}{\to} \Gamma(N\mathcal{L})$, with $\alpha_N(s)_A = s$ for any $s \in \Gamma(N\mathcal{L}_A)$.

Since $\Gamma(N\mathcal{L} - A) \cong \Gamma((N-1)\mathcal{L})$, we see by a descending induction using the exact long cohomology sequence

$$0 \to \Gamma(N\mathcal{L} - A) \to \Gamma(N\mathcal{L}) \to \Gamma(N\mathcal{L}_A) \to H^1(X, N\mathcal{L} - A),$$

that we get the isomorphism $\Gamma(N\mathcal{L}) \cong \oplus_{t=0}^{N}\Gamma(t\mathcal{L}_A)s^{N-t}$ compatible with the mappings $\Gamma(a\mathcal{L}) \otimes \Gamma(b\mathcal{L}) \to \Gamma((a+b)\mathcal{L})$. Now use the criterion, (5.4.1), for X to be a cone. □

In the smooth case and for $\dim X \geq 3$ we have the following result proved by Fujita in [245].

Theorem 5.4.3. *Let L be an ample line bundle on a normal irreducible projective variety X with $\dim X \geq 3$. Assume that there is a smooth $A \in |L|$ such that $H^1(\mathcal{T}_A \otimes (-tL_A)) = (0)$ for all $t \geq 1$. Then (X, L) is the cone on (A, L_A).*

Proof. Let us give a sketch of the proof. Since $H^1(\mathcal{T}_A \otimes (-L_A)) = (0)$, $\mathcal{T}_X \cong \mathcal{T}_A \oplus \mathcal{N}_{A/X}$. From this it follows that $(\mathcal{T}_X \otimes (-L))_A$ has a nowhere vanishing section, s, corresponding to a nowhere vanishing section of $\mathcal{N}_{A/X} \otimes (-L_A) \cong \mathcal{O}_A$. The cohomological vanishing hypothesis implies that $H^1((\mathcal{T}_X \otimes (-tL))_\mu) = (0)$ for all $t \geq 1$ and all $\mu \geq 0$, where $(\mathcal{T}_X \otimes (-tL))_\mu$ denotes the restriction of $\mathcal{T}_X \otimes (-tL)$ to the μ-th infinitesimal neighborhood of A in X (see §2.7). By (2.7.2) we conclude that there is an extension, s_U, of s to a neighborhood U of A. Note that s_U as a section of $\mathcal{T}_U$ vanishes to the first order on A. Integrating this vector field we obtain a holomorphic retraction of some possibly smaller open neighborhood of A onto A. Now use (5.4.2). □

Corollary 5.4.4. *Let A be a product $\prod_{i=1}^{k} A_i$ of connected projective manifolds, A_i. Assume that A is an ample Cartier divisor on a normal projective variety, X. If $k \geq 3$ or if $k = 2$ and $\dim A_i \geq 2$ for $i = 1, 2$, then X is a cone on A.*

Proof. Let $L = \mathcal{O}_X(A)$ and let L_A be the restriction of L to A. Let $p_i : A \to A_i$ denote the projection onto the i-th factor. Note that

$$\mathcal{T}_A \otimes (-tL_A) \cong \oplus_i p_i^* \mathcal{T}_{A_i} \otimes (-tL_A).$$

Note that by hypothesis the fibers of p_i are at least 2-dimensional. Thus by the Kodaira vanishing theorem $p_{i(j)}(-tL_A) = 0$ for $t \geq 1$ and $j \leq 1$. Using this, the projection formula, $p_{i(j)}(p_i^*\mathcal{T}_{A_i} \otimes (-tL_A)) \cong \mathcal{T}_{A_i} \otimes p_{i(j)}(-tL_A)$, and the Leray spectral sequence for p_i gives that $h^1(p_i^*\mathcal{T}_{A_i} \otimes (-tL_A)) = 0$. Thus $h^1(\mathcal{T}_A \otimes (-tL_A)) = 0$ for $t \geq 1$. Now use (5.4.3). □

The following result of Mori-Sumihiro and Wahl [611] was proved in the course of the proof of (5.4.3) in the special case of it when A is smooth. In Remark (5.4.6) we show that (5.4.3) follows from (5.4.5).

Theorem 5.4.5. *Let L be an ample line bundle on a normal, irreducible, projective variety X with* $\dim X \geq 2$. *Assume that* $h^0(\mathcal{T}_X \otimes (-L)) > 0$. *Then there is an effective divisor A on X, which is a normal projective variety, such that $L \cong \mathcal{O}_X(A)$ and (X, L) is the cone on (A, L_A).*

Remark 5.4.6. Let us show here how Theorem (5.4.3) also easily follows from Theorem (5.4.5). The first author would like to thank Bădescu for pointing out the simple argument below. Consider the exact sequence

$$0 \to \mathcal{T}_A \to \mathcal{T}_{X|A} \to L_A \to 0.$$

Write $\mathcal{T}_A(-i) := \mathcal{T}_A \otimes (-iL_A)$, $\mathcal{T}_X(-iA) := \mathcal{T}_X \otimes (-iL)$, for $i \geq 1$. By tensoring with $-iL_A$, $i \geq 2$, we get the exact sequence

$$0 \to \mathcal{T}_A(-i) \to \mathcal{T}_X(-iA)_{|A} \to (1-i)L_A \to 0. \tag{5.8}$$

By assumption $h^1(\mathcal{T}_A(-i)) = 0$, for $i \geq 1$. Furthermore, since $\dim A \geq 2$, $h^1((1-i)L_A) = 0$ for $i \geq 2$. Thus we find, for $i \geq 2$, $h^1(\mathcal{T}_X(-iA)_{|A}) = 0$. Then the exact sequence

$$0 \to \mathcal{T}_X(-(i+1)A) \to \mathcal{T}_X(-iA) \to \mathcal{T}_X(-iA)_{|A} \to 0 \tag{5.9}$$

gives a surjection, for $i \geq 2$,

$$H^1(\mathcal{T}_X(-(i+1)A)) \to H^1(\mathcal{T}_X(-iA)). \tag{5.10}$$

Since $\mathcal{T}_X$ is a reflexive sheaf and X is normal with $\dim X \geq 2$, we have that $h^1(\mathcal{T}_X(-iA)) = 0$ for $i \gg 0$. Thus from (5.10) we conclude that $h^1(\mathcal{T}_X(-2A)) = 0$ and therefore the exact sequence (5.9) yields a surjection

$$H^0(\mathcal{T}_X(-A)) \to H^0(\mathcal{T}_X(-A)_{|A}).$$

Since $h^1(\mathcal{T}_A(-1)) = 0$ by assumption, from the exact sequence (5.8), with $i = 1$, we find a surjection

$$H^0(\mathcal{T}_X(-A)_{|A}) \to H^0(\mathcal{O}_A).$$

The two surjections above imply $h^0(\mathcal{T}_X(-A)) > 0$. Therefore Theorem (5.4.5) applies to give the result.

Bădescu gives in [46] two new different proofs of the following theorem, which generalizes a result due to L'vovsky [418] and Zak [630].

Theorem 5.4.7. ([46]) *Let A be a normal projective variety in $\mathbb{P}^r$. Assume that* $\operatorname{cod}_{\mathbb{P}^r} A \geq 2$ *and that the map (5.7) for $A \subset \mathbb{P}^r$ is surjective. Let X be an irreducible projective subvariety of $\mathbb{P}^{r+1}$ with* $\dim X = \dim A + 1$, *such that $A = X \cap \mathbb{P}^r$ scheme-theoretically. Then $(X, \mathcal{O}_{\mathbb{P}^{r+1}}(1)_{|X})$ is a cone over $(A, \mathcal{O}_{\mathbb{P}^r}(1)_{|A})$.*

Remark 5.4.8. Zak [630] and L'vovsky [418] proved the result above under the assumption that A is smooth. Note that the surjectivity assumption in (5.4.7) implies that $h^0(\mathcal{N}_{A/\mathbb{P}^r}(-1)) = r+1$, the minimal possible value. Note also that for a hypersurface A of degree d in $\mathbb{P}^r$, $h^0(\mathcal{N}_{A/\mathbb{P}^r}(-1)) = h^0(\mathcal{O}_A(d-1)) = r+1$ if and only if A is a quadric hypersurface in $\mathbb{P}^r$, i.e., $d=2$.

The first proof of Theorem (5.4.7) applies, with minor changes, to yield the following result (also due to L'vovsky [418] and Zak [630]).

Theorem 5.4.9. *Let A be an irreducible and nondegenerate projective submanifold of $\mathbb{P}^r$ of dimension ≥ 2. Assume that $h^0(\mathcal{N}_{A/\mathbb{P}^r}(-1)) = r+t+1$, $t \geq 0$, and $h^0(\mathcal{N}_{A/\mathbb{P}^r}(-2)) = 0$. Let X be a closed irreducible subvariety of $\mathbb{P}^{r+m}$, with $m \geq t+1$, of dimension $\dim X = \dim A + m$ and such that $A = X \cap H$ (scheme-theoretically) for some linear subspace $H = \mathbb{P}^r$ of $\mathbb{P}^{r+m}$. Then $(X, \mathcal{O}_{\mathbb{P}^{r+m}}(1)_{|X})$ is a cone over $(Y, \mathcal{O}_{\mathbb{P}^{r+m-1}}(1)_{|Y})$ for a subvariety Y of $\mathbb{P}^{r+m-1}$.*

There is also a specialized criterion of ours [98] for a variety to be a cone, that settles a conjecture of Conte and Murre [168] on Gorenstein varieties with ample anticanonical bundle.

We give a second argument classifying pairs (X, A) where $A \cong \mathbb{P}^n$ with $n \geq 2$ is an ample Cartier divisor on a normal space, X. The smooth version of this was proven in modern times by Ramanujam [508] and Sommese [567]. The general case is due to Bădescu [42].

Theorem 5.4.10. *Let $A \cong \mathbb{P}^n$ be an ample Cartier divisor on a normal projective variety, X. If $n \geq 2$, then X is the cone $\mathcal{C}(\mathbb{P}^n, \mathcal{O}_{\mathbb{P}^n}(k))$ on $(\mathbb{P}^n, \mathcal{O}_{\mathbb{P}^n}(k))$ where $\mathcal{N}_{A/X} \cong \mathcal{O}_{\mathbb{P}^n}(k)$.*

Proof. Let $L = \mathcal{O}_X(A)$. Assume first that $L_A \cong \mathcal{O}_{\mathbb{P}^n}(1)$. Consider

$$0 \to -L \to \mathcal{O}_X \to \mathcal{O}_{\mathbb{P}^n} \to 0. \tag{5.11}$$

From the fact that $n \geq 2$ and X is normal we conclude that $h^1(-L) = 0$ by (2.2.7) which combined with $h^1(\mathcal{O}_{\mathbb{P}^n}) = 0$ implies that $h^1(\mathcal{O}_X) = 0$. Using this and the sequence (5.11) tensored with L we see that $h^0(L) = n+2$ and that L is spanned by global sections. Thus we get a map, $\phi : X \to \mathbb{P}^{n+1}$, with degree of ϕ equal to $L^{n+1} = L^n_{\mathbb{P}^n} = 1$. Since L is ample, ϕ is finite and generically one-to-one. By the Zariski main theorem, we conclude that $X \cong \mathbb{P}^{n+1}$. This proves the theorem when $L_{\mathbb{P}^n} \cong \mathcal{O}_{\mathbb{P}^n}(1)$.

If $L_{\mathbb{P}^n} \cong \mathcal{O}_{\mathbb{P}^n}(k)$, then we use the Rossi extension theorem, (2.6.1), to conclude that $X \cong \mathbb{P}^{n+1}/G$ where $G \cong \mathbb{Z}/k\mathbb{Z}$ acts on $\mathbb{P}^{n+1}$ with $\mathbb{P}^n$ as a set of fixed points. From this and the description of the automorphism group of $\mathbb{P}^{n+1}$ as the group of projective linear transformations, we see that a generator of G acts as

$$\begin{pmatrix} I_{n+1} & 0 \\ 0 & e^{\frac{2\pi\sqrt{-1}}{k}} \end{pmatrix}$$

where $\mathbb{P}^n$ is identified with the hyperplane of equation $z_{n+1} = 0$ and I_{n+1} is the $(n+1) \times (n+1)$ identity matrix. From this we see that the full fixed point set of G consists of $\mathbb{P}^n$ given by $z_{n+1} = 0$ and the point $\{0, \dots, 0, 1\}$. Using the identification of $\mathbb{P}^{n+1} - \{0, \dots, 0, 1\}$ as $\mathcal{O}_{\mathbb{P}^n}(1)$, it is straightforward to check that X is the cone over $(\mathbb{P}^n, \mathcal{O}_{\mathbb{P}^n}(k))$. Indeed if $\mathcal{L}$ is a line bundle on a manifold X and $\mathbb{Z}_k$ acts, $\rho : \mathbb{Z}_k \times \mathcal{L} \to \mathcal{L}$, on fibers of $\mathcal{L} \to X$ by multiplication of k-th roots of unity, then $\mathcal{L}/\mathbb{Z}_k \cong k\mathcal{L}$. From this it can be seen that the quotient of the cone on $(X, \mathcal{L})$ by the extension to the cone of the action, ρ, is the cone on $(X, k\mathcal{L})$. □

5.5 $\mathbb{P}^d$-bundles as ample divisors

Smooth $\mathbb{P}^d$-bundles as ample divisors on local complete intersections come up in many cases. The following conjecture describes all known examples.

Conjecture 5.5.1. *Let L be an ample line bundle on a projective local complete intersection variety, X, of dimension $n \geq 3$. Assume that there is a smooth $A \in |L|$ such that A is a $\mathbb{P}^d$-bundle, $p : A \to B$, over a manifold, B, of dimension b. Then $d \geq b - 1$ and it follows that $(X, L) \cong (\mathbb{P}(\mathscr{E}), \xi_{\mathscr{E}})$ for an ample vector bundle, $\mathscr{E}$, on B with p equal to the restriction to A of the induced projection $\mathbb{P}(\mathscr{E}) \to B$, except if either:*

1. *$X \subset \mathbb{P}^4$ is a quadric and $L \cong \mathcal{O}_{\mathbb{P}^4}(1)_X$; or*
2. *$(X, L) \cong (\mathbb{P}^3, \mathcal{O}_{\mathbb{P}^3}(2))$; or*
3. *$A \cong \mathbb{P}^1 \times \mathbb{P}^{n-2}$, p is the product projection onto the second factor, $(X, L) \cong (\mathbb{P}(\mathscr{E}), \xi_{\mathscr{E}})$ for an ample vector bundle, $\mathscr{E}$, on $\mathbb{P}^1$ with the product projection of A onto the first factor equal to the induced projection $\mathbb{P}(\mathscr{E}) \to \mathbb{P}^1$.*

The conjecture has been shown except when $d = 1$, $b \geq 3$, and the base B does not map finite-to-one into its Albanese variety. The case when either $d \geq 2$ or B is a submanifold of an Abelian variety follows from Sommese's extension theorems (see (5.2.1), Sommese [567], and Fujita [241]), and was one of the motivations for those theorems. This argument works also in the case when B maps finite-to-one into its Albanese variety.

Theorem 5.5.2. *Conjecture* (5.5.1) *is true if $d \geq 2$.*

Proof. Since the result is trivial if B is a point we can assume without loss of generality that $\dim B \geq 1$ and thus that $n \geq d + 2 \geq 4$.

Since A is smooth and ample it follows that the singularities of X are finite. By (2.2.8) it follows that the singularities of X are rational. It thus follows from

(5.2.1) and (5.2.3) that $p : A \to B$ extends to a morphism, $\overline{p} : X \to B$. Note that $d \geq b - 1$ by (5.2.5). All fibers of $\overline{p}$ are equal dimensional. Indeed the fibers of $\overline{p}$ are all at least $(d+1)$-dimensional, and since any fiber of $\overline{p}$ meets A in a $\mathbb{P}^d$, the fibers of $\overline{p}$ are at most $(d+1)$-dimensional. For a general fiber, F, of $\overline{p}$, the divisor $A_F = A \cap F$ is a smooth ample divisor which is isomorphic to $\mathbb{P}^d$. Note that the condition $d \geq 2$ implies $n \geq 4$. Therefore Theorem (5.4.10) implies that $F \cong \mathbb{P}^{d+1}$ and $L^{d+1} \cdot F = 1$. The result now follows from Proposition (3.2.1). □

The conjecture is also known in the case when $d = 1$ and $b \leq 2$. The proof when $b = 1$ is due to Bădescu [39, 40, 41], and the proof when $b = 2$ is due to Fania and Sommese [228], Fania, Sato, and Sommese [226], Sato and Spindler [541, 542], and Sato [539]. These results follow as consequences of the results in Chapter 7. We include the proof when $d = 1 = b$ following the argument of Bădescu. The central part of the proof is an analysis of how the proof of the second author's extension theorem, (5.2.1), can fail in the concrete case of a $\mathbb{P}^1$-bundle over $\mathbb{P}^1$.

Theorem 5.5.3. (Bădescu) *Conjecture* (5.5.1) *is true if* $d = 1$ *and* $b = 1$.

Proof. Since A is smooth and ample it follows that the singularities of X are finite. Note that by (2.2.8) it follows that the singularities of X are rational.

Choose a section E of $p : A \to B$ of minimal self-intersection, $-r$, and let f denote a fiber of p. By the Lefschetz theorem on ample divisors, (2.3.4), it happens that either the restriction map $\alpha : \mathrm{Pic}(X) \to \mathrm{Pic}(A) = \mathbb{Z}[E] \oplus \mathbb{Z}[f]$ is an isomorphism or α has as image $\mathbb{Z}(xE + yf)$ for two relatively prime integers, x, y.

We claim that we will be done if we show that p extends to a morphism $\overline{p} : X \to B$. In this case $\mathrm{rank} H^2(X, \mathbb{Z}) \geq 2$, and therefore $\mathrm{Pic}(X) \cong \mathrm{Pic}(A) = \mathbb{Z}[E] \oplus \mathbb{Z}[f]$. Let F be a general fiber of the extension $\overline{p}$. Since a fiber of p, which is isomorphic to $\mathbb{P}^1$, is an ample divisor on F we know by (3.1.3) and (3.1.2) that either $F \cong \mathbb{F}_a$ for some $a \geq 0$ or $F \cong \mathbb{P}^2$. The class $F \in H^2(X, \mathbb{Z})$ goes to $f \in H^2(A, \mathbb{Z})$ and $0 \in H^2(F, \mathbb{Z})$. Thus there is a class $D \in H^2(X, \mathbb{Z})$ which must go to $E \in H^2(A, \mathbb{Z})$. Note that F and D go to a basis of $H^2(A, \mathbb{Z})$, and thus are a basis of $H^2(X, \mathbb{Z})$. Since F goes to 0 in $H^2(F, \mathbb{Z})$ and since L goes to a nontrivial element of $H^2(F, \mathbb{Z})$, it follows that $L_F \sim tD_F$ for some $t \neq 0$. Since

$$t^2 D_F^2 = L_F^2 = tD_F \cdot L_F = tD \cdot L \cdot F = tD_A \cdot F_A = tE \cdot f = t$$

we see that $t = 1$ and $L_F^2 = 1$. Thus $(F, L_F) \cong (\mathbb{P}^2, \mathcal{O}_{\mathbb{P}^2}(1))$, and by (3.2.1) we are done.

If $B \not\cong \mathbb{P}^1$, it follows from (5.2.3) that $p : A \to B$ extends to a morphism, $\overline{p} : X \to B$. Therefore we can assume without loss of generality that $B \cong \mathbb{P}^1$.

We know that $A \cong \mathbb{F}_r$ for some $r \geq 0$ with p the induced projection $\mathbb{F}_r \to \mathbb{P}^1$ by (3.2.9). Let E denote a section of $\mathbb{F}_r \to \mathbb{P}^1$ satisfying $E^2 = -r$ and let f

denote a fiber of p. Assume we are in the case where α is an isomorphism and the curves E and f give a basis for $H^2(X, \mathbb{Z})$. By (3.2.9) we know that $L_A \cong aE + bf$ with $b \geq ar + 1$ and $a \geq 1$.

Let H be the line bundle on X corresponding to the bundle associated to f. By the proof of (5.2.1), the map p will extend to $\overline{p} : X \to \mathbb{P}^1$ if we show that $h^1((H - tL)_A) = 0$ for all $t \geq 1$. This is equivalent to $h^1(K_{\mathbb{F}_r} + tL_A - f) = 0$ for $t \geq 1$. This happens if $tL_A - f$ is nef and big, which happens by (3.2.9) unless $b = 1$, $r = 0$. In this case we can use the other projection, $q : A \to \mathbb{P}^1$, of $\mathbb{F}_0 \cong \mathbb{P}^1 \times \mathbb{P}^1$. If q extends then we will be in case 3) of (5.5.1) and done. If q does not extend to $\overline{q} : X \to \mathbb{P}^1$, then we have, by the same argument that we just used to try to construct an extension of p, that $a = 1$ also. Therefore $-K_X \approx 3L$ by the adjunction formula. This implies that we are in case 1) by (3.1.6).

Now we will do the case when $\operatorname{Pic}(X) \not\cong \operatorname{Pic}(A)$. Since $L_A \cong aE + bf$, with $a \geq 1$ and $b \geq ar + 1$, we know that we can choose a generator of the image of $\operatorname{Pic}(X)$ of the form $xE + yf$ with $x \geq 1$ and $y \geq xr + 1$. Since $-K_A \approx 2E + (2+r)f$ must be a positive multiple of $xE + yf$ we have that $2 + r \geq 2r + 1$, i.e., $r = 0, 1$.

If $r = 0$ this gives $-K_X \approx tH$ for some $H \in \operatorname{Pic}(X)$ with $H_A \cong E + f$. Here $t \geq 3$. By (3.1.6) this happens only in the cases 1) and 2) of (5.5.1).

The case $r = 1$ does not occur. Note that in this case we conclude from (3.1.6) that $-K_X \approx 2L$ with $L_A = 2E + 3f$. We give a sketch of the argument that we leave as an exercise to the reader to complete (e.g., Fujita's argument [257, (8.5)] showing that E is the restriction of some element of $\operatorname{Pic}(X)$ applies here). To see that $r = 1$ does not occur note that by the vanishing theorem, (2.2.4), we have $h^1(-L) = 0$. Thus from the sequence

$$0 \to -L \to \mathcal{O}_X \to \mathcal{O}_A \to 0$$

we conclude that $h^1(\mathcal{O}_X) = 0$. Thus from

$$0 \to \mathcal{O}_X \to L \to L_A \to 0$$

we conclude that $\Gamma(L) \to \Gamma(L_A) \to 0$ is onto. Since L_A is spanned we conclude that $\Gamma(L)$ spans L on A. Since $A \in |L|$ we see that L is spanned off A and thus on all of X. The general $A' \in |L|$ is smooth and since $h^1(\mathcal{T}_{\mathbb{F}_1}) = 0$ we have that $A' \cong \mathbb{F}_1$. Choosing a general linear $\mathbb{P}^1 \subset |L|$, we can define an effective divisor, $\mathcal{D}$, as the union of the curves E from the fibers of the pencil that are isomorphic to $\mathbb{F}_1$. We have the contradiction that the divisor $\mathcal{D} \in \operatorname{Pic}(X)$ goes to E which is not a multiple of any $xE + yf$ with $x \geq 1$ and $y \geq rx + 1$. □

Chapter 6

Families of unbreakable rational curves

Recall that an irreducible, reduced curve is said to be *rational* if the normalization of the curve is isomorphic to $\mathbb{P}^1$.

The varieties arising naturally in adjunction theory are of negative Kodaira dimension, and by basic results of Mori, covered with rational curves, which satisfy a strong nonbreaking condition. Often these curves are in fact lines (made precise below), and these families which are intrinsic to the adjunction theory of the varieties are of classical interest.

In §6.1 we give some definitions and some classical examples of families of lines.

In §6.2 we first make precise what we mean by a family of rational curves, and show that there exist families satisfying a strong maximality property. This maximality property lets us use deformation theory when X is smooth to give lower bounds on the size of the family in terms of the *length* of a curve C in the family, $-K_X \cdot C$, or equivalently in terms of the *normal degree*, $\nu := -K_X \cdot C - 2$, of a curve, C, in the family. Note that if C was smooth ν is simply the degree of the normal bundle of C in X. Our main debt in this section is to Mori.

In §6.3 we prove the complex version of Mori's nonbreaking lemma, which guarantees that the locus of unbreakable rational curves through a point cannot be mapped to any lower dimensional variety except a point. As one consequence we prove an important bound due to Ionescu [335] relating the size of the locus of such curves on projective manifolds to the normal degree of the curves. We also show that on manifolds, equality happens in the Kawamata rationality theorem, (1.5.2), if and only if the nefvalue morphism gives to X the structure of $\mathbb{P}(\mathcal{E})$ for a vector bundle, $\mathcal{E}$, over a projective manifold.

In §6.4 we show that given a morphism of a variety, X, which is covered by unbreakable rational curves, to a variety, Y, then either the curves must map to points or the morphism must satisfy stringent conditions. In particular if the morphism is a nefvalue morphism, we show that either the morphism is a contraction of an extremal ray (represented by curves in the family), or the nefvalue and the invariants of the rational curves are small.

In §6.4 we also show that if L is an ample line bundle on a connected, projective manifold, X, with K_X nonnef, then if the nefvalue is $> \dim X/2 + 1$, then the nefvalue morphism of the pair is either birational or the contraction of an extremal ray. We also classify what happens in the boundary case when the nefvalue of the pair is equal to $\dim X/2 + 1$.

In §6.5 we make an extended application of the methods of this chapter to the classical problem of the classification of projective manifolds covered by lines. We give an application of the results to projective manifolds of positive defect, i.e., whose dual varieties are of codimension > 1.

In §6.6 we prove a number of results on spannedness of adjoint bundles.

After this book was written we received Kollár's paper [360] which we refer the reader to for interesting related material.

6.1 Examples

Let X denote a projective variety. We say that an irreducible and reduced curve, $C \subset X$, is *unbreakable* if the integer homology class, $[C]$, in $H_2(X, \mathbb{Z})$ associated to C cannot be written as a nontrivial sum of the integer homology classes of effective curves. For us there are two important examples of unbreakable curves.

1. If L is an ample line bundle on X and if C is a curve such that $L \cdot C = 1$, then C is unbreakable. Such a curve, if rational, is called a *line on X relative to L*, or a *line on* (X, L), or simply *a line on* X when no confusion will result. If L is ample and spanned then by Corollary (1.4.3) lines on X are smooth $\mathbb{P}^1$'s.

2. If X is smooth and C is an extremal rational curve (see Theorem (4.2.5) for the definition) then C is unbreakable. Note that by the fundamental results of Mori there are extremal rational curves on any smooth n-fold with nonnef canonical sheaf.

There are many other examples, e.g., smooth rational curves of self-intersection -2 on surfaces with nef canonical bundle.

Examples of varieties with many lines are very common in adjunction theory.

Example 6.1.1. Let $X = \mathbb{P}^n$. Then (X, L) with $L := \mathcal{O}_{\mathbb{P}^n}(k)$ has lines only if $k = 1$. In this case the family of lines, which covers X, is naturally isomorphic to $\mathbb{P}(\mathcal{T}^*_{\mathbb{P}^n}) \subset \mathrm{Grass}(2, n+1) \times X$. In particular the family is of dimension $2n - 2$. Note that, in the notation of §6.3, D_x is $\mathbb{P}^n$, $\mathcal{P}_x$ is $\mathbb{P}^{n-1}$, $\mathcal{P}$ is $\mathrm{Grass}(2, n+1)$, and $\mathcal{F}_x$ is isomorphic to $\mathbb{P}(\mathcal{O}_{\mathbb{P}^{n-1}} \oplus \mathcal{O}_{\mathbb{P}^{n-1}}(1))$.

Example 6.1.2. Let X denote an irreducible quadric of dimension n, i.e., $X \subset \mathbb{P}^{n+1}$ and $L := \mathcal{O}_{\mathbb{P}^{n+1}}(1)_X$. If $n = 1$, then there are no lines on X, but if $n \geq 2$, then

X is covered by lines. If X is smooth and odd dimensional (respectively even dimensional) then X is covered by one family (respectively two distinct families) of lines of dimension $2n-3$. See Hodge and Pedoe [325] and Zak [630] for more on these important families. See also Segre [550] for a classification of n-dimensional varieties covered by a family of lines of dimension $2n-3$.

Example 6.1.3. Let (X, L) denote an irreducible degree 3 variety of dimension n, where L is a very ample line bundle on X such that $L^n = 3$. If $n = 1$ then X is not a line. If $n = 2$ and X is smooth, then as we will see later in §8.10 either:

1. X is a hypersurface of degree 3 in $\mathbb{P}^3$ with $L := \mathcal{O}_{\mathbb{P}^{n+1}}(1)_X$; or

2. $X := \mathbb{F}_1$ with $L := E + 2f$.

In the first case X has exactly 27 lines (see Mumford [466, §8.5] or Hartshorne [318, Chap. V, §4]) for a survey of the results on the lines in this case), and in the second there are two families of lines, one consisting only of E, and the other 1-dimensional family of lines consists of fibers of the map $\mathbb{F}_1 \to \mathbb{P}^1$.

If $n \geq 3$ then X is covered by lines. This will follow from an easy general result we give following this example. We note that if X is a smooth cubic hypersurface, the family of lines has a smooth surface, $\mathcal{P}$, that embeds in its 5-dimensional Albanese variety, as its parameter space. The surface, $\mathcal{P}$, which is called the Fano family of lines, has been much studied, and plays a major role (see Clemens and Griffiths [165]) in the proof that X is not birational to $\mathbb{P}^3$.

Theorem 6.1.4. *Let X be an irreducible variety of $\mathbb{P}^N$ of degree d and dimension n. Then X is covered by a family of lines of dimension at least $2n-d-1$ if $d \leq n$.*

Proof. First assume that X is a hypersurface, and $N = n+1$. Let $f = 0$ be a defining equation of X. Choose $x \in X$ and a hyperplane $H \subset \mathbb{P}^{n+1}$ with $x \notin H$. Restricting f to the Euclidean coordinate chart, $\mathbb{P}^{n+1} - H$ of $\mathbb{P}^{n+1}$, we can assume that x is at the origin of $\mathbb{C}^{n+1}$ and $f = f_1 + \cdots + f_d$ where the f_i are homogeneous polynomials of degree i in the coordinates of $\mathbb{C}^{n+1}$. If $d \leq n$ then the system of the f_i's defines a subvariety of dimension at least $n-d$ of $\mathbb{P}^n$ which equals the union of the lines on X containing x. If $d \leq n$ then this subvariety is nonempty proving the theorem for hypersurfaces.

Choose a general projection, φ, of X from a $\mathbb{P}^{N-n-1}$ to $\mathbb{P}^{n+1}$ which is generically one-to-one and such that $\varphi_{\varphi^{-1}(U)} : \varphi^{-1}(U) \to U$ is an isomorphism for some affine neighborhood U. Let $X' := \varphi(X)$. Then X' is a hypersurface of degree d in $\mathbb{P}^{n+1}$ and, by the above, is covered by a $(2n-d-1)$-dimensional family of lines. Hence for any $x \in \varphi^{-1}(U)$ there is an $(n-d)$-dimensional family of lines ℓ through $\varphi(x)$. Take the proper transform ℓ' of ℓ under φ. Since φ is generically one-to-one it follows that $L \cong \varphi^*\mathcal{O}_{\mathbb{P}^{n+1}}(1)_{|X'}$ and therefore $L \cdot \ell' = \mathcal{O}_{\mathbb{P}^{n+1}}(1) \cdot \ell = 1$. Therefore we find an $(n-d)$-dimensional family of lines for (X, L) through x. □

Motivated by results of Clemens and Griffiths [165] on the Fano surface of lines on a cubic hypersurface in $\mathbb{P}^4$ and classical results of Van der Waerden [605] on lines on hypersurfaces, Barth and Van de Ven make a study in [73] of the family of lines on hypersurfaces in $\mathbb{P}^N$. They show the following result.

Theorem 6.1.5. (Barth-Van de Ven [73, Theorem 8]) *The parameter space of the family of lines on a hypersurface of degree, d, in* $\mathbb{P}^N$ *is:*

1. *nonempty for every such hypersurface if* $d+3 \leq 2N$*;*
2. *nonsingular of dimension* $2N-d-3$ *for the general hypersurface if* $d+3 \leq 2N$*; and*
3. *connected for every hypersurface if* $d+5 \leq 2N$*, or if* $N = d = 4$.

6.2 Families of unbreakable rational curves

"Families of unbreakable rational curves" play an important role in adjunction theory. In this section we make precise what we mean by such a family, and show that maximal families exist. We need some minimal information on the Chow variety in this section. For more details on Chow varieties, we refer the reader to the helpful discussion in Harris [311], and the detailed exposition in Barlet [67].

If L is a very ample line bundle on a smooth projective manifold, X, containing unbreakable curves and the degree of the unbreakable curves is 1 or 2, then the curves are all smooth. This makes it straightforward to use the Hilbert scheme to construct maximal flat families with good properties, containing a given curve. Indeed given such a curve, C, let W be the connected component of $\mathrm{Hilb}(X)$ containing C, and let $Z_W \to W$ be the pullback of the universal flat family $Z \to \mathrm{Hilb}(X)$ to W. It is in fact true that $Z_W \to W$ is a $\mathbb{P}^1$-bundle and has all the properties one can hope for.

Unfortunately it cannot be assumed that unbreakable rational curves are smooth. Such a curve, if singular, will have arithmetic genus > 0, and the flat families obtained from the universal flat family over the Hilbert scheme will contain curves we do not want, and miss the curves that we do want. For example, consider the flat family containing the smooth twisted cubics in $\mathbb{P}^3$. Though certainly not unbreakable curves, this example shows the key difficulty. There is in the flat family of rational curves (beautifully illustrated by Hartshorne [318, Chap. I, Ex. 2.10]) a singular rational curve, C, which is generically reduced, but of necessity nonreduced. To see that C is nonreduced, note that the arithmetic genus, $p_a(C)$, is positive. On the other hand, since the family is flat, $\chi(\mathcal{O}_C) = 1$, which contradicts $p_a(C) > 0$ unless C is nonreduced. If we started with the reduction of that singular curve, the flat family we would get from the Hilbert scheme could contain none of the smooth twisted cubics. To get them from the singular rational curve, we would have to put the nonreduced structure on the singular curve. The prospect of putting nonreduced structures on rational curves so that things work out is daunting at best.

This leads us to define a *family of rational curves on a quasi-projective variety*, X, to be a quadruple, $(\mathcal{F}, \mathcal{P}, p, q)$, consisting of a reduced subvariety, $\mathcal{F} \subset \mathcal{P} \times X$, with $\mathcal{P}$ a reduced variety, the *parameter space*, and where the maps $p : \mathcal{F} \to \mathcal{P}$, $q : \mathcal{P} \to X$, are induced by the product projections, and where p is proper, algebraic, with all fibers generically reduced with reductions being irreducible, rational curves. Such a family is *irreducible* (respectively *compact*, respectively *connected*) if $\mathcal{F}$ or equivalently $\mathcal{P}$ is *irreducible* (respectively *compact*, respectively *connected*). A family, $(\mathcal{F}', \mathcal{P}', p', q')$, is said to contain a family $(\mathcal{F}, \mathcal{P}, p, q)$ if there is a holomorphic map, $i : \mathcal{P} \to \mathcal{P}'$, such that the image of $\mathcal{F}$ in $\mathcal{P}' \times X$ under (i, id_X) is contained in $\mathcal{F}'$. A family, $(\mathcal{F}', \mathcal{P}', p', q')$, is said to be *comparably equal* to a family $(\mathcal{F}, \mathcal{P}, p, q)$ if there is a surjective holomorphic map, $i : \mathcal{P} \to \mathcal{P}'$, such that the image of $\mathcal{F}$ in $\mathcal{P}' \times X$ under (i, id_X) is contained in $\mathcal{F}'$. Two families, $(\mathcal{F}', \mathcal{P}', p', q')$ and $(\mathcal{F}, \mathcal{P}, p, q)$, are said to be *equal* if they are contained in the same equivalence class of the equivalence relation generated by the relation of being comparably equal.

If any two distinct fibers of p go to distinct curves in X under q, we say the family is a *Chow family of rational curves*. Given any family of rational curves, $(\mathcal{F}, \mathcal{P}, p, q)$, it follows because the fibers of p are irreducible, generically reduced, curves that there is a holomorphic map from $\mathcal{P}$ to the Chow variety of X. If $\mathcal{P}$ is compact, then the image in the Chow variety of X (with its reduced structure) will be a variety, and in this case we can replace our family by a Chow family which contains all the curves in the family.

A family of rational curves, $(\mathcal{F}, \mathcal{P}, p, q)$, is *maximal* if there is no family, $(\mathcal{F}', \mathcal{P}', p', q')$, containing $(\mathcal{F}, \mathcal{P}, p, q)$ and not equal to it. Let L be an ample line bundle on X. We say that a family of rational curves, $(\mathcal{F}, \mathcal{P}, p, q)$, is a *family of lines* if $L \cdot C = 1$ for any curve, C, from the family.

6.2.1. Maximal Chow families of unbreakable rational curves. Assume X is compact. To study families of rational curves, we initially regard a singular, rational, irreducible curve, $C \subset X$, as a map, $\phi : \overline{C} \to X$, of the normalization, $\overline{C} (\cong \mathbb{P}^1)$ of C, into the variety X, and then deform the pair $(\overline{C}, \phi)$. Since the normalization of a rational curve is isomorphic to $\mathbb{P}^1$, and has no deformations, this comes down to studying the deformations of the graph, $G(\phi)$, of ϕ in $\mathbb{P}^1 \times X$. Using the Hilbert scheme of $\mathbb{P}^1 \times X$ we obtain a connected universal flat family of rational curves, $\mathcal{G} \subset \mathcal{M} \times \mathbb{P}^1 \times X$ of $\mathbb{P}^1 \times X$, $\mathcal{M}$ the parameter space, containing the graph, $G(\phi)$, we started with. Let $\mathcal{G}'$ denote the image under the product projection of the universal flat family, $\mathcal{G}$, in $\mathcal{M} \times X$. To show that $\mathcal{G}' \subset \mathcal{M} \times X$ is a family of rational curves, we need that any fiber of $\mathcal{G}' \to \mathcal{M}$ is irreducible and reduced. To see this when C is unbreakable, note that all fibers of the flat map $\mathcal{G} \to \mathcal{M}$ represent the same homology class in $\mathbb{P}^1 \times X$. By construction each fiber of $\mathcal{G}' \to \mathcal{M}$ is the image of a fiber of $\mathcal{G} \to \mathcal{M}$, and thus they also have the same homology classes. Thus if a fiber of $\mathcal{G}' \to \mathcal{M}$ was not irreducible and generically reduced, we would have that the homology class of C is the nontrivial sum of homology classes of effective curves, contradicting the definition of unbreakable curve. Replacing the family $\mathcal{G}' \to \mathcal{M}$ with its associated Chow family gives the family that we want.

We will need the following technical result in the proof of Proposition (6.2.4).

Lemma 6.2.2. *Let X be an irreducible, projective variety, $\Delta \subset \mathbb{C}$ the unit disk, and $\mathcal{U} \subset \Delta \times X$ an irreducible and reduced variety. Assume that the proper map, $u : \mathcal{U} \to \Delta$, induced by the product projection has irreducible and generically reduced curves as fibers. Assume that each fiber has $\mathbb{P}^1$ as the normalization of its reduction. Let $\mu : \overline{\mathcal{U}} \to \mathcal{U}$ be the normalization of $\mathcal{U}$, and let $\overline{u} : \overline{\mathcal{U}} \to \Delta$ be the composition of μ with u. Then there is a biholomorphism, $\Delta \times \mathbb{P}^1 \to \overline{\mathcal{U}}$, whose composition with $\overline{u}$ is the product projection. In particular this biholomorphism to $\Delta \times \mathbb{P}^1$ gives a family of graphs of maps of $\mathbb{P}^1 \to X$, $\widetilde{\mathcal{U}} \subset \Delta \times \mathbb{P}^1 \times X$, which projects to $\mathcal{U}$ under the product projection.*

Proof. Using Corollary (4.1.3), this result is clear. □

Remark 6.2.3. If the genus of the normalizations of the fibers of $\mathcal{U} \to \Delta$ were constant, but not 0, the map $\overline{u} : \overline{\mathcal{U}} \to \Delta$ would be of maximal rank, and therefore we would get a similar conclusion allowing not only the map but the smooth curve which is the domain of the map to deform.

The following result shows that the family constructed contains the curves it should and satisfies a good maximality property.

Proposition 6.2.4. *Let X be an irreducible projective variety. Let $(\mathcal{F}, \mathcal{P}, p, q)$ be an irreducible, not necessarily compact Chow family of unbreakable rational curves. There exists a compact, maximal Chow family of rational curves of the type constructed in* (6.2.1) *that contains $(\mathcal{F}, \mathcal{P}, p, q)$.*

Proof. Choose any rational curve, C, arising as a fiber of the product projection $p : \mathcal{F} \to \mathcal{P}$. Let $\mathcal{C} \subset \mathcal{T} \times X$ be the universal family over the open subvariety, $\mathcal{T}$, of the Chow variety of X parameterizing irreducible, generically reduced curves. Let $\mathcal{V}$ be the parameter space containing C, arising in the construction (6.2.1). We can regard $\mathcal{P}$ and $\mathcal{V}$ as being contained in $\mathcal{T}$. If $\mathcal{P}$ is contained in $\mathcal{V}$, then the result is proven. If $\mathcal{P}$ is not contained in $\mathcal{V}$, then there exists a map of a disk Δ

into $\mathcal{P}$ with 0 going to the point representing C, and $\Delta - \{0\}$ going to $\mathcal{P} - \mathcal{V}$. This is impossible by Lemma (6.2.2). □

Let us explicitly point out the following consequence of the proposition above.

Remark 6.2.5. From Proposition (6.2.4) it follows in particular that given a single unbreakable rational curve, C, on an irreducible projective variety X, there exists a compact, maximal Chow family of rational curves that contains C.

In the case when L is a very ample line bundle on a projective variety, and we are dealing with a line, C, with respect to L it is possible to work directly with the Hilbert scheme.

We often deal with the restriction of a maximal family to an irreducible component of the parameter space. Such a family is a maximal irreducible family.

6.3 The nonbreaking lemma

Throughout this section X is an irreducible projective variety, and $(\mathcal{F}, \mathcal{P}, p, q)$ is an irreducible, compact, maximal Chow family of unbreakable rational curves on X. We define the *locus of the family*, $(\mathcal{F}, \mathcal{P}, p, q)$, to be the variety, $\mathcal{E}(\mathcal{F}, \mathcal{P}, p, q) := q(\mathcal{F})$. Often we make the convenient abuse of referring to the family and to the locus of the family as $\mathcal{F}$, $\mathcal{E}(\mathcal{F})$ respectively. We say the family is a *covering family* if $\mathcal{E}(\mathcal{F}) = X$. We call $\dim \mathcal{P}$ the *dimension of the family*.

The dimension, $\dim \mathcal{F} - \dim \mathcal{E}(\mathcal{F})$, of the general fiber of the map $q : \mathcal{F} \to \mathcal{E}(\mathcal{F})$ is denoted δ. Note that

$$\dim \mathcal{F} = \dim \mathcal{P} + 1 \tag{6.1}$$

and hence that

$$\delta + \dim \mathcal{E}(\mathcal{F}) = \dim \mathcal{P} + 1. \tag{6.2}$$

Note that the fiber $F_x := q^{-1}(x)$ over a point $x \in \mathcal{E}(\mathcal{F})$ of $q : \mathcal{F} \to \mathcal{E}(\mathcal{F})$ maps homeomorphically under p to $\mathcal{P}_x$, where $\mathcal{P}_x$ is the subvariety of $\mathcal{P}$ corresponding to curves from the family that contain x. We denote the dimension of $\mathcal{P}_x$ by δ_x. By upper semi-continuity of dimensions of the fibers of the map q we conclude that

$$\delta_x \geq \delta \text{ for all } x \in \mathcal{E}(\mathcal{F}). \tag{6.3}$$

For $x \in \mathcal{E}(\mathcal{F})$, we denote $p^{-1}(\mathcal{P}_x)$ by $\mathcal{F}_x$, and let D_x denote the set $q(\mathcal{F}_x)$, which we will refer to as the locus of curves from the family containing x.

Note that if a projective manifold has a rational curve going through a general point then the canonical bundle is nonnef by a theorem of Miyaoka and Mori [444].

Let X be a projective manifold and let $\rho : X \to Z$ be the contraction of an extremal ray, R. We often use $\mathcal{E}(\rho)$ for the exceptional set of the contraction. Let us point out the following fact, used in the proof of Theorem (6.3.6). Let C be an extremal rational curve in R. Let E be any irreducible component of the not necessarily connected algebraic set, $\mathcal{E}(\rho)$. Then E contains the maximal irreducible compact family of unbreakable rational curves containing C.

The following is essentially Mori's nonbreaking lemma over the complex numbers.

Theorem 6.3.1. *Notation as above. Fix $x \in \mathcal{E}(\mathcal{F})$. Given any connected variety, $B \subset \mathcal{P}_x$, let $f : p^{-1}(B) \to Z$ be an holomorphic map which takes $F_x \cap p^{-1}(B) := q^{-1}(x) \cap p^{-1}(B)$ to a point. Either $f(p^{-1}(B))$ is a single point, or the restriction map $f_{p^{-1}(B)-F_x \cap p^{-1}(B)}$ is finite-to-one. In particular, any holomorphic function, $h : q(p^{-1}(B)) \to Z$, is either finite-to-one or takes $q(p^{-1}(B))$ to a point.*

Proof. Assume that f is not finite-to-one on $p^{-1}(B) - F_x \cap p^{-1}(B)$. Let $R \subset p^{-1}(B)$ be an irreducible curve going to a point under f, but with $R \not\subset F_x \cap p^{-1}(B)$. Let $Y := p(R) \subset B$ and let $W := p^{-1}(Y)$. Let $\overline{W} \to W$ and $\overline{Y} \to Y$ be the normalization maps of W and Y respectively. The induced mapping $\overline{W} \to \overline{Y}$ is a $\mathbb{P}^1$-bundle by Corollary (4.1.3) or Lemma (6.2.2). Thus the composition of $\overline{W} \to W$ with the restriction $f_W : W \to Z$ of f to W has an image of dimension ≤ 1 by Corollary (3.2.15). Since a fiber of $W \to Y$ goes to a rational curve, the whole of W goes to a rational curve. This contradicts the Chow property of the family that no curve is represented by more than one fiber of the family. □

From the above we conclude that $\dim D_x = \dim \mathcal{F}_x (= \dim \mathcal{P}_x + 1)$ and hence

$$\dim D_x = \delta_x + 1 \text{ for all } x \in \mathcal{E}(\mathcal{F}). \tag{6.4}$$

Note also that, by taking $B = \mathcal{P}_x$ in the theorem above, we have that any holomorphic function $h : D_x \to Z$ is either finite-to-one or takes D_x to a point.

Given a rational curve on a $\mathbb{Q}$-Gorenstein variety, X, we define its *normal degree*, $\nu(C)$, to be $-K_X \cdot C - 2$, or ν when no confusion will result. Note that, if C is smooth, $\nu(C) = \deg \mathcal{N}_{C/X}$. In particular $\nu(C) \geq 0$ if $\mathcal{N}_{C/X}$ is spanned. Given a family of rational curves, $(\mathcal{F}, \mathcal{P}, p, q)$, it follows from Corollary (4.1.3), that the normal degree of any curve of the family is independent of the curve in the family; this number, $\nu(\mathcal{F})$, is called *the normal degree of the family*. The *length*, length(C), of the curve C is defined to be $\nu(C)+2$, i.e., $-K_X \cdot C$. Similarly the *length of the family* $\mathcal{F}$, length$(\mathcal{F})$, is defined to be $\nu(\mathcal{F})+2$. We sometimes use the notation $(\mathcal{F}, \mathcal{P}, p, q, \nu)$ to mean a family of normal degree $\nu = \nu(\mathcal{F})$. We also say that an extremal rational curve, C, is of *minimal degree* if length$(C) =$ length(R) where R is the extremal ray $R := \mathbb{R}_+[C]$.

Theorem 6.3.2. *Notation as above. Let $(\mathcal{F}, \mathcal{P}, p, q)$ be a maximal connected family of unbreakable rational curves on a connected projective manifold, X, of dimension n. Let $\nu := \nu(\mathcal{F})$ be the normal degree of the family. Then for each irreducible component, $P \subset \mathcal{P}$, we have that* $\dim P \geq n + \nu(\mathcal{F}) - 1$, *and* $\delta_x \geq \nu(\mathcal{F})$ *for $x \in \mathcal{E}(\mathcal{F})$. Moreover $\nu(\mathcal{F}) \leq n - 1$.*

Proof. Fix an arbitrary curve C in the family. Let $\phi : \mathbb{P}^1 \to C \subset X$ be the normalization of C and let $G(\phi)$ be the graph of ϕ. By (4.1.9) applied to $G(\phi)$ we know that the dimension of any component, H_ϕ, of the Hilbert scheme containing $G(\phi)$ is $\geq \chi(\phi^*(\mathcal{T}_X)) = -K_X \cdot C + n = \nu + 2 + n$.

Any two graphs, $G(\phi_1)$, $G(\phi_2)$, for generically one-to-one maps $\phi_i : \mathbb{P}^1 \to X$, $i = 1, 2$, going to the same C satisfy $\phi_1 = \phi_2 \circ g$ for some $g \in \mathrm{Aut}(\mathbb{P}^1)$. It suffices to check this for generically one-to-one maps $\phi_i : \mathbb{P}^1 \to C$, which upon normalizing C reduces to checking it for two generically one-to-one maps of $\mathbb{P}^1$ to itself. Since such maps are biholomorphic by the Zariski main theorem, the result is immediate. Let P_ϕ be the irreducible component of $\mathcal{P}$ corresponding to H_ϕ. Since $\dim P_\phi \geq \dim H_\phi - \dim \mathrm{Aut}(\mathbb{P}^1)$ and since $\dim \mathrm{Aut}(\mathbb{P}^1) = 3$, we conclude that the bound $\dim P \geq n + \nu - 1$ holds for every irreducible component, P, of $\mathcal{P}$.

Fix a point $x \in C$ and a generically one-to-one map $\phi : \mathbb{P}^1 \to X$ with $\phi(0) = x$ and $\phi(\mathbb{P}^1) = C$. We have that the dimension of each irreducible component, H_ϕ, of the Hilbert scheme of $\mathbb{P}^1 \times X$ containing ϕ, or the graph $G(\phi)$, is bounded below by $\chi(\phi^*\mathcal{T}_X \otimes \mathcal{I}_0) = -K_X \cdot C - n + n = \nu + 2$, where $\mathcal{I}_0$ denotes the ideal sheaf of the point 0 in $\mathbb{P}^1$. Noting that the subgroup, A_0, of the automorphisms of $\mathbb{P}^1$ that fixes 0 is 2-dimensional, the same argument as above lets us conclude that $\delta_x = \dim \mathcal{P}_x \geq \dim H_\phi - \dim A_0 \geq \nu$.

Finally choose two distinct points x, x' on C and a generically one-to-one map $\phi : \mathbb{P}^1 \to X$ with $\phi(\mathbb{P}^1) = C$, $\phi(0) = x$, and $\phi(\infty) = x'$. By (4.1.9), the dimension of each irreducible component, H_ϕ, of the Hilbert scheme containing ϕ is bounded below in dimension by $\chi(\phi^*\mathcal{T}_{\mathbb{P}^1} \otimes \mathcal{I}_0 \otimes \mathcal{I}_\infty) = -K_X \cdot C - 2n + n = \nu + 2 - n$, where $\mathcal{I}_\infty$ is the ideal sheaf of ∞ in $\mathbb{P}^1$. Since the automorphisms of $\mathbb{P}^1$ taking 0 to 0 and ∞ to ∞ form a 1-dimensional group, $A_{(0,\infty)}$, and since $\dim(\mathcal{P}_x \cap \mathcal{P}_{x'}) \geq \dim H_\phi - \dim A_{(0,\infty)}$, we conclude that $\dim(\mathcal{P}_x \cap \mathcal{P}_{x'}) \geq \nu + 1 - n$. If $\nu + 1 - n \geq 1$ then we get $\dim(\mathcal{P}_x \cap \mathcal{P}_{x'}) \geq 1$. Take a curve C in $\mathcal{P}_x \cap \mathcal{P}_{x'}$ and look at the surface $S := p^{-1}(C)$. Then $S \subset \mathcal{F}_x$. The fibers, S_x, $S_{x'}$, over x, x' of the restriction map $q : S \to X$ are 1-dimensional. Note that $S_x = S \cap F_x$, $S_{x'} = S \cap F_{x'}$. Therefore the restriction of q to $S - S_x \subset \mathcal{F}_x - F_x$ contracts $S_{x'}$. This is in contradiction to the fact that $q_{\mathcal{F}_x - F_x}$ is finite-to-one by Theorem (6.3.1). Therefore we conclude that $\nu + 1 \leq n$. □

Let us consider the following particular case when the family of unbreakable rational curves covers X.

Corollary 6.3.3. *Let $(\mathcal{F}, \mathcal{P}, p, q)$ be a maximal connected family of unbreakable rational curves on a connected projective manifold, X, of dimension n. Let $\nu := \nu(\mathcal{F})$ be the normal degree of $\mathcal{F}$. Assume that $\mathcal{F}$ covers X, i.e., $\mathcal{E}(\mathcal{F}) = X$. Let C be a general curve from the family and let P_C be any irreducible component of $\mathcal{P}$ containing the point corresponding to C. Then* $\dim P_C = n + \nu - 1$ *and $\delta_x = \nu$ for a general point x in X.*

Proof. Let C be a general irreducible curve in the family. Let $\phi : \mathbb{P}^1 \to X$ be the morphism from the normalization of C into the variety X and let $G(\phi)$ be the graph of ϕ. As in (6.2.1), let $\mathcal{G} \subset \mathcal{M} \times \mathbb{P}^1 \times X$ be a connected flat family of curves of $\mathbb{P}^1 \times X$ containing the graph $G(\phi)$. Since $\mathcal{F}$ covers X we know that there is a graph, $\gamma \in \mathcal{G}$, of a map $\phi : \mathbb{P}^1 \to X$ corresponding to a general curve C with $\phi(\mathbb{P}^1)$ in the family $\mathcal{F}$, such that small deformations of γ in $\mathbb{P}^1 \times X$ have image in X containing an open set U of X. Note that if γ is in the family $\mathcal{G}$, then the graph of $\phi \circ g$ belongs to $\mathcal{G}$ for $g \in \mathrm{Aut}(\mathbb{P}^1)$. Thus the image, $\mathcal{E}(\mathcal{G})$, of $\mathcal{G}$ in $\mathbb{P}^1 \times X$ contains $\mathbb{P}^1 \times U$. This implies that the family $\mathcal{G}$ covers $\mathbb{P}^1 \times X$.

Thus by combining Lemma (4.1.11) and Remark (4.1.12) we see that at all points, $h^0(\mathcal{N}_{G(\phi)/\mathbb{P}^1 \times X})$ spans $\mathcal{N}_{G(\phi)/\mathbb{P}^1 \times X}$. Thus the dimension of any component, H_ϕ, of the Hilbert scheme containing $G(\phi)$ is $\chi(\phi^*\mathcal{T}_X) = -K_X \cdot C + n = \nu + 2 + n$ and $\dim P_C = \dim H_\phi - \dim \mathrm{Aut}(\mathbb{P}^1)$, where P_C is the irreducible component of $\mathcal{P}$ containing (the point corresponding to) C and corresponding to H_ϕ.

Fix now a general point $x \in X$ and a generically one-to-one map $\phi : \mathbb{P}^1 \to X$ with $\phi(0) = x$ and $\phi(\mathbb{P}^1) = C$. Since x is general, we may assume that $x \in C$. The same argument as in the proof of Theorem (6.3.2), showing $\delta_x \geq \nu$, gives now $\delta_x = \nu$ because the spannedness condition for $\mathcal{N}_{G(\phi)/\mathbb{P}^1 \times X}$ implies that all the inequalities " $\geq$" in that proof are now equalities. □

Remark 6.3.4. Let $(\mathcal{F}, \mathcal{P}, p, q, \nu)$ be a maximal connected family of unbreakable rational curves on a connected projective manifold, X. Assume that $\mathcal{E}(\mathcal{F})$ is irreducible. Let δ, δ_x be as above. Then one has $\delta = \min_{x \in \mathcal{E}(\mathcal{F})}\{\delta_x\} = \delta_{x_0}$, for a general point $x_0 \in \mathcal{E}(\mathcal{F})$. In particular, by (6.3.2), it follows that $\delta \geq \nu$.

If moreover $\mathcal{F}$ covers X, i.e., $\mathcal{E}(\mathcal{F}) = X$, then by (6.3.3), we conclude that $\delta = \nu = \delta_{x_0}$ for a general point x_0 in X.

Following the reasoning of Ionescu [335, (0.4)] (see also Wiśniewski [619]) we use the nonbreaking lemma to give lower bounds for the locus of deformations of rational curves.

Theorem 6.3.5. *Notations as above. Let X be an irreducible, projective variety, and let $(\mathcal{F}, \mathcal{P}, p, q)$ be an irreducible, compact, maximal family of unbreakable rational curves. Let $\nu := \nu(\mathcal{F})$ be the normal degree of $\mathcal{F}$. Let $\phi : X \to Y$ be an holomorphic mapping, and let Δ denote any fiber of the restriction $\phi_{\mathcal{E}(\mathcal{F})}$. Let $\mathcal{F}_x := p^{-1}(\mathcal{P}_x)$, $D_x := q(\mathcal{F}_x)$.*

1. *If* $\dim \phi(C) = 1$ *for some curve, C, in the family, then for any* $x \in \mathscr{E}(\mathscr{F})$, $\dim(\Delta \cap D_x) = 0$ *and* $\dim \phi(\mathscr{E}(\mathscr{F})) \geq \delta_x + 1$.

2. *If* $\phi(C)$ *is a point for some curve, C, in the family, then* ϕ *takes every curve in the family to a point and if X is smooth, then* $\dim \mathscr{E}(\mathscr{F}) + \dim \Delta \geq \dim X + \nu(\mathscr{F}) + 1$.

Proof. For a given $x \in \mathscr{E}(\mathscr{F})$ take a fiber Δ through x. If $\dim(\Delta \cap D_x) \geq 1$ there exists a curve C_x contained in $\Delta \cap D_x$. Then $\dim \phi(C_x) = 0$, contradicting the fact that no curves from the family get contracted since $\dim \phi(C) = 1$. Thus $\dim(\Delta \cap D_x) = 0$. Since no curves from the family get contracted by ϕ we have that $\phi(D_x)$ is not a point. Therefore $\phi(q(\mathscr{F}_x))$ is not a point. Let $f := \phi \circ q$. Then by Theorem (6.3.1) we know that $f_{\mathscr{F}_x - F_x}$ is finite-to-one, or $f_{\mathscr{F}_x}$ is generically finite-to-one. Thus, since $D_x = q(\mathscr{F}_x)$, ϕ_{D_x} is generically finite-to-one. Therefore

$$\dim \phi(\mathscr{E}(\mathscr{F})) \geq \dim \phi(D_x) = \dim D_x = \delta_x + 1.$$

This proves the first statement.

To see the second part of the theorem, note that since $\dim \phi(C) = 0$, ϕ takes every curve of the family to a point. Then clearly $D_x \subset \Delta$ if $x \in \Delta$. Therefore $\dim \Delta \geq \dim D_x$. By (6.3), (6.4) we conclude that $\dim \Delta \geq \delta + 1$, where δ denotes the dimension of the general fiber of the map $q : \mathscr{F} \to \mathscr{E}(\mathscr{F})$. Thus

$$\dim \mathscr{E}(\mathscr{F}) + \dim \Delta \geq \dim \mathscr{E}(\mathscr{F}) + \delta + 1.$$

Now using (6.2) we are reduced to showing $\dim \mathscr{P} \geq n + \nu(C) - 1$, which is just Theorem (6.3.2). □

The following very useful consequence of Theorem (6.3.5) is essentially due to Ionescu [335, (0.4)] (see Wiśniewski [621, (1.1)]).

Theorem 6.3.6. *Let X be a smooth projective variety of dimension n. Assume that* K_X *is nonnef and let R be an extremal ray on X of length* $\ell(R)$. *Let* ρ *be the contraction of R and let E be any irreducible component of the locus of R. Let* Δ *be any irreducible component of any fiber of the restriction,* ρ_E, *of* ρ *to E. Then*

$$\dim E + \dim \Delta \geq n + \ell(R) - 1.$$

Proof. Let C be a rational curve, $[C] \in R$, such that $-K_X \cdot C$ reaches the minimum, i.e., $-K_X \cdot C = \ell(R)$, or $\ell(R) = \nu(C) + 2$, where $\nu(C)$ denotes the normal degree of C. Note that the curve C is unbreakable by the minimality condition. Take a maximal irreducible compact family $(\mathscr{F}, \mathscr{P}, p, q, \nu)$, $\nu = \nu(C)$, of unbreakable rational curves containing C. Let $\mathscr{E}(\mathscr{F}) := q(\mathscr{F})$. Then $\mathscr{E}(\mathscr{F}) \subset E$. Let $\Delta_{\mathscr{F}} :=$

$\Delta \cap \mathscr{E}(\mathscr{F})$. Then by Theorem (6.3.5) we have

$$\dim E + \dim \Delta \geq \dim \mathscr{E}(\mathscr{F}) + \dim \Delta_{\mathscr{F}} \geq n + \nu(C) + 1$$

or $\dim E + \dim \Delta \geq n + \ell(R) - 1$. □

Remark 6.3.7. Ionescu [335, (0.4)] stated the bound in (6.3.6) for the set, E, as a whole, i.e., $2 \dim E \geq n + \ell(R) - 1$. Wiśniewski [621, (1.1)] made the very useful observation that Ionescu's argument actually showed $\dim E + \dim \Delta \geq n + \ell(R) - 1$. Indeed the key point of the strengthened inequality is the fact that $\dim \Delta \geq \dim D_x$, which is implicit in Ionescu's argument, plus the inequality [335, p. 460, line 7]. It has also been pointed out (we believe by Kawamata) that the inequality, $\dim E + \dim \Delta \geq n + \ell(R) - 1$, reduces by simply slicing with general hyperplane sections on $\rho(X)$ to the inequality $2 \dim E \geq n + \ell(R) - 1$.

We refer to Kawamata [345] for a version of the result above in the singular case.

The following result from Andreatta, Ballico, and Wiśniewski [20, (1.1)] is a useful fact in adjunction theory. This result, which is not used in this book, describes the boundary case in the theorem above, in the special case when the contraction is the nefvalue morphism associated to a polarized pair. See also our paper [110] for a similar result covering the cases when the upper bound allowed by the Kawamata rationality theorem is taken on.

Proposition 6.3.8. *Let X be a smooth projective variety of dimension n and let L be an ample line bundle on X. Let ϕ be the nefvalue morphism of (X, L). Assume that ϕ is a contraction of an extremal ray, R. Let E be an irreducible component of the exceptional locus of ϕ and let Δ be an irreducible component of any fiber of the restriction ϕ_E of ϕ to E. Let $d = \dim \Delta$ and let $\alpha : \Delta' \to \Delta$ be the normalization of Δ. Further assume that $\dim E + \dim \Delta = n + \ell(R) - 1$. Then $(\Delta', \alpha^* L_\Delta) \cong (\mathbb{P}^d, \mathcal{O}_{\mathbb{P}^d}(1))$.*

Let us prove the following consequences of Theorem (6.3.6) and the contraction theorem, (4.3.1), which will be used in the sequel. The first of the lemmas below is the analogous, in the smooth case, of Lemma (4.2.17).

Lemma 6.3.9. *Let X be a smooth connected n-dimensional variety. Let R_1, R_2 be distinct extremal rays on X with $\text{length}(R_1) = \ell_1$, $\text{length}(R_2) = \ell_2$. Let E_1, E_2 be the loci of R_1, R_2 respectively. If $\ell_1 + \ell_2 \geq n + 3$, then E_1 and E_2 are disjoint.*

Furthermore if R_1, R_2 are nonnef and $\ell_1 + \ell_2 \geq n + 1$ then $E_1 \cap E_2 = \emptyset$.

Proof. Assume that $E_1 \cap E_2 \neq \emptyset$ and take a point $x \in E_1 \cap E_2$. Let ρ_i be the contractions of R_i and let Δ_i be an irreducible component of a fiber of ρ_i with $x \in \Delta_i$, $i = 1, 2$. If $\dim(\Delta_1 \cap \Delta_2) \geq 1$, there exists a curve, ℓ, contained in $\Delta_1 \cap \Delta_2$

which contracts to a point under ρ_1, ρ_2. Therefore $[C] \in R_1$, $[C] \in R_2$. This leads to the contradiction $R_1 = R_2$. Thus we have

$$0 = \dim(\Delta_1 \cap \Delta_2) \geq \dim \Delta_1 + \dim \Delta_2 - n$$

and hence

$$\dim \Delta_1 + \dim \Delta_2 \leq n. \tag{6.5}$$

By (6.3.6) we have, for $i = 1, 2$,

$$n + \dim \Delta_i \geq \dim E_i + \dim \Delta_i \geq n + \ell_i - 1.$$

This gives $\dim \Delta_1 + \dim \Delta_2 \geq \ell_1 + \ell_2 - 2 \geq n + 1$, which contradicts (6.5).

If ρ_1, ρ_2 are birational, Theorem (6.3.5) yields for $i = 1, 2$

$$n - 1 + \dim \Delta_i \geq \dim E_i + \dim \Delta_i \geq n + \ell_i - 1.$$

This gives $\dim \Delta_1 + \dim \Delta_2 \geq \ell_1 + \ell_2 \geq n + 1$, the same contradiction as above. □

Remark 6.3.10. Notation as in (6.3.9). The above proof shows that $E_1 \cap E_2 = \emptyset$ if $\ell_1 + \ell_2 \geq n + 3 - \mathrm{cod}_X E_1 - \mathrm{cod}_X E_2$. Note that in this case the assumption that R_1, R_2 are nonnef is not needed.

Lemma 6.3.11. *Let L be an ample line bundle on a smooth projective variety X of dimension n. Let R be an extremal ray on X and let $\varphi : X \to Y$ be the contraction of R. Assume that $K_X + (n-2)L$ is nef and big and $K_X + (n-2))L \approx \varphi^* H$ for some line bundle H on Y. Then the locus E of R is a prime divisor and the restriction $\varphi_E : E \to \varphi(E)$ maps E to a point or to a curve.*

Proof. Note that φ is a birational morphism since $\varphi^* H$ is nef and big, so in particular R is nonnef. Let Δ be an irreducible component of a fiber of φ_E and let $R = \mathbb{R}_+[\ell]$, ℓ an extremal rational curve. Assume $\dim E \leq n - 2$. Then, by (6.3.6),

$$\dim \Delta \geq \mathrm{length}(R) + 1.$$

Since $(K_X + (n-2)L) \cdot \ell = 0$ we have $-K_X \cdot \ell = (n-2)L \cdot \ell \geq n - 2$, and hence $\mathrm{length}(R) \geq n - 2$. So we conclude that $\dim \Delta \geq n - 1$, which contradicts the assumption $\dim E \leq n - 2$.

Thus $\dim E = n - 1$. As in the proof of (4.3.1), since R is nonnef, there exists a prime divisor, D, such that $D \cdot R < 0$. Therefore $D \cdot C < 0$ and hence $D \supset C$ for every curve C in X, $[C] \in R$. This implies that D is unique and, since $\dim E = n - 1$, D coincides with E. Then from Theorem (6.3.6) we get

$$\dim \Delta \geq \mathrm{length}(R) \geq n - 2.$$

Thus φ_E has either a curve or a point as image. □

Lemma 6.3.12. *Let X be a smooth projective variety of dimension n. Assume that there exists an extremal rational curve, C, on X.*

1. *If $\nu(C) = n-1$, i.e., $-K_X \cdot C = n+1$, then* $\mathrm{Pic}(X) \cong \mathbb{Z}$ *and $-K_X$ is ample.*
2. *If $\nu(C) = n-2$, i.e., $-K_X \cdot C = n$, then either* $\mathrm{Pic}(X) \cong \mathbb{Z}$ *and $-K_X$ is ample or X has Picard number 2 and the contraction $\rho = \mathrm{cont}_R : X \to Z$ associated to the extremal ray $R = \mathbb{R}_+[C]$ is a morphism onto a smooth curve, whose general fiber is a smooth $(n-1)$-fold, F, with* $\mathrm{Pic}(F) \cong \mathbb{Z}$ *and $-K_F$ is ample.*

Proof. Let $\rho : X \to Z$ be the contraction associated to the extremal ray $R = \mathbb{R}_+[C]$. Take a maximal irreducible compact family $(\mathcal{F}, \mathcal{P}, p, q, \nu)$, $\nu = \nu(C)$, of unbreakable rational curves containing C. Let $\mathcal{E}(\mathcal{F}) := q(\mathcal{F})$, and $\Delta = \rho^{-1}(z) \cap \mathcal{E}(\mathcal{F})$ for any $z \in \rho(\mathcal{E}(\rho))$. Then $\mathcal{E}(\mathcal{F}) \subset \mathcal{E}(\rho)$, where $\mathcal{E}(\rho)$ denotes the exceptional locus of the contraction ρ.

Assume $\nu(C) = n-1$. If we show that $\dim Z = 0$, then by the basic exact sequence associated to a contraction of an extremal ray

$$0 \to \mathrm{Pic}(Z) \to \mathrm{Pic}(X) \overset{\cdot C}{\to} \mathbb{Z},$$

with the cokernel of the map $\mathrm{Pic}(X) \overset{\cdot C}{\to} \mathbb{Z}$ being finite, we will conclude that $\mathrm{Pic}(X) \cong \mathbb{Z}$ and that $-K_X$ is ample. By Theorem (6.3.5) we see that for any $z \in \rho(\mathcal{E}(\rho))$,

$$\dim \mathcal{E}(\rho) + \dim \rho^{-1}(z) \geq \dim \mathcal{E}(\mathcal{F}) + \dim \Delta \geq n + \nu(C) + 1 = 2n.$$

This implies that $\dim \mathcal{E}(\rho) = \dim \rho^{-1}(z) = n$ which proves 1).

Assume $\nu(C) = n-2$. The same argument as above, by using Theorem (6.3.5), shows that $\dim \mathcal{E}(\mathcal{F}) = n$, i.e., $\mathcal{E}(\mathcal{F}) = X$. Hence in particular $\mathcal{E}(\rho) = X$ and $\dim \rho^{-1}(z) \geq n-1$. Then either ρ maps X to a point or onto a smooth curve Z. In the first case we fall again in the case 1) above. If $\dim \rho^{-1}(z) = n-1$, X has Picard number 2 since Z has Picard number 1 and $-K_F$ is ample for a general fiber, F, of $\rho : X \to Z$ by the contraction theorem, (4.3.1). Since $\mathcal{E}(\mathcal{F}) = X$, deformations, C', of C meet the general fiber, F, of ρ. Since ρ contracts C to a point, C' is contained in F. Since $\nu(C) = \nu(C')$, we have $-K_X \cdot C = -K_X \cdot C' = n = \dim F + 1$. So we conclude that $\mathrm{Pic}(F) \cong \mathbb{Z}$ by the same argument as in case 1). □

Remark 6.3.13. Notation as in Lemma (6.3.12). Note that the same conclusions as in (6.3.12), 1), 2) hold true if instead of the assumption on the normal degree of C, we assume $\ell(R) = n+1$ or $\ell(R) = n$ respectively, where R is the extremal ray generated by C (see Wiśniewski [619, (2.4)]). To see that (6.3.12) gives the statement in terms of the length, take a rational curve, γ, in the ray R such that $-K_X \cdot \gamma = \ell(R)$ and note that γ is extremal (see also (4.2.15)). Then apply (6.3.12) to γ.

Recently, Cho and Miyaoka [156] gave the following new numerical characterization of complex projective spaces and quadric hypersurfaces. The result is stronger than Lemma (6.3.12). Though we do not use it through the book, we include it for completeness, referring to [156] for the proof.

Theorem 6.3.14. *Let X be a smooth Fano variety of dimension n.*

1. *If $\nu(C) \geq n-1$ for every effective rational curve, $C \subset X$, then $X \cong \mathbb{P}^n$.*
2. *$X \cong Q$, Q quadric hypersurface in $\mathbb{P}^{n+1}$, if and only if*

$$\min\{\nu(C) \mid C \subset X \text{ is a rational curve}\} = n-2.$$

The following result characterizes the cases where the nefvalue takes the maximal value allowed by the Kawamata rationality theorem, (1.5.2).

Theorem 6.3.15. *Let L be an ample line bundle on X, a connected projective manifold of dimension n. Assume that K_X is nonnef and let $\phi : X \to Y$ be the nefvalue morphism of the pair (X, L). Assume that $vK_X + uL \cong \phi^*H$ for some ample line bundle H on Y and coprime positive integers u and v. If $u = \max_{y\in Y}\{\dim \phi^{-1}(y)\} + 1$, then ϕ expresses X as a $\mathbb{P}^{u-1}$-bundle over Y. Moreover there exists an ample line bundle $\mathcal{L}$ on X and a positive integer b such that $K_X + u\mathcal{L} \cong b\phi^*H$, and in particular $X \cong \mathbb{P}(\phi_*\mathcal{L})$.*

Proof. By Lemma (1.5.6) we can find an ample line bundle $\mathcal{L}$ on X with $K_X+u\mathcal{L} \cong b\phi^*H$ for some positive integer b.

Let R be an extremal ray such that $(K_X + u\mathcal{L}) \cdot R = 0$. Let C be an extremal rational curve generating the ray, and let $\rho : X \to Z$ be the contraction associated to the ray. Let $\mathcal{E}(\rho)$ be the exceptional locus of the contraction, ρ. We have $\phi = \mu \circ \rho$ for a morphism with connected fibers, $\mu : Z \to Y$. Let Δ be a positive dimensional fiber of ρ. By the same argument as in the proof of Theorem (6.3.6), by using Theorem (6.3.5), and since $\nu(C) = -K_X \cdot C - 2 = u\mathcal{L} \cdot C - 2$, we have

$$\begin{aligned} \dim \Delta &\geq \nu(C) + 1 + (n - \dim \mathcal{E}(\rho)) \qquad (6.6)\\ &= u\mathcal{L} \cdot C - 1 + (n - \dim \mathcal{E}(\rho)) \geq u - 1 + (n - \dim \mathcal{E}(\rho)). \end{aligned}$$

Since $\rho(\Delta)$ is a point, it follows that $\phi(\Delta)$ is a point. Thus

$$\dim \Delta \leq \max_{y\in Y}\{\dim \phi^{-1}(y)\} = u - 1.$$

This implies that $n = \mathcal{E}(\rho)$ and thus that $\dim Z < \dim X$, and by inequality (6.6) that $\dim \rho^{-1}(z) = u-1$ for all $z \in Z$. Let F be a fiber of ρ. Since $(K_X+u\mathcal{L})_F \approx \mathcal{O}_F$ we have $K_X+u\mathcal{L} \approx \rho^*\mathcal{H}$ for some line bundle $\mathcal{H}$ on Z. By replacing $\mathcal{H}$ by $\mathcal{H}+uM$ and $\mathcal{L}$ by $\mathcal{L} + \rho^*M$ for some ample line bundle M on Z, we may assume that

$\mathcal{H}$ is ample. Hence $(X, \mathcal{L})$ is a scroll over Z under ρ. Then, for a general fiber F of ρ, $(F, \mathcal{L}_F) \approx (\mathbb{P}^{u-1}, \mathcal{O}_{\mathbb{P}^{u-1}}(1))$. Thus we see from Proposition (3.2.1) that $\rho : X \to Z$ is in fact a $\mathbb{P}^{u-1}$-bundle. Note also that μ is an isomorphism. If not, then it would follow by Zariski's main theorem that there is a positive dimensional fiber, $A = \mu^{-1}(y)$, $y \in Y$. Therefore $\dim \phi^{-1}(y) = \dim A + u - 1 \geq u$. This contradicts the fact that $\dim \phi^{-1}(y) \leq \max_{y \in Y}\{\dim \phi^{-1}(y)\} = u - 1$. Thus we conclude that $\phi : X \to Y$ is a $\mathbb{P}^{u-1}$-bundle. Since $K_X + u\mathcal{L} \cong \phi^*(bH)$, we have $X \cong \mathbb{P}(\phi_* \mathcal{L})$. □

The following result will be useful (see Ye and Zhang [626] and also Fujita [259]).

Theorem 6.3.16. (Ye-Zhang) *Let E be an ample vector bundle on a connected projective manifold, X, of dimension n. If* rank$E \geq n + 1$, *then $K_X + \det E$ is ample unless* rank$E = n + 1$ *and* $(X, E) \cong (\mathbb{P}^n, \mathcal{O}_{\mathbb{P}^n}(1)^{\oplus(n+1)})$.

Proof. Since $K_X + \det E$ is not ample there exists an effective curve, B, such that $(K_X + \det E) \cdot B \leq 0$. By the cone theorem, (4.2.5), we can write B in $NE(X)$ as a finite sum $B = \sum_i \lambda_i C_i + \gamma$ where $\lambda_i \in \mathbb{R}_+$, $R_i := \mathbb{R}_+[C_i]$ are extremal rays and γ satisfies the condition $K_X \cdot \gamma \geq 0$. Since $(K_X + \det E) \cdot \gamma > 0$, if γ is not zero, it must be $(K_X + \det E) \cdot C_i \leq 0$ for some index i, say $i = 1$. Then there exists a rational curve, C, which generates R_1 and such that $-K_X \cdot C = \text{length}(R_1)$, i.e., C is of minimal degree, $(K_X + \det E) \cdot C \leq 0$. Since the normalization $\overline{C}$ is biholomorphic to $\mathbb{P}^1$, the pullback, $\overline{E}$, under the normalization map $\overline{C} \to C$ splits as a direct sum of line bundles, $\oplus_{i=1}^{\text{rank}E} \mathcal{O}_{\mathbb{P}^1}(a_i)$. Since E is ample, the pullback, $\overline{E}$, is ample and thus $a_i > 0$ for all i. Since the normalization map is generically one-to-one, we conclude that

$$\det E \cdot C = \det \overline{E} \cdot \overline{C} = \deg \overline{E}_{\overline{C}} = \sum_{i=1}^{\text{rank}E} a_i \geq \text{rank}E.$$

Thus $-K_X \cdot C \geq \det E \cdot C \geq \text{rank}E$. Since C is extremal, $-K_X \cdot C \leq n + 1$. Therefore we conclude that rank$E = n + 1$ and $(K_X + \det E) \cdot C = 0$. By Lemma (6.3.12), we conclude that $\text{Pic}(X) \cong \mathbb{Z}$, and thus that $K_X + \det E \cong \mathcal{O}_X$.

Consider $M \cong \mathbb{P}(E)$. Note that $\dim M = n + \text{rank}E - 1 = 2n$. Since $\text{Pic}(X) \cong \mathbb{Z}$, we have that $\text{Pic}(M) \cong \mathbb{Z} \oplus \mathbb{Z}$. The cone of effective curves in $\overline{NE}(M)$ is 2-dimensional. Note that $K_M \cong -(n+1)\xi$ where ξ is the tautological line bundle on $\mathbb{P}(E)$. Since E is ample, ξ and hence $-K_M$ is ample. Since $-K_M$ is ample, we conclude that there are 2 extremal rays generating the cone, that is, M has Picard number 2.

Choose an extremal ray, R, which is not generated by any curve in a fiber of $M \to X$. Let ℓ be an extremal rational curve of minimal degree generating the ray. Let $\rho : M \to Y$ be the contraction associated to this ray. Note that

$-K_M \cdot \ell = (n+1)\xi \cdot \ell \geq n+1$. Hence $\nu(\ell) = \text{length}(R) - 2 \geq n-1$. Therefore by Theorem (6.3.5) we see that given any fiber $\Delta = \rho^{-1}(y)$ where $y \in \rho(\mathscr{E}(\rho))$, it follows that $\dim \mathscr{E}(\rho) + \dim \Delta \geq \dim M + \nu(\ell) + 1 \geq 3n$. Noting that

$$0 \geq \dim(\Delta \cap F) \geq \dim \Delta + \dim F - \dim M \geq \dim \Delta - n$$

for any fiber, F, of $M \to X$, we see that $\dim \Delta \leq n$. Thus $\dim \mathscr{E}(\rho) = 2n$, i.e., $\dim Y < \dim M$. It also follows that $\dim \Delta = n$. Thus for a general fiber, Δ, of ρ we have $K_\Delta \cong -(\dim \Delta + 1)\xi_\Delta$. Since, by the above, ρ is equidimensional, by Theorem (3.1.6) and Proposition (3.2.1), we conclude that ρ is a $\mathbb{P}^n$-bundle. The map $M \to X$ restricted to Δ is finite-to-one. By Theorem (3.1.7), we conclude that $X \cong \mathbb{P}^n$. Hence in particular $-K_X \cdot C = -K_{\mathbb{P}^n} \cdot C = n+1$, so that C is a line. Thus E restricted to any line in X is a direct sum of $n+1$ copies of $\mathscr{O}_{\mathbb{P}^1}(1)$. From this we conclude by Okonek, Schneider, and Spindler [493, (3.2.1)] that E splits as a direct sum of $n+1$ copies of $\mathscr{O}_{\mathbb{P}^n}(1)$. □

6.4 Morphisms of varieties covered by unbreakable rational curves

The following result is basic.

Proposition 6.4.1. *Let X be an n-dimensional connected projective manifold with nonnef canonical bundle. Let $\rho : X \to Z$ be the contraction associated to an extremal ray, R. Let C be an extremal rational curve of minimal degree. Let $(\mathscr{F}, \mathscr{P}, p, q)$ be a maximal connected family of unbreakable rational curves with normal degree $\nu(\mathscr{F})$. Assume that $\mathscr{E}(\rho) \cap \mathscr{E}(\mathscr{F}) \neq \emptyset$, and that $\nu(C) + \nu(\mathscr{F}) \geq n-2$. If ρ does not take curves in the family $\mathscr{F}$ to points, then all the following hold:*

1. *$\nu(C) + \nu(\mathscr{F}) = n-2$;*
2. *$\mathscr{E}(\rho) = X$;*
3. *all fibers of ρ are $(\nu(C)+1)$-dimensional;*
4. *any morphism $\mu : Z \to W$ with a positive dimensional image is finite-to-one. In particular if Z is smooth, then $-K_Z$ is ample, and* Pic(Z) $\cong \mathbb{Z}$.

Proof. Let Δ be a fiber of the restriction map $\rho_{\mathscr{E}(\rho)}$ which meets $\mathscr{E}(\mathscr{F})$. Let $x \in \Delta \cap \mathscr{E}(\mathscr{F})$. Let D_x denote the locus of curves in the family $\mathscr{F}$ which contain x. Using the smoothness of X, Theorem (6.3.6) and Theorem (6.3.2) we have

$$\begin{aligned} \dim(\Delta \cap D_x) &\geq \dim \Delta + \dim D_x - n \\ &\geq n - \dim \mathscr{E}(\rho) + \operatorname{length}(R) - 1 + \dim D_x - n \\ &\geq n - \dim \mathscr{E}(\rho) + \nu(C) + 1 + \nu(\mathscr{F}) + 1 - n \\ &\geq \nu(C) + \nu(\mathscr{F}) + 2 - n. \end{aligned} \tag{6.7}$$

If $\nu(C) + \nu(\mathscr{F}) \geq n - 1$, then Δ meets D_x in a positive dimensional set, $\rho(D_x)$ is a point by Theorem (6.3.1), and hence ρ takes all curves in the family to points. Thus if ρ does not take each curve in the family to a point, it follows that $\nu(C) + \nu(\mathscr{F}) = n - 2$ and $n = \dim \mathscr{E}(\rho)$, or $X = \mathscr{E}(\rho)$. Hence $\dim Z < n$. Note that, by Theorem (6.3.1), $\dim \rho(D_x) = \dim D_x$. Thus by Theorem (6.3.2) we have $\dim Z \geq \dim D_x \geq \nu(\mathscr{F}) + 1$. Therefore

$$n - \dim Z \leq n - \nu(\mathscr{F}) - 1 = \nu(C) + 1. \tag{6.8}$$

Since $\mathscr{E}(\rho) = X$ we may assume that C is contained in a general fiber, F, of ρ. Note that C is unbreakable since C is extremal. Take the maximal compact Chow family, $(\mathscr{F}', \mathscr{P}', p', q')$, of rational curves that contains C (see Remark (6.2.5)) and let $D'_x := q'(\mathscr{F}'_x)$. Then $\dim D'_x \geq \nu(C) + 1$ by Theorem (6.3.2) so that, since $D'_x \subset F$, $\dim F = n - \dim Z \geq \nu(C) + 1$. Therefore by (6.8) we conclude that $\dim Z = \nu(\mathscr{F}) + 1 = n - \nu(C) - 1$. Thus by Theorem (6.3.5.1) all fibers are equal dimensional.

To see the last point let $\mu : Z \to W$ be a morphism with a positive dimensional image. Note that by the above it follows that $\rho(D_x) = Z$ since $\dim Z = \nu(\mathscr{F}) + 1$ and $\dim \rho(D_x) = \dim D_x \geq \nu(\mathscr{F}) + 1$. Assume that μ has a positive dimensional fiber. Then $\mu \circ \rho$ is not finite-to-one on D_x, so that by Theorem (6.3.1), $\mu \circ \rho(D_x)$ is a point, contradicting $\dim \mu(Z) > 0$. Thus μ is finite-to-one. If Z is smooth, then since $\rho(D_x) = Z$, we conclude that Z is covered by rational curves. Thus K_Z is nonnef by a result of Miyaoka and Mori [444]. Thus there exists a contraction $\sigma : Z \to Y$ of an extremal ray. By the last paragraph, since ρ is not finite-to-one, $\dim \sigma(Z) = 0$ which implies that $\operatorname{Pic}(Z) \cong \mathbb{Z}$ by using the exact sequence of (4.3.1). Since K_Z is nonnef, this implies that $-K_Z$ is ample. □

Very often we will be interested in analyzing a nefvalue morphism, and in particular asking whether it is a contraction of an extremal ray. The following is very useful.

Lemma 6.4.2. *Let X be a projective variety with terminal singularities and a canonical sheaf which is not nef. Let L be an ample line bundle on X and let $\phi : X \to Y$ be the nefvalue morphism associated to (X, L) with nefvalue τ. There is an extremal ray, R, such that $(K_X + \tau L) \cdot R = 0$. Let $\rho_R : X \to Z$ be the contraction from X onto the normal projective variety, Z, associated to R. The map ϕ factors, $\phi = \mu \circ \rho_R$, where $\mu : Z \to Y$ has connected fibers.*

Proof. Let C be any effective curve that ϕ takes to a point. Fix an $\varepsilon \in \mathbb{R}$ such that $\tau > \varepsilon > 0$. Then, by the cone theorem, we can write $[C]$, the homology class of C, as a sum, $\sum_{i\in I} \lambda_i \ell_i + D$, where I is a finite set, ℓ_i are disjoint extremal curves, each $\lambda_i > 0$, and D is an element of the closure of the cone of curves with $(K_X + \varepsilon L) \cdot D \geq 0$. Since $K_X + \tau L$ is nef, we conclude that

$$0 = (K_X + \tau L) \cdot C = (K_X + \tau L) \cdot (\sum_{i\in I} \lambda_i \ell_i) + (K_X + \varepsilon L) \cdot D + (\tau - \varepsilon) L \cdot D \geq 0.$$

From this we conclude that $(\tau - \varepsilon) L \cdot D = 0$ and $(K_X + \tau L) \cdot (\sum_{i\in I} \lambda_i \ell_i) = 0$. The former implies $D = 0$ in $\overline{NE}(X)$, and the later $(K_X + \tau L) \cdot \ell_i = 0$ for each i. Take R to be $R := \mathbb{R}_+[\ell_i]$ for any i. □

We say that the contractions ρ_R with $(K_X + \tau L) \cdot R = 0$ are *subordinate* to ϕ. To show that ϕ is the contraction associated to an extremal ray it suffices to show that the map μ in Lemma (6.4.2) is an isomorphism.

The following two results give some circumstances under which nefvalue morphisms are contractions of extremal rays.

Theorem 6.4.3. *Let L be an ample line bundle on a connected, projective manifold, X, of dimension n. Assume that there is a connected, covering family of unbreakable rational curves, $(\mathcal{F}, \mathcal{P}, p, q)$, on X with normal degree $\nu := \nu(\mathcal{F})$. Let τ and $\phi : X \to Y$ be the nefvalue and nefvalue morphism of (X, L) respectively. Write $\tau = u/v$ with u, v coprime positive integers. Assume that $u + \nu \geq n$. Then one of the following holds:*

1. *ϕ is the contraction of an extremal ray, R, of the form $R = \mathbb{R}_+[C]$ for a curve C in the family and* $\dim Y \leq n - \nu - 1$;
2. *(ϕ is not as in* 1) *and) $u + \nu = n$. In this case either*
   a) *ϕ takes curves from the family to points,* $\dim Y = 0$, *and* $(X, L) \cong (\mathbb{P}^{u-1} \times \mathbb{P}^{\nu+1}, \mathcal{O}_{\mathbb{P}^{u-1}\times\mathbb{P}^{\nu+1}}(v, (\nu+2)v/u))$, *or*
   b) *ϕ does not take curves from the family to points, ϕ is the contraction of an extremal ray, Y is smooth, ϕ expresses X as $\mathbb{P}(E)$ for some ample vector bundle E, of rank u,* $\mathrm{Pic}(Y) \cong \mathbb{Z}$, *$-K_Y$ is ample and the restriction $\phi_{D_x} : D_x \to Y$ is a finite-to-one map, where $D_x := q(p^{-1}(\mathcal{P}_x))$ and $\mathcal{P}_x$ is the subvariety of the parameter space $\mathcal{P}$ corresponding to curves from the family that contains x.*

Proof. By Lemma (1.5.6), we can find an ample line bundle $\mathcal{L}$ on X with $K_X + u\mathcal{L} \cong b\phi^* H$ for some positive integer b and where H is the ample line bundle on Y such that $vK_X + uL \cong \phi^* H$.

Let $\rho : X \to Z$ be a contraction associated to an extremal ray, and choose a rational curve, ℓ, of minimal degree, $-K_X \cdot \ell$, in the ray, and such that $(K_X + u\mathcal{L}) \cdot \ell = 0$. We can do this by Remark (4.2.15). Then ρ is a contraction subordinate to ϕ.

Note that $\nu(\ell) = -K_X \cdot \ell - 2 = u\mathcal{L} \cdot \ell - 2 \geq u - 2$. Therefore $\nu(\ell) + \nu(\mathcal{F}) \geq u - 2 + \nu(\mathcal{F})$. Since $\mathcal{E}(\mathcal{F}) = X$, $\mathcal{E}(\rho) \cap \mathcal{E}(\mathcal{F}) \neq \emptyset$. Therefore if $u + \nu(\mathcal{F}) \geq n + 1$, we have $\nu(\ell) + \nu(\mathcal{F}) \geq n - 1$ and hence we conclude by Proposition (6.4.1), that ρ takes curves in the family $\mathcal{F}$ to points. In particular ρ is the contraction of an extremal ray of the form $\mathbb{R}_+[C]$ for C in the family. This would imply that ρ was the unique contraction subordinate to ϕ and therefore that $\phi = \rho$ (recall that since ϕ has connected fibers, if ϕ factors through ρ, $\phi = \mu \circ \rho$, then either μ is an isomorphism or μ contracts some curve). We are in case 1) after proving that $\dim Y \leq n - \nu - 1$. To see this take a point $x \in C$. Since all curves in the family are contracted to points and since $\mathcal{F}$ covers X we have that for any fiber, F, of ϕ, $\dim F \geq \dim D_x$. Hence $\dim F \geq \nu + 1$ by Theorem (6.3.2). Therefore $\dim Y \leq n - \nu - 1$.

We are therefore reduced to the case when $u + \nu(\mathcal{F}) = n$. We may assume that ρ is not the contraction of an extremal ray of the form $\mathbb{R}_+[C]$ for C in the family, since otherwise we fall again in case 1). Then by Proposition (6.4.1), $\mathcal{E}(\rho) = X$ and all fibers of ρ are $u - 1 = (\nu(\ell) + 1)$-dimensional. Note that $K_F + u\mathcal{L}_F \cong \mathcal{O}_F$ for a general fiber F of ρ. Thus by Theorem (3.1.6) and by Proposition (3.2.1) we conclude that X is $\mathbb{P}(E)$ for a rank u ample vector bundle, $E := \rho_*\mathcal{L}$, on Z, and that Z is smooth. By Proposition (6.4.1), we see that $\mathrm{Pic}(Z) \cong \mathbb{Z}$, $-K_Z$ is ample, and $\rho_{D_x} : D_x \to Z$ is finite-to-one.

Write $\phi = \mu \circ \rho$. If $\phi = \rho$, then we are in case 2b). If $\phi \neq \rho$, then, since $\mathrm{Pic}(Z) \cong \mathbb{Z}$, we conclude that $Y = \mu(Z)$ is a point by Lemma (3.1.8). Thus $K_X + u\mathcal{L} \cong \mathcal{O}_X$. Since $K_X + u\mathcal{L} \cong \rho^*(K_Z + \det E) \cong \mathcal{O}_X$, we conclude that $K_Z + \det E \cong \mathcal{O}_Z$. By Theorem (6.3.16) we conclude that $(Z, E) \cong (\mathbb{P}^{\nu(\mathcal{F})+1}, \mathcal{O}_{\mathbb{P}^{\nu(\mathcal{F})+1}}(1)^{\oplus u})$. This implies that

$$(X, L) \cong (\mathbb{P}^{u-1} \times \mathbb{P}^{\nu(\mathcal{F})+1}, \mathcal{O}_{\mathbb{P}^{u-1} \times \mathbb{P}^{\nu(\mathcal{F})+1}}(v, \frac{v(\nu(\mathcal{F}) + 2)}{u}))$$

and we are in case 2a). To check the last isomorphism recall that by Lemma (1.5.6) we have $\mathcal{L} = bK_X + aL$ for integers a, b, such that $av - bu = 1$. Therefore

$$\begin{aligned} \frac{1 + bu}{v} L &\cong \mathcal{O}_{\mathbb{P}^{u-1} \times \mathbb{P}^{\nu+1}}(1, \frac{\nu + 2}{u}) \otimes \mathcal{O}_{\mathbb{P}^{u-1} \times \mathbb{P}^{\nu+1}}(bu, b(\nu + 2)) \\ &\cong \mathcal{O}_{\mathbb{P}^{u-1} \times \mathbb{P}^{\nu+1}}(1 + bu, (\nu + 2)\frac{1 + bu}{u}). \end{aligned}$$

Thus $L \cong \mathcal{O}_{\mathbb{P}^{u-1} \times \mathbb{P}^{\nu+1}}(v, v(\nu + 2)/u)$. □

The following consequence of Mori's nonbreaking lemma is at the heart of the results in [114].

In [457], Mukai conjectured the following. Let X be a smooth complex connected Fano variety of dimension n. If $\operatorname{index}(X) > n/2+1$ then $\operatorname{Pic}(X) \cong \mathbb{Z}$. This was proved by Wiśniewski in [620]. The following theorem can be also viewed as the "relative version" of the Mukai conjecture, which indeed can be deduced from it when the variety Y (from the theorem) is a point. Let us also point out that in the theorem below we do not assume *a priori* the existence of a covering family of lines.

Theorem 6.4.4. (Relative Mukai conjecture) *Let L be an ample line bundle on a connected n-dimensional projective manifold, X. Assume that K_X is nonnef and that the nefvalue, τ, of the pair (X, L) satisfies $\tau \geq n/2+1$. Let $\phi : X \to Y$ be the nefvalue morphism of the pair, (X, L), and assume that* $\dim Y < \dim X$. *Then there exists a covering family of lines $(\mathcal{F}, \mathcal{P}, p, q)$ for (X, L). Furthermore either ϕ is the contraction associated to an extremal ray and* $\operatorname{Pic}(X) \cong \phi^*\operatorname{Pic}(Y) \oplus \mathbb{Z}[L]$, *or $\tau = n/2+1$, $(X, L) \cong (\mathbb{P}^{n/2} \times \mathbb{P}^{n/2}, \mathcal{O}_{\mathbb{P}^{n/2}\times\mathbb{P}^{n/2}}(1,1))$ and* $\dim Y = 0$.

Proof. Let F be a smooth fiber of ϕ. Since $-K_F$ is ample, it is a well-known general fact that we have seen attributed to Kollár, that there is a family of extremal rational curves, ℓ, covering F, and satisfying $-K_F \cdot \ell \leq \dim F + 1$ (see, e.g., Mori [452] and Miyaoka and Mori [444]). Thus $L \cdot \ell \leq (\dim F+1)/\tau \leq (n+1)/\tau$. Since this is true for all smooth fibers, by finiteness of the number of components of the Hilbert scheme parameterizing curves whose intersection with L is bounded by the same bound, we conclude that there is a maximal connected family of unbreakable rational curves, $(\mathcal{F}, \mathcal{P}, p, q)$, covering X, and with $\nu := \nu(\mathcal{F}) \leq \dim F - 1$. Letting ℓ be a curve in the family we have that $(K_X + \tau L) \cdot \ell = 0$, and therefore $L \cdot \ell (n/2+1) \leq \tau L \cdot \ell = -K_X \cdot \ell = -K_F \cdot \ell \leq \dim F + 1$. Thus we see that $L \cdot \ell = 1$. Therefore $\nu + 2 = -K_X \cdot \ell = \tau$ is integral. Moreover $\tau + \nu(\mathcal{F}) = \tau + \tau - 2 \geq n$. Thus by Theorem (6.4.3), if ϕ is not the contraction of an extremal ray, we conclude that (X, L), ϕ are as in (6.4.3.2a). Hence in particular $\tau + \nu(\mathcal{F}) = n$, $\tau = n/2 + 1 = \nu + 2$. This gives the conclusions of the theorem.

To see that $\operatorname{Pic}(X) \cong \phi^*\operatorname{Pic}(Y) \oplus \mathbb{Z}[L]$ when ϕ is the contraction of an extremal ray use the exact sequence from (4.3.1). Note that the morphism $\operatorname{Pic}(X) \xrightarrow{\cdot \ell} \mathbb{Z}$ is surjective since $L \cdot \ell = 1$. □

Let us recall now two lemmas from [114], that we need to prove the main result of the next section. For completeness, we give here the proof of these lemmas.

Lemma 6.4.5. ([114, (1.4.4)]) *Let L be an ample line bundle on a connected, projective manifold, X, of dimension n. Assume that there is a covering family of unbreakable rational curves $(\mathcal{F}, \mathcal{P}, p, q)$ for (X, L) with normal degree $\nu := \nu(\mathcal{F})$. Assume that there exists a morphism $\varphi : X \to Y$ with connected fibers onto a projective normal variety Y. If φ contracts any curve, C, from the family then* $\dim Y \leq n - \nu - 1$.

Further assume that $\mathcal{F}$ is a family of lines for (X, L), that φ is the nefvalue morphism of (X, L) and that $\dim Y = n - \nu - 1$. *Then (X, L) is a scroll under φ.*

Proof. First note that φ contracts all the curves from the family. Fix a point $x \in C$ and let $D_x := q(p^{-1}(\mathcal{P}_x))$ be the locus of curves from the family containing x. Since $\mathcal{F}$ covers X we have that for any fiber, F, of φ, $\dim F \geq \dim D_x$ and hence $\dim F \geq \nu + 1$ by Theorem (6.3.2). Therefore $\dim Y \leq n - \nu - 1$.

Let τ be the nefvalue of (X, L). To show the second part of the statement note that $(K_F + \tau L_F) \cdot C = 0$ and hence $\tau = -K_X \cdot C = \nu + 2$ since we are now assuming that C is a line. On the other hand $\dim F = \nu + 1 = \tau - 1$. Therefore $K_F \approx -(\dim F + 1)L_F$, so that $(F, L_F) \cong (\mathbb{P}^{\nu+1}, \mathcal{O}_{\mathbb{P}^{\nu+1}}(1))$ by the Kobayashi-Ochiai result (see (3.1.6)). Since $K_X + \tau L \approx \varphi^* H$ for some ample line bundle H on Y, and $\tau = \nu + 2 = \dim F + 1$, we see that (X, L) is a scroll under φ. □

Lemma 6.4.6. ([114, (1.4.5)]) *Let $(\mathcal{F}, \mathcal{P}, p, q, \nu)$ be a connected family of unbreakable rational curves on a connected, projective manifold, X, of dimension n. Assume that $\mathcal{F}$ covers X. Assume that there is a morphism $\varphi : X \to Y$ with connected fibers onto a projective normal variety Y. Further assume that φ does not contract curves from the family to points. For any point $x \in X$, let $\mathcal{P}_x$ be the subvariety of the parameter space $\mathcal{P}$ corresponding to curves from the family that contain x, and let $\delta := \min_{x \in X}\{\delta_x\}$ where $\delta_x = \dim \mathcal{P}_x$. Then any fiber of φ is of dimension $\leq n - \delta - 1$ $(\leq n - \nu - 1)$ and in particular* $\dim Y \geq \delta + 1$. *Moreover, if there exists a fiber, F, of φ such that* $\dim F = n - \nu - 1$, *then*

$$NE(X) = NE(F) + \mathbb{R}_+[C],$$

where $NE(F)$ denotes the subcone of $NE(X)$ generated by curves contained in F, and C is any curve from the family $\mathcal{F}$.

Proof. Fix a point $x \in X$ such that a curve in the family containing x meets a positive dimensional fiber, F, of φ. Let $D_x := q(p^{-1}(\mathcal{P}_x))$ be the locus of the curves from the family that contain x. Since $\mathcal{F}$ covers X, we have $\mathcal{E}(\mathcal{F}) = X$. Therefore, by (6.4), $\dim D_x = \delta_x + 1$ for each $x \in X$. Since φ does not contract curves from the family we have $\dim(F \cap D_x) \leq 0$ and hence

$$0 \geq \dim F + \dim D_x - n = \dim F + \delta_x + 1 - n.$$

In particular $\dim Y \geq \delta_x + 1$. This gives the first part of the statement.

Let δ be the dimension of the general fiber of q. By Remark (6.3.4) we have $\delta = \nu$. Thus we can assume that there exists a fiber, F, of φ such that $\dim F = n - \delta - 1$.

Let $\mathcal{P}(F)$ be the subvariety of $\mathcal{P}$ parameterizing curves meeting F. Note that $\mathcal{P}(F) = p(q^{-1}(F))$. Since no curves from the family belong to F, we see that the restriction map $p_{q^{-1}(F)}$ is finite-to-one onto $\mathcal{P}(F)$, i.e., $q^{-1}(F)$ is a multi-section of the map $p_{p^{-1}(\mathcal{P}(F))}$. Thus $\dim q^{-1}(F) = \dim \mathcal{P}(F)$ and since p has one dimensional fibers we have

$$\dim p^{-1}(\mathcal{P}(F)) = \dim \mathcal{P}(F) + 1 = \dim q^{-1}(F) + 1.$$

Since all fibers of q are of dimension $\geq \delta$ we see that $\dim q^{-1}(F) \geq \dim F + \delta$. Then, by the above,

$$\dim p^{-1}(\mathcal{P}(F)) \geq \dim F + \delta + 1. \tag{6.9}$$

We claim that the restriction, $q' := q_{p^{-1}(\mathcal{P}(F)) - q^{-1}(F)}$, is a finite map. To see this assume otherwise. Then there exists an irreducible curve, $\Gamma \subset p^{-1}(\mathcal{P}(F)) - q^{-1}(F)$, such that $q'(\Gamma)$ is a point, x, and $x \notin F$ since no curves from the family are contracted to points. Let $B := p(\Gamma)$. Note that B is in fact a subvariety of $\mathcal{P}_x$. Let $h := \varphi \circ q : p^{-1}(B) \to Y$. Then h takes Γ to x. By construction, $q^{-1}(F)$ meets all fibers of $p_{p^{-1}(B)}$ and $h(q^{-1}(F))$ is a point on Y. Therefore h is not finite-to-one on $p^{-1}(B) - \Gamma$. Thus by the nonbreaking lemma, (6.3.1), we conclude that $h(p^{-1}(B))$ is a single point. This is not possible, since, by construction, $\varphi(x)$, $\varphi(F) \in h(p^{-1}(B))$ and $x \notin F$.

The above contradiction shows that q' is a finite map. Since $q^{-1}(F)$ is a plurisection of $p_{p^{-1}(\mathcal{P}(F))}$ we see that

$$\dim(p^{-1}(\mathcal{P}(F)) - q^{-1}(F)) = \dim p^{-1}(\mathcal{P}(F)).$$

Since q' is a finite map we have that the restriction $q_{p^{-1}(\mathcal{P}(F))}$ is generically finite-to-one on $p^{-1}(\mathcal{P}(F))$ and hence $\dim p^{-1}(\mathcal{P}(F)) = \dim q(p^{-1}(\mathcal{P}(F)))$. Therefore (6.9) leads to

$$\dim F + \delta + 1 \leq \dim q(p^{-1}(\mathcal{P}(F)) \leq n.$$

Since $\dim F = n - \delta - 1$, we thus conclude that the map $q_F := p^{-1}(\mathcal{P}(F)) \to X$ is onto and hence the family $(\mathcal{F}_F, \mathcal{P}(F), p_F, q_F)$ dominates X, where $\mathcal{F}_F := p^{-1}(\mathcal{P}(F))$ and p_F, q_F are the restrictions of p, q to $\mathcal{F}_F$ respectively.

Let γ be an effective irreducible curve on X. We have to show that $\gamma \sim aC + bf$ for some curve C in the family $\mathcal{F}$, some effective 1-cycle f contained in F and nonnegative real numbers a, b. Let $\mathcal{C}$ be an irreducible curve in $p^{-1}(\mathcal{P}(F))$ such that $q(\mathcal{C}) = \gamma$. Let $B := p(\mathcal{C})$. Then B is an irreducible curve, inside $\mathcal{P}(F)$, which parameterizes curves meeting both F and γ. We normalize B to get a smooth curve $\overline{B}$ and by base change we obtain a family S of rational curves parameterized by $\overline{B}$ whose normalization is a smooth ruled surface $\pi : \overline{S} \to \overline{B}$ (see also Wiśniewski [619, (1.14)]). Let $\psi : \overline{S} \to X$ be the composition of q with the base change map $S \to \mathcal{F}$ and the normalization. The morphism ψ maps any line ℓ of the ruling π birationally onto a curve, C, from the family $\mathcal{F}$ which, by the above construction, meets both F and γ. Therefore $\overline{S}$ contains two irreducible curves, γ_0, f_0, such that

$\psi(\gamma_0) = \gamma$, $\psi(f_0)$ is contained in F, so that f_0 goes to a point on Y and hence f_0 is a section isomorphic to $\overline{B}$. Moreover neither γ_0, nor f_0 are numerically equivalent to ℓ. On the surface $\overline{S}$ one has $\gamma_0 \sim a\ell + bf_0$, for some real numbers a, b, and then, by taking the image under ψ, $\gamma \sim aC + bf$ on X, where $f := \psi(f_0)$. Thus we are done after proving that a, b are nonnegative. Take a line ℓ on $\overline{S}$. Since $\ell \cdot \ell = 0$, $\ell \cdot \gamma_0 > 0$, $\ell \cdot f_0 > 0$ we find $b = (\ell \cdot \gamma_0)/(\ell \cdot f_0) > 0$. Let $H = (\varphi \circ \psi)^* A$ for some ample line bundle A on Y. Then $H \cdot \gamma_0 \geq 0$, $H \cdot \ell = A \cdot \varphi(\psi(\ell)) = A \cdot \varphi(C) > 0$, since $\varphi(C)$ is not a point by assumption, and $H \cdot f_0 = A \cdot \varphi(\psi(f_0)) = 0$ since $\psi(f_0) \subset F$. Therefore $a = (H \cdot \gamma_0)/(H \cdot \ell) \geq 0$. □

Corollary 6.4.7. *Let E be an ample vector bundle of rank r on a smooth connected variety Y. Let $X = \mathbb{P}(E)$, $\pi : X \to Y$. Assume that there is a covering family of unbreakable rational curves $(\mathcal{F}, \mathcal{P}, p, q)$ on X, with normal degree $\nu \geq -1$. Assume that π does not take curves from the family to points and $r = n - \nu$. Then X is a Fano variety of Picard number $\rho(X) = 2$ (and hence $\rho(Y) = 1$).*

Proof. Note that the condition $r = n - \nu$ is equivalent to the fact that the fibers of π are of dimension $n - \nu - 1$. Then, by the previous lemma, we have $NE(X) = NE(F) + \mathbb{R}_+[C]$, where $F = \mathbb{P}^{r-1}$ is a fiber of π and C is a curve from the family. Then $NE(X) = R + \mathbb{R}_+[C]$ where $R := NE(\mathbb{P}^{r-1})$. Note that $K_X \cdot R < 0$ and, since $\nu \geq -1$, $K_X \cdot C < 0$. It thus follows that R, $\mathbb{R}_+[C]$ are the two extremal rays generating the closure of the cone $\overline{NE}(X)$. □

The following useful consequence of Theorem (6.4.3) is a result of the authors and of Wiśniewski [114, (2.1)].

Corollary 6.4.8. *Let L be an ample line bundle on a connected, projective manifold, X. Assume that there is a connected, covering family of lines $(\mathcal{F}, \mathcal{P}, p, q)$ for (X, L) with normal degree $\nu := \nu(\mathcal{F})$. Let τ and $\phi : X \to Y$ be the nefvalue and the nefvalue morphism of (X, L) respectively. Then either*

1. *$\nu = \tau - 2$, ϕ takes lines from the family to points and* $\dim Y \leq n - \nu - 1$*; or*
2. *$\nu < \tau - 2$, $\tau + \nu \leq n$ and in particular $\nu < (n-2)/2$.*

Proof. Let ℓ be a line from the family $\mathcal{F}$. From $0 \leq (K_X + \tau L) \cdot \ell = K_X \cdot \ell + \tau = -\nu - 2 + \tau$ it follows $\tau \geq \nu + 2$.

If $\nu = \tau - 2$, then $(K_X + \tau L) \cdot \ell = 0$, so $\dim \phi(\ell) = 0$. Note that $\dim Y \leq n - \nu - 1$ by Lemma (6.4.5).

Let $\nu < \tau - 2$. It is enough to show that $\tau + \nu \leq n$. Thus, assume $\tau + \nu \geq n + 1$. Let $\tau = u/v$, u, v coprime positive integers. Then $u + \nu \geq \tau + \nu \geq n + 1$, so Theorem (6.4.3) applies to say that ϕ is the contraction of an extremal ray $R = \mathbb{R}_+[\ell]$, for a line ℓ from the family. Hence in particular $(K_X + \tau L) \cdot \ell = 0$, contradicting the present assumption $\nu > \tau - 2$. □

6.5 The classification of projective manifolds covered by lines

In this section we make an extended application of the methods of this chapter to the classification of projective manifolds covered by lines. The main result of this section is the following structure theorem.

Theorem 6.5.1. *Let L be an ample line bundle on a connected, projective manifold, X, of dimension n. Assume that there is a connected covering family of lines $(\mathcal{F}, \mathcal{P}, p, q)$ for (X, L) with normal degree $\nu := \nu(\mathcal{F})$. Let τ and $\phi : X \to W$ be the nefvalue and the nefvalue morphism of (X, L) respectively. If the dimension of any maximal irreducible component, P, of $\mathcal{P}$ is $\geq (3n-5)/2$ (or, equivalently, if $\nu \geq (n-3)/2$) then either $\nu < \tau - 2$ and $X \cong \mathbb{P}^{\tau-1} \times \mathbb{P}^{\nu+1}$, or $\tau = \nu + 2$ and $\phi : X \to W$ takes lines of the given family to points. The dimension, m, of W is bounded by $m \leq n - \nu - 1 \leq (n+1)/2$.*

Proof. First note that by Corollary (6.3.3), the condition $\dim P \geq (3n-5)/2$ is equivalent to $\nu \geq (n-3)/2$. By Corollary (6.4.8) we have $\tau \geq \nu + 2$ and $\tau = \nu + 2$ if and only if ϕ takes lines from the family to points. Furthermore we have $\dim W \leq n - \nu - 1 \leq (n+1)/2$.

The following argument runs parallel to that of [114, (2.2), (2.5)]. Assume now $\nu < \tau - 2$. Let $\tau = u/v$, u, v coprime positive integers. Then $u + \nu \geq \tau + \nu > 2\nu + 2 \geq n - 1$ so that $u + \nu \geq n$. By Theorem (6.4.3) we conclude that $u + \nu = n$, otherwise ϕ takes lines from the family to points, and we are as in (6.4.3.2b). Note that since

$$\frac{u+1}{v} \geq \tau = \frac{u}{v} > \nu + 2 \geq \frac{n+1}{2},$$

we have $v = 1$, $\tau = u$ and hence

$$u + \nu = \tau + \nu = n. \tag{6.10}$$

We may also assume $\tau = \nu + 3$. Indeed $\tau \geq \nu + 3$ and if $\tau \geq \nu + 4$, then we would get $u + \nu \geq n + 1$, contradicting (6.10). Therefore, from (6.10), one has $\nu = (n-3)/2$, $\tau = (n+3)/2$.

Thus we know that $\phi : X \to W$ is a contraction of an extremal ray, $R = \mathbb{R}_+[\mathfrak{l}]$, which expresses X as $X = \mathbb{P}(E)$ for an ample vector bundle E of rank u, and all fibers, F, of ϕ are $\mathbb{P}^{u-1} = \mathbb{P}^{(n+1)/2}$.

Take a line, ℓ, from the family $\mathcal{F}$. Since $\dim F = n - \nu - 1$, we conclude by Lemma (6.4.6) that $NE(X) = NE(F) + \mathbb{R}_+[\ell] = R + \mathbb{R}_+[\ell]$. Since $K_X \cdot R < 0$ and $K_X \cdot \ell < 0$, we see that ℓ is an extremal rational curve and the contraction, $\alpha : X \to V$, associated to $R' := \mathbb{R}_+[\ell]$ is of fiber type since $\mathcal{F}$ covers X. Since $\mathrm{Pic}(W) \cong \mathbb{Z}$, we have $\rho(X) = 2$. It follows that V is not a point since $\rho(V) = \rho(X) - 1$ by the exact sequence of (4.3.1) associated to α.

Let F' be any fiber of α. By Theorem (6.3.6) we have

$$n + \dim F' \geq n + \ell(R') - 1 = n + \nu + 1,$$

that is, $\dim F' \geq \nu + 1$. Let $\mathfrak{T}$ be the family of deformations of $\mathfrak{l}$, the curve generating the ray R. Note that since $\tau = (n+3)/2$, we conclude from $(K_X + \tau L) \cdot R = 0$ that $L \cdot \mathfrak{l} = 1$ and $\ell(R) = -K_X \cdot \mathfrak{l} = \tau$. Therefore $\nu(\mathfrak{l}) = -K_X \cdot \mathfrak{l} - 2 = \tau - 2$. Since α does not contract curves from the family $\mathfrak{T}$, Lemma (6.4.6) applies to say that

$$\dim F' \leq n - \nu(\mathfrak{l}) - 1 = n - \tau + 1 = \nu + 1.$$

Thus $\dim F' = \nu + 1$. Since $(K_X + (\nu+2)L) \cdot \ell = K_X \cdot \ell + \nu + 2 = 0$, we know that $K_X + (\nu + 2)L$ is trivial on F'. Therefore we conclude that a general fiber of α is $(\mathbb{P}^{\nu+1}, \mathcal{O}_{\mathbb{P}^{\nu+1}}(1))$. From Proposition (3.2.1) we conclude that α is a $\mathbb{P}^{\nu+1}$-bundle.

We claim that X is biholomorphic to $\mathbb{P}^{\tau-1} \times \mathbb{P}^{\nu+1}$. To see this, consider the product map $g = (\alpha, \phi) : X \to V \times W$. Note that the map is finite-to-one. Indeed if it is not then there would be an irreducible curve, C, that went to a point under g. Such a curve C would be contained in a fiber $\phi^{-1}(w)$ for some $w \in W$. Since $\mathrm{Pic}(\mathbb{P}^{\tau-1}) \cong \mathbb{Z}$, it would follow that $\alpha(\phi^{-1}(w)) = \alpha(C) = z$, which would imply that $\phi^{-1}(w)$ is contained in $\alpha^{-1}(z)$ for some $z \in V$. This is absurd for dimension reasons.

We have just shown that $\alpha : \phi^{-1}(w) \to V$ for $w \in W$ and $\phi : \alpha^{-1}(z) \to W$ for $z \in V$ are finite-to-one surjections. Since W, V are smooth, and since $\phi^{-1}(w) \cong \mathbb{P}^{\tau-1}$ and $\alpha^{-1}(z) \cong \mathbb{P}^{\nu+1}$ we conclude by a theorem of Lazarsfeld [402, (4.1)] that $W \cong \mathbb{P}^{\nu+1}$ and $V \cong \mathbb{P}^{\tau-1}$. Using this we see that it suffices by the Zariski main theorem to show that the product map $(\alpha, \phi) : X \to \mathbb{P}^{\tau-1} \times \mathbb{P}^{\nu+1}$ is one-to-one. To see this assume otherwise. Then we would have a fiber $F (\cong \mathbb{P}^{\tau-1})$ of ϕ which α maps t-to-one onto $V (\cong \mathbb{P}^{\tau-1})$ where $t > 1$. If we can show that $\alpha_F^* \mathcal{O}_V(1) \approx L_F \approx \mathcal{O}_{\mathbb{P}^{\tau-1}}(1)$ we will have a contradiction to $t > 1$. To show this note that by the canonical bundle formula for the bundle α we have

$$K_X + (\nu + 2)L \approx \alpha^*(K_V + \det \mathcal{G})$$

where $\mathcal{G} = \alpha_* L$. Since L is ample, $\mathcal{G}$ is an ample vector bundle of rank $\nu + 2$. By restriction to a line in V we see that $\det \mathcal{G} \approx \mathcal{O}_V(b)$ where $b \geq \nu + 2$. Thus

$$\mathcal{O}_F(\nu + 2 - \tau) \approx K_F + (\nu + 2)L_F \approx \alpha_F^* \mathcal{O}_V(b - \tau).$$

So $\nu + 2 - \tau = \lambda(b - \tau)$ where $\alpha_F^* \mathcal{O}_V(1) \approx \mathcal{O}_F(\lambda)$. Since $b - \tau \geq \nu + 2 - \tau$ we conclude that $\lambda = 1$ unless $\nu + 2 - \tau = 0$. This contradicts the assumption $\nu < \tau - 2$. □

Let F be a general fiber of the nefvalue morphism, $\phi : X \to W$, as in (6.5.1). Then F is a Fano variety with $K_F \approx -\tau L_F$. Since $L \cdot \ell = 1$ for a line ℓ from the family, we see that τ is the index of F. The theorem above shows that such Fano varieties, F, are the "building blocks" for the projective manifolds covered by lines, in the sense that the study of manifolds covered by lines can be reduced to the study of these Fano varieties. The following application of the general adjunction theory results of Chapter 7 gives an explicit example of this situation. Note that the Mukai varieties occurring in the case 5) below are not completely classified. These Fano varieties are the first "building blocks" to be studied in order to understand projective manifolds covered by lines. In that case $\tau = n - 2$, $\nu = \tau - 2 = n - 4$ (see Mukai [458, 459, 460] for results on these varieties).

For the definitions of first and second reduction of a polarized pair we use in the paragraph below we refer to Chapter 7. We also keep the same notations as in that chapter.

6.5.2. Coarse classification of smooth varieties of dimension $n \geq 6$, containing a family of lines of dimension $\geq 2n - 5$. Let $\mathcal{M}$ be a smooth connected variety of dimension $n \geq 6$. Let $\mathcal{L}$ be a very ample line bundle on $\mathcal{M}$. Let $(\mathcal{F}, \mathcal{P}, p, q)$ be a connected family of lines for $(\mathcal{M}, \mathcal{L})$. Assume that $\mathcal{F}$ covers $\mathcal{M}$ and that $\mathcal{F}$ is of dimension $\geq 2n - 5$ (i.e., $\dim \mathcal{P} \geq 2n - 5$). Then one of the following holds:

1. $(\mathcal{M}, \mathcal{L})$ is either $(\mathbb{P}^n, \mathcal{O}_{\mathbb{P}^n}(1))$, a $\mathbb{P}^{n-1}$-bundle over a curve, or a quadric Q in $\mathbb{P}^{n+1}$ with $\mathcal{L}_Q \cong \mathcal{O}_Q(1)$;
2. $(\mathcal{M}, \mathcal{L})$ is a Del Pezzo variety, i.e., $K_{\mathcal{M}} \approx -(n-1)\mathcal{L}$;
3. $(\mathcal{M}, \mathcal{L})$ is a quadric fibration over a smooth curve;
4. $(\mathcal{M}, \mathcal{L})$ is a $\mathbb{P}^{n-2}$-bundle over a smooth surface;
5. $(\mathcal{M}, \mathcal{L})$ is a Mukai variety, i.e., $K_{\mathcal{M}} \approx -(n-2)\mathcal{L}$;
6. $(\mathcal{M}, \mathcal{L})$ is a Del Pezzo fibration over a smooth curve;
7. $(\mathcal{M}, \mathcal{L})$ is a quadric fibration over a smooth surface; or
8. $(\mathcal{M}, \mathcal{L})$ is a scroll over a normal 3-fold.

To see this, let $\ell (\cong \mathbb{P}^1)$ be a general line from the family. Let $\mathcal{P}_\ell$ be any maximal irreducible component of $\mathcal{P}$ containing the point corresponding to ℓ. By Corollary (6.3.3) we have $\dim \mathcal{P}_\ell = \nu + n - 1$. Therefore $\nu \geq n - 4$ by the assumption on the dimension of $\mathcal{P}$. From the results of Chapter 7 we may assume that the first reduction, (M, L), of $(\mathcal{M}, \mathcal{L})$ exists (since otherwise $(\mathcal{M}, \mathcal{L})$ would be as in one of the cases 1)—4) above) and also that the second reduction, $(X, \mathcal{D})$, of $(\mathcal{M}, \mathcal{L})$ exists (since otherwise (M, L) would be as in one of the cases 5)—8) above). Since $n \geq 6$, we have $\nu \geq n - 4 \geq 2$ and hence by Theorem (7.6.6) we conclude that $\mathcal{M} \cong M \cong X$.

Since $\nu = -K_{\mathcal{M}} \cdot \ell - 2 \geq n - 4$ we find $K_{\mathcal{M}} \cdot \ell < 3 - n$ and hence $(K_{\mathcal{M}} + (n - 3)\mathcal{L}) \cdot \ell < 0$. Then $K_{\mathcal{M}} + (n-3)\mathcal{L}$ is nonnef, so the results of Chapter 7 say that $(\mathcal{M}, \mathcal{L})$ is as in one of the cases listed above. Note the fact that the base surface is smooth in case 7) follows from Theorem (14.2.3).

We refer to Rogora [523] for the classification of n-dimensional varieties covered by a family of lines of dimension $2n - 4$ and for further classical references on these topics.

The following result gives a useful condition insuring that two maximal families have a line in common. It is a question if it holds under the assumption that L is merely ample (see our paper [110, (0.3.3)] for a weaker form of this result).

Proposition 6.5.3. *Let L be an ample line bundle on a smooth connected projective manifold, X, of dimension n. Let $(\mathcal{F}_1, \mathcal{P}_1, p_1, q_1)$ and $(\mathcal{F}_2, \mathcal{P}_2, p_2, q_2)$ be two irreducible maximal families of lines on (X, L) with normal degrees ν_1 and ν_2 respectively. Assume that $\mathcal{E}(\mathcal{F}_1) \cap \mathcal{E}(\mathcal{F}_2) \neq \emptyset$, and that $\nu_1 + \nu_2 \geq n - 1$. Further assume that either*

a) *L is very ample, or*

b) *L is ample and there exists a line ℓ from one of the families, say $\mathcal{F}_1$, such that $R := \mathbb{R}_+[\ell]$ is an extremal ray.*

Then there exists at least one line common to both families.

Proof. Assume a). Take a point $x \in \mathcal{E}(\mathcal{F}_1) \cap \mathcal{E}(\mathcal{F}_2)$. Let $\mathcal{P}_{ix}$ be the subvariety of the parameter space $\mathcal{P}_i$ corresponding to lines from the family $\mathcal{F}_i$ that contain x, $i = 1, 2$. Let $\delta_{ix} = \dim \mathcal{P}_{ix}$, $i = 1, 2$. Then by Theorem (6.3.2) we have $\delta_{1x} \geq \nu_1$, $\delta_{2x} \geq \nu_2$. Consider the projective space $\mathbb{P}^{n-1} = \mathbb{P}(\mathcal{T}_x^*) = (\mathcal{T}_x - \{0\})/\mathbb{C}^*$. Then $\mathcal{P}_{ix}$ gives rise to a subvariety V_i of $\mathbb{P}^{n-1}$ of dimension δ_{ix}, $i = 1, 2$. Thus by the above

$$\dim(V_1 \cap V_2) \geq \dim V_1 + \dim V_2 - (n-1) \geq \nu_1 + \nu_2 - n + 1 \geq 0.$$

Therefore there exists a point, $v \in V_1 \cap V_2$, which corresponds to a line, ℓ_1, from $\mathcal{F}_1$ and to a line, ℓ_2, from $\mathcal{F}_2$, such that ℓ_1, ℓ_2 meet in x and are tangent at x. Since L is very ample it thus follows that $\ell_1 = \ell_2$.

Assume b). Let $\rho : X \to Y$ be the contraction of the extremal ray R and let E be the locus of R. Then $\mathcal{E}(\mathcal{F}_1) \subset E$. Let Δ be any irreducible component of a fiber of the restriction, ρ_E, of ρ to E, passing through $x \in \mathcal{E}(\mathcal{F}_1) \cap \mathcal{E}(\mathcal{F}_2)$. Let $D_x \subset \mathcal{E}(\mathcal{F}_2)$ be the locus of lines from the family $\mathcal{F}_2$ containing x. Recall that $\dim D_x = \delta_x + 1 \geq \nu_2 + 1$ by Theorem (6.3.1). It suffices to show $\dim(\Delta \cap D_x) \geq 1$. Assume otherwise. Then

$$\dim \Delta + \nu_2 + 1 - n \leq \dim \Delta + \dim D_x - n \leq 0.$$

Therefore by Theorem (6.3.6), $\ell(R) - 1 - \dim E + \nu_2 + 1 \le 0$. Since $\ell(R) = \nu_1 + 2$ we find $\nu_1 + \nu_2 + 2 - \dim E \le 0$, which gives the contradiction $n - 1 \le \nu_1 + \nu_2 \le \dim E - 2$. □

6.6 Some spannedness results

In this section we prove the existence of global sections of adjunction bundles that span the adjunction bundles on linear subspaces. We refer to our paper with Schneider [97] for further developments of the results of this section.

We need a special case of the following general lemma which follows from the results in the introduction of Mumford [465].

Lemma 6.6.1. *Let L be a very ample line bundle on an irreducible projective variety, X. Let $Y \subset X$ be an irreducible subvariety of degree d relative to L, i.e., $d = L^{\dim Y} \cdot Y$. If either Y is smooth or $Y \subset \mathrm{reg}(X)$ and $\mathrm{cod}_X Y = 1$ then $(dL) \otimes \mathcal{J}_Y$ is spanned by global sections, where $\mathcal{J}_Y$ denotes the ideal sheaf of Y in X.*

Proof. We give here the proof in the simple case we need, when X, Y are smooth and $d = 1$, referring to [465] for the general case.

In our case Y is a linear $\mathbb{P}^{\dim Y}$. Look at the embedding $X \hookrightarrow \mathbb{P}^N$ given by $\Gamma(L)$, $N = h^0(L) - 1$. Let $\mathcal{J}_Y$ be the ideal sheaf defining Y in $\mathbb{P}^N$. If $\mathcal{P}$ is the set of hyperplanes in $\mathbb{P}^N$ that contain Y, one has the scheme-theoretic intersection $Y = \cap_{H \in \mathcal{P}} H$. From this it thus follows that $\Gamma(\mathcal{O}_{\mathbb{P}^N}(1) \otimes \mathcal{J}_Y)$ spans $\mathcal{O}_{\mathbb{P}^N}(1) \otimes \mathcal{J}_Y$. By looking at the restriction to X we see that $\Gamma(L \otimes \mathcal{J}_Y)$ spans $L \otimes \mathcal{J}_Y$ on X. □

The following consequence of the lemma above was one basic starting observation for [97].

Corollary 6.6.2. *Let $\mathcal{E}$ be a rank r very ample vector bundle on a smooth projective variety, X, i.e., the tautological bundle $\mathcal{O}_{\mathbb{P}(\mathcal{E})}(1)$ is very ample. Let $\pi : \overline{X} \to X$ denote the blowup of X at a point x. Let $\mathcal{P} := \pi^{-1}(x)$. Then $\pi^*\mathcal{E} \otimes \mathcal{J}_{\mathcal{P}}$ is a spanned vector bundle on $\overline{X}$, where $\mathcal{J}_{\mathcal{P}}$ denotes the ideal sheaf of $\mathcal{P}$ in $\overline{X}$.*

Proof. We have a commutative diagram

$$\begin{array}{ccc} \mathbb{P}(\mathcal{E}) & \xleftarrow{\overline{\pi}} & \mathbb{P}(\pi^*\mathcal{E}) \\ p \downarrow & & \downarrow \overline{p} \\ X & \xleftarrow{\pi} & \overline{X} \end{array}$$

where p, $\overline{p}$ are the bundle projections and $\overline{\pi}$ is the blowingup of $\mathbb{P}(\mathcal{E})$ along the fiber $F := p^{-1}(x)$. Let $\overline{\mathcal{P}} := \overline{\pi}^{-1}(F)$ and letting $\xi = \mathcal{O}_{\mathbb{P}(\mathcal{E})}(1)$ denote the tautological bundle of $\mathbb{P}(\mathcal{E})$. Let $\overline{\xi} := \overline{\pi}^*\xi$ be a tautological bundle on $\mathbb{P}(\pi^*\mathcal{E})$. Since F is a linear $\mathbb{P}^{r-1}$ relative to ξ, by Lemma (6.6.1) we conclude that $\xi \otimes \mathcal{J}_F$ is spanned, $\mathcal{J}_F$ denoting the ideal sheaf of F in $\mathbb{P}(\mathcal{E})$. Since $\overline{\xi}\otimes\mathcal{J}_{\overline{\mathcal{P}}} \cong \pi^*(\xi\otimes\mathcal{J}_F)$, we conclude that $\overline{\xi}\otimes\mathcal{J}_{\overline{\mathcal{P}}}$ is spanned. Note that $\overline{\xi}\otimes\mathcal{J}_{\overline{\mathcal{P}}}$ is a tautological bundle on $\mathbb{P}(\pi^*\mathcal{E})$. It thus follows that $\overline{p}_*(\overline{\xi}\otimes\mathcal{J}_{\overline{\mathcal{P}}})$ is spanned. Now $\overline{p}_*(\overline{\xi}\otimes\mathcal{J}_{\overline{\mathcal{P}}}) \cong \overline{p}_*(\overline{\xi}\otimes\overline{p}^*\mathcal{J}_{\mathcal{P}}) \cong \overline{p}_*\overline{\xi}\otimes\mathcal{J}_{\mathcal{P}} \cong \pi^*\mathcal{E}\otimes\mathcal{J}_{\mathcal{P}}$. □

Lemma 6.6.3. *Let L be a very ample line bundle on an n-dimensional normal Cohen-Macaulay projective variety, X. Assume that the locus,* $\mathrm{Irr}(X)$, *of nonrational singularities of X is at most 0-dimensional. Given a k-dimensional linear subspace, $P \subset \mathrm{reg}(X)$, relative to L it follows that the restriction map*

$$\Gamma(K_X + tL) \to \Gamma((K_X + tL)_P) \to 0$$

is onto if $t \geq n - k + 1$. If the rational map associated to $|L - P|$ has an n-dimensional image, then the conclusion holds with $t \geq n - k$.

Proof. The conclusion of the lemma will follow if we prove that $H^1(X, K_X \otimes L^{\otimes t} \otimes \mathfrak{I}_P) = (0)$ for the hypothesized t where $\mathfrak{I}_P$ is the ideal sheaf associated to P.

Let $\psi : \overline{X} \to X$ denote the blowup of X at P. Then $K_{\overline{X}} \cong \psi^*K_X + (n-k-1)\mathcal{P}$ where $\mathcal{P} := \psi^{-1}(P)$. Since P is a linear $\mathbb{P}^k$ relative to L, Lemma (6.6.1) applies to say that $L \otimes \mathfrak{I}_P$ is spanned by global sections and hence it follows that $\mathcal{L} := \psi^*L - \mathcal{P}$ is spanned by global sections. Thus $h^1(K_{\overline{X}} + s\mathcal{L} + (t-s)\psi^*L) = 0$ and $\psi_{(1)}(K_{\overline{X}} + s\mathcal{L} + (t-s)\psi^*L) = 0$ for all $t \geq s \geq 1$ if $\mathcal{L}$ is big, and for all $t > s \geq 0$, in general.

Thus by the Leray spectral sequence and the projection formula, we conclude that $h^1(K_X + tL + \psi_*((n-k-1-s)\mathcal{P})) = 0$ for all $t \geq s \geq 1$ if $\mathcal{L}$ is big, and for all $t > s \geq 0$ in general. Since $\psi_*\mathcal{O}_{\overline{X}}(-\mathcal{P}) = \mathfrak{I}_P$, choosing $s = n - k$ proves the lemma. □

Remark 6.6.4. Letting a be such that $\det\mathcal{N}_{P/X} = \mathcal{O}_{\mathbb{P}^k}(a)$ in the above lemma, we see that global sections of $\Gamma(K_X + tL)$ span $K_X + tL$ in a neighborhood of P if $t \geq \max\{k + 1 + a, n - k + 1\}$. In fact

$$(K_X + tL)_P \approx K_P - \det\mathcal{N}_{P/X} + tL_P \approx \mathcal{O}_{\mathbb{P}^k}(-k - 1 - a + t)$$

is spanned if $t \geq k + 1 + a$, so that sections lift by the above lemma as soon as it is also $t \geq n - k + 1$.

If the rational map associated to $|L - P|$ has an n-dimensional image, then we get the spannedness of $K_X + tL$ with $t \geq \max\{k + 1 + a, n - k\}$.

Corollary 6.6.5. *Let X be an n-dimensional smooth connected variety. Let L be a very ample line bundle on X. Assume that (X, L) is a scroll, $p : X \to Y$, over* *a variety Y of dimension m. Let $k = n - m$. Then $K_X + (k+1)L$ is spanned by global sections if $k \geq n/2$. The same is true if (X, L) is a $\mathbb{P}^k$-bundle.*

Proof. If (X, L) is a $\mathbb{P}^k$-bundle, then X is covered by linear $\mathbb{P}^k$'s, so we conclude by Lemma (6.6.3) and Remark (6.6.4) that $K_X+(k+1)L$ is spanned if $k+1 \geq n-k+1$, or $k \geq n/2$.

Let (X, L) be a scroll. Fix a point $x \in X$. We can choose a sequence F_i of general fibers of p such that the point x is contained in the limit of the F_i's. Let P be the limit of the F_i's in the Hilbert scheme of the F_i's. The fibers F_i are linear $\mathbb{P}^k$'s, so P is a linear $\mathbb{P}^k$ and $\det \mathcal{N}_{P/X} \cong \mathcal{O}_P$. Thus Lemma (6.6.3) and Remark (6.6.4) apply again to give the result. □

Corollary 6.6.6. *Let $\mathcal{E}$ be a very ample vector bundle on a smooth projective variety, X, of dimension n. Then $K_X + \det \mathcal{E}$ is spanned by global sections if* $\operatorname{rank}\mathcal{E} \geq n+1$, *or if* $\operatorname{rank}\mathcal{E} = n$ *and* $c_1(\mathcal{E})^n \neq (\operatorname{rank}\mathcal{E})^n$.

Proof. Let ξ be the tautological bundle of $\mathbb{P}(\mathcal{E})$. Let $r = \operatorname{rank}\mathcal{E}$. Then $\mathbb{P}(\mathcal{E})$ is covered by linear $\mathbb{P}^{r-1}$'s such that $\mathcal{N}_{\mathbb{P}^{r-1}/\mathbb{P}(\mathcal{E})} \cong \mathcal{O}_{\mathbb{P}^{r-1}}$. Let $r \geq n+1$. By Lemma (6.6.3) and Remark (6.6.4), applied to $(\mathbb{P}(\mathcal{E}), \xi)$, $\mathbb{P}^{r-1}$, it follows that $K_{\mathbb{P}(\mathcal{E})} + r\xi$ is spanned by global sections if $r \geq \dim \mathbb{P}(\mathcal{E}) - (r-1) + 1 = n+1$. Let $p : \mathbb{P}(\mathcal{E}) \to X$ be the bundle projection. Since $K_{\mathbb{P}(\mathcal{E})} + r\xi \approx p^*(K_X + \det \mathcal{E})$ the first part of the statement is proved.

Assume now $r = n$. Let $\pi : \overline{X} \to X$ be the blowup of X at a point x. Let $\mathcal{J}_{\mathcal{P}}$ be the ideal sheaf of $\mathcal{P} := \pi^{-1}(x)$ in $\overline{X}$. By Corollary (6.6.2) we know that $\pi^*\mathcal{E} \otimes \mathcal{J}_{\mathcal{P}}$ is spanned and therefore $\det(\pi^*\mathcal{E} \otimes \mathcal{J}_{\mathcal{P}}) \cong \pi^* \det \mathcal{E} - n\mathcal{P}$ is spanned. By Leray's spectral sequence and projection formula we have

$$H^1(K_{\overline{X}} + \pi^* \det \mathcal{E} - n\mathcal{P}) \cong H^1(\pi^* K_X + \pi^* \det \mathcal{E} - \mathcal{P}) \cong H^1(K_X \otimes \det \mathcal{E} \otimes \mathfrak{m}_x),$$

where $\mathfrak{m}_x$ is the ideal sheaf of x in X. Thus by the Kawamata-Viehweg vanishing theorem, we conclude that

$$h^1(K_X \otimes \det \mathcal{E} \otimes \mathfrak{m}_x) = 0 \tag{6.11}$$

as soon as $\pi^* \det \mathcal{E} - n\mathcal{P}$ is nef and big. By the above, this is the case unless $(\pi^* \det \mathcal{E} - n\mathcal{P})^n = 0$, which is equivalent to $c_1(\mathcal{E})^n = (\operatorname{rank}\mathcal{E})^n$. From (6.11) we have that the restriction map $\Gamma(K_X + \det \mathcal{E}) \to \Gamma((K_X + \det \mathcal{E})_x) \cong \mathbb{C}$ is onto and hence $K_X + \det \mathcal{E}$ is spanned in a neighborhood of x. Since this is true for all $x \in X$, $K_X + \det \mathcal{E}$ is spanned. □

Corollary 6.6.7. *Let L be a very ample line bundle on a projective manifold, X, of dimension n. Let $p : X \to Y$ be a quadric fibration, i.e., p is surjective with connected fibers onto a normal space, Y, of dimension m and $K_X + (n-m)L \approx p^*H$*

for an ample line bundle H on Y. Then setting $f = n - m$, $K_X + (n-m)L$ is spanned for $f \geq 2n/3 + 1$.

Proof. Let F be a general fiber of p. Then either $\dim F = f = 2k$ or $\dim F = f = 2k + 1$ and in both cases F contains a linear $\mathbb{P}^k$.

Let $f = 2k$. By the adjunction formula we get

$$\det \mathcal{N}_{\mathbb{P}^k/X} \cong K_{\mathbb{P}^k} - K_{X|\mathbb{P}^k} \cong K_{\mathbb{P}^k} - K_{F|\mathbb{P}^k} \cong \mathcal{O}_{\mathbb{P}^k}(k-1).$$

Fix a point $x \in X$. We can choose a sequence of such $\mathbb{P}^k$'s, contained in general fibers of p, such that the point x is contained in the limit, P, of the $\mathbb{P}^k$'s in the Hilbert scheme of the $\mathbb{P}^k$'s. Then $\det \mathcal{N}_{P/X} \cong \mathcal{O}_P(k-1)$. By Lemma (6.6.3) and Remark (6.6.4) we know that $K_X + fL$ is spanned if $f = 2k \geq \max\{n-k+1, 2k\}$, which is equivalent to $2k \geq n - k + 1$, or $k \geq (n+1)/3$, that is $f \geq 2(n+1)/3$.

In the case $f = 2k + 1$ we have $\det \mathcal{N}_{\mathbb{P}^k/X} \cong \mathcal{O}_{\mathbb{P}^k}(k)$ and exactly the same argument as above, shows that $K_X + fL$ is spanned if $k \geq n/3$ or $f \geq 2n/3 + 1$. □

Corollary 6.6.8. *Let L be a very ample line bundle on a projective manifold, X, of dimension n. Let $k := \operatorname{def}(X, L)$ be the defect of (X, L). Assume $k > 0$. Let τ be the nefvalue of (X, L). Then $\tau = (n+k)/2 + 1$ and $m(K_X + ((n+k)/2 + 1)L)$, for $m \gg 0$, is spanned by global sections and defines a morphism which is a fiber type contraction of an extremal ray. If $k \geq n/3$ this is true for $m = 1$.*

Proof. Since $k \geq 1$ it is a classical fact that there is a linear $\mathbb{P}^k$ through a general point of X and $K_{X|\mathbb{P}^k} \cong \mathcal{O}_{\mathbb{P}^k}((-n-k-2)/2)$ or $\det \mathcal{N}_{\mathbb{P}^k/X} \cong \mathcal{O}_{\mathbb{P}^k}((n-k)/2)$ (see, e.g., [87, (0.3)], [203, (0.4)] and §14.4). By a standard limit argument, as in the proof of Corollary (6.6.5), we see that there is a linear $\mathbb{P}^k$ through each point of X and the determinant of the normal bundle of this $\mathbb{P}^k$ is $\mathcal{O}_{\mathbb{P}^k}((n-k)/2)$. Thus Lemma (6.6.3) and Remark (6.6.4) apply to say that $K_X + ((n+k)/2 + 1)L$ is spanned by global sections if $(n+k)/2 + 1 \geq n - k + 1$, or $k \geq n/3$.

To show the general case, take a line, ℓ, through every point $x \in X$. Then $K_X \cdot \ell = -(n+k)/2 - 1$. Let $\nu = -K_X \cdot \ell - 2$ be the normal degree of ℓ. Then $\nu = (n+k)/2 - 1 > n/2 - 1$ since $k > 0$. Therefore, by Corollary (6.4.8), $\tau = \nu + 2 = (n+k)/2 + 1$ and $m(K_X + ((n+k)/2 + 1)L)$, for $m \gg 0$, defines a morphism, $\phi : X \to Y$, the nefvalue morphism of (X, L), which is a contraction of an extremal ray. Furthermore $\dim Y \leq n - \nu - 1 = (n-k)/2$, hence ϕ is a contraction of fiber type. □

Let us conclude this section mentioning that the important new criteria of Ein and Lazarsfeld [208] prove a famous conjecture of Fujita [253] about spannedness of adjoint line bundles in the 3-fold case.

Conjecture 6.6.9. *Let A be an ample line bundle on a smooth projective variety, X, of dimension n. Then $K_X + (n+1)A$ is spanned by its global sections and $K_X + (n+2)A$ is very ample.*

To go on, we need the following definitions. Let L be a nef line bundle on an irreducible projective manifold, X. Let x be a fixed point on X. The *Seshadri constant* of L at x is defined to be the real number (cf., Ein and Lazarsfeld [209] and Demailly [185])

$$s(L, x) := \inf_{x \in C} \left\{ \frac{L \cdot C}{m_x(C)} \right\},$$

where the infimum is taken over all irreducible curves, C, passing through x and $m_x(C)$ is the multiplicity of C at x. For example, if $L = \mathcal{O}_X(D)$ for an ample effective divisor D, $x \in D$, which is smooth at x, e.g., if L is very ample, then $s(L, x) \geq 1$.

Let $\pi : \overline{X} \to X$ be the blowing up of X at x and let $E = \pi^{-1}(x)$ be the exceptional divisor. Then one has

$$s(L, x) := \sup\{t \in \mathbb{R} \mid t > 0, \pi^* L - tE \text{ is nef }\}.$$

The *global Seshadri constant* is defined as $s(L) := \inf_{x \in X}\{s(L, x)\}$. Note that the Seshadri criterion for ampleness states that $s(L) > 0$ if and only if L is ample. We refer to [209] for results on Seshadri constants on smooth surfaces and to [185, §6] for applications of Seshadri constants for producing sections of adjoint bundles $K_X + tL$.

In [208], Ein and Lazarsfeld prove the following general numerical criterion.

Theorem 6.6.10. *Let L be a nef and big line bundle on a smooth complex projective threefold, X, and let $x \in X$ be a fixed point. Assume that $\sigma_1, \sigma_2, \sigma_3$ are rational numbers such that $L^3 > \sigma_3^3 > 3^3$, $L \cdot S > \sigma_2^2$ and $L \cdot C > \sigma_1$ for all surfaces S, and curves C on X such that $x \in S$, $x \in C$. If*

$$\sigma_2 \geq \max\left\{ \frac{2\sigma_3}{\sigma_3 - 1}, \frac{\sqrt{2}\sigma_3}{\sigma_3 - 2} \right\} \; ; \; \sigma_1 \geq \max\left\{ \frac{2\sigma_3}{\sigma_3 - 1}, \frac{\sigma_3 - s(L, x)}{\sigma_3 - 3} \right\}$$

then $K_X + L$ is free at x.

Note that the assumption $L^3 > 27$ is sharp as the Remark-Example (11.2.1) shows.

As a consequence of the result above they obtain the following.

Theorem 6.6.11. *Let X, L, $x \in X$ be as in* (6.6.10). *Assume that $L^3 \geq 92$, $L^2 \cdot S \geq 7$ for each irreducible surface $S \subset X$ with $x \in S$ and $L \cdot C \geq 3$ for each irreducible curve $C \subset X$* with *$x \in C$. Then $K_X + L$ is spanned at x. If $L^3 \gg 0$ (e.g., $L^3 \geq 1000$) then the same conclusion holds with the second inequality replaced by $L^2 \cdot S \geq 5$.*

A major consequence of Theorem (6.6.10) is the spannedness part of Fujita's conjecture for threefolds. The first part of the following theorem is due to Ein and Lazarsfeld [208, §6]. By using the results of [208], Fujita [266] recently proved a stronger version of Conjecture (6.6.9) as in 2) below.

Theorem 6.6.12. *Let A be an ample line bundle on a smooth complex projective threefold X. Then*

1. $K_X + 4A$ *is globally generated;*

2. $K_X + 3A$ *is globally generated if* $A^3 > 1$.

Proof. To prove 1), take $L = 4A$ and apply Theorem (6.6.10). In [266], Fujita improves Theorem (6.6.11) above showing that the assumption $L^3 \geq 92$ can be replaced by $L^3 \geq 51$. From this he derives 2). We refer to [266] for the proof and full details. □

We refer to our paper with Schneider [95] for some applications of the criteria of Ein and Lazarsfeld for the spannedness of the adjoint bundle, L, on a smooth complex projective 3-fold, X.

Let us only discuss here the following consequence of Theorem (6.6.10). Assume that L is a very ample line bundle on a smooth variety, X, of dimension 3, with $L^3 \geq 850$. Assume that X contains no rational curves, i.e., curves whose desingularization is $\mathbb{P}^1$. For example, this happens if X is hyperbolic in the sense of Kobayashi, or if the cotangent bundle of X is nef. Then the condition $L \cdot C \geq 3$ of Theorem (6.6.11) is immediate, since $L \cdot C \leq 2$ implies that C is a rational curve. Lemma 14 of [95] implies that any irreducible reduced projective surface, S, of degree ≤ 4 contains a rational curve. Therefore the condition $L^2 \cdot S \geq 5$ is also satisfied. Thus by Theorem (6.6.11) we conclude that $K_X + L$ is spanned. By using the Kodaira vanishing theorem we immediately extend the result in higher dimension $n \geq 4$. Thus by the above we have the following [95, Theorem 15].

Theorem 6.6.13. *Let X be a smooth projective n-fold,* $n \geq 3$, *with no rational curves. Let L be a very ample line bundle on X such that* $L^n \geq 850$. *Then* $K_X + (n-2)L$ *is spanned by global sections.*

Chapter 7

General adjunction theory

Adjunction theory typically uses the adjoint $\mathbb{Q}$-Cartier divisors, $K_{\mathcal{M}} + t\mathcal{L}$, $t \in \mathbb{Q}$, to study pairs, $(\mathcal{M}, \mathcal{L})$, where $\mathcal{L}$ is an appropriately ample line bundle on a $\mathbb{Q}$-Gorenstein projective variety, $\mathcal{M}$, with appropriate singularities. There is a spectrum of hypotheses on $(\mathcal{M}, \mathcal{L})$. At one end we make assumptions on the existence of sections of $\mathcal{L}$, and often can significantly relax the smoothness assumptions of $\mathcal{M}$. At the other extreme we relax the spannedness assumptions of $\mathcal{L}$, but restrict singularities so that the theory of extremal rays can be used. Some results need hypotheses on the spannedness of $\mathcal{L}$, e.g., results on the very ampleness of $K_{\mathcal{M}} + (n-1)\mathcal{L}$, or results on inequalities of Chern numbers. We often refer to the collection of those results which hold with the assumption of ampleness and little more as *general adjunction theory*.

In this chapter we carry out general adjunction theory. Developing this part of theory first gives a clearer understanding of classical adjunction theory. Occasionally we will prove a result twice with slightly different hypotheses, when both proofs are important.

Let us describe the structure of this chapter.

In §7.1 we discuss the notion of spectral value, $u(V, D)$, of a pair (V, D) where V is a normal r-Gorenstein variety and D is a $\mathbb{Q}$-Cartier divisor on V. The spectral value plays an important role in the classification of the polarized varieties (see Sommese [583] and Beltrametti and Sommese [111]).

Let $\mathcal{L}$ denote an ample line bundle on an n-dimensional projective normal variety, $\mathcal{M}$, with terminal, $\mathbb{Q}$-factorial singularities, and $n \geq 2$.

In §7.2 we classify the pairs, $(\mathcal{M}, \mathcal{L})$, with $K_{\mathcal{M}} + (n-1)\mathcal{L}$ nonnef. We end with a number of numerical characterizations of these special pairs.

Assume further that $\mathcal{M}$ has at worst terminal factorial singularities. In §7.3 we show that, except for a short list of exceptions:

1. $\mathcal{M}$ is the blowup, $\pi : \mathcal{M} \to M$, of a finite set $B \subset \mathrm{reg}(M)$ with M a projective variety with no worse singularities that $\mathcal{M}$;
2. $L := (\pi_*\mathcal{L})^{**}$ is an ample line bundle such that $K_{\mathcal{M}} + (n-1)\mathcal{L} \cong \pi^*(K_M + (n-1)L)$ with $K_M + (n-1)L$ ample;
3. if $n \geq 3$ then $K_M + (n-2)L$ is nef and big.

The new pair (M, L) is called the *first reduction* of $(\mathcal{M}, \mathcal{L})$. If $n \geq 3$, then by this result and the Kawamata-Shokurov basepoint free theorem, there exists a birational morphism, $\varphi : M \to X$, with X normal and $K_M + (n-2)L \cong \varphi^*\mathcal{H}$ for an ample line bundle $\mathcal{H}$ on X. Let $\mathcal{D} := (\varphi_* L)^{**}$. The pair $(X, \mathcal{D})$ is called the second reduction of $(\mathcal{M}, \mathcal{L})$.

In §7.4 we discuss some useful properties of the first reduction.

Assume now that $\mathcal{M}$ is smooth. The structure of $(X, \mathcal{D})$, the second reduction of $(\mathcal{M}, \mathcal{L})$, is studied when $n \geq 4$ in §7.5. It is shown that X has isolated terminal singularities, and is moreover Gorenstein in even dimensions and 2-Gorenstein in odd dimensions. Further X is 2-factorial for $n \geq 4$ and $\mathcal{D} := (\varphi_* L)^{**}$ is 2-Cartier. We show that $K_M+(n-2)L \cong \varphi^*(K_X+(n-2)\mathcal{D})$, and that $(n-2)(K_X+(n-3)\mathcal{D}) \approx K_X + (n-3)\mathcal{H}$. The more involved structure of the second adjunction when $n = 3$ is developed in Chapter 12 under the added hypothesis that $\mathcal{L}$ is spanned by global sections.

In §7.6 we work out some useful properties of the first and second reductions—especially the relations between sections of the adjoint bundles on $(\mathcal{M}, \mathcal{L})$ and the corresponding adjoint bundles on (M, L) and $(X, \mathcal{D})$.

In §7.7 we prove the strong result from our paper [106] that $K_X + (n-3)\mathcal{D}$ is nef for $n \geq 6$ unless $(X, \mathcal{D}) \cong (\mathbb{P}^6, \mathcal{O}_{\mathbb{P}^6}(2))$. This can be considered as the main result of the chapter. We also give weaker results for $n = 4$, and in §7.8, for $n = 3$. In the cases $n = 3, 4$ and partially in the case $n = 5$, the results are due to Fujita [261, 263]. Most of the results for projective manifolds with $n \geq 3$ and $\mathcal{L}$ very ample were done by Fania and the authors of this book [86].

The above results have many consequences. For example it follows that if $\mathcal{M}$ is a submanifold of a projective space with $\dim \mathcal{M} \geq 4$, then either $\mathcal{M}$ is degenerate with a very special structure (e.g., a scroll, a quadric fibration, a Del Pezzo fibration, a $(\mathbb{P}^2, \mathcal{O}_{\mathbb{P}^2}(2))$-fibration over a curve,...) or any smooth 3-fold section, V of $\mathcal{M}$, is of general type and there is a very simple birational morphism, $\varphi \circ \pi : \mathcal{M} \to X$, such that $(\varphi \circ \pi)_V : V \to (\varphi \circ \pi)(V)$ is the map of V onto a minimal model $V' = (\varphi \circ \pi)(V)$. Moreover V' is 2-Gorenstein (see (7.9.1)).

In §7.9 we give some applications of the results above. One of these is a coarse classification of smooth threefolds, $\mathcal{M}$, carrying an ample divisor, $\mathcal{L}$, such that $|\mathcal{L}|$ contains a smooth surface, S, of given Kodaira dimension.

Our main technique is Mori theory, and in particular the precise form of the Kawamata rationality theorem.

Earlier versions of the results of §7.2 and §7.3 are due to Fujita [252], Ionescu [335], and Sommese [583]. In the case when $\mathcal{L}$ is very ample or ample and spanned, the structure of the second reduction was studied in Sommese [579], Fania [219], Fania and Sommese [229], and Beltrametti, Fania, and Sommese [86]. We studied the second reduction with $n \geq 5$ in [106], the main reference for this chapter. Here we slightly improve the results of [106], adding a complementary part about properties and applications of the structure of the first and the second reduction of a polarized pair. We also use the results of Fujita [261, 264] on the structure of the second reduction when $n = 3, 4$ and L is merely ample.

7.1 Spectral values

In this section we give the definition of spectral value associated with a polarized pair and we recall a few of its properties that we use in the sequel.

Definition 7.1.1. Let V be a normal, r-Gorenstein, projective variety of dimension n and let D be a $\mathbb{Q}$-Cartier divisor on V. Recall that the Kodaira dimension of D, $\kappa(D)$, equals the Kodaira dimension of mD, $\kappa(mD)$, for any positive rational number m. Thus $\kappa(D) = -\infty$ if and only if $\kappa(mD) = -\infty$ for some positive integer m.

Now let D be a $\mathbb{Q}$-Cartier divisor on V such that $\kappa(D) = n$. Let us define the *unnormalized spectral value*,

$$u(V, D) := \sup\{t \in \mathbb{Q} \mid \kappa(K_V + tD) = -\infty\},$$

that is

$$u(V, D) := \sup\{t \in \mathbb{Q} \mid h^0(N(K_V + tD)) = 0 \text{ for all } N > 0 \text{ such that } N(K_V + tD) \text{ is an integral Cartier divisor}\}.$$

We are taking the sup as a real number. Often, when the space is obvious from the context, we denote $u(V, D)$ by $u(D)$. Note that $u(V, D) < +\infty$ since $\kappa(D) = n$. It is an open question whether it is rational (see the paper [583] of Sommese). In this direction, it is known that if V is a projective manifold of dimension ≤ 3, and D is an ample line bundle such that $u(V, D) > 0$ then $u(V, D)$ is rational. For $\dim V = 1$, rationality is trivial, while for surfaces and threefolds this has been proved by Sakai [532] and Batyrev [74] respectively. This real number, $u(V, D)$, is called the *unnormalized spectral value* of (V, D). Note also that the *spectral value*, $\sigma(V, \mathcal{L})$, defined in [583, §1] for a line bundle $\mathcal{L}$, is $\sigma(V, \mathcal{L}) := n + 1 - u(V, \mathcal{L})$.

Remark 7.1.2. Let $\mathcal{L}$ be an ample line bundle on an irreducible normal variety V. Fujita [261] defines the *Kodaira energy* of $(V, \mathcal{L})$ as $\kappa\epsilon(V, \mathcal{L}) := -\inf\{t \in \mathbb{Q} \mid \kappa(K_V + t\mathcal{L}) \geq 0\}$. Note that $\kappa\epsilon(V, \mathcal{L}) = -u(V, \mathcal{L})$.

Lemma 7.1.3. *If $\mathcal{L}$ is a nef and big line bundle on an irreducible normal n-dimensional projective variety V, then $u(V, \mathcal{L}) \leq n + 1$, i.e., $\sigma(V, \mathcal{L}) \geq 0$.*

Proof. The simple argument of the second author [583] goes as follows. Let $p : \overline{V} \to V$ be a desingularization of V, and let $\overline{\mathcal{L}} := p^*\mathcal{L}$. Since $p_*(K_{\overline{V}} + j\overline{\mathcal{L}}) \cong$

$p_* K_{\overline{V}} + j\mathcal{L}$ is a subsheaf of $K_V + j\mathcal{L}$ it suffices to show that $h^0(K_{\overline{V}} + j\overline{\mathcal{L}}) > 0$ for some positive integer $j \le n+1$. To see this note that since $\overline{\mathcal{L}}$ is nef and big, $q(j) := \chi(K_{\overline{V}} + j\overline{\mathcal{L}})$ is a nontrivial polynomial of degree n. Since $\overline{\mathcal{L}}$ is nef and big we have $\chi(K_{\overline{V}} + j\overline{\mathcal{L}}) = h^0(K_{\overline{V}} + j\overline{\mathcal{L}})$ for $j \ge 1$ by the Kawamata-Viehweg vanishing theorem. Thus since $q(j)$ can have at most n zeroes, it is not possible that $q(j) = 0$ for $1 \le j \le n+1$. □

Lemma 7.1.4. *Let D be a big $\mathbb{Q}$-Cartier divisor on an irreducible n-dimensional normal projective variety, V. Assume that K_V is $\mathbb{Q}$-Cartier. If $aK_V + bD$ is big for rational numbers a, b with $a > 0$, then $u(V, D) < b/a$.*

Proof. Since $a > 0$ we have that $\kappa(K_V + (b/a)D) = \kappa(aK_V + bD)$. Thus we can assume without loss of generality that $a = 1$. By Lemma (2.5.5), it follows that given any ample divisor H there exists an integer $N > 0$ and an effective divisor, D', such that $N(K_V + bD) \approx D' + H$. Since H is ample it follows, for all sufficiently small $\epsilon > 0$, that $H - \epsilon D$ is represented by an effective $\mathbb{Q}$-Cartier divisor D''. Thus $NK_V + (Nb - \epsilon)D \approx D' + D''$ is $\mathbb{Q}$-effective. Therefore $u(V, D) \le b - \epsilon/N$. □

Lemma 7.1.5. *Let D be a big $\mathbb{Q}$-Cartier divisor on an irreducible n-dimensional normal projective variety, V. Assume that K_V is $\mathbb{Q}$-Cartier and $u(V, D) \ge 0$. If $aK_V + bD$ is big for rational numbers a, b with $a \ge 0$, then*

$$u(aK_V + bD) = \frac{u(D)}{b - au(D)}; \quad u(D) = \frac{bu(aK_V + bD)}{1 + au(aK_V + bD)}.$$

Proof. We can assume that $a > 0$ since if $a = 0$ the result follows immediately from the definition. Since $n = \kappa(aK_V + bD) = \kappa(K_V + (b/a)D)$ it follows from (7.1.4) that $u := u(V, D) < b/a$. It suffices therefore to prove the first assertion since the second follows from the first by simple algebra.

We can assume without loss of generality that $u > 0$ since otherwise the relation follows immediately from the definition. Thus $u = \sup\{t \in \mathbb{Q} \mid t > 0, \kappa(K_V + tD) = -\infty\}$. Note that for any rational number $t' > 0$

$$\kappa(K_V + t'(aK_V + bD)) = \kappa(K_V + \frac{t'b}{1 + t'a} D). \tag{7.1}$$

Note that for t' on $[0, \infty)$, $(t'b)/(1 + t'a)$ is a monotonically increasing continuous function from 0 to b/a. Since $0 < u < b/a$ it follows that there is a $u' \in (0, \infty)$ such that $u = (u'b)/(1 + u'a)$. Thus using formula (7.1) we see that for positive fractions $t' < u'$ we have $\kappa(K_V + t'(aK_V + bD)) = -\infty$ and, for positive fractions $t' > u'$, $\kappa(K_V + t'(aK_V + bD)) = n$. Thus $u(V, aK_V + bD) = u'$. □

Lemma 7.1.6. ([111, (0.4.3)]) *Let $\mathcal{L}$ be an ample line bundle on a normal irreducible projective variety V with terminal singularities. Assume that K_V is nonnef* *and let τ, $u := u(V, \mathcal{L})$ be the nefvalue and the unnormalized spectral value of $(V, \mathcal{L})$ respectively. Then $u \le \tau$ with equality if and only if the nefvalue morphism, ϕ, of $(V, \mathcal{L})$ is not birational.*

Proof. Since the Kawamata-Shokurov basepoint free theorem implies that there are positive integers N with $N(K_V + \tau\mathcal{L})$ Cartier and spanned, we see that $u(V, \mathcal{L}) \le \tau$.

If the nefvalue morphism is not birational, then given a general fiber F of ϕ, $(K_V + t\mathcal{L})_F = (t - \tau)\mathcal{L}_F$. Thus given any $t \in \mathbb{Q}$ with $t < \tau$, no positive power of $(K_V + t\mathcal{L})_F$ can have any sections. Since F is a general fiber this implies that no positive power of $K_V + t\mathcal{L}$ has any sections and hence that $t \le u(V, \mathcal{L}) \le \tau$ for any $t < \tau$. Thus $u(V, \mathcal{L}) = \tau$.

Now assume that $u(V, \mathcal{L}) = \tau$ and that ϕ is birational. Thus $K_V + \tau\mathcal{L}$ is big, and by Lemma (2.5.5), there exists an integer $N > 0$ and an effective divisor, D, such that $N(K_V + \tau\mathcal{L}) = D + \mathcal{L}$. Therefore $u(V, \mathcal{L}) \le (N\tau - 1)/N < \tau$. □

The following result applies in a number of situations to show that when the nefvalue morphism is birational, spectral values transform well.

Theorem 7.1.7. *Let $\mathcal{L}$ be an ample line bundle on a normal irreducible projective variety V with terminal singularities. Assume that K_V is nonnef and let τ, $u := u(V, \mathcal{L})$ be the nefvalue and the unnormalized spectral value of $(V, \mathcal{L})$ respectively. Assume that the nefvalue morphism, $\phi : V \to W$, is birational, W has terminal singularities, and $D := (\phi_*\mathcal{L})^{**}$ is $\mathbb{Q}$-Cartier. Then D is big and if $u \ge 0$ and $u(W, D) \ge 0$ then $u(W, D) = u(V, \mathcal{L})$.*

Proof. Since we will not use this result we refer the reader to our paper [111, Theorem (0.4.6)]. That result combined with Lemma (7.1.5) gives the theorem. □

The following variant of the conjecture in our appendix to [114] (compare also with Fujita [261]) was a motivating conjecture for our paper [111].

Conjecture 7.1.8. *Let $\mathcal{L}$ be an ample line bundle on $\mathcal{M}$, a connected projective manifold of dimension n. If $u := u(\mathcal{M}, \mathcal{L}) > (n + 1)/2$ then $u \in \mathbb{Q}$ and $K_{\mathcal{M}} + u\mathcal{L}$ is nef.*

Note that if $n \le 7$ or $n \ge 8$ with $u \ge n - 3$ this conjecture follows from the results of this chapter and our appendix to [114].

Remark 7.1.9. Another invariant of a pair, (V, D), where D is a big $\mathbb{Q}$-Cartier divisor on a normal irreducible $\mathbb{Q}$-Gorenstein projective variety, V, is the *cospectral value* of (V, D), defined as

$$\alpha(V, D) := \sup\{t \in \mathbb{Q} \mid \kappa(-K_V + tD) = -\infty\}.$$

Note that $\alpha(V, D)+u(V, D) \geq 0$. This invariant and the unnormalized spectral value allow us to define an interesting measure of complexity of the pair, (V, D). We define the *spectral diameter* of the pair (V, D) to be $\delta(V, D) := |\alpha(V, D)-u(V, D)|$. Note that $u(V, D) > 0 > \alpha(V, D)$ if $-K_V$ is big and $\alpha(V, D) > 0 > u(V, D)$ if K_V is big. Thus if the spectral diameter is 0, we have that neither K_V nor $-K_V$ is big. It is not hard to check that if the spectral diameter is 0 then $\kappa(K_V) \leq 0 \geq \kappa(-K_V)$. The spectral diameter gives a measure in terms of D of how much the negative part of K_V differs from the negative part of $-K_V$. For example if $V \cong \mathbb{P}^1 \times R$, where R is a smooth curve of positive genus, and D is a line bundle which is of degree 1 on fibers of the product fibrations, then the spectral diameter is 1.

7.2 Polarized pairs $(\mathcal{M}, \mathcal{L})$ with nefvalue $> \dim \mathcal{M} - 1$ and $\mathcal{M}$ singular

Throughout this section $\mathcal{M}$ will denote an n-dimensional normal projective variety with terminal singularities of index $r = r(\mathcal{M})$ and with $K_{\mathcal{M}}$ nonnef. Let $\mathcal{L}$ be an ample line bundle on $\mathcal{M}$. We will denote by τ the nefvalue of $(\mathcal{M}, \mathcal{L})$ and by $\phi : \mathcal{M} \to W$ the nefvalue morphism of $(\mathcal{M}, \mathcal{L})$. Recall that τ is the unique rational number such that $K_{\mathcal{M}} + \tau\mathcal{L}$ is nef but not ample. Even if we are mainly interested in the nonsingular case, terminal singularities occur naturally in the theory (see §7.5), so we must deal with the singular case also.

The following result has been proved by Maeda [422, (2.1)] in the more general case when $\mathcal{M}$ has log-terminal singularities.

Theorem 7.2.1. (Maeda) *Let $\mathcal{M}$ be an n-dimensional irreducible normal projective variety with terminal singularities. Let $\mathcal{L}$ be an ample line bundle on $\mathcal{M}$. Then $K_{\mathcal{M}} + n\mathcal{L}$ is nef unless $(\mathcal{M}, \mathcal{L}) \cong (\mathbb{P}^n, \mathcal{O}_{\mathbb{P}^n}(1))$. In particular $K_{\mathcal{M}} + (n+1)\mathcal{L}$ is always nef and is ample unless $(\mathcal{M}, \mathcal{L}) \cong (\mathbb{P}^n, \mathcal{O}_{\mathbb{P}^n}(1))$.*

Proof. Let r denote the index of $\mathcal{M}$, i.e., the smallest positive integer such that $rK_{\mathcal{M}}$ is a Cartier divisor. We can assume that $K_{\mathcal{M}}$ is nonnef since otherwise there is nothing to prove. Let τ be the nefvalue of the pair, $(\mathcal{M}, \mathcal{L})$, and let $\phi : \mathcal{M} \to Y$ be the nefvalue morphism (see (1.5.3)) of the pair $(\mathcal{M}, \mathcal{L})$. Note that the nefvalue, τ, of the pair $(\mathcal{M}, \mathcal{L})$ coincides with the ϕ-nefvalue of the pair $(\mathcal{M}, \mathcal{L})$ by Remark (1.5.4).

By the Kawamata rationality theorem, (1.5.2), applied to $\mathcal{L}$ and the morphism ϕ we see that we can write $r\tau = u/v$ with u and v coprime positive integers such that

$$u \leq r(1 + \max_{y \in Y}\{\dim \phi^{-1}(y)\}) \leq r(n+1). \tag{7.2}$$

We conclude that $\tau = u/(rv) \le u/r \le 1+n$ with equality happening only if $u = r(n+1)$, $v = 1$, and ϕ maps $\mathcal{M}$ to a point. In this case we have by (1.5) that $r(K_{\mathcal{M}} + (n+1)\mathcal{L}) \cong \mathcal{O}_{\mathcal{M}}$, and thus by Theorem (3.1.6) that $(\mathcal{M}, \mathcal{L}) \cong (\mathbb{P}^n, \mathcal{O}_{\mathbb{P}^n}(1))$.

If $n < \tau < n+1$ then by the Kawamata rationality theorem we conclude that $n < \tau = u/(rv) < n+1$. By the inequality (7.2) we see that

$$n < \frac{u}{rv} \le \frac{(1 + \max_{y \in Y}\{\dim \phi^{-1}(y)\})}{v}$$

which implies that $v = 1$, $n < u/r < n+1$, and ϕ maps $\mathcal{M}$ to a point. Thus we conclude by (1.5) that $rK_{\mathcal{M}} + u\mathcal{L} \cong \mathcal{O}_{\mathcal{M}}$. Consideration of the Hilbert polynomial, $p(t) := \chi(t\mathcal{L})$, leads to a contradiction. Note that since $t\mathcal{L} - K_{\mathcal{M}} = (t + u/r)\mathcal{L}$ is ample for $t > -u/r > -n-1$ we have $p(t) = h^0(t\mathcal{L})$ for $t \ge -n$ by Theorem (2.2.1). Thus we conclude that $p(0) = 1$ and $p(t) = 0$ for $-1 \ge t \ge -n$. Thus since $p(t)$ is a polynomial of degree n we conclude that $p(t) = (t+1)\cdots(t+n)/n!$. We further conclude that $\mathcal{L}^n = 1$ and $h^0(\mathcal{L}) = n+1$ which implies that $(\mathcal{M}, \mathcal{L}) \cong (\mathbb{P}^n, \mathcal{O}_{\mathbb{P}^n}(1))$ by Theorem (3.1.4). This gives the contradiction $\tau = n+1$. □

As an immediate consequence, we have the following general fact we use in the sequel (see also Fujita [255]).

Proposition 7.2.2. *Let $\mathcal{L}$ be an ample line bundle on $\mathcal{M}$, an n-dimensional irreducible normal projective variety with terminal singularities. Let τ be the nefvalue of $(\mathcal{M}, \mathcal{L})$ and $\phi : \mathcal{M} \to W$ the nefvalue morphism of $(\mathcal{M}, \mathcal{L})$. Then either*

1. *$\tau = n+1$ and $(\mathcal{M}, \mathcal{L}) \cong (\mathbb{P}^n, \mathcal{O}_{\mathbb{P}^n}(1))$; or*
2. *$\tau = n$ and $(\mathcal{M}, \mathcal{L}) \cong (Q, \mathcal{O}_Q(1))$, Q is a hyperquadric in $\mathbb{P}^{n+1}$; or*
3. *$\tau = n$ and $(\mathcal{M}, \mathcal{L})$ is a $(\mathbb{P}^{n-1}, \mathcal{O}_{\mathbb{P}^{n-1}}(1))$-bundle over a smooth curve under ϕ; or*
4. *$\tau \le n$ and $K_{\mathcal{M}} + n\mathcal{L}$ is nef and big.*

Proof. We can assume without loss of generality that $n \ge 2$. Indeed, if $n = 1$, then $K_{\mathcal{M}}$ is nef except for $\mathcal{M} \cong \mathbb{P}^1$. So either $\mathcal{M} \cong \mathbb{P}^1$ or we are in case (7.2.2.4). If $\mathcal{M} \cong \mathbb{P}^1$ and $K_{\mathcal{M}} + \mathcal{L}$ is not nef and big we have $\deg \mathcal{L} \le 2$. Then we are in case (7.2.2.1) if $\deg \mathcal{L} = 1$ and in case (7.2.2.2) if $\deg \mathcal{L} = 2$.

From (7.2.1) we can assume that $K_{\mathcal{M}} + n\mathcal{L}$ is nef and hence $\tau \le n$. Otherwise $(\mathcal{M}, \mathcal{L}) \cong (\mathbb{P}^n, \mathcal{O}_{\mathbb{P}^n}(1))$, $\tau = n+1$, and we are in case (7.2.2.1). If $\tau < n$, then $K_{\mathcal{M}} + n\mathcal{L}$ is ample and we are in case (7.2.2.4).

Thus we can assume $\tau = n$ and that $K_{\mathcal{M}} + n\mathcal{L}$ is not nef and big. Thus $\dim W < n$. Let F be a general fiber of ϕ. Therefore $K_F + n\mathcal{L}_F \sim \mathcal{O}_F$ and hence we see that $\dim F \ge n-1$ (see, e.g., (3.3.2)). If $\dim F = n$ then $\mathcal{M} = F$, W is a point and $(\mathcal{M}, \mathcal{L}) \cong (Q, \mathcal{O}_Q(1))$, Q is a hyperquadric in $\mathbb{P}^{n+1}$, by the Kobayashi-Ochiai result, (3.1.6). So we are in case (7.2.2.2). If $\dim F = n-1$, W is a smooth curve, $(F, \mathcal{L}_F) \cong (\mathbb{P}^{n-1}, \mathcal{O}_{\mathbb{P}^{n-1}}(1))$ and all fibers of ϕ are of dimension $n-1$. Then ϕ is a $\mathbb{P}^{n-1}$-bundle by (3.2.1) and we are in case (7.2.2.3). □

By the above we can assume that $\tau \leq n$ and $K_{\mathcal{M}} + n\mathcal{L}$ is nef and big. Here we make the assumption that $n \geq 2$. The smooth version of the next argument is due to Ionescu [335].

Theorem 7.2.3. *Let $\mathcal{L}$ be an ample line bundle on $\mathcal{M}$, an n-dimensional irreducible normal projective variety with terminal singularities. Assume further that $\mathcal{M}$ is $\mathbb{Q}$-factorial and that $K_{\mathcal{M}} + n\mathcal{L}$ is nef and big. Then $K_{\mathcal{M}} + n\mathcal{L}$ is ample.*

Proof. Let $\phi : \mathcal{M} \to W$ be the nefvalue morphism of $(\mathcal{M}, \mathcal{L})$. Let us assume that $K_{\mathcal{M}} + n\mathcal{L}$ is not ample. Hence in particular $\tau = n$ (see (1.5.5)) and ϕ is a birational morphism which contracts some curve. From (4.2.14) we know that there exists an extremal ray R such that $(K_{\mathcal{M}} + n\mathcal{L}) \cdot R = 0$. Let $\rho = \mathrm{cont}_R : \mathcal{M} \to Y$ be the contraction of R. Then ϕ factors through ρ, $\phi = \alpha \circ \rho$.

Let $\mathcal{A} := \mathcal{L} + \rho^* A$ for some ample line bundle A on Y. Then $\mathcal{A}$ is ample and we claim that $K_{\mathcal{M}} + n\mathcal{A} \approx \rho^* H$ for some ample $\mathbb{Q}$-Cartier divisor H on Y. Take a large integer m such that $m(K_{\mathcal{M}} + n\mathcal{L}) \approx \phi^* \mathcal{H}$ for some ample line bundle $\mathcal{H}$ on W. Then $K_{\mathcal{M}} + n\mathcal{L} \approx \rho^* D$ for some nef and big $\mathbb{Q}$-Cartier divisor D on Y (in fact $mD = \alpha^* \mathcal{H}$). Thus $K_{\mathcal{M}} + n\mathcal{A} = K_{\mathcal{M}} + n\mathcal{L} + n\rho^* A \approx \rho^*(D + nA)$, so that the claim is proved by taking $H = D + nA$.

By replacing $\mathcal{L}$ with $\mathcal{A}$ and renaming, we can assume without loss of generality that ρ is the nefvalue morphism of the pair, $(\mathcal{M}, \mathcal{L})$. It follows that, for some large integer m, $|m(K_{\mathcal{M}} + n\mathcal{L})|$ gives the morphism ρ and $K_{\mathcal{M}} + n\mathcal{L}$ is nef and big but not ample since $(K_{\mathcal{M}} + n\mathcal{L}) \cdot R = 0$. Thus $\tau = n$ is the ρ-nefvalue of $(\mathcal{M}, \mathcal{L})$. Therefore by the rationality theorem, (1.5.2), we know that

$$rn = \frac{u}{v} \leq u \leq r(\max_{y \in Y}\{\dim \rho^{-1}(y)\} + 1)$$

where r is the index of $\mathcal{M}$ and u, v are coprime positive integers. Then we see that ρ has a fiber of dimension $n-1$ and hence the locus E of R is an irreducible divisor on $\mathcal{M}$ (see the contraction theorem, (4.3.1), and compare also with the proof of (4.2.18)). We are going to show that this leads to a contradiction, and hence $K_{\mathcal{M}} + n\mathcal{L}$ is ample. For any integer t, consider the exact sequence of sheaves (cf., proof of (4.2.18))

$$0 \to t\mathcal{L} - E \to t\mathcal{L} \to t\mathcal{L}_E \to 0.$$

Recall that $K_{\mathcal{M}} + n\mathcal{L}$ is trivial on E. Then we have

$$(t\mathcal{L} - K_{\mathcal{M}})_E \sim (t+n)\mathcal{L}_E \quad \text{and} \quad (t\mathcal{L} - E - K_{\mathcal{M}})_E \sim ((t+n)\mathcal{L} - E)_E.$$

Therefore, since $-E$ is ρ-ample, we have that $t\mathcal{L} - K_{\mathcal{M}}$ is ρ-ample (i.e., $t\mathcal{L} - K_{\mathcal{M}}$ is ample on E) for $t \geq -n+1$ and $t\mathcal{L} - K_{\mathcal{M}} - E$ is ρ-ample, (i.e., $t\mathcal{L} - K_{\mathcal{M}} - E$ is ample on E) for $t \geq -n$. Consider the exact sequence for direct image sheaves

$$\cdots \to \rho_{(i)}(t\mathcal{L}) \to \rho_{(i)}(t\mathcal{L}_E) \to \rho_{(i+1)}(t\mathcal{L} - E) \to \cdots$$

where $\rho_{(i)}(\mathcal{F}) = R^i\rho_*(\mathcal{F})$ is the i-th direct image of a sheaf, $\mathcal{F}$. By using the relative vanishing theorem, (2.2.1), we have by the above that for $i > 0$, $\rho_{(i)}(t\mathcal{L}) = 0$ for $t \geq -n+1$ and $\rho_{(i)}(t\mathcal{L} - E) = 0$ for $t \geq -n$. Therefore $\rho_{(i)}(t\mathcal{L}_E) = 0$ for $t \geq -n+1$ and $i > 0$. Note that, for each i, $\rho_{(i)}(t\mathcal{L}_E) \cong H^i(t\mathcal{L}_E)$ since $t\mathcal{L}_E$ is supported on E. Let $p(t) := \chi(t\mathcal{L}_E)$. By the above $p(t) = h^0(t\mathcal{L}_E)$ for $t \geq 1-n$. Note also that $h^0(t\mathcal{L}_E) = 0$ if $t < 0$ since $t\mathcal{L}_E$ is ample. Then $p(t) = 0$ for $-n+1 \leq t \leq -1$, and since $p(0) = h^0(\mathcal{O}_E) = 1$ we find $p(t) = (t+1)\cdots(t+n-1)/(n-1)!$. Hence $\mathcal{L}_E^{n-1} = 1$ and $p(1) = h^0(\mathcal{L}_E) = n$, and therefore $(E, \mathcal{L}_E) \cong (\mathbb{P}^{n-1}, \mathcal{O}_{\mathbb{P}^{n-1}}(1))$ by (3.1.4). This is not possible. To see this note that, since $\mathcal{M}$ is $\mathbb{Q}$-factorial, there exists some integer $m > 0$ such that $m(K_{\mathcal{M}} + E)$ is a Cartier divisor. Since $\mathcal{M}$ has terminal singularities, $\mathrm{cod}_{\mathcal{M}}\mathrm{Sing}(\mathcal{M}) \geq 3$ by (1.3.1), and thus $\mathrm{cod}_E(\mathrm{Sing}(\mathcal{M}) \cap E) \geq 2$. Thus since E is smooth,

$$\mathcal{O}_{E-\mathrm{Sing}(\mathcal{M})\cap E}(m(K_{\mathcal{M}} + E)) \cong mK_{E-\mathrm{Sing}(\mathcal{M})\cap E}$$

extends to a unique $\mathcal{O}_{\mathbb{P}^{n-1}}(-mn)$. Note also that $K_{\mathcal{M}|E} \cong -n\mathcal{L}_E \cong \mathcal{O}_{\mathbb{P}^{n-1}}(-n)$ and $-E_E$ is an ample $\mathbb{Q}$-Cartier divisor by the contraction theorem. Thus the line bundle, $-E_{E-\mathrm{Sing}(\mathcal{M})\cap E}$, extends to a unique $\mathcal{O}_{\mathbb{P}^{n-1}}(e)$ for some $e > 0$, and hence $(K_{\mathcal{M}} + E)_{E-\mathrm{Sing}(\mathcal{M})\cap E}$ extends to a unique $\mathcal{O}_{\mathbb{P}^{n-1}}(-n-e)$. Therefore we have $m(-n-e) = -mn$, which gives the contradiction $e = 0$. □

In view of the result above we can assume from now on that $K_{\mathcal{M}} + n\mathcal{L}$ is ample and $\tau < n$. The following theorem, which in this generality is essentially due to Fujita, [255], takes care of the range $n-1 < \tau < n$. We give a reduction of the theorem to results of Fujita. We give a second proof in the smooth case.

Theorem 7.2.4. *Let $\mathcal{L}$ be an ample line bundle on $\mathcal{M}$, an n-dimensional irreducible normal projective variety with terminal singularities. Assume further that $\mathcal{M}$ is $\mathbb{Q}$-factorial, $n \geq 2$, and that $K_{\mathcal{M}} + n\mathcal{L}$ is ample. Then the nefvalue τ of $(\mathcal{M}, \mathcal{L})$ is $\tau \leq n-1$ unless $\tau = n - 1/2$ and $(\mathcal{M}, \mathcal{L}) = C_n(\mathbb{P}^2, \mathcal{O}_{\mathbb{P}^2}(2))$ is a generalized cone over $(\mathbb{P}^2, \mathcal{O}_{\mathbb{P}^2}(2))$. In particular if $n = 2$, then $(\mathcal{M}, \mathcal{L}) \cong (\mathbb{P}^2, \mathcal{O}_{\mathbb{P}^2}(2))$.*

Proof. Note that $\tau \leq n-1$ is equivalent to $K_{\mathcal{M}} + (n-1)\mathcal{L}$ being nef.

From Fujita [255, (3.4)], we know that $K_{\mathcal{M}} + (n-1)\mathcal{L}$ is nef unless either $\Delta(\mathcal{M}, \mathcal{L}) = 0$ or $\mathcal{M}$ is a $\mathbb{P}^{n-1}$-bundle over a smooth curve and $\mathcal{L}_F \cong \mathcal{O}_{\mathbb{P}^{n-1}}(1)$ for any fiber, F. The second case is excluded since $K_{\mathcal{M}} + n\mathcal{L}$ is ample. If $\Delta(\mathcal{M}, \mathcal{L}) = 0$ we know from Theorem (3.1.2) that either $(\mathcal{M}, \mathcal{L}) \cong (\mathbb{P}^n, \mathcal{O}_{\mathbb{P}^n}(1))$,

$(\mathcal{M}, \mathcal{L}) \cong (Q, \mathcal{O}_Q(1))$, Q a hyperquadric in $\mathbb{P}^{n+1}$, $\mathcal{M}$ is a $\mathbb{P}^{n-1}$-bundle over $\mathbb{P}^1$ and the restriction of $\mathcal{L}$ to any fiber is $\mathcal{O}_{\mathbb{P}^{n-1}}(1)$, $(\mathcal{M}, \mathcal{L}) \cong (\mathbb{P}^2, \mathcal{O}_{\mathbb{P}^2}(2))$, or $(\mathcal{M}, \mathcal{L})$ is a generalized cone over a submanifold V with $\Delta(V, \mathcal{L}_V) = 0$ and $K_V + (\dim V)\mathcal{L}_V$ ample, where $\mathcal{L}_V$ denotes the restriction of $\mathcal{L}$ to V. The first three cases are excluded since $K_{\mathcal{M}} + n\mathcal{L}$ is ample. Thus either $(\mathcal{M}, \mathcal{L}) \cong (\mathbb{P}^2, \mathcal{O}_{\mathbb{P}^2}(2))$ or $(\mathcal{M}, \mathcal{L})$ is a nondegenerate generalized cone. If $\dim V \geq 2$, then either $(V, \mathcal{L}_V) \cong (\mathbb{P}^2, \mathcal{O}_{\mathbb{P}^2}(2))$ and $(\mathcal{M}, \mathcal{L}) \cong C_n(\mathbb{P}^2, \mathcal{O}_{\mathbb{P}^2}(2))$ or V is a $\mathbb{P}^d$-bundle over $\mathbb{P}^1$ with $K_V + (\dim V)\mathcal{L}_V$ ample. The latter is not possible. Therefore $\dim V = 1$, $(V, \mathcal{L}_V) \cong (\mathbb{P}^1, \mathcal{O}_{\mathbb{P}^1}(a))$, where by the above we can assume $a \geq 3$, and we have that the singular set of $\mathcal{M}$ is of codimension 2 contradicting the fact that terminal singularities are of codimension ≥ 3.

We leave it as an exercise to the reader to check that $\tau = n - 1/2$ if $(\mathcal{M}, \mathcal{L}) \cong C_n(\mathbb{P}^2, \mathcal{O}_{\mathbb{P}^2}(2))$. □

Here is a second proof of Theorem (7.2.4) in the smooth case.

Proof. (smooth case) Since $K_{\mathcal{M}} + n\mathcal{L}$ is ample, we have $\tau < n$. Then we can assume that $n - 1 < \tau < n$. Let $\phi : \mathcal{M} \to Y$ be the nefvalue morphism of the pair $(\mathcal{M}, \mathcal{L})$. Writing $\tau = u/v$, where u, v are coprime positive integers, we have by the Kawamata rationality theorem, (1.5.2),

$$u \leq 1 + \max_{y \in Y}\{\dim \phi^{-1}(y)\} \leq n + 1. \tag{7.3}$$

First assume that $n = 2$. By (7.3) we have that

$$v\tau = u \leq 1 + \max_{y \in Y}\{\dim \phi^{-1}(y)\} \leq 3.$$

Since $\tau > n - 1 = 1$, we conclude that if $v > 1$ then $2 \leq v < u \leq 3$, and therefore that $v = 2$, $u = 3$, ϕ is the constant map. Thus $2K_{\mathcal{M}} + 3\mathcal{L} \cong \mathcal{O}_{\mathcal{M}}$ and there is an ample line bundle, H on $\mathcal{M}$, with $3H \approx -K_{\mathcal{M}}$, $\mathcal{L} \approx 2H$. From this and the Kobayashi-Ochiai theorem, (3.1.6), we conclude that $(\mathcal{M}, \mathcal{L}) \cong (\mathbb{P}^2, \mathcal{O}_{\mathbb{P}^2}(2))$. Therefore we are reduced to the case $v = 1$. This gives the absurdity that $1 < u < 2$.

Now assume that $n \geq 3$. Then from (7.3) we have

$$(n-1)v < u \leq 1 + \max_{y \in Y}\{\dim \phi^{-1}(y)\} \leq n + 1,$$

which implies that

$$(n-1)v \leq \max_{y \in Y}\{\dim \phi^{-1}(y)\} \leq n.$$

Hence $n(v - 1) \leq v$. Since $n \geq 3$ we have that $v = 1$ and $n - 1 < u < n$. □

Table 7.1. Polarized pairs with nefvalue $> n-1$ or spectral value < 2

Let $\mathcal{L}$ be an ample line bundle on a normal projective variety, $\mathcal{M}$, of dimension $n \geq 2$ with $\mathbb{Q}$-factorial terminal singularities. We give the spectral value, $\sigma := \sigma(\mathcal{M}, \mathcal{L}) = n + 1 - u$ (where $u := u(\mathcal{M}, \mathcal{L})$ denotes the unnormalized spectral value), the nefvalue $\tau := \tau(\mathcal{M}, \mathcal{L})$, and the sectional genus $g := g(\mathcal{L})$. In all cases $u = \tau$.

σ	τ	g	*Structure of* $(\mathcal{M}, \mathcal{L})$
0	$n+1$	0	$(\mathcal{M}, \mathcal{L}) \cong (\mathbb{P}^n, \mathcal{O}_{\mathbb{P}^n}(1))$
1	n	0	$\mathcal{M} \subset \mathbb{P}^{n+1}$ is a quadric and $\mathcal{L} \cong \mathcal{O}_{\mathbb{P}^{n+1}}(1)_{\mathcal{M}}$
		$h^1(\mathcal{O}_{\mathcal{M}})$	$(\mathcal{M}, \mathcal{L}) \cong (\mathbb{P}(\mathcal{E}), \xi_{\mathcal{E}})$ for an ample rank n vector bundle, $\mathcal{E}$, on a smooth curve, C. Furthermore $h^1(\mathcal{O}_{\mathcal{M}}) = g(\mathcal{L}) = h^1(\mathcal{O}_C)$
$\frac{3}{2}$	$n - \frac{1}{2}$	0	a possibly degenerate generalized cone over $(\mathbb{P}^2, \mathcal{O}_{\mathbb{P}^2}(2))$; the index of $\mathcal{M}$ is 2 if $n \geq 3$, and $2K_{\mathcal{M}} + (2n-1)\mathcal{L} \cong \mathcal{O}_{\mathcal{M}}$

For the reader's convenience we summarize in Table (7.1) the results given in Proposition (7.2.2), Theorem (7.2.3), and Theorem (7.2.4). In the trivial case when $\dim \mathcal{M} = 1$ (with $\mathcal{M}$ as in the table), $K_{\mathcal{M}}$ nonnef implies that $\mathcal{M} \cong \mathbb{P}^1$, $\mathcal{L} \cong \mathcal{O}_{\mathbb{P}^1}(d)$ for some $d > 0$, and hence that $K_{\mathcal{M}} + (2/d)\mathcal{L} \approx \mathcal{O}_{\mathcal{M}}$.

Let $\mathcal{L}$ be an ample line bundle on a normal irreducible projective variety, $\mathcal{M}$, of dimension n with at worst terminal singularities. If $K_{\mathcal{M}} + (n-1)\mathcal{L}$ is nef then it follows that $(K_{\mathcal{M}} + (n-1)\mathcal{L})^k \cdot \mathcal{L}^{n-k} \geq 0$ for all $k \geq 0$. The following numerical result which follows from the results proven so far shows that this almost characterizes the pairs, $(\mathcal{M}, \mathcal{L})$, with $K_{\mathcal{M}} + (n-1)\mathcal{L}$ nef. Here we use $g(\mathcal{L})$ to denote the sectional genus of a line bundle $\mathcal{L}$.

Theorem 7.2.5. *Let $\mathcal{L}$ be an ample line bundle on a normal irreducible projective variety, $\mathcal{M}$, of dimension $n \geq 2$ with at worst $\mathbb{Q}$-factorial, terminal singularities. If $K_{\mathcal{M}} + (n-1)\mathcal{L}$ is nonnef then $(-1)^k(K_{\mathcal{M}} + (n-1)\mathcal{L})^k \cdot \mathcal{L}^{n-k} < 0$ for $2 \leq k \leq n$ unless $g(\mathcal{L}) = 0$ and $(\mathcal{M}, \mathcal{L})$ is one of the following pairs:*

1. $(\mathbb{P}(\mathcal{E}), \xi_{\mathcal{E}})$ *where $\mathcal{E}$ is a rank n ample vector bundle on $\mathbb{P}^1$ with $n \leq \deg \mathcal{E} \leq 4$;*
2. $(\mathbb{P}^n, \mathcal{O}_{\mathbb{P}^n}(1))$*;*
3. $(Q, \mathcal{O}_Q(1))$*, Q a quadric hypersurface in $\mathbb{P}^{n+1}$;*
4. *a generalized cone over* $(\mathbb{P}^2, \mathcal{O}_{\mathbb{P}^2}(2))$.

In particular $K_{\mathcal{M}} + (n-1)\mathcal{L}$ is nef if and only if

$$(K_{\mathcal{M}} + (n-1)\mathcal{L})^k \cdot \mathcal{L}^{n-k} \geq 0 \ \ \textit{for} \ \ 0 \leq k \leq n.$$

If $g(\mathcal{L}) > 0$, then $K_{\mathcal{M}} + (n-1)\mathcal{L}$ is nef if and only if $(K_{\mathcal{M}} + (n-1)\mathcal{L})^2 \cdot \mathcal{L}^{n-2} \geq 0$.

Proof. If $K_{\mathcal{M}} + (n-1)\mathcal{L}$ is nonnef, then the nefvalue of $(\mathcal{M}, \mathcal{L})$ is bigger than $n-1$. Then the proof is simply a matter of going down the list in Table (7.1).

If $(\mathcal{M}, \mathcal{L}) \cong (\mathbb{P}^n, \mathcal{O}_{\mathbb{P}^n}(1))$, then

$$K_{\mathcal{M}} + (n-1)\mathcal{L} \cong \mathcal{O}_{\mathbb{P}^n}(-2), \quad (K_{\mathcal{M}} + (n-1)\mathcal{L})^k \cdot \mathcal{L}^{n-k} = (-2)^k,$$

so that

$$(-1)^k (K_{\mathcal{M}} + (n-1)\mathcal{L})^k \cdot \mathcal{L}^{n-k} = 2^k > 0.$$

If $(\mathcal{M}, \mathcal{L}) \cong (Q, \mathcal{O}_Q(1))$, Q a quadric in $\mathbb{P}^{n+1}$, then $K_{\mathcal{M}} + (n-1)\mathcal{L} \cong \mathcal{O}_Q(-1)$ and $(K_{\mathcal{M}} + (n-1)\mathcal{L})^k \cdot \mathcal{L}^{n-k} = (-1)^k$.

If $(\mathcal{M}, \mathcal{L})$ is a generalized cone over $(\mathbb{P}^2, \mathcal{O}_{\mathbb{P}^2}(2))$, then $2(K_{\mathcal{M}} + (n-1)\mathcal{L}) \cong -\mathcal{L}$ and

$$(K_{\mathcal{M}} + (n-1)\mathcal{L})^k \cdot \mathcal{L}^{n-k} = 4(-\frac{1}{2})^k.$$

The fraction is not surprising since $K_{\mathcal{M}}$ is 2-Cartier.

If $(\mathcal{M}, \mathcal{L}) \cong (\mathbb{P}(\mathcal{E}), \xi_{\mathcal{E}})$ where $\mathcal{E}$ is a rank n ample vector bundle on a curve B of genus $q := h^1(\mathcal{O}_{\mathcal{M}})$, then, by the canonical bundle formula, $K_{\mathcal{M}} + (n-1)\mathcal{L} \cong -\mathcal{L} + p^*(K_B + \det \mathcal{E})$ where $p : \mathbb{P}(\mathcal{E}) \to B$ is the natural projection. Thus

$$(-1)^k (K_{\mathcal{M}} + (n-1)\mathcal{L})^k \cdot \mathcal{L}^{n-k} = \mathcal{L}^n - k \deg(K_B + \det \mathcal{E}) = (1-k) \deg \mathcal{E} - k(2q-2).$$

If $2q - 2 \geq 0$ and $k \geq 2$, this is negative. Thus we can assume that $q = 0$, hence in particular $g(\mathcal{L}) = 0$. Therefore we can assume that $(1-k)\deg \mathcal{E} + 2k \geq 0$ for some $k \geq 2$. Thus $\deg \mathcal{E} \leq 2k/(k-1)$. So $\deg \mathcal{E} \leq 4$. Noting that the rank r, ample vector bundles on $\mathbb{P}^1$ are precisely the direct sums, $\oplus_{i=1}^r \mathcal{O}_{\mathbb{P}^1}(a_i)$, with $a_i > 0$ for all i, we have $\deg \mathcal{E} = \sum_i a_i \geq \operatorname{rank} \mathcal{E} = n$. □

Now we specialize down to the case when $n := \dim \mathcal{M} = 2$. Since terminal singularities have codimension 3, we are in the smooth case. Sakai [532] showed the following for normal surfaces. We follow the presentation in our paper with Andreatta [21, Lemma (0.6)].

Theorem 7.2.6. *Let $\mathcal{L}$ be an ample line bundle on a smooth connected projective surface, $\mathcal{M}$. The following are equivalent:*

1. $h^0(K_{\mathcal{M}} + \mathcal{L}) > 0$;
2. $h^0(N(K_{\mathcal{M}} + \mathcal{L})) > 0$ *for some integer* $N > 0$;
3. $K_{\mathcal{M}} + \mathcal{L}$ *is nef;*

4. *$K_{\mathcal{M}} + \mathcal{L}$ is semiample;*

5. *$g(\mathcal{L}) \geq 1$ and $(K_{\mathcal{M}} + \mathcal{L})^2 \geq 0$.*

Proof. Condition 1) immediately implies condition 2).

Going down Table (7.1) we see that when $K_{\mathcal{M}} + \mathcal{L}$ is nonnef, then $h^0(N(K_{\mathcal{M}} + \mathcal{L})) = 0$ for all integers $N > 0$. Thus condition 2) implies the condition 3).

Condition 3) implies the condition 4) by the Kawamata-Shokurov basepoint free theorem, (1.5.1).

By the genus formula, condition 4) immediately implies condition 5).

Therefore we can assume that $g(\mathcal{L}) \geq 1$, $(K_{\mathcal{M}} + \mathcal{L})^2 \geq 0$, but $h^0(K_{\mathcal{M}} + \mathcal{L}) = 0$. By the Kodaira vanishing theorem $h^i(K_{\mathcal{M}} + \mathcal{L}) = 0$ for $i > 0$ and therefore $\chi(K_{\mathcal{M}} + \mathcal{L}) = 0$. By the Riemann-Roch theorem applied to $\chi(K_{\mathcal{M}} + \mathcal{L}) = 0$, we have that $\chi(\mathcal{O}_{\mathcal{M}}) = 1 - g(\mathcal{L})$.

First assume that $g(\mathcal{L}) = 1$. Then we have $\mathcal{L} \cdot (K_{\mathcal{M}} + \mathcal{L}) = 0$ and $(K_{\mathcal{M}} + \mathcal{L})^2 \geq 0$. By the Hodge index theorem we conclude that $K_{\mathcal{M}} \sim -\mathcal{L}$ and therefore $h^i(\mathcal{O}_{\mathcal{M}}) = 0$ for $i > 0$, so that we find the contradiction $1 = \chi(\mathcal{O}_{\mathcal{M}}) = 1 - g(\mathcal{L}) = 0$.

Therefore we can assume that $g(\mathcal{L}) \geq 2$. This gives $\chi(\mathcal{O}_{\mathcal{M}}) = 1 - g(\mathcal{L}) < 0$, which implies that $\mathcal{M}$ is birationally ruled. Thus $h^2(\mathcal{O}_{\mathcal{M}}) = 0$ and we conclude from $\chi(\mathcal{O}_{\mathcal{M}}) = 1 - g(\mathcal{L})$ that $g(\mathcal{L}) = h^1(\mathcal{O}_{\mathcal{M}})$. Thus we have from $(K_{\mathcal{M}} + \mathcal{L})^2 \geq 0$ and the genus formula, $2g(\mathcal{L}) - 2 = (K_{\mathcal{M}} + \mathcal{L}) \cdot \mathcal{L}$, that

$$0 < \mathcal{L}^2 \leq K_{\mathcal{M}}^2 + 4h^1(\mathcal{O}_{\mathcal{M}}) - 4.$$

Since $\mathcal{M}$ is birationally ruled we know that $K_{\mathcal{M}}^2 \leq 8 - 8h^1(\mathcal{O}_{\mathcal{M}})$. This gives the absurdity $0 < \mathcal{L}^2 \leq 4(1 - h^1(\mathcal{O}_{\mathcal{M}})) = 4(1 - g(\mathcal{L})) < 0$. □

Using the bundle $K_{\mathcal{M}} + (n-1)\mathcal{L}$ when $n \geq 3$ in place of $K_{\mathcal{M}} + \mathcal{L}$ when $n = 2$, it follows from the results of this section that conditions 2), 3), 4), and 5) of Theorem (7.2.6) are equivalent for pairs $(\mathcal{M}, \mathcal{L})$ with $n \geq 3$. Condition 1) is more subtle. We have the following conjecture.

Conjecture 7.2.7. *If $\mathcal{L}$ is an ample line bundle on a connected projective manifold, $\mathcal{M}$, of dimension n, and $K_{\mathcal{M}} + (n-1)\mathcal{L}$ is nef then $h^0(K_{\mathcal{M}} + (n-1)\mathcal{L}) > 0$.*

We have the following result in the direction of the conjecture.

Corollary 7.2.8. *Let $\mathcal{L}$ be an ample line bundle on an n-dimensional normal projective variety, $\mathcal{M}$, with at worst terminal singularities and $n \geq 2$. Assume that there exist $n-2$ elements $H_i \in |\mathcal{L}|$ for $1 \leq i \leq n-2$, such that the H_i intersect transversally in a smooth surface S. Then the following conditions are equivalent:*

1. *$h^0(K_{\mathcal{M}} + (n-1)\mathcal{L}) > 0$;*

2. *$h^0(N(K_{\mathcal{M}} + (n-1)\mathcal{L})) > 0$ for some integer $N > 0$;*

3. $K_{\mathcal{M}} + (n-1)\mathcal{L}$ *is nef;*

4. $K_{\mathcal{M}} + (n-1)\mathcal{L}$ *is semiample;*

5. $g(\mathcal{L}) \geq 1$ *and* $(K_{\mathcal{M}} + (n-1)\mathcal{L})^k \cdot \mathcal{L}^{n-k} \geq 0$ *for* $k \geq 2$.

Proof. This result in the case $n = 2$ is Theorem (7.2.6). The proofs of the implications 1) $\Rightarrow$ 2) $\Rightarrow$ 3) $\Rightarrow$ 4) $\Rightarrow$ 5) are exactly as in the proof of (7.2.6). We prove the last implication 5) $\Rightarrow$ 1) by induction from the case $n = 2$. Since the sections of $\mathcal{L}$ corresponding to the H_i span $\mathcal{L}$ outside S, we conclude from Theorem (1.7.1) that the intersections $H_1 \cap \ldots \cap H_i$ are irreducible, normal, and have terminal singularities for all i.

The condition 5) implies the corresponding condition on H_1. Let $\mathcal{L}_{H_1}$ be the restriction of $\mathcal{L}$ to H_1. Thus by the induction hypothesis $h^0(K_{H_1} + (\dim H_1 - 1)\mathcal{L}_{H_1}) > 0$. From the vanishing theorem, (2.2.1), we conclude that $h^1(K_{\mathcal{M}} + (n-2)\mathcal{L}) = 0$ and thus, noting that X has Cohen-Macaulay singularities, from the exact sequence

$$0 \to K_{\mathcal{M}} + (n-2)\mathcal{L} \to K_{\mathcal{M}} + (n-1)\mathcal{L} \to K_{H_1} + (\dim H_1 - 1)\mathcal{L}_{H_1} \to 0$$

we conclude that $h^0(K_{\mathcal{M}} + (n-1)\mathcal{L}) > 0$. □

We have the following corollary of Theorem (7.2.6).

Corollary 7.2.9. *Let $\mathcal{L}$ be an ample line bundle on a smooth connected projective surface, $\mathcal{M}$. Then $h^2(\mathcal{O}_{\mathcal{M}}) + g(\mathcal{L}) \geq h^1(\mathcal{O}_{\mathcal{M}})$ with equality only if $h^2(\mathcal{O}_{\mathcal{M}}) = 0$ and $K_{\mathcal{M}} + \mathcal{L}$ is nonnef.*

Proof. As came out in the proof of Theorem (7.2.6) the inequality $h^2(\mathcal{O}_{\mathcal{M}}) + g(\mathcal{L}) \geq h^1(\mathcal{O}_{\mathcal{M}})$ is just the inequality $h^0(K_{\mathcal{M}} + \mathcal{L}) \geq 0$. Thus equality implies $h^0(K_{\mathcal{M}} + \mathcal{L}) = 0$, which implies that $K_{\mathcal{M}} + \mathcal{L}$ is nonnef by Theorem (7.2.6). The fact that $h^2(\mathcal{O}_{\mathcal{M}}) = 0$ follows from Table (7.1). □

The following result is useful. The characterization of the equality $h^1(\mathcal{O}_{\mathcal{M}}) = g(\mathcal{L})$ for surfaces under the minimal condition $h^0(\mathcal{L}) \geq 2$ is due to Lanteri [364].

Theorem 7.2.10. *Let $\mathcal{L}$ be an ample line bundle on an n-dimensional normal projective variety, $\mathcal{M}$, with at worst terminal singularities and $n \geq 2$. Assume that $h^0(\mathcal{L}) \geq n-1$ and that there exist $n-2$ elements $H_i \in |\mathcal{L}|$ for $1 \leq i \leq n-2$, such that the H_i intersect transversally in a smooth surface S. Then $g(\mathcal{L}) \geq h^1(\mathcal{O}_{\mathcal{M}})$. If $h^0(\mathcal{L}) \geq n$ then the equality $g(\mathcal{L}) = h^1(\mathcal{O}_{\mathcal{M}})$ implies that $K_{\mathcal{M}} + (n-1)\mathcal{L}$ is nonnef, and therefore that $(\mathcal{M}, \mathcal{L})$ is as in* Table (7.1).

Proof. Let $\mathcal{L}_S$ be the restriction of $\mathcal{L}$ to S. Since $h^1(\mathcal{O}_S) = h^1(\mathcal{O}_{\mathcal{M}})$, $g(\mathcal{L}) = g(\mathcal{L}_S)$ and since $(K_{\mathcal{M}} + (n-1)\mathcal{L})_S \approx K_S + \mathcal{L}_S$, it suffices to prove the result in the case $n = 2$.

By Corollary (7.2.9) we have that $h^2(\mathcal{O}_{\mathcal{M}}) + g(\mathcal{L}) \geq h^1(\mathcal{O}_{\mathcal{M}})$. If $h^2(\mathcal{O}_{\mathcal{M}}) = 0$ then we have the desired inequality. In this case when $h^2(\mathcal{O}_{\mathcal{M}}) = 0$, it follows from (7.2.9) that the inequality $g(\mathcal{L}) \geq h^1(\mathcal{O}_{\mathcal{M}})$ is an equality only if $K_{\mathcal{M}} + \mathcal{L}$ is nonnef. Therefore without loss of generality we can assume that $h^2(\mathcal{O}_{\mathcal{M}}) > 0$.

Since $h^0(\mathcal{L}) > 0$ then

$$h^0(K_{\mathcal{M}} + \mathcal{L}) \geq h^0(K_{\mathcal{M}}) + h^0(\mathcal{L}) - 1 \geq h^0(K_{\mathcal{M}}),$$

and since $h^0(K_{\mathcal{M}}) = h^2(\mathcal{O}_{\mathcal{M}})$ we conclude from the equality $h^0(K_{\mathcal{M}} + \mathcal{L}) = h^2(\mathcal{O}_{\mathcal{M}}) + g(\mathcal{L}) - h^1(\mathcal{O}_{\mathcal{M}})$ noted in the proof of Theorem (7.2.6), that

$$h^2(\mathcal{O}_{\mathcal{M}}) + g(\mathcal{L}) = h^1(\mathcal{O}_{\mathcal{M}}) + h^0(K_{\mathcal{M}} + \mathcal{L}) \geq h^1(\mathcal{O}_{\mathcal{M}}) + h^2(\mathcal{O}_{\mathcal{M}})$$

which gives the desired inequality.

Finally we are reduced to the case of a surface $\mathcal{M}$ with $h^0(K_{\mathcal{M}}) > 0$, $h^1(\mathcal{O}_{\mathcal{M}}) = g(\mathcal{L})$, and $h^0(\mathcal{L}) \geq 2$. Since $h^0(K_{\mathcal{M}}) > 0$ and $h^0(\mathcal{L}) \geq 2$, $h^0(K_{\mathcal{M}} + \mathcal{L}) \geq h^0(K_{\mathcal{M}}) + 1$ and we conclude by the same reasoning as in the last paragraph the contradiction that $g(\mathcal{L}) \geq h^1(\mathcal{O}_{\mathcal{M}}) + 1$. □

Note that when $g(\mathcal{L}) = h^1(\mathcal{O}_{\mathcal{M}})$ and $h^0(\mathcal{L}) = 1$ there exist a number of examples with $K_{\mathcal{M}} + (n-1)\mathcal{L}$ nef. One simple example is given by a smooth genus 2 curve embedded in its Jacobian, i.e., its Albanese variety. The following example related to this is given in the interesting article by Lanteri [364] which studies these exceptional varieties. Let C be a smooth curve of genus g. Let $\mathcal{M} := C^{(n)}$, where $C^{(n)}$ is the n-th symmetric product of C. It is a standard fact that $\mathcal{M}$ is smooth and $K_{\mathcal{M}}$ is nef for $n \leq g$. Fulton and Lazarsfeld [269, Lemma (2.7)] showed that $C^{(n-1)}$ is an ample divisor on $C^{(n)}$. Let $\mathcal{L} := \mathcal{O}_{\mathcal{M}}(C^{(n-1)})$. Then in particular $K_{\mathcal{M}} + (n-1)\mathcal{L}$ is ample and the pair $(\mathcal{M}, \mathcal{L})$ with $n \leq g$ gives the example since it can be shown that $g(\mathcal{L}) = h^1(\mathcal{O}_{\mathcal{M}}) = g$, and $h^0(\mathcal{L}) = 1$.

The above results suggest the following question.

Question 7.2.11. *Let $\mathcal{L}$ be an ample line bundle on an n-dimensional projective manifold, $\mathcal{M}$.*

1. *Is $g(\mathcal{L}) \geq h^1(\mathcal{O}_{\mathcal{M}})$?*

2. *What are the pairs, $(\mathcal{M}, \mathcal{L})$, with $K_{\mathcal{M}} + (n-1)\mathcal{L}$ nef and $g(\mathcal{L}) \leq h^1(\mathcal{O}_{\mathcal{M}})$?*

7.3 The first reduction of a singular variety

In this section we start considering a polarized pair $(\mathcal{M}, \mathcal{L})$, allowing on $\mathcal{M}$ some mild singularities.

7.3.1. Assumptions and notation. Through this section, $\mathcal{M}$ will denote an n-dimensional projective normal variety with terminal singularities and with $n \geq 2$. Let $\mathcal{L}$ be an ample line bundle on $\mathcal{M}$. We will denote by $\tau(\mathcal{L})$ the nefvalue of $(\mathcal{M}, \mathcal{L})$ and by $\phi_{\mathcal{L}} : \mathcal{M} \to M$ the nefvalue morphism of $(\mathcal{M}, \mathcal{L})$.

By the results of the previous section we can assume that $\tau(\mathcal{L}) \leq n-1$. Here we make the blanket assumption that $(\mathcal{M}, \mathcal{L})$ is not one of the special pairs described in §7.2. The following result has been proved, in the smooth case, by Fujita [252] and Ionescu [335]. There is a partial result of Fujita [252] for $\mathcal{M}$ with rational, Gorenstein singularities, but does not identify the exceptional divisors in the case (7.3.2.4). For $\mathcal{M}$ with codimension three Gorenstein singularities, $\dim \operatorname{Irr}(\mathcal{M}) \leq 0$, and an ample, spanned line bundle, $\mathcal{L}$, this result was proved by Sommese [583], where the exceptional divisors in (7.3.2.4) were identified by using the main result of Lipman and Sommese [408].

Theorem 7.3.2. *Let $\mathcal{M}$ be an n-dimensional normal projective variety with terminal singularities. Let $\mathcal{L}$ be an ample line bundle on $\mathcal{M}$. Let $\phi_{\mathcal{L}}$ be the nefvalue morphism of $(\mathcal{M}, \mathcal{L})$. Let r be the index of $\mathcal{M}$. Assume $\tau(\mathcal{L}) \leq n-1$. Then $K_{\mathcal{M}} + (n-1)\mathcal{L}$ is ample unless $\tau(\mathcal{L}) = n-1$ and either*

1. *$rK_{\mathcal{M}} \approx -r(n-1)\mathcal{L}$, i.e., $(\mathcal{M}, \mathcal{L})$ is a Del Pezzo variety; or*
2. *$(\mathcal{M}, \mathcal{L})$ is a quadric fibration over a smooth curve under $\phi_{\mathcal{L}}$; or*
3. *$(\mathcal{M}, \mathcal{L})$ is a scroll over a normal surface under $\phi_{\mathcal{L}}$; or*
4. *the nefvalue morphism $\phi_{\mathcal{L}} : \mathcal{M} \to M$ is birational. In this case, if further $\mathcal{M}$ is factorial, then $\phi_{\mathcal{L}} : \mathcal{M} \to M$ is the simultaneous contraction to distinct smooth points of divisors $E_i \cong \mathbb{P}^{n-1}$ such that $E_i \subset \operatorname{reg}(\mathcal{M}), \mathcal{O}_{E_i}(E_i) \cong \mathcal{O}_{\mathbb{P}^{n-1}}(-1)$ and $\mathcal{L}_{E_i} \cong \mathcal{O}_{\mathbb{P}^{n-1}}(1)$ for $i = 1, \dots, t$. Furthermore $L := (\phi_{\mathcal{L}*}\mathcal{L})^{**}$ and $K_M + (n-1)L$ are ample and $K_{\mathcal{M}} + (n-1)\mathcal{L} \cong \phi_{\mathcal{L}}^*(K_M + (n-1)L)$.*

Proof. For simplicity, write $\tau = \tau(\mathcal{L}), \phi = \phi_{\mathcal{L}}$. We assume that $K_{\mathcal{M}} + (n-1)\mathcal{L}$ is not ample. Note that $K_{\mathcal{M}} + (n-1)\mathcal{L}$ is nef since $\tau \leq n-1$. It thus follows that $\tau = n-1$ (see (1.5.5)) and ϕ is the morphism associated to $|mr(K_{\mathcal{M}} + (n-1)\mathcal{L})|$ for $m \gg 0$.

First, assume that $\phi : \mathcal{M} \to M$ has a lower dimensional image and let F be a general fiber of ϕ. Note that F has terminal singularities. We have $r(K_F + (n-1)\mathcal{L}_F) \approx \mathcal{O}_F$ so that, by (3.3.2), $\dim F \geq n-2$, i.e., $\dim M \leq 2$. If $\dim M = 0$, $rK_{\mathcal{M}} \approx -r(n-1)\mathcal{L}$ and $(\mathcal{M}, \mathcal{L})$ is a Del Pezzo variety as in (7.3.2.1). If $\dim M = 1$, $(\mathcal{M}, \mathcal{L})$ is a quadric fibration over a smooth curve as in (7.3.2.2). If $\dim M = 2$, $(\mathcal{M}, \mathcal{L})$ is a scroll over M, and we are in case (7.3.2.3).

Thus we can assume that $K_{\mathcal{M}} + (n-1)\mathcal{L}$ is nef and big, i.e., ϕ is a birational morphism which contracts some curve. Assume now that $\mathcal{M}$ has terminal factorial singularities. From (4.2.14) we know that there exists an extremal ray $R = \mathbb{R}_+[C]$ such that $(K_{\mathcal{M}} + (n-1)\mathcal{L}) \cdot R = 0$. Let $\rho : \mathcal{M} \to Y$ be the contraction of the extremal ray R. Then ϕ factors through ρ, $\phi = \alpha \circ \rho$. Hence in particular $K_{\mathcal{M}} + (n-1)\mathcal{L} \approx \rho^* H$ where $H = \alpha^*\mathcal{H}$ and $\mathcal{H}$ is an ample line bundle on M such that $K_{\mathcal{M}} + (n-1)\mathcal{L} \approx \phi^*\mathcal{H}$. Let E be the locus of R. Therefore Proposition (4.2.18) applies to say that $(E, \mathcal{L}_E) \cong (\mathbb{P}^{n-1}, \mathcal{O}_{\mathbb{P}^{n-1}}(1))$. Thus since E is smooth and Cartier, $E \subset \operatorname{reg}(\mathcal{M})$. Note that $(K_{\mathcal{M}} + E)_E \approx K_E \approx \mathcal{O}_E(-n)$.

Since $K_{\mathcal{M}} + (n-1)\mathcal{L}$ is trivial on E we also have $K_{\mathcal{M}|E} \approx -(n-1)\mathcal{L}_E \approx \mathcal{O}_{\mathbb{P}^{n-1}}(1-n)$. Therefore we find $E_E \approx \mathcal{O}_E(-1)$.

We have shown that $\mathcal{M}$ contains a finite number of divisors $E_i \cong \mathbb{P}^{n-1}$ such that $E_i \subset \operatorname{reg}(\mathcal{M})$, $\mathcal{O}_{E_i}(E_i) \cong \mathcal{O}_{\mathbb{P}^{n-1}}(-1)$, $\mathcal{L}_{E_i} \cong \mathcal{O}_{\mathbb{P}^{n-1}}(1)$ and E_i can be contracted to a smooth point, $i = 1, \dots, t$.

Let us show now that all the exceptional divisors E_i above, corresponding to all the extremal rays R_i such that $(K_{\mathcal{M}} + (n-1)\mathcal{L}) \cdot R_i = 0$, are disjoint. Let R_1, R_2 be two distinct extremal rays as above. If $n \geq 3$, Corollary (4.2.17) applies to say that E_1, E_2 are disjoint. If $n = 2$, assume that $E_1 = \ell_1$, $E_2 = \ell_2$ are two exceptional lines meeting in a point. Then $\ell_i \cong \mathbb{P}^1$, $\ell_i^2 = -1$ for $i = 1, 2$, and since the intersection matrix $E_i \cdot E_j$ is negative definite, we know that $(\ell_1 + \ell_2)^2 < 0$. This leads to the contradiction $\ell_1 \cdot \ell_2 < 1$.

Since the E_i are disjoint, we have the simultaneous contraction $\sigma : \mathcal{M} \to V$ of all the E_i. Note that σ agrees in some complex neighborhood of E_i with the contraction associated to R_i for each i. Then V is a normal, compact analytic variety and ϕ factors through σ, $\phi = \beta \circ \sigma$. Note that β has connected fibers since ϕ, σ have connected fibers. To prove that β is an isomorphism, assume the converse. Then there exists a curve contained in V, say γ, such that $\beta(\gamma)$ is a point. Let γ' be an irreducible curve in $\mathcal{M}$, such that γ' goes onto γ under σ. Since γ' is contracted to a point under ϕ, we have $(K_{\mathcal{M}} + (n-1)\mathcal{L}) \cdot \gamma' = 0$. Hence by (4.2.14) we can write in $NE(\mathcal{M})$, $\gamma' = \sum_i \lambda_i C_i$ where $\lambda_i \in \mathbb{R}_+$ and $R_i = \mathbb{R}_+[C_i]$ are the extremal rays which correspond to the E_i that gave rise to the morphism σ. Recall that $E_j \cdot C_i = 0$ since $E_i \cap E_j = \emptyset$ for $i \neq j$ and $E_i \cdot C_i < 0$ for all i. Therefore for each index j, $E_j \cdot \gamma' = \sum_i \lambda_i E_j \cdot C_i = \lambda_j E_j \cdot C_j < 0$, so that γ' is contained in E_j for all j. This leads to a contradiction unless $\gamma' = \lambda_j C_j$ for some index j. In this case $[\gamma'] \in R_j$ and hence $\sigma(\gamma')$ is a point. This contradicts the fact that $\sigma(\gamma') = \gamma$. Thus we conclude that β is an isomorphism.

Now, let $L := (\phi_*\mathcal{L})^{**}$, the double dual. Since $\mathcal{L}$ is ample and ϕ is an isomorphism outside of a finite set of points, L is ample. The fact that $K_{\mathcal{M}} + (n-1)\mathcal{L} \approx \phi^*(K_M + (n-1)L)$ simply follows by noting that ϕ expresses $\mathcal{M}$ as the blowing up of M along a finite set, B, of points with $B \subset \operatorname{reg}(M)$. □

Table (7.2) summarizes the conclusions of Theorem (7.3.2).

Table 7.2. Polarized pairs with nefvalue $= n - 1$ or spectral value $= 2$
In this table $\mathcal{L}$ is an ample line bundle on a normal projective variety, $\mathcal{M}$, of dimension $n \geq 2$ with terminal singularities. Let r be the index of $\mathcal{M}$. We give the spectral value, $\sigma := \sigma(\mathcal{M}, \mathcal{L}) = n+1-u$ (where $u := u(\mathcal{M}, \mathcal{L})$ denotes the unnormalized spectral value), the nefvalue $\tau := \tau(\mathcal{M}, \mathcal{L})$, and the sectional genus $g := g(\mathcal{L})$.

<table>
<tr><th>σ</th><th>τ</th><th>g</th><th>Structure of $(\mathcal{M}, \mathcal{L})$</th></tr>
<tr><td rowspan="3">2</td><td rowspan="3">$n-1$</td><td>1</td><td>$(\mathcal{M}, \mathcal{L})$ is a Del Pezzo variety, i.e., $rK_{\mathcal{M}} \approx -r(n-1)\mathcal{L}$</td></tr>
<tr><td>$\deg H + 1$</td><td>$(\mathcal{M}, \mathcal{L})$ is a quadric fibration, $p : \mathcal{M} \to C$, over a smooth curve, C, i.e., $r(K_{\mathcal{M}} + (n-1)\mathcal{L}) \approx p^*H$ for some ample line bundle H on C. Furthermore $h^1(\mathcal{O}_C) = h^1(\mathcal{O}_{\mathcal{M}})$</td></tr>
<tr><td>≥ 2</td><td>$(\mathcal{M}, \mathcal{L})$ is a scroll over a normal surface S</td></tr>
<tr><td>> 2</td><td>$\leq n-1$</td><td>≥ 2</td><td>$K_{\mathcal{M}}+(n-1)\mathcal{L}$ is semiample and big for these pairs, and the first reduction (M, L) exists</td></tr>
</table>

We can now define the first reduction of $(\mathcal{M}, \mathcal{L})$.

Definition 7.3.3. (First reduction) Let $\mathcal{M}$ be an n-dimensional normal projective variety with terminal singularities. Let r be the index of $\mathcal{M}$. Let $\mathcal{L}$ be an ample line bundle on $\mathcal{M}$. Assume that $K_{\mathcal{M}} + (n-1)\mathcal{L}$ is nef and big. By the Kawamata-Shokurov basepoint free theorem, we have a birational morphism, $\pi : \mathcal{M} \to M$ with connected fibers and normal image, associated to $|mr(K_{\mathcal{M}} + (n-1)\mathcal{L})|$ for $m \gg 0$. Let $L = (\pi_*\mathcal{L})^{**}$, the double dual. The pair (M, L) together with the morphism π is called the *first reduction* of $(\mathcal{M}, \mathcal{L})$. Note that if $K_{\mathcal{M}} + (n-1)\mathcal{L}$ is nef and big but not ample, then π coincides with the nefvalue morphism $\phi_{\mathcal{L}}$ of $(\mathcal{M}, \mathcal{L})$. Moreover if in addition the singularities of $\mathcal{M}$ are factorial, then (M, L) satisfies the conditions in (7.3.2.4). In particular, in this case, M has terminal factorial singularities since $\phi_{\mathcal{L}}$ is an isomorphism in a neighborhood of $\mathrm{Sing}(\mathcal{M})$ which expresses $\mathcal{M}$ as a blowing up of M at a finite set contained in $\mathrm{reg}(M)$.

We denote by $\tau(L)$ the nefvalue and by $\phi_L : M \to X$ the nefvalue morphism of (M, L). Note that $\tau(L) < n-1$ since $K_M + (n-1)L$ is ample by (7.3.2.4).

Note that if $\mathcal{M}$ is smooth and $\mathcal{L}$ is ample and spanned then there is a one-to-one correspondence between smooth divisors of $|L|$ which contain the set of points blown up under $\phi_{\mathcal{L}}$ and smooth divisors of $|\mathcal{L}|$.

We work now on the first reduction (M, L) of $(\mathcal{M}, \mathcal{L})$ and we make the assumption that $n \geq 3$. The following theorem takes care of the range $n-2 < \tau(L) < n-1$.

Theorem 7.3.4. *Let $\mathcal{M}$ be an n-dimensional normal projective variety with $n \geq 3$ and with terminal factorial singularities. Let $\mathcal{L}$ be an ample line bundle on $\mathcal{M}$. Let (M, L) be the first reduction of $(\mathcal{M}, \mathcal{L})$ as in (7.3.3). Let $\tau(L)$ be the nefvalue* *of* (M, L) *and* $\phi_L : M \to X$ *be the nefvalue morphism of* (M, L). *Assume that* $n-2 < \tau(L) < n-1$. *Then either*

1. $n = 4$, $\tau(L) = 5/2$, $(M, L) \cong (\mathbb{P}^4, \mathcal{O}_{\mathbb{P}^4}(2))$*; or*
2. $n = 3$, $\tau(L) = 3/2$, $(M, L) \cong (Q, \mathcal{O}_Q(2))$, Q *a hyperquadric in* $\mathbb{P}^4$*; or*
3. $n = 3$, $\tau(L) = 4/3$, $(M, L) \cong (\mathbb{P}^3, \mathcal{O}_{\mathbb{P}^3}(3))$*; or*
4. $n = 3$, $\tau(L) = 3/2$, ϕ_L *has a smooth curve as image and* $(F, L_F) \cong (\mathbb{P}^2, \mathcal{O}_{\mathbb{P}^2}(2))$ *for a general fiber,* F, *of* ϕ_L.

Proof. We will use over and over the rationality theorem, (1.5.2), and Lemma (3.3.2) without explicitly referring to them. For simplicity, write $\tau = \tau(L)$, $\phi = \phi_L$. We have $\tau = u/v$ where u, v are coprime positive integers and

$$u \leq \max_{x \in X}\{\dim \phi^{-1}(x)\} + 1 \leq n + 1.$$

Note that τ is not an integer by the assumption $n-2 < \tau < n-1$ and hence $v \geq 2$. Therefore from $n - 2 < \tau = u/v \leq (n+1)/2$ we find $2n - 4 < n + 1$, or $n < 5$.

Let $n = 4$. Then $2 < \tau < 3$. If $\tau = u/v \neq 5/2$ we see that $v \geq 3$, so we find $3\tau \leq v\tau = u \leq n + 1 = 5$, i.e., $\tau \leq 5/3$. This contradicts $\tau > 2$. Therefore $\tau = 5/2$. Hence in particular ϕ contracts M to a point, so we have $2K_M + 5L \approx \mathcal{O}_M$. Thus there exists an ample line bundle A on M such that $K_M \approx -5A$, $L \approx 2A$ and $(M, L) \cong (\mathbb{P}^4, \mathcal{O}_{\mathbb{P}^4}(2))$ as in case (7.3.4.1).

Let $n = 3$. Then $1 < \tau < 2$. From $1 < \tau = u/v$ and $u \leq n + 1 = 4$ we see that $v = 2, 3$. If $v = 2$, $1 < u/2 < 2$ gives $u = 3$. If $v = 3$, $1 < u/3 < 2$ and $u \leq 4$ give $u = 4$.

Let $u = 4$, $v = 3$, $\tau = 4/3$. In this case ϕ contracts M to a point, so we have $3K_M + 4L \approx \mathcal{O}_M$. Hence $K_M \approx -4A$, $L \approx 3A$ for some ample line bundle A on M and $(M, L) \cong (\mathbb{P}^3, \mathcal{O}_{\mathbb{P}^3}(3))$ as in (7.3.4.3).

Let $u = 3$, $v = 2$, $\tau = 3/2$. First, assume that $\phi : M \to X$ has lower dimensional image. If $\dim X = 0$, then $2K_M + 3L \approx \mathcal{O}_M$. Hence there exists an ample line bundle A on M such that $K_M \approx -3A$, $L \approx 2A$ and $(M, L) \cong (Q, \mathcal{O}_Q(2))$, Q a hyperquadric in $\mathbb{P}^4$, as in (7.3.4.2). If $\dim X = 1$, let F be a general fiber of ϕ. Then $2K_F + 3L_F \approx \mathcal{O}_F$, so that $K_F \approx -3A$, $L_F \approx 2A$ for some ample line bundle A on F and $(F, L_F) \cong (\mathbb{P}^2, \mathcal{O}_{\mathbb{P}^2}(2))$. We are in case (7.3.4.4). Note that $\dim X \leq 1$. Otherwise we would have a 1-dimensional fiber, f, with $K_f \approx -3A$ for some ample line bundle A on f and this is not possible.

Thus we can assume that $2K_M + 3L$ is nef and big and let R be an extremal ray on M such that $(2K_M + 3L) \cdot R = 0$ (see (4.2.14)). Write

$$K_M + 3(K_M + 2L) = 2(2K_M + 3L)$$

and note that $H := K_M + 2L$ is ample. For any real number $0 < \epsilon \ll 1$ we have $(K_M+(3-\epsilon)H)\cdot R < 0$. Let $\rho : M \to Y$ be the contraction of R and let $\rho^{-1}(y)$ be a fiber of ρ, $y \in Y$. Then Lemma (4.2.16) applies to say that $\dim \rho^{-1}(y) > 3 - \epsilon$, i.e., $\dim \rho^{-1}(y) = 3$. Since ϕ factors through ρ, this contradicts the fact that $2K_M + 3L$ is nef and big. □

Remark 7.3.5. Let $(\mathcal{M}, \mathcal{L})$, (M, L), ϕ_L be as in (7.3.4.4). Note that if $\mathcal{L}$ is very ample, than $(F, L_F) \cong (\mathbb{P}^2, \mathcal{O}_{\mathbb{P}^2}(2))$ for any fiber F of ϕ_L. Indeed in this case, $K_{\mathcal{M}} + 2\mathcal{L}$ is spanned by its global sections (see Theorem (9.2.2)). Hence $H := K_M+2L$ is spanned. Moreover H is ample since $2H \approx (2K_M+3L)+L \approx \phi_L^*\mathcal{H}+L$ for some ample line bundle $\mathcal{H}$. The restriction of H to any general fiber ($\cong \mathbb{P}^2$) is $\mathcal{O}_{\mathbb{P}^2}(1)$. So $H^2 \cdot F = 1$ for any fiber F. Hence $(F, H_F) \cong (\mathbb{P}^2, \mathcal{O}_{\mathbb{P}^2}(1))$. This implies $(F, L_F) \cong (\mathbb{P}^2, \mathcal{O}_{\mathbb{P}^2}(2))$.

7.4 The polarization of the first reduction

The following facts are useful.

Proposition 7.4.1. *Let $\mathcal{L}$ be an ample line bundle on $\mathcal{M}$, a normal projective variety of dimension $n \geq 2$ with terminal factorial singularities. Assume that the first reduction, (M, L), as in (7.3.3) exists. Let $\pi : \mathcal{M} \to M$ be the first reduction map and assume that π is not an isomorphism. Hence in particular π coincides with the nefvalue morphism, $\phi_{\mathcal{L}}$, of $(\mathcal{M}, \mathcal{L})$. Let $y_1, \dots, y_\gamma$ be the points of M with $\phi_{\mathcal{L}}^{-1}(y_i)$ positive dimensional and let $\mathfrak{m}_i$ be the ideal sheaf of y_i in M, $i = 1, \dots, \gamma$. The following holds.*

1. *If there exist global sections of $\mathcal{L}$ spanning $\mathcal{L}$ on the positive dimensional fibers of $\phi_{\mathcal{L}}$, then the restriction map $\Gamma(L) \to \Gamma(L \otimes \mathfrak{m}_i/\mathfrak{m}_i^2))$ is onto for each $i = 1, \dots, \gamma$.*

2. *If $h^1(\mathcal{L}) = 0$, then the points y_i, $i = 1, \dots, \gamma$, are not in the base locus of $|L|$ and the image of $\{y_1, \dots, y_\gamma\}$ under the map given by $\Gamma(L)$ spans a $\mathbb{P}^{\gamma-1}$;*

3. *If $h^1(\mathcal{L}) = 0$ and $\mathcal{L}$ is spanned, then L is spanned and $|L|$ separates a point $y \in M$ and any of the points y_i, $i = 1, \dots, \gamma$, distinct from y.*

Proof. Write $\phi := \phi_{\mathcal{L}}$. To prove 1) we have to show that $\Gamma(L)$ separates infinitely near points (v, w) where v, w are nonzero elements of $\mathcal{T}_{M,y_i}$, $i = 1, \dots, \gamma$. Fix an index i, say $i = 1$. Let p_v, p_w be the points on $E_1 := \phi^{-1}(y_1)$ which correspond to v, w. By assumption, there exist sections s, t in $\Gamma(\mathcal{L})$ which span $\mathcal{L}$ on E_1, i.e., such that $s(p_v) = 0$, $s(p_w) \neq 0$, $t(p_v) \neq 0$, $t(p_w) = 0$. Let s', t' be the pushforwards under ϕ of s, t respectively. Thus we have $s'(y_1) = t'(y_1) = 0$ and, after choosing a trivialization of L, we see that $ds'(v) = dt'(w) = 0$ and $ds'(w) \neq 0$, $dt'(v) \neq 0$, where ds', $dt' \in \mathcal{T}^*_{M,y_1}$ are the exterior derivatives of s', t' at y_1.

To show 2), consider the exact sequence

$$0 \to L \otimes \mathfrak{m}_1 \otimes \cdots \otimes \mathfrak{m}_\gamma \to L \to \oplus_{i=1}^{\gamma} \mathbb{C}_{y_i} \to 0.$$

Let $E_i := \phi^{-1}(y_i)$, $i = 1, \ldots, \gamma$. By the projection formula we have

$$\phi_* \mathcal{L} \approx \phi_*(\phi^* L - \sum_i E_i) \approx L \otimes \mathfrak{m}_1 \otimes \cdots \otimes \mathfrak{m}_\gamma.$$

From Leray's spectral sequence we get $h^1(\phi_* \mathcal{L}) = h^1(\mathcal{L}) = 0$. Therefore we conclude that $h^1(L \otimes \mathfrak{m}_1 \otimes \cdots \otimes \mathfrak{m}_\gamma) = 0$. From the long cohomology sequence associated to the exact sequence above it thus follows that the map

$$\Gamma(L) \to \oplus_{i=1}^{\gamma} \mathbb{C}_{y_i} \to 0 \tag{7.4}$$

is onto. This implies that L is spanned at the points y_i and that the image of the points $y_1, \ldots, y_\gamma$ under the map given by $\Gamma(L)$ spans a linear $\mathbb{P}^{\gamma-1}$.

In particular if $\mathcal{L}$ is spanned, then L is spanned on $M - \{y_1, \ldots, y_\gamma\}$ and therefore L is spanned.

To finish showing 3), take a point $y \in M$ such that $y \notin \{y_1, \ldots, y_\gamma\}$. Since $\mathcal{L}$ is spanned and $\Gamma(\mathcal{L}) \cong \Gamma(\phi_* \mathcal{L}) \cong \Gamma(L \otimes \mathfrak{m}_1 \otimes \cdots \otimes \mathfrak{m}_\gamma)$ we can find a section $s \in \Gamma(L)$ such that $s(y) \neq 0$, $s(y_i) = 0$ for each $i = 1, \ldots, \gamma$. Moreover, given any point y_i and by using the surjection (7.4), we have a section $t \in \Gamma(L)$ with $t(y_i) \neq 0$. Therefore $t' := t - cs$ for some constant $c \neq 0$ has the property that $t'(y_i) \neq 0$ and $t'(y) = 0$. Then $\{s, t'\}$ separate $\{y, y_i\}$, $i = 1, \ldots, \gamma$. □

Corollary 7.4.2. *Let $(\mathcal{M}, \mathcal{L})$, (M, L) be as in (7.3.3) with $n \geq 2$ and $\mathcal{M}$ factorial. Assume that $h^1(\mathcal{L}) = 0$. If $\mathcal{L}$ is very ample, then L is very ample.*

Proof. We can clearly assume that $\mathcal{M}$, M are not isomorphic. Let $\phi_{\mathcal{L}} : \mathcal{M} \to M$ be the first reduction map. Since $\mathcal{L}$ is very ample, then $\Gamma(L)$ gives an embedding on the subset of M where $\phi_{\mathcal{L}}$ is an isomorphism. Therefore by Lemma (7.4.1) we conclude that L is very ample. □

It is natural to ask the following general question, suggested by the results above.

Question 7.4.3. *Let $(\mathcal{M}, \mathcal{L})$, (M, L) be as in (7.3.3) with $n \geq 2$. Assume that $\mathcal{L}$ is very ample and that $\mathcal{M}$ is smooth. Is L very ample?*

In explicit computations the following results are often useful.

Lemma 7.4.4. *Let S be a smooth connected projective surface of nonnegative Kodaira dimension. Let $\pi : S \to Y$ be the map of S onto its minimal model. Denote $\nu := K_Y^2 - K_S^2$. Let L be an ample line bundle on S and let $L_Y := (\pi_* L)^{**}$. Then $L \cdot K_S \geq \nu + \pi^* K_Y \cdot L$. Moreover if the equality holds, then π expresses S as Y with a finite set blown up and (Y, L_Y) is the first reduction of (S, L).*

Proof. We can factor $\pi : S \to Y$ into simple blowups

$$S = S_0 \xrightarrow{\pi_1} S_1 \to \cdots \xrightarrow{\pi_\nu} S_\nu = Y.$$

Clearly $K_S \approx \pi^* K_Y + \sum_{i=1}^{\nu} \lambda_i E_i$, where the E_i's are the irreducible, reduced components of the positive dimensional fibers of π and the λ_i's are positive integers. Therefore we have

$$K_S \cdot L = \pi^* K_Y \cdot L + \sum_{i=1}^{\nu} \lambda_i E_i \cdot L \geq \pi^* K_Y \cdot L + \nu.$$

If $K_S \cdot L = \pi^* K_Y \cdot L + \nu$, then $\lambda_i = E_i \cdot L = 1$ for all $i = 1, \cdots, \nu$. If we show that the E_i's are disjoint, then we conclude that (Y, L_Y), $\pi : S \to Y$, is the reduction of (S, L).

To show disjointness, let E_1, E_2 be two distinct exceptional curves of π, $E_1^2 = E_2^2 = -1$. Assume that $K_S + L$ is nef and big. Then, since $(K_S + L) \cdot (E_1 + E_2) = 0$ and $E_1 + E_2$ is not numerically trivial, the Hodge index theorem gives $(E_1 + E_2)^2 < 0$. Therefore we find $E_1 \cdot E_2 < 1$, implying $E_1 \cap E_2 = \emptyset$.

Thus it remains to prove that $K_S + L$ is nef and big. If $K_S + L$ is nonnef, the nefvalue of (S, L) is bigger than 1 (compare with §1.5). Therefore, by combining (7.2.2) and (7.2.4) (see also (7.2.5)), we see that S is either $\mathbb{P}^2$, a quadric in $\mathbb{P}^3$, or a scroll over a curve. All such cases are not possible since $\kappa(S) \geq 0$. Then $K_S + L$ is nef and hence $\Gamma(m(K_S + L))$ defines a morphism, say ψ, for $m \gg 0$. Clearly $\dim \psi(S) = 2$, so $K_S + L$ is also big. Indeed, if $\dim \psi(S) = 0$, (S, L) would be a Del Pezzo surface and, if $\dim \psi(S) = 1$, (S, L) would be a conic fibration, this contradicting again the assumption $\kappa(S) \geq 0$. □

Corollary 7.4.5. *Let S be a smooth connected projective surface of nonnegative Kodaira dimension. Let $\pi : S \to Y$ be the map of S onto its minimal model. Denote $\nu := K_Y^2 - K_S^2$. Let L be an ample line bundle on S and let $L_Y := (\pi_* L)^{**}$. Then we have $L \cdot K_S \geq -K_S^2$ with equality only if $K_Y \sim \mathcal{O}_Y$. In particular $p_g(S) \geq 2$ implies that $L \cdot K_S \geq -K_S^2 + 1$.*

Proof. By (7.4.4) we have $L \cdot K_S \geq -K_S^2 + K_Y^2 + \pi^* K_Y \cdot L$. Since $\kappa(S) \geq 0$, the canonical bundle K_Y is nef, so that $K_Y^2 \geq 0$, $\pi^* K_Y \cdot L \geq 0$. Then $L \cdot K_S \geq -K_S^2$.

Assume the equality holds. Hence in particular $\pi^* K_Y \cdot L = K_Y^2 = 0$. Since K_Y is nef we conclude by the Hodge index theorem that $K_Y \sim \mathcal{O}_Y$.

Let $p_g(S) = p_g(Y) \geq 2$. Then by the above, either we are done or $L \cdot K_S = -K_S^2$ and $K_Y \sim \mathcal{O}_Y$. This leads to the contradiction $h^0(K_Y) \leq 1$. □

7.5 The second reduction in the smooth case

In this section we start considering a smooth polarized variety $(\mathcal{M}, \mathcal{L})$.

7.5.1. Assumptions and notation. Throughout this section $\mathcal{M}$ will denote an n-dimensional smooth projective variety with $n \geq 3$. Let $\mathcal{L}$ be an ample line bundle on $\mathcal{M}$. Let (M, L), $\pi : \mathcal{M} \to M$ be the first reduction of $(\mathcal{M}, \mathcal{L})$ as in (7.3.3). Let $\tau(L)$ be the nefvalue of (M, L) and let $\phi_L : M \to X$ be the nefvalue morphism of (M, L).

By the results of the previous sections we can assume that $\tau(L) \leq n - 2$. Next we study the case when $\tau(L) = n - 2$.

Let us first state, in the case we need, one of the main results from Andreatta and Wiśniewski [28, §4]. This result allows us to describe every fiber, F, from Theorem (7.5.3.5a). Note that without using Proposition (7.5.2) we obtain the same conclusion for a general fiber F. We do not include the proof of the following result, since we have nothing to add to it and since we do not need the extra information it gives in any essential way.

Proposition 7.5.2. ([28]) *Let X be a projective normal variety with canonical singularities and let L be an ample line bundle on X. Assume that K_X is nonnef. Let t be a positive integer such that $K_X + tL$ is nef, and let ψ be the morphism associated to $|m(K_X + tL)|$ for $m \gg 0$. Assume that ψ is the birational contraction of an extremal ray R. Let F' be any irreducible component of a fiber, F, of ψ. Assume that* $\dim F' = t$ *and that X is smooth along F. Then F is irreducible and L is spanned on a neighborhood, U, of F.*

Theorem 7.5.3. *Let $\mathcal{M}$ be an n-dimensional smooth projective variety with $n \geq 3$. Let $\mathcal{L}$ be an ample line bundle on $\mathcal{M}$. Let (M, L) be the first reduction of $(\mathcal{M}, \mathcal{L})$. Let $\tau(L)$ be the nefvalue of (M, L) and let $\phi_L : M \to X$ be the nefvalue morphism of (M, L). Assume that $\tau(L) = n - 2$. Then one of the following holds:*

1. *$K_M \approx -(n-2)L$, i.e., (M, L) is a Mukai variety;*
2. *(M, L) is a Del Pezzo fibration over a smooth curve under ϕ_L;*
3. *(M, L) is a quadric fibration over a normal surface under ϕ_L;*

4. *(M, L) is a scroll over a normal threefold under ϕ_L;*

5. *$K_M + (n-2)L$ is nef and big.*

   Assume $n \geq 4$. Then $\phi_L : M \to X$ is an isomorphism outside of $\phi_L^{-1}(Z)$, where $Z \subset X$ is an algebraic subset of X, such that $\dim Z \leq 1$. We have the following.

   a) *The 1-dimensional component, Z_1, of Z is the disjoint union of smooth curves and it is contained in the regular set of X. Let C be a curve in Z_1 and $E := \phi_L^{-1}(C)$. Let F be any fiber of the restriction, ϕ_E, of ϕ_L to E. Then $(F, L_F) \cong (\mathbb{P}^{n-2}, \mathcal{O}_{\mathbb{P}^{n-2}}(1))$ and $\mathcal{N}_{E/M|F} \cong \mathcal{O}_{\mathbb{P}^{n-2}}(-1)$.*

   b *If x is a 0-dimensional component of Z then $\phi_L^{-1}(x)$ is an irreducible reduced divisor, E, and either*

      i) *$(E, L_E) \cong (\mathbb{P}^{n-1}, \mathcal{O}_{\mathbb{P}^{n-1}}(1))$ with $\mathcal{N}_{E/M} \cong \mathcal{O}_{\mathbb{P}^{n-1}}(-2)$, or*

      ii) *$(E, L_E) \cong (Q, \mathcal{O}_Q(1))$, Q (possibly singular) hyperquadric in $\mathbb{P}^n$, with $\mathcal{N}_{E/M} \cong \mathcal{O}_Q(-1)$.*

Proof. For simplicity, write $\tau = \tau(L)$, $\phi = \phi_L$. First, assume that ϕ has lower dimensional image and let F be a general fiber of ϕ. Then $K_F + (n-2)L_F \approx \mathcal{O}_F$. Hence in particular $\dim F \geq n-3$, i.e., $\dim X \leq 3$, by (3.3.2). If $\dim X = 0$, $K_M \approx -(n-2)L$ and (M, L) is a Mukai variety. If $\dim X = 1$, (M, L) is a Del Pezzo fibration over a smooth curve. If $\dim X = 2$, (M, L) is a quadric fibration over a normal surface. If $\dim X = 3$, (M, L) is a scroll over a normal threefold. So we find the cases (7.5.3.1) — (7.5.3.4).

Thus we can assume that $n \geq 4$ and $K_M + (n-2)L$ is nef and big, i.e., since $K_M+(n-2)L$ is not ample, ϕ is a birational morphism which contracts some curve. From (4.2.14) there exists an extremal ray R such that $(K_M + (n-2)L) \cdot R = 0$. Let $\rho : M \to Y$ be the contraction of R. Then ϕ factors through ρ, $\phi = \alpha \circ \rho$. Hence in particular ρ is birational. Let E be the locus of R and let Δ be a general fiber of the restriction $\rho_E : E \to \rho(E)$. By Lemma (6.3.11) we see that two cases are possible. Either $E = \Delta$ and ρ contracts E to a point or ρ contracts E to a curve, say B, and Δ is an $(n-2)$-dimensional fiber of the restriction $\rho_E : E \to B$.

First, consider the case when E contracts to a point. From the contraction theorem we know that $E \cdot \ell = -k$ for some positive integer k, where $R = \mathbb{R}_+[\ell]$.

Claim 7.5.4. The line bundle $tL + E$ is nef and big for $t \geq k$.

Proof. To prove nefness let γ be an irreducible curve on M. If γ is not contained in E, then $E \cdot \gamma \geq 0$ so that $(tL + E) \cdot \gamma > 0$. If $\gamma \subset E$, then $\gamma \sim \mu\ell$ in M where μ is a positive real number, and $(tL + E) \cdot \gamma = (tL + E) \cdot (\mu\ell) = \mu(t-k) \geq 0$.

To prove bigness, write, for $t > k$, $tL+E = (t-k)L+(kL+E)$. Since $kL+E$ is nef we have that $tL+E$ is ample. So we can assume $t = k$. Since $(kL+E) \cdot \ell = 0$

we have $kL + E \approx \rho^* H$ for some line bundle H on Y (see (4.3.1.1)). Let m be a large integer such that mkL is very ample. From $\rho_*(mkL + mE) \approx mH$ we see that mH is spanned outside of the point $\rho(E)$ and gives an embedding outside of that point. It thus follows that H is big and hence $kL + E$ is big. □

Consider the exact sequence

$$0 \to K_M + tL \to K_M + tL + E \to K_E + tL_E \to 0.$$

By the Kodaira and Kawamata-Viehweg vanishing theorems we get $h^i(K_E + tL_E) = 0$ for $i \geq 1$ and $t \geq k$, so that $\chi(K_E + tL_E) = h^0(K_E + tL_E)$ for $t \geq k \geq 1$. Note that $(K_M + (n-2)L)_E \approx \mathcal{O}_E$ since E is contracted to a point by ϕ. Moreover $E + kL \in \rho^*\mathrm{Pic}(Y)$ since $(E + kL) \cdot \ell = 0$ and therefore $\mathcal{O}_E(E) \approx -kL_E$. Thus

$$K_E + tL_E \approx (K_M + E + tL)_E \approx (t - n + 2 - k)L_E.$$

Then we have

$$h^0(K_E + tL_E) = \left\{ \begin{array}{lll} 0 & \text{if} & t < n - 2 + k \\ 1 & \text{if} & t = n - 2 + k \end{array} \right\}. \tag{7.5}$$

It follows that the polynomial $p(t) := \chi(K_E + tL_E)$ has $n-2$ roots $t = k, \ldots, n - 3 + k$. Now, by Serre duality and by the above, we see that

$$p(t) = (-1)^{n-1}\chi(-tL_E) = (-1)^{n-1}\chi(K_E + (k + n - 2 - t)L_E).$$

Hence $p(t) = (-1)^{n-1}p(k + n - 2 - t)$. Let c be the $(n-1)$-th zero of $p(t)$. Note that the map $t \longmapsto k + n - 2 - t$ sends zeroes of $p(t)$ in zeroes of $p(t)$. Precisely, the zeroes

$$t = k, k + 1, \ldots, k + n - 3, c \tag{7.6}$$

correspond to the zeroes

$$t = n - 2, n - 3, \ldots, 1, k + n - 2 - c. \tag{7.7}$$

This implies $k \leq 2$. Indeed, if e.g., $k = 3$ we would have $1, \ldots, n - 2$ as zeroes from equation (7.7) and $3, \ldots, n$ as zeroes from equation (7.6). Hence $p(t)$ would have n zeroes, a contradiction. Similarly we exclude $k \geq 4$. Thus either $k = 1$ or $k = 2$.

Let $k = 1$. The first $n - 2$ zeroes from equation (7.6) coincide with the first $n-2$ zeroes from equation (7.7) and hence $c = k + n - 2 - c$, i.e., $c = (n-1)/2$. By equation (7.5) we know that $p(n - 2 + k) = p(n - 1) = 1$. Therefore

$$p(t) = \frac{2(t-1)\cdots(t+2-n)(t-(n-1)/2)}{(n-1)!}.$$

Then the coefficient $L_E^{n-1}/(n-1)!$ of the leading term is $2/(n-1)!$, or $L_E^{n-1} = 2$. Since $(K_M + (n-2)L)_E \approx \mathcal{O}_E$ and $\mathcal{O}_E(E) \approx -L_E$ we have

$$K_E \approx -(n-2)L_E - L_E = -(n-1)L_E$$

and hence $h^0(L_E) = h^0(K_E + nL_E) = p(n) = n+1$. In particular (E, L_E) has Δ-genus $\Delta(E, L_E) = 0$ (see equation (3.1)). Therefore we conclude by Proposition (3.1.2) that $(E, L_E) \cong (Q, \mathcal{O}_Q(1))$, Q irreducible, reduced hyperquadric in $\mathbb{P}^n$, and $\mathcal{N}_{E/M} \cong \mathcal{O}_E(-1)$. We are in case (7.5.3.5bii).

Let $k = 2$. In this case $p(t)$ vanishes for $t = 2, \ldots, n-1$ from equation (7.6) and for $t = 1, \ldots, n-2$ from equation (7.7), so $p(t)$ has exactly the $n-1$ zeroes $t = 1, \ldots, n-1$. By equation (7.5) we know that $p(n-2+k) = p(n) = 1$. Thus

$$p(t) = (t-1)\cdots(t-(n-1))/(n-1)!.$$

Then $L_E^{n-1} = 1$. Since $(K_M + (n-2)L)_E \approx \mathcal{O}_E$ and $\mathcal{O}_E(E) \approx -2L_E$ we have $K_E \approx -(n-2)L_E - 2L_E = -nL_E$ and hence

$$h^0(L_E) = h^0(K_E + (n+1)L_E) = p(n+1) = n.$$

In particular (E, L_E) has Δ-genus $\Delta(E, L_E) = 0$. Therefore we conclude that $(E, L_E) \cong (\mathbb{P}^{n-1}, \mathcal{O}_{\mathbb{P}^{n-1}}(1))$ by (3.1.2) and $\mathcal{N}_{E/M} \cong \mathcal{O}_E(-2)$. We are in case (7.5.3.5bi).

We consider now the case when $\rho = \mathrm{cont}_R : M \to Y$ contracts E to a curve, B. Let F be a general fiber of the restriction $\rho_E : E \to B$. Recall the $\phi : M \to X$ factors through ρ as $\phi = \alpha \circ \rho$.

We claim that for a suitable very ample line bundle $\mathcal{A}$ on Y the linear system $|K_M + (n-2)L + \rho^*\mathcal{A}|$ defines the contraction ρ. To see this note that $K_M + (n-2)L \approx \phi^*\mathcal{H} = \rho^*\alpha^*\mathcal{H}$, where $\mathcal{H}$ is an ample divisor on X. Then the claim follows from the facts that $\alpha^*\mathcal{H} + \mathcal{A}$ is very ample if $\mathcal{A}$ is very ample enough on Y and that $K_M + (n-2)L + \rho^*\mathcal{A} = \rho^*(\alpha^*\mathcal{H} + \mathcal{A})$.

Let A be a general element of $|\mathcal{A}|$. By Bertini's theorem we can assume that $A' := \rho^{-1}(A)$ is a smooth divisor on M. By the above, and by noting that $K_{A'} \approx (K_M + \rho^*\mathcal{A})_{A'}$ we have that the restriction $\rho_{A'}$ of ρ to A' is given by the linear system associated to the divisor

$$(K_M + (n-2)L + \rho^*\mathcal{A})_{A'} \approx K_{A'} + (n-2)L_{A'} = K_{A'} + (\dim A' - 1)L_{A'},$$

i.e., $\rho_{A'}$ is the morphism associated to the adjoint line bundle of $(A', L_{A'})$. Therefore we can apply Theorem (7.3.2) to the pair $(A', L_{A'})$ to conclude that $\rho_{A'}$ has disjoint exceptional divisors $D_i \cong \mathbb{P}^{n-2}$ with $L_{A'|D_i} \cong \mathcal{O}_{\mathbb{P}^{n-2}}(1)$ for all indices i. It thus follows that the general fiber, F, of $\rho_E : E \to B$ is a linear $\mathbb{P}^{n-2}$ with $L_F \cong \mathcal{O}_{\mathbb{P}^{n-2}}(1)$. The fact that this is true for any fiber, F, of ρ_E follows from a more general result proved by Andreatta and Wiśniewski [28], §4, by using two results by Fujita [261,

(1.5)] and Ando [7, (2.3)]. Let us summarize here, in our case, the nice argument of Andreatta and Wiśniewski [28].

We may assume that the nefvalue morphism, $\phi_L : M \to X$, is the contraction of a single extremal ray, R, such that the locus, E, of R is an irreducible divisor which is contracted by ϕ_L onto a curve, B. That is, ϕ_L is an elementary Mori contraction supported by $K_M + (n-2)L$. Let b be a point of B and let Δ be any irreducible component, of dimension $n-2$, of the fiber $F := \phi_L^{-1}(b)$ over b. Let $\nu : \Delta' \to \Delta$ be the normalization of Δ and let L_Δ be the restriction of L to Δ. From Fujita [261, (1.5)] we know that $(\Delta', \nu^* L_\Delta) \cong (\mathbb{P}^{n-2}, \mathcal{O}_{\mathbb{P}^{n-2}}(1))$. From Proposition (7.5.2) we know that the line bundle L is spanned on a neighborhood, U, of F.

It thus follows that $(\Delta, L_\Delta) \cong (\mathbb{P}^{n-2}, \mathcal{O}_{\mathbb{P}^{n-2}}(1))$. To see this, look at the morphism, σ, given by $\Gamma(L_\Delta)$. Since L_Δ is ample and spanned, σ is a finite morphism and, by dimension reasons, $h^0(L_\Delta) \geq n-1$. On the other hand $h^0(L_\Delta) \leq h^0(\nu^* L_\Delta) = h^0(\mathcal{O}_{\mathbb{P}^{n-2}}(1)) = n-1$. Therefore we have a finite morphism $\sigma \circ \nu : \Delta' \to \mathbb{P}^{n-2}$ which is generically one-to-one. Since Δ' is normal we conclude by the Zariski main theorem that $\sigma \circ \nu$ is an isomorphism, and hence ν is an isomorphism.

By using again the fact that L is spanned in a neighborhood of the fiber F, we can choose an irreducible section, A, of L which does not contain any irreducible component Δ of F. Take a point $y \in \Delta \cap A$. We claim that y is a smooth point of A. Note that in the following argument we need the fact that M is smooth. Assume that y is a singular point of A. Then for any curve $C \subset \Delta$, $y \in C$, C not contained in A, we have $\operatorname{mult}_y(A \cdot C) > 1$. Since $(\Delta, L_\Delta) \cong (\mathbb{P}^{n-2}, \mathcal{O}_{\mathbb{P}^{n-2}}(1))$, there exists a rational curve ℓ in Δ, $y \in \ell$, such that $\operatorname{mult}_y(A \cdot \ell) = 1$. Therefore, by the above, ℓ is contained in A. Since any two points of Δ can be joined by such rational curves, ℓ, we would get the contradiction that $\Delta \subset A$. Thus $\Delta \cap A \subset \operatorname{reg}(A)$ for each irreducible component Δ of F. Hence $F \subset \operatorname{reg}(A)$.

Tensoring the standard exact sequence

$$0 \to \mathcal{O}_M(-A) \to \mathcal{O}_M \to \mathcal{O}_A \to 0$$

by $m(K_M + (n-2)L)$, we get the exact sequence

$$0 \to mK_M + (m(n-2)-1)L \to mK_M + m(n-2)L \to mK_A + m(n-3)L_A \to 0.$$

Writing

$$mK_M + (m(n-2)-1)L = K_M + ((n-3)L + (m-1)(K_M + (n-2)L)),$$

we see by the Kawamata-Viehweg vanishing theorem that $h^1(mK_M + (m(n-2) - 1)L) = 0$. Then we conclude that the restriction, φ, of ϕ_L to A is the morphism given by $\Gamma(m(K_A + (n-3)L_A))$, for $m \gg 0$. Hence in particular φ has connected fibers and normal image.

We claim that the fiber F is irreducible. Assume otherwise. Then, by using the fact that $F \subset \text{reg}(A)$ and by iterating the procedure to slicing with suitable $n-2$ divisors $A_1, \ldots, A_{n-2}$, $A_1 := A$, of $|L|$, we reduce to the case when the intersection $F \cap A_1 \cap \ldots \cap A_{n-2}$ coincides with a finite number of distinct points. This contradicts the connectedness argument above.

By using the same induction argument and Ando's result [7, (2.3)] it thus follows that the curve B is smooth and it is contained in the regular set of X (see [28, (4.1),(4.2)]). This shows the first statement of (7.5.3.5a).

To show that $\mathcal{N}_{E/M|F} \cong \mathcal{O}_{\mathbb{P}^{n-2}}(-1)$ note that $(K_M + (n-2)L)_F \approx \mathcal{O}_F$. Then $K_{M|F} \approx \mathcal{O}_{\mathbb{P}^{n-2}}(2-n)$ and hence the adjunction formula yields

$$\mathcal{O}_{\mathbb{P}^{n-2}}(-n+1) \cong K_F \cong K_{E|F} \cong K_{M|F} + \mathcal{N}_{E/M|F},$$

whence $\mathcal{N}_{E/M|F} \cong \mathcal{O}_{\mathbb{P}^{n-2}}(-1)$.

To conclude the proof of (7.5.3.5) we have to show that all the exceptional divisors E_i we found as in (7.5.3.5), a), bi) or bii) and corresponding to all the extremal rays R_i such that $(K_M + (n-2)L) \cdot R_i = 0$, are disjoint and that the nefvalue morphism ϕ coincides, up to isomorphism in a complex neighborhood of each E_i, with the contraction of the extremal rays R_i. Let R_1, R_2 be two distinct extremal rays on M such that $(K_M + (n-2)L) \cdot R_i = 0, i = 1, 2$. If $n \geq 5$ we have $n - 2 \geq (n+1)/2$ and hence Corollary (4.2.17) applies to say that the loci E_1, E_2 of R_1, R_2 are disjoint.

If $n = 4$ a different argument due to Fujita [261, (2.4)] shows that the loci, E_1, E_2 of R_1, R_2 are disjoint. The argument goes as follows. Assume that $S := E_1 \cap E_2 \neq \emptyset$. Then $\dim S \geq n - 2 = 2$. Let $\rho_i : M \to Y_i$ be the contraction maps of the R_i, $i = 1,\ 2$. Suppose that ρ_1 is of type (7.5.3.5a) and $\rho_2(E_2)$ is a point. Since $\rho_1(E_1)$ is a curve, there is a curve, $C \subset S$, such that $\rho_1(C)$ is a point. This contradicts $R_1 \neq R_2$. If both $\rho_1(E_1)$ and $\rho_2(E_2)$ are points we get a contradiction by a similar and easier argument. Hence we may assume that $B_i := \rho_i(E_i)$ is a curve for each $i = 1,\ 2$.

Let C be a general fiber of $S \to B_2$. If $\rho_1(C)$ is a point we get a contradiction as above. So $\rho_1(C) = B_1$. Since C is proportional to R_2 and $(K_M + 2L) \cdot R_2 = 0$, we have $0 = \phi^*\mathcal{H} \cdot C = \deg(C \to B_1) \deg \mathcal{H}_{B_1}$, where $\mathcal{H}$ is an ample line bundle on X, so B_1 is mapped to a point x by the map $\alpha_1 : Y_1 \to X$, where $\phi = \alpha_1 \circ \rho_1$. Clearly $x = \phi(S)$ and similarly B_2 is mapped to x.

Take a hyperplane H on X such that $x \in H$. Then $\phi^*H \approx aE_1 + bE_2 + N$ for some positive integers $a,\ b$, where N is the sum of the components other than the E_i's. Let F_i be a general fiber of $E_i \to B_i$, $i = 1,\ 2$. Then the restriction of $\mathcal{O}_M(E_j)$, $j \neq i$, to $F_i \cong \mathbb{P}^2$ is $\mathcal{O}_{\mathbb{P}^2}(\delta_i)$ for some $\delta_i > 0$ since $\rho_i(S) = B_i$. Set $\mathcal{O}_{F_i}(N) := \mathcal{O}_{\mathbb{P}^2}(\nu_i)$. Then $\nu_i \geq 0$ and $0 = -a + b\delta_1 + \nu_1 = a\delta_2 - b + \nu_2$. From this we infer that $\delta_1 = \delta_2 = 1$, $a = b$, and $\nu_1 = \nu_2 = 0$. On the other hand, ϕ^*H is nef and big, so $h^1(\mathcal{O}_M(-\phi^*H)) = 0$. This implies that $h^0(\mathcal{O}_{\phi^*H}) = 1$, and hence that ϕ^*H is connected. Therefore N meets E_i for some i. The intersection $N \cap E_i$

consists of several fibers since $\nu_i = 0$. But then the components of $N \cap S$ are in R_i, hence $\rho_j(N \cap S) = B_j$, so $\nu_j > 0$, for the other j. This contradicts $\nu_1 = \nu_2 = 0$.

Thus we can consider the contraction $\sigma : M \to V$ which is a biholomorphism in the complement of all the E_i and which agrees in some complex neighborhood of E_i with the contraction associated to R_i for each i. Then V is a normal, compact, analytic variety and ϕ factors through σ, $\phi = \beta \circ \sigma$. Note that β has connected fibers since ϕ, σ have connected fibers. Exactly the same argument as in the proof of (7.3.2.4) applies to say that β is an isomorphism. □

Remark 7.5.5. Assume that (M, L) is as in Theorem (7.5.3.1)—(7.5.3.4) with $K_M + (n-2)L$ nef but not big. If the nefvalue morphism of (M, L), $\phi : M \to X$, has a lower dimensional image, with $\dim M \geq 7$, then it follows from [114, Theorem A.3] that the first reduction morphism is an isomorphism.

We study now in detail the singularities which occur on the variety X in Theorem (7.5.3).

Proposition 7.5.6. *Let $\mathcal{M}$ denote an n-dimensional smooth projective variety with $n \geq 4$. Let $\mathcal{L}$ be an ample line bundle on $\mathcal{M}$. Let (M, L), $\pi : \mathcal{M} \to M$ be the first reduction of $(\mathcal{M}, \mathcal{L})$ as in (7.3.3). Assume that $\tau(L)$, the nefvalue of (M, L), equals $n-2$, and let $\phi_L : M \to X$ be the nefvalue morphism of (M, L). Then X has isolated terminal singularities. Furthermore X is 2-factorial, and in even dimensions is Gorenstein.*

Proof. Write $\phi = \phi_L$. From (7.5.3.5) we know that ϕ is an isomorphism outside of $\phi^{-1}(Z)$, where $Z \subset X$ is an algebraic subset such that $\dim Z \leq 1$. Furthermore we know that X is singular at those points x which are 0-dimensional components of Z (case (7.5.3.5b)). From the proof of (7.5.3) we also know that ϕ coincides with a simultaneous contraction of a finite number of extremal rays of divisorial type. By the contraction theorem, (4.3.1), this implies that X has terminal singularities.

Let $R = \mathbb{R}_+[C]$ be an extremal ray contracted by ϕ and let E be the locus of R. If E is either as in (7.5.3.5a) or (7.5.3.5bii), then $E \cdot C = -1$ while if E is as in (7.5.3.5bi), then $E \cdot C = -2$. Therefore the contraction theorem applies again to say that X is factorial along $\phi(E)$ in the first two cases and 2-factorial at $\phi(E)$ in the third case. For the reader's convenience let us review the proof of this fact. Let D be any prime Weil divisor on X and let D' be the proper transform of D under ϕ. Put $-\alpha = (E \cdot C)$, $\beta = (D' \cdot C)$. Then $(\alpha D' + \beta E) \cdot C = 0$, so that by (4.3.1.1), $\alpha D' + \beta E \approx \phi^* \mathcal{F}$, $\mathcal{F} \in \operatorname{Pic}(X)$. Therefore we have a section, s, of $\mathcal{F}$ over $X - \phi(E)$ which defines αD on $X - \phi(E)$. Since X is normal and $\phi(E)$ is of codimension at least two, we conclude by Levi's extension theorem that s extends to a global section of $\mathcal{F}$. This means that αD is a Cartier divisor and hence X is $\alpha = -(E \cdot C)$-factorial.

Thus, to conclude the proof, we can assume that $\phi : M \to X$ is the contraction of an extremal ray $R = \mathbb{R}_+[C]$ whose locus E is as in (7.5.3.5bi), i.e., $(E, L_E) \cong (\mathbb{P}^{n-1}, \mathcal{O}_{\mathbb{P}^{n-1}}(1))$, $\mathcal{N}_{E/M} \cong \mathcal{O}_{\mathbb{P}^{n-1}}(-2)$ and we have to show that X is Gorenstein in even dimension. Note that $2K_X$ is a Cartier divisor since X is 2-factorial. Since ϕ is an isomorphism on $M - E$ we have $2K_M \approx \phi^*(2K_M) + \lambda E$ for some integer λ. From $(K_M + (n-2)L) \cdot C = 0$ and $L \cdot C = 1$ we get $K_M \cdot C = -(n-2)$. Therefore, since $\phi^*(2K_X) \cdot C = 0$ and $E \cdot C = -2$, we find $\lambda = (2K_M \cdot C)/(E \cdot C) = n - 2$. Then $2K_M \approx \phi^*(2K_X) + (n-2)E$. By definition this means that $K_M \approx \phi^*(K_X) + ((n-2)/2)E$. Since $n = 2n'$ is even, $(n-2)E/2 = (n'-1)E$ is an integral divisor and $K_M \approx \phi^*(K_X) + (n'-1)E$. Since $(K_M - (n'-1)E) \cdot R = 0$ we also have $K_M - (n'-1)E \approx \phi^*\mathcal{G}$ for some line bundle $\mathcal{G}$ on X by (4.3.1.1). Let $x = \phi(E)$, $U = X - \{x\}$, and $j : U \to X$ be the open embedding. Note that K_X and $\mathcal{G}$ agree on U since $\phi^*(K_X) \approx \phi^*\mathcal{G}$ and ϕ is isomorphism outside of x. This means that $K_U \approx \mathcal{G}_U$ where K_U is the canonical divisor of U and $\mathcal{G}_U$ is the restriction of $\mathcal{G}$ to U. Therefore $j_*K_U \approx j_*\mathcal{G}_U$. Note also that $j_*K_U = K_X$ since $U = \operatorname{reg}(X)$ and $j_*\mathcal{G}_U \approx \mathcal{G}$ by Levi's extension theorem. Thus we conclude that $K_X \approx \mathcal{G}$ and hence in particular X is Gorenstein. □

We can now define the second reduction of $(\mathcal{M}, \mathcal{L})$.

7.5.7. Definition-Notation (Second reduction) Let $\mathcal{M}$ be an n-dimensional smooth connected variety with $n \geq 3$ and $\mathcal{L}$ an ample line bundle on $\mathcal{M}$. Let (M, L), $\pi : \mathcal{M} \to M$ be the first reduction of $(\mathcal{M}, \mathcal{L})$ as in (7.3.3). Assume that $K_M + (n-2)L$ is nef and big. By the Kawamata-Shokurov basepoint free theorem we can choose $m \gg 0$ such that the morphism, $\varphi : M \to X$, associated to $|m(K_M + (n-2)L)|$ has connected fibers and normal image. Note that if $K_M + (n-2)L$ is nef but not ample, i.e., the nefvalue, $\tau(L)$, of (M, L) is $\tau(L) = n - 2$, then φ coincides with the nefvalue morphism, ϕ_L, of (M, L). We have $K_M + (n-2)L \approx \varphi^*\mathcal{K}$ for some ample line bundle $\mathcal{K}$ on X. Let $\mathcal{D} := (\varphi_*L)^{**}$. The pair $(X, \mathcal{D})$ together with the morphism $\varphi : M \to X$ is called the *second reduction* of $(\mathcal{M}, \mathcal{L})$. If $n \geq 4$, and $\varphi = \phi_L$, the structure of ϕ_L and the singularities of X are described in (7.5.3.5) and (7.5.6) above.

We say that $(\mathcal{M}, \mathcal{L})$ is of *log-general type* if $K_M + (n-2)L$ is nef and big, i.e., if the second reduction, $(X, \mathcal{D})$, exists. Note that if $\kappa(\mathcal{M}) \geq 0$, then $(\mathcal{M}, \mathcal{L})$ is of *log-general type*. Indeed if $h^0(tK_M) > 0$ for some positive integer, t, then $t(K_M + (n-2)L)$ gives a birational embedding, given on a Zariski open set by sections of $\Gamma(L)$, and thus $\kappa(K_M + (n-2)L) = n$.

We will denote by $\tau(\mathcal{K})$ the nefvalue of $(X, \mathcal{K})$ and by $\phi_{\mathcal{K}}$ the nefvalue morphism of $(X, \mathcal{K})$, defined by $|m(K_X + \tau(\mathcal{K})\mathcal{K})|$ for $m \gg 0$.

Note that for $n \geq 4$ the second reduction map $\varphi : M \to X$ can be factored as $M \to M^\# \to X$ where $\varphi^\# : M \to M^\#$ consists of contractions of types (7.5.3.5a) and (7.5.3.5bii) and $M^\# \to X$ consists of contractions of the type (7.5.3.5bi).

Following [261, (2.6)], $M^{\#}$ will be called the *factorial stage* since $\varphi(E)$ is a factorial singularity of X if E is as in (7.5.3.5bii). In particular every divisor on $M^{\#}$ is Cartier. Hence $L^{\#} := (\varphi_*^{\#} L)^{**}$ is invertible.

We also recall the following property of the second reduction map φ. Factor φ as $\varphi = \alpha \circ \beta$ with $\alpha : M \to V, \beta : V \to X, V$ normal variety, where α agrees with φ away from fibers F of φ of the form $F \cong \mathbb{P}^{n-1}, \mathcal{N}_{F/M} \cong \mathcal{O}_{\mathbb{P}^{n-1}}(-2)$ as in (7.5.3.5bi) and α is an isomorphism in a neighborhood of such fibers F. The morphism β is a partial resolution of those points $x \in X$ such that $\varphi^{-1}(x)$ is a fiber F as above. Then one has $\beta_* L_V \approx \mathcal{D}$, where $L_V := (\alpha_* L)^{**}$. For a proof we refer to [112], (2.6).

Lemma 7.5.8. *Notation and assumptions as in* (7.5.7). *Let* $\mathcal{D} := (\varphi_* L)^{**}$, *the double dual. Then we have*

1. $\mathcal{D}$ *is a* 2*-Cartier divisor and* $\mathcal{K} \approx K_X + (n-2)\mathcal{D}$*;*
2. $\mathcal{D}$ *is* $\phi_{\mathcal{K}}$*-ample;*
3. *the* $\phi_{\mathcal{K}}$*-nefvalue of* $2\mathcal{D}$ *is* $\tau(2\mathcal{D}) = \dfrac{\tau(\mathcal{K})(n-2)}{2(\tau(\mathcal{K})+1)}$.

Proof. We may assume that φ is not an isomorphism, i.e., φ coincides with the nefvalue morphism, $\phi_L : M \to X$, of (M, L). To show (7.5.8.1), note that $\mathcal{D}$ is a 2-Cartier divisor since X is 2-factorial. By the above we know that ϕ_L induces an isomorphism on an open subset U of X such that $\operatorname{cod}_X(X - U) \geq 2$. Furthermore the restriction $\mathcal{K}_U$ of $\mathcal{K}$ to U satisfies the condition

$$\mathcal{K}_U \cong (K_X + (n-2)\mathcal{D})_U \cong (K_M + (n-2)L)_{\phi_L^{-1}(U)}.$$

Since $K_X + (n-2)\mathcal{D}$ is a rank 1 reflexive sheaf it thus follows that $\mathcal{K} \approx K_X + (n-2)\mathcal{D}$.

The assertion in (7.5.8.2) follows from (7.5.8.1). It suffices to note that $-K_X$ is $\phi_{\mathcal{K}}$-ample and $\mathcal{K}$ is ample.

Finally, (7.5.8.3) follows from the equality

$$\frac{1}{\tau+1}(K_X + \tau\mathcal{K}) = K_X + \frac{\tau(n-2)}{2(\tau+1)}(2\mathcal{D})$$

where $\tau = \tau(\mathcal{K})$. □

Remark 7.5.9. Let $\mathcal{M}$ be an n-dimensional smooth connected variety with $n \geq 3$ and $\mathcal{L}$ an ample line bundle on $\mathcal{M}$. Let (M, L), $\pi : \mathcal{M} \to M$ be the first reduction of $(\mathcal{M}, \mathcal{L})$ as in (7.3.3). Let $\tau(L)$ be the nefvalue of (M, L) and assume $\tau(L) = n-2$, i.e., the nefvalue morphism $\varphi_L : M \to X$ is the second reduction map, φ. Let $\mathcal{D} := (\varphi_* L)^{**}$. Let D_1 be the union of all irreducible divisors, D on M, with

$\phi_L(D)$ not a divisor, and with $(D, L_D) \not\cong (\mathbb{P}^{n-1}, \mathcal{O}_{\mathbb{P}^{n-1}}(1))$. Let D_2 be the union of all irreducible divisors D on M with $\phi_L(D)$ not a divisor, and with $(D, L_D) \cong (\mathbb{P}^{n-1}, \mathcal{O}_{\mathbb{P}^{n-1}}(1))$. It is straightforward to show that $2L \approx \phi_L^*(2\mathcal{D}) - 2D_1 - D_2$ (see also our paper with Fania [86, (0.2.4)]).

Remark 7.5.10. Recently Andreatta [10] extended the structure Theorems (7.3.2.4) and (7.5.3) for the first and the second reduction by relaxing the assumptions on the singularities. Precisely, in [10], the statement (7.3.2.4) is proved under the assumption that $\mathcal{M}$ has terminal Gorenstein singularities and Theorem (7.5.3) is proved under the assumption that $\mathcal{M}$—and hence M—has terminal factorial singularities.

The arguments in [10] follow along the same lines as in [8]. Essentially, the above extensions are proved by using the methods of Andreatta and Wiśniewski [28] and the main result of Lipman and Sommese [408], as well as results from Cutkowsky [175].

7.6 Properties of the first and the second reduction

In this section we prove some general facts about the behavior of global sections of adjoint line bundles and of the unnormalized spectral value under reductions: these results are true in the case of terminal factorial singularities with no change. We also prove a useful result from our appendix to [114] which shows how the first and the second reduction maps are affected by the presence of coverings families of lines.

Proposition 7.6.1. *Let $\mathcal{M}$ be an n-dimensional smooth connected variety with $n \geq 3$ and let $\mathcal{L}$ be an ample line bundle on $\mathcal{M}$. Let (M, L), $\pi : \mathcal{M} \to M$, be the first reduction of $(\mathcal{M}, \mathcal{L})$. Let $(X, \mathcal{D})$, $\varphi : M \to X$, $\mathcal{D} = (\varphi_* L)^{**}$, be the second reduction of $(\mathcal{M}, \mathcal{L})$. Given integers a, b it follows that:*

1. *π induces a canonical isomorphism of $\Gamma(aK_{\mathcal{M}} + b\mathcal{L})$ and $\Gamma(aK_M + bL)$ for $b \leq a(n-1)$;*

2. *$\varphi \circ \pi$ induces a canonical injection of $\Gamma(aK_{\mathcal{M}} + b\mathcal{L})$ into $\Gamma(aK_X + b\mathcal{D})$. If $b \leq a(n-2)$ and $a(n-2) - b \equiv 0 \mod 2$, then $\varphi \circ \pi$ induces a canonical isomorphism of $\Gamma(aK_{\mathcal{M}} + b\mathcal{L})$ and $\Gamma(aK_X + b\mathcal{D})$.*

Proof. The statement 1) is proved by the second author in [584, (0.3.1)]. Simply note that

$$\pi_* : H^0(aK_{\mathcal{M}} + b\mathcal{L}) \to H^0(aK_M + bL)$$

is an injection by Hartogs's theorem. Note that

$$\pi^*(aK_M + bL) \cong aK_{\mathcal{M}} + b\mathcal{L} - (a(n-1) - b)\pi^{-1}(B)$$

where $B \subset M$ is the set such that π expresses $\mathcal{M}$ as M blown up at B. Thus if $b \leq a(n-1)$ we have a natural injective map $\pi^* : H^0(aK_M + bL) \to H^0(aK_{\mathcal{M}} + b\mathcal{L})$. Since $\pi_* \circ \pi^*$ is an isomorphism and since $H^0(aK_M + bL)$ and $H^0(aK_{\mathcal{M}} + b\mathcal{L})$ are finite dimensional we have that π_* is an isomorphism.

To show 2), in view of the last paragraph it suffices to prove that φ induces a canonical injection of $\Gamma(aK_M + bL)$ into $\Gamma(aK_X + b\mathcal{D})$, and if $b \leq a(n-2)$ and $a(n-2) - b \equiv 0 \bmod 2$, then φ induces a canonical isomorphism of $\Gamma(aK_M + bL)$ and $\Gamma(aK_X + b\mathcal{D})$.

Note that φ gives a biholomorphism of $X - Z$ and $M - \varphi^{-1}(Z)$ with $\operatorname{cod}_X Z \geq 2$, and

$$\varphi^*(aK_X + b\mathcal{D})_{X-Z} \cong (aK_M + bL)_{M-\varphi^{-1}(Z)}.$$

Since $aK_X + b\mathcal{D}$ is reflexive, the injectivity is an immediate consequence of Hartogs's theorem. Let

$$\Delta' := \{y \in M \mid \varphi \text{ is not an isomorphism at } y\}.$$

Then $L \approx \varphi^*\mathcal{D} - \Delta$, where Δ is a $\mathbb{Q}$-effective (indeed 2-effective) divisor such that $\lceil \Delta \rceil = \Delta'$. Thus, since $K_M + (n-2)L \approx \varphi^*(K_X + (n-2)\mathcal{D})$, one has $K_M \approx \varphi^* K_X + (n-2)\Delta$. Therefore

$$aK_M + bL \approx \varphi^*(aK_X + b\mathcal{D}) + (a(n-2) - b)\Delta.$$

Note that $(a(n-2) - b)\Delta$ is an integral divisor because $a(n-2) - b \equiv 0 \bmod 2$. Therefore $\Gamma(aK_X + b\mathcal{D})$ injects in $\Gamma(aK_M + bL)$ under φ^* as soon as $a(n-2) - b \geq 0$. The rest of the argument is exactly as in the first paragraph of the proof. □

Corollary 7.6.2. *Let $\mathcal{M}$ be an n-dimensional smooth connected variety with $n \geq 3$ and let $\mathcal{L}$ be an ample line bundle on $\mathcal{M}$. Let (M, L), $\pi : \mathcal{M} \to M$, be the first reduction of $(\mathcal{M}, \mathcal{L})$. Let $(X, \mathcal{D})$, $\varphi : M \to X$, $\mathcal{D} = (\varphi_* L)^{**}$, be the second reduction of $(\mathcal{M}, \mathcal{L})$. Let $\mathcal{H}$ be the ample line bundle on X defined in (7.5.7). Then $(n-2)(K_X + (n-3)\mathcal{D}) \cong K_X + (n-3)\mathcal{H}$ and*

1. *$\kappa(K_{\mathcal{M}} + (n-2)\mathcal{L}) = \kappa(K_M + (n-2)L)$;*

2. *$\kappa(K_{\mathcal{M}} + (n-3)\mathcal{L}) = \kappa(K_X + (n-3)\mathcal{D}) = \kappa(K_X + (n-3)\mathcal{H})$.*

Proof. The relation

$$(n-2)(K_X + (n-3)\mathcal{D}) \cong K_X + (n-3)\mathcal{H}$$

follows from (7.5.8.1). From (7.6.1.1) we have, for any positive integer m,

$$H^0(m(K_{\mathcal{M}} + (n-2)\mathcal{L})) \cong H^0(m(K_M + (n-2)L)).$$

This gives 1). From (7.6.1.2) we have, for any positive even integer m,

$$H^0(m(K_{\mathcal{M}} + (n-3)\mathcal{L})) \cong H^0(m(K_X + (n-3)\mathcal{D})).$$

This gives the first equality in 2). The second equality follows from $(n-2)(K_X + (n-3)\mathcal{D}) \approx K_X + (n-3)\mathcal{K}$. □

Remark 7.6.3. The fact that $(n-2)(K_X + (n-3)\mathcal{D}) \cong K_X + (n-3)\mathcal{K}$ is quite important. It implies that $K_X + (n-3)\mathcal{D}$ is nef if and only if $K_X + (n-3)\mathcal{K}$ is nef. Thus even though $\mathcal{D}$ might not be nef and big, $\mathcal{K}$ is and therefore by the basepoint free theorem some multiple of $K_X + (n-3)\mathcal{K}$ and hence of $K_X + (n-3)\mathcal{D}$ is spanned by global sections.

Also it is sometimes more convenient to describe the pair $(X, \mathcal{K})$ rather than the pair $(X, \mathcal{D})$. Using $(n-2)(K_X + (n-3)\mathcal{D}) \cong K_X + (n-3)\mathcal{K}$ it is an easy exercise, which we will often leave to the reader, to switch between the equivalent formulations.

We have the following result. Precursors are papers of the second author [584, (0.3.1)], Fania and the authors of this book [86, (0.2.7)], Fujita [261, (1.11), (2.7)], and our paper [111, (0.4.7)].

Proposition 7.6.4. *Let $\mathcal{L}$ be an ample line bundle on $\mathcal{M}$, an irreducible normal projective variety with terminal singularities of dimension n. If the first reduction, (M, L), of $(\mathcal{M}, \mathcal{L})$ exists, then $u(\mathcal{M}, \mathcal{L}) = u(M, L)$. If the second reduction, $(X, \mathcal{D})$, $\mathcal{K} \approx K_X + (n-2)\mathcal{D}$, of $(\mathcal{M}, \mathcal{L})$ exists, and $\mathcal{M}$ is smooth, then*

1. $u(\mathcal{M}, \mathcal{L}) = u(M, L) = u(X, \mathcal{K})$; *and*
2. $u(X, \mathcal{D}) = \frac{u(X,\mathcal{K})(n-2)}{1+u(X,\mathcal{K})}$.

Proof. Since we do not use this result in the rest of the book we do not include the complete proof. If $\mathcal{L}$ is very ample the result is proved for smooth $\mathcal{M}$ in [111, (0.4.7)]. The same proof using (7.1.7) in place of [111, (0.4.6)] gives the above result. □

The following theorem from our paper [111] is not used in any essential way throughout the book. We only refer to the last statement of it in Remark (7.7.11) below. Thus we only give a sketch of the proof of the theorem.

Theorem 7.6.5. *Let $\mathcal{M}$ be an n-dimensional smooth connected variety with $n \geq 3$ and let $\mathcal{L}$ be an ample line bundle on $\mathcal{M}$. Let (M, L), $\pi : \mathcal{M} \to M$ and $(X, \mathcal{D})$, $\varphi : M \to X$ be the first and the second reduction of $(\mathcal{M}, \mathcal{L})$ respectively. Let u be the unnormalized spectral value of $(\mathcal{M}, \mathcal{L})$. Then we have:*

1. *If $u > \lceil (n+1)/2 \rceil$ (or $u > \lceil n/2 \rceil$ if $\mathcal{L}^n \neq 2^n - 1$) then $n \geq 4$ and $(\mathcal{M}, \mathcal{L}) \cong (M, L)$;*

2. *If* $u > \lceil (n+2)/3 \rceil$ *(or* $u > \lceil n/3 \rceil$ *if* $\mathcal{L}^n \neq (3^n-1)/2$*) then* $n \geq 4$ *and* X *is factorial;*

3. *If* $u > \lceil (n+1)/2 \rceil$ *then* $n \geq 4$ *and* $\mathcal{D}$ *is ample and the only positive dimensional fibers of* φ *are* $(n-1)$*-dimensional irreducible quadrics,* Q*, with* $\mathcal{N}_{Q/M} \cong \mathcal{O}_Q(-1)$*;*

4. *If* $u > 2\lceil (n-1)/3 \rceil$ *then* $n \geq 5$ *and* $(\mathcal{M}, \mathcal{L}) \cong (M, L) \cong (X, \mathcal{D})$*.*

Proof. First note that the assumptions in each case 1)—4) imply that $n \geq 4$. Indeed, for $n = 3$,

$$\lceil (n+1)/2 \rceil = \lceil (n+2)/3 \rceil = 2\lceil (n-1)/3 \rceil = 2.$$

Hence in particular $h^0(N(K_{\mathcal{M}} + 2\mathcal{L})) = 0$ for each $N > 0$. This is clearly impossible, since, by assumption, there exists the first reduction, M, of $\mathcal{M}$.

Furthermore in case 4) we have $n \geq 5$. Indeed, for $n = 4$, the assumption on u gives $u > 2$. Hence in particular $h^0(N(K_{\mathcal{M}} + 2\mathcal{L})) = 0$ for each $N > 0$. From (7.6.1.1) we know that, for any $N > 0$, $h^0(N(K_{\mathcal{M}}+2\mathcal{L})) = h^0(N(K_M+2L))$. Since, by assumption, there exists the second reduction, X, we have $h^0(N(K_M+2L)) > 0$ for some $N \gg 0$. This leads to a contradiction.

To prove 1), note that if $\mathcal{M}$ is not isomorphic to M, there exists $D \cong \mathbb{P}^{n-1}$ in $\mathcal{M}$ with $\mathcal{N}_{D/\mathcal{M}} \cong \mathcal{O}_{\mathbb{P}^{n-1}}(-1)$. Then [111, (2.2.1)] applies to $\mathcal{M}$, $\mathcal{L}$, D to say that $h^0(K_{\mathcal{M}} + q\mathcal{L}) > 0$ if $q \geq \lceil (n+1)/2 \rceil$. This contradicts the assumption that $u > \lceil (n+1)/2 \rceil$.

Note that in 4), since $n \geq 5$, the assumption $u > 2\lceil (n-1)/3 \rceil$ implies $u > \lceil (n+1)/2 \rceil$, so that the isomorphism $(\mathcal{M}, \mathcal{L}) \cong (M, L)$ follows from 1).

The proof of the remaining statements is similar. We refer to [111, (3.1)] for the complete proof, which makes use of the results in [111, §2]. □

For some further results in the direction of (7.6.5) we refer to Nakamura [472, 473, 474].

The following is [114, Theorem A.1]. We keep the same notation as in Chapter 6. In the theorem below we say that a rational curve ℓ on $\mathcal{M}$ (respectively on M or X) is a *line* if $\mathcal{L} \cdot \ell = 1$ (respectively $L \cdot \ell = 1$ or $\mathcal{D} \cdot \ell = 1$). Furthermore, let B denote the finite set of points blown up under the first reduction map π and let Z denote the algebraic subset of X where the second reduction map φ is not an isomorphism.

Theorem 7.6.6. *Let* $\mathcal{M}$ *be an* n*-dimensional smooth connected variety with* $n \geq 3$ *and let* $\mathcal{L}$ *be an ample line bundle on* $\mathcal{M}$*. Let* (M, L)*,* $\pi : \mathcal{M} \to M$*, be the first reduction of* $(\mathcal{M}, \mathcal{L})$*. Let* $(X, \mathcal{D})$*,* $\varphi : M \to X$ *be the second reduction of* $(\mathcal{M}, \mathcal{L})$ *with* $\mathcal{D} = (\varphi_* L)^{**}$*.*

1. *If $\mathcal{M}$ is not isomorphic to M, no lines on M can meet B.*

2. *Assume that $\mathcal{L}$ is spanned. If there is a connected covering family of lines $(\mathcal{F}, \mathcal{P}, p, q)$ for $(\mathcal{M}, \mathcal{L})$ of normal degree $\nu = \nu(\mathcal{F})$, then either $\mathcal{M} \cong M$ or $\nu = 0$ and the morphism π is a blowing up at a single point.*

3. *If M is not isomorphic to X, no lines on X can meet Z.*

4. *Assume that $\mathcal{L}$ is spanned. If there is a connected covering family of lines $(\mathcal{G}, \mathcal{P}, p, q)$ for (M, L) of normal degree $\nu(\mathcal{G})$, then either $M \cong X$, or $\nu(\mathcal{G}) = 1$ and X is smooth, or $\nu(\mathcal{G}) = 0$. In the latter case there is at most one divisorial fiber of $\varphi : M \to X$.*

Proof. Assume that there is a line ℓ on M containing a point $b \in B$. Let γ be the proper transform of ℓ under $\pi : \mathcal{M} \to M$. Then one has

$$\mathcal{L} \cdot \gamma \leq \pi^* L \cdot \gamma - \pi^{-1}(b) \cdot \gamma = L \cdot \ell - \pi^{-1}(b) \cdot \gamma = 0,$$

which contradicts the ampleness of $\mathcal{L}$. This shows 1).

To prove 2), let $\mathcal{M}$ be filled out by the family $\mathcal{F}$ and assume that $\mathcal{M}$ is not isomorphic to M. Then there is a linear $\mathbb{P}^{n-1} \subset \mathcal{M}$ which contracts to a point in M. We can choose a general line ℓ from the family $\mathcal{F}$ such that ℓ is not contained in $\mathbb{P}^{n-1}$ (since otherwise π would have a lower dimensional image) and $\ell \cap \mathbb{P}^{n-1} \neq \emptyset$. Let $x \in \ell$ be a smooth point of ℓ such that $x \notin \mathbb{P}^{n-1}$. Note that since $\mathcal{L}$ is ample and spanned, ℓ is smooth so that since ℓ is general, $\mathcal{N}_{\ell/\mathcal{M}}$ is spanned and $\nu = \deg \mathcal{N}_{\ell/\mathcal{M}} \geq 0$. Let D_x be the locus of the lines from the family containing x. Recall that $\dim D_x \geq \nu + 1$ by Theorem (6.3.2). Then $\dim(D_x \cap \mathbb{P}^{n-1}) \geq \nu + 1 + n - 1 - n = \nu$. Recall that, as a consequence of the nonbreaking lemma, (6.3.1), we have that the restriction $\pi_{D_x} : D_x \to M$ is either finite-to-one or takes D_x to a point. Thus we see that $\dim(D_x \cap \mathbb{P}^{n-1}) = 0$ and therefore we conclude that $\nu = 0$.

Assume that $\mathcal{M} \ncong M$ and let $P_1 \cong \mathbb{P}^{n-1}$ be a linear $\mathbb{P}^{n-1}$ contracted under π. Since $\mathcal{F}$ covers $\mathcal{M}$ we can take a limit, $\mathfrak{l}$, of a sequence $\{\ell_i\}_i$ of lines from the family such that $\mathfrak{l} \cap P_1 \neq \emptyset$. Clearly, neither $\mathfrak{l}$, nor the lines ℓ_i can be contained in P_1 since otherwise all lines from the family would be contracted. Then we conclude that $\mathfrak{l} \cap P_1$ is a finite set and hence $\ell \cap P_1 \neq \emptyset$ for each line ℓ from the family.

Assume now that B contains at least two distinct points, b_1, b_2, and let $P_1 = \pi^{-1}(b_1)$, $P_2 = \pi^{-1}(b_2)$. Take a curve $C \subset \mathcal{P}$ and let $S := p^{-1}(C)$. Consider the ruled surface $q(S) \subset \mathcal{M}$. Then by the above, $q(S)$ meets P_1, P_2 in two divisors D_1, D_2. Let $g : S' \to S$ be the normalization of S and let E_i denote the pullback of D_i, $i = 1, 2$, under the composition map $q \circ g : S' \to q(S)$. Therefore S' is a $\mathbb{P}^1$-bundle (over the normalization of C). Under the composition map $\pi' = \pi \circ q \circ g : S' \to M$ the two curves E_1, E_2 are contracted to the points b_1, b_2, though $\pi'(S')$ is a surface.

This contradicts the fact that at most one section of a $\mathbb{P}^1$-bundle can be contracted to a point, as shown in Lemma (3.2.13).

To show 3), assume that there is a line, ℓ, on X containing a point $x \in Z$. Let γ be the proper transform of ℓ under $\varphi : M \to X$. Recall that $L \approx \varphi^*\mathcal{D} - \Delta$ where Δ is a 2-effective Cartier divisor with $\operatorname{supp}(\Delta) = \varphi^{-1}(Z)$. Then

$$L \cdot \gamma = \varphi^*\mathcal{D} \cdot \gamma - \Delta \cdot \gamma = 1 - \Delta \cdot \gamma < 1,$$

which contradicts the ampleness of L.

To prove 4), let M be filled out by a family of lines, $\mathcal{G}$, of normal degree $\nu(\mathcal{G})$ and assume that M is not isomorphic to X. Then $Z \neq \emptyset$.

Assume that there exists an isolated point $x \in Z$. Therefore $\varphi^{-1}(x)$ belongs to the support of Δ. The same argument as in the proof of 2) applies to say that $M \cong X$ if $\nu(\mathcal{G}) \geq 2$, there are no divisorial fibers if $\nu(\mathcal{G}) = 1$, and there exists at most one divisorial fiber of φ if $\nu(\mathcal{G}) = 0$. By Theorem (7.5.3.5) we conclude that X is smooth in the case above when $\nu(\mathcal{G}) = 1$.

Thus, by using again Theorem (7.5.3.5), we can assume that Z is a disjoint union of smooth curves, C, and X is smooth. Furthermore the restriction $\varphi_P : P := \varphi^{-1}(C) \to C$ is a $\mathbb{P}^{n-2}$-bundle. Fix a fiber $\mathbb{P}^{n-2}$ of φ_P. As usual, we can choose a line from the family such that $\ell \cap \mathbb{P}^{n-2} \neq \emptyset$ and ℓ is not contained in $\mathbb{P}^{n-2}$. Let x be a smooth point of ℓ, $x \notin \mathbb{P}^{n-2}$. Let D_x be the locus of lines from the family $\mathcal{G}$ containing x. As in the proof of 2) we get $\dim(D_x \cap \mathbb{P}^{n-2}) \geq \nu(\mathcal{G}) - 1$. If $\nu(\mathcal{G}) > 1$ we contradict again the nonbreaking lemma, (6.3.1). Thus we fall in one of the previous possible cases when either $\nu(\mathcal{G}) = 1$ and X is smooth or $\nu(\mathcal{G}) = 0$ and there is at most one divisorial fiber. □

The following corollary is an immediate consequence of Theorem (7.6.6) (compare with (7.6.5)).

Corollary 7.6.7. *Let $\mathcal{M}$ be a smooth connected n-fold with $n \geq 3$ and $\mathcal{L}$ an ample line bundle on $\mathcal{M}$. Assume that $(\mathcal{M}, \mathcal{L})$ admits a first reduction (M, L), $\pi : \mathcal{M} \to M$.*

1. *If (M, L) is a scroll, then $(\mathcal{M}, \mathcal{L}) \cong (M, L)$;*
2. *If (M, L) is a quadric fibration, $p : M \to Y$, over a normal variety Y and $n - \dim Y \geq 2$, then $(\mathcal{M}, \mathcal{L}) \cong (M, L)$.*

Example 7.6.8. (Effective construction of the second reduction map) Let $\mathcal{A}$, $\mathcal{H}$ be two spanned line bundles on an n-dimensional projective manifold X, $n \geq 3$. Assume that $\mathcal{H}$ is very ample and $\mathcal{A}^t \neq 0$ as an element of $H^{2t}(X, \mathbb{Q})$. Then there exist t elements $A_1, \dots, A_t$ of $|\mathcal{A}|$ meeting transversally along a smooth connected codimension t submanifold Z of X. Let $\varphi : M \to X$ be the blowing up of X along Z, denote $E := \varphi^{-1}(Z)$, $\mathcal{F} := \varphi^*\mathcal{A} - E$ and $L := \mathcal{F} + \varphi^*\mathcal{H}$. The line bundle L

is very ample. To see this, note that $\mathcal{F}$ is spanned by global sections and $\Gamma(\mathcal{F})$ gives a map $q : M \to \mathbb{P}^{t-1}$ with connected fibers and with $\mathcal{F} = q^*\mathcal{O}_{\mathbb{P}^{t-1}}(1)$ (see also (1.7.7)). Furthermore $K_M \approx \varphi^* K_X + (t-1)E$. Thus

$$K_M + (t-1)L \approx \varphi^*(K_X + (t-1)(\mathcal{A} + \mathcal{H})).$$

Now let $t = n - 1$. Then

$$K_M + (n-2)L \approx \varphi^*(K_X + (n-2)(\mathcal{A} + \mathcal{H})).$$

Note that $\mathcal{D} := (\varphi_* L)^{**} = \mathcal{A} + \mathcal{H}$ and $\mathcal{K}_X := K_X + (n-2)\mathcal{D}$ is ample on X if $\mathcal{H}$ is very ample enough on X. This shows that $(X, \mathcal{D})$, $\varphi : M \to X$, is the second reduction of (M, L) and φ is simply the blowing up along a smooth connected curve.

We will use this construction in Example (13.1.2).

Let us conclude this section, proving the following useful fact.

Proposition 7.6.9. *Let $\mathcal{M}$ be an n-dimensional, smooth, connected variety and $\mathcal{L}$ an ample line bundle on $\mathcal{M}$, $n \geq 3$. Then $\kappa(K_{\mathcal{M}} + (n-2)\mathcal{L}) \geq 0$ (respectively $= n$) if and only if the first reduction, (M, L), of $(\mathcal{M}, \mathcal{L})$ exists and $K_M + (n-2)L$ is nef (respectively nef and big).*

Proof. By combining Proposition (7.2.2) and Theorems (7.2.3), (7.2.4) and (7.3.2), we see that either the first reduction, (M, L), of $(\mathcal{M}, \mathcal{L})$ exists or $\kappa(K_{\mathcal{M}} + (n-2)\mathcal{L}) < 0$. Assume now $\kappa(K_{\mathcal{M}} + (n-2)\mathcal{L}) \geq 0$, so that the first reduction (M, L) exists and we are as in (7.3.2.4). In this case $K_M + (n-1)L$ is ample, and hence the nefvalue, $\tau(L)$, of (M, L) satisfies the condition that $\tau(L) < n - 1$.

To show that $K_M + (n-2)L$ is nef, assume otherwise. Then $n-2 < \tau(L) < n-1$ and therefore (M, L) is as in one of the cases listed in Theorem (7.3.4). In all these cases an easy check shows that there are no positive integers, t, such that $t(K_M + (n-2)L)$ is effective. From Corollary (7.6.2) we have $\kappa(K_{\mathcal{M}} + (n-2)\mathcal{L}) = \kappa(K_M + (n-2)L)$. Therefore, since $\kappa(K_{\mathcal{M}} + (n-2)\mathcal{L}) \geq 0$, there exists a positive integer m such that $m(K_M + (n-2)L)$ is effective. This gives a contradiction. Also, $\kappa(K_{\mathcal{M}} + (n-2)\mathcal{L}) = n$ implies that $K_M + (n-2)L$ is nef and big.

To show the converse, assume that $K_M + (n-2)L$ is nef (respectively nef and big). Then by (1.5.1) we know that, for $m \gg 0$, $m(K_M + (n-2)L)$ is spanned (respectively $\Gamma(m(K_M + (n-2)L))$ gives a birational morphism). Therefore, by using again Corollary (7.6.2.1), we have $\kappa(K_{\mathcal{M}} + (n-2)\mathcal{L}) \geq 0$ (respectively $= n$). □

Remark 7.6.10. Notation as in Proposition (7.6.9). Assume that $\mathcal{L}$ is spanned by its global sections and let $\widehat{S}$ be a smooth surface obtained as transversal intersection of $n-2$ general members of $|\mathcal{L}|$. Let S be the corresponding smooth surface on

M (see (7.3.3)). Assume that $\kappa(K_{\mathcal{M}}+(n-2)\mathcal{L}) \geq 0$. Then from (7.6.9) it follows that K_S is nef. Therefore the restriction, $\pi_{\widehat{S}} : \widehat{S} \to S$, of the first reduction map to $\widehat{S}$ maps $\widehat{S}$ onto its minimal model.

Note that if $K_M+(n-2)L$ is nef and $\kappa(K_M+(n-2)L) \geq 2$ then S is a surface of general type. To see this, let φ be the morphism associated to $\Gamma(m(K_M+(n-2)L))$ for $m \gg 0$. Let φ_S be the restriction of φ to S. From (5.1.3) we know that φ_S has a 2-dimensional image. Furthermore $K_S \approx \varphi_S^* H$ for some ample line bundle H, so that K_S is nef and big. Note also that if $K_M + (n-2)L$ is nef and big then $K_{\widehat{S}}^2 > 0$ easily follows from $K_{\widehat{S}}^2 = (K_M + (n-2)L)^2 \cdot L^{n-2} > 0$.

In particular if $\kappa(\mathcal{M}) \geq 0$, then $\kappa(K_{\mathcal{M}}+(n-2)\mathcal{L}) = n$ and $\widehat{S}$ is of general type.

7.7 The second reduction $(X, \mathfrak{D})$ with $K_X + (n-3)\mathfrak{D}$ nef

In this section we work on the second reduction $(X, \mathfrak{D})$. Let $\mathcal{K}$ be the ample line bundle on X defined in (7.5.7). Using the relation

$$(n-2)(K_X + (n-3)\mathfrak{D}) \cong K_X + (n-3)\mathcal{K}$$

from (7.6.3) we find it convenient to usually work with $K_X + (n-3)\mathcal{K}$. By using the rationality theorem systematically, we show that $K_X + (n-3)\mathcal{K}$ is nef if $n \geq 6$ except for one exceptional case with $n = 6$. If $n = 5$ we will show that $K_X + (n-3)\mathcal{K} = K_X + 2\mathcal{K}$ is nef except for three exceptional cases. Finally we include the $n = 4$ case results due to Fujita [261]. Let us make the following assumptions.

7.7.1. Assumptions and notation. Let $\mathcal{M}$ be a n-dimensional smooth connected projective variety with $n \geq 4$ and $\mathcal{L}$ an ample line bundle on $\mathcal{M}$. Assume that $K_{\mathcal{M}} + (n-1)\mathcal{L}$ is nef and big, and let (M, L), $\pi : \mathcal{M} \to M$ be the first reduction of $(\mathcal{M}, \mathcal{L})$ as in (7.3.3). Assume that $K_M+(n-2)L$ is nef and big, and let $(X, \mathfrak{D})$, $\mathfrak{D} = (\varphi_* L)^{**}$, $\varphi : M \to X$ denote the second reduction of $(\mathcal{M}, \mathcal{L})$ as in (7.5.7). Recall that $K_M + (n-2)L \approx \varphi^*\mathcal{K}$, where $\mathcal{K} \approx K_X + (n-2)\mathfrak{D}$ by (7.5.8). We will denote by $\tau(\mathcal{K})$ the nefvalue of $(X, \mathcal{K})$ and by $\phi_{\mathcal{K}}$ the nefvalue morphism of $(X, \mathcal{K})$, defined by $|m(K_X + \tau(\mathcal{K})\mathcal{K})|$, for $m \gg 0$.

We first consider the case when n is even.

Theorem 7.7.2. *Let $(X, \mathfrak{D})$, $\mathcal{K}$, $\tau(\mathcal{K})$, and $\phi_{\mathcal{K}}$ be as in (7.7.1) with n even, $n \geq 6$. Then $\tau(\mathcal{K}) \leq n-3$ unless $n = 6$, $\tau(\mathcal{K}) = 7$, and $(X, \mathcal{K}) \cong (\mathbb{P}^6, \mathcal{O}_{\mathbb{P}^6}(1))$.*

Proof. For simplicity, write $\tau = \tau(\mathcal{K})$, $\phi = \phi_{\mathcal{K}}$. By using Proposition (7.2.2) we know that $\tau \leq n+1$.

If $\tau = n+1$, then $(X, \mathcal{H}) \cong (\mathbb{P}^n, \mathcal{O}_{\mathbb{P}^n}(1))$. Therefore $\mathcal{D} \approx \mathcal{O}_{\mathbb{P}^n}(a)$ for some integer a and from $\mathcal{H} \approx K_X + (n-2)\mathcal{D} \approx \mathcal{O}_{\mathbb{P}^n}(1)$ we get $1 = (n-2)a - n - 1$, i.e., $(n-2)a = n+2$. Since $n \geq 6$, this implies $n = 6$ so we find the exceptional case.

If $\tau = n$, then either $(X, \mathcal{H}) \cong (Q, \mathcal{O}_Q(1))$ with Q a hyperquadric in $\mathbb{P}^{n+1}$, or $(X, \mathcal{H})$ is a $\mathbb{P}^{n-1}$-bundle over a smooth curve under ϕ, or $K_X + n\mathcal{H}$ is nef and big.

In the first case $\mathcal{D} \approx \mathcal{O}_Q(a)$ for some integer a, hence from $\mathcal{H} \approx K_X+(n-2)\mathcal{D} \approx \mathcal{O}_Q(1)$ we get $1 = (n-2)a - n$. Then $(n-2)a = n+1$ and this is not possible for $n \geq 6$.

In the second case, let F be a general fiber of ϕ. Let $\mathcal{H}_F$ and $\mathcal{D}_F$ be the restrictions of $\mathcal{H}$ and $\mathcal{D}$ to F. Then $\mathcal{D}_F \approx \mathcal{O}_{\mathbb{P}^{n-1}}(a)$ for some integer a and from $\mathcal{H}_F \approx K_F + (n-2)\mathcal{D}_F \approx \mathcal{O}_{\mathbb{P}^{n-1}}(1)$ we get $1 = (n-2)a - n$, the same contradiction as above.

In the third case we know from Theorem (7.2.3) that $K_X + n\mathcal{H}$ is in fact ample since it is nef and big. This contradicts the assumption $\tau = n$.

Thus from now on we can assume $n - 3 < \tau < n$. By the rationality theorem, (1.5.2), we can write $\tau = u/v$ where u, v are coprime positive integers and $u \leq n+1$. Note that $v \leq 2$. Otherwise

$$n - 3 < u/v \leq u/3 \leq (n+1)/3$$

leads to the contradiction $n < 5$. If $v = 1$, then either $u = n-2$ or $u = n-1$. If $v = 2$, then $n - 3 < u/2 \leq (n+1)/2$ gives $n = 6$ and $u = 7$. The following case by case analysis proves the theorem. We will use over and over the rationality theorem, (7.5.8.3) and (3.3.2) without explicitly refer to them.

Case $v = 1$, $u = n-1$. We have $\tau = n-1$ and the ϕ-nefvalue of $2\mathcal{D}$ is

$$\tau(2\mathcal{D}) = \frac{\tau(n-2)}{2(\tau+1)} = \frac{(n-1)(n-2)}{2n}.$$

Then $(n-1)(n-2)/(2n) = a/b$ where a, b are coprime positive integers such that $a \leq n+1$ and either

$$a = (n-1)(n-2)/2 \text{ if } ((n-1)(n-2)/2, n) = 1, \text{ i.e., if } n \not\equiv 2(4),$$

or

$$a = (n-1)(n-2)/4 \text{ if } ((n-1)(n-2)/4, n/2) = 1, \text{ i.e., if } n \equiv 2(4).$$

Since $n \geq 6$, the first case contradicts the bound $a \leq n+1$. In the second case, $(n-1)(n-2) \leq 4(n+1)$ leads to $n \leq 7$ and hence $n = 6, \tau = 5$. We claim that this does not occur.

Subcase $n = 6$, $\tau = 5$. Assume first that the nefvalue morphism $\phi : X \to W$ of $(X, \mathcal{H})$ has lower dimensional image. If $\dim W = 0$, then $K_X \approx -5\mathcal{H}$ and $(X, \mathcal{H})$ is a Gorenstein Del Pezzo variety. Fujita [261, (3.15.2)] used the following argument to show that this case does not occur. From [258] we know that $\mathrm{Bs}|\mathcal{H}| = \emptyset$ unless $\mathcal{H}^6 = 1$. Moreover when $\mathcal{H}^6 = 1$, $\mathrm{Bs}|\mathcal{H}|$ is a single point at which any general member $A \in |\mathcal{H}|$ is smooth. Hence, in any case, A is smooth by Bertini's theorem and $\mathcal{D}$ is invertible on A. Note that $4\mathcal{D} = \mathcal{H} - K_X = 6\mathcal{H}$ is a Cartier divisor. Set $B := \mathcal{D}_A - \mathcal{H}_A \in \mathrm{Pic}(A)$ where $\mathcal{D}_A$, $\mathcal{H}_A$ denote the restrictions of $\mathcal{D}$, $\mathcal{H}$ to A. Notice that $-K_A \approx 4\mathcal{H}_A$ is ample and therefore $\mathrm{Pic}(A)$ is torsion free. Therefore we have $2B = \mathcal{H}_A$ and $\omega_A = (K_X+\mathcal{H})_A = -4\mathcal{H}_A = -8B$. This contradicts (3.3.2) since B is ample.

Now assume that $\dim W > 0$. Let F be a general fiber of ϕ. Note that F is smooth since $\dim W > 0$ and since X has isolated singularities. Let $\mathcal{H}_F$ denote the restriction of $\mathcal{H}$ to F. Note also that $K_F + 5\mathcal{H}_F \approx \mathcal{O}_F$ implies $\dim F \geq 4$ and hence $\dim W \leq 2$. If $\dim W = 1$, we know from (3.1.6) that F is a hyperquadric in $\mathbb{P}^6$ and $\mathcal{H}_F \approx \mathcal{O}_F(1)$. Hence in particular $\mathcal{D}_F \approx \mathcal{O}_F(a)$ for some integer a. From $\mathcal{O}_F(1) \approx \mathcal{H}_F \approx K_F + 4\mathcal{D}_F$ we find $1 = -5 + 4a$, which is a contradiction. If $\dim W = 2$, we know by (3.1.6) that $(F, \mathcal{H}_F) \cong (\mathbb{P}^4, \mathcal{O}_{\mathbb{P}^4}(1))$. Again $\mathcal{D} \approx \mathcal{O}_{\mathbb{P}^4}(a)$ for some positive integer a and we find the same numerical contradiction as above.

Thus we can assume that $K_X+5\mathcal{H}$ is nef and big, i.e., ϕ is a birational morphism which contracts some curve. Let R be an extremal ray contracted by ϕ (see (4.2.14)). Let $\rho : X \to Y$ be the contraction of R and let E be the locus of R. Then, since ϕ factors through ρ, $K_X + 5\mathcal{H} \approx \rho^* H$ for some line bundle H on Y. Recall that X is Gorenstein since $n = 6$ and 2-factorial. Therefore Proposition (4.2.18) applies to say that $(E, \mathcal{H}_E) \cong (\mathbb{P}^5, \mathcal{O}_{\mathbb{P}^5}(1))$ where $\mathcal{H}_E$ denotes the restriction of $\mathcal{H}$ to E. Let $\mathcal{D}_E$ be the restriction of $\mathcal{D}$ to E. Then $\mathcal{D}_E \approx \mathcal{O}_{\mathbb{P}^5}(a)$ for some integer a and $K_{X|E} \approx \mathcal{O}_{\mathbb{P}^5}(-5)$ since $(K_X + 5\mathcal{H})_E \approx \mathcal{O}_{\mathbb{P}^5}$. Hence from $\mathcal{O}_{\mathbb{P}^5}(1) \approx \mathcal{H}_E \approx K_{X|E}+4\mathcal{D}_E$ we find again the numerical contradiction $1 = -5+4a$.

Case $v = 1$, $u = n - 2$. We have $\tau = n - 2$ and the ϕ-nefvalue of $2\mathcal{D}$ is

$$\tau(2\mathcal{D}) = \frac{(n-2)^2}{2(n-1)}.$$

Then $(n-2)^2/(2(n-1)) = a/b$, where a, b are coprime positive integers with $a \leq n+1$. Since $((n-2)^2/2, (n-1)) = 1$ we get $a = (n-2)^2/2$ and hence $(n-2)^2 \leq 2(n+1)$. This leads to the contradiction $n \leq 5$.

Case $n = 6$, $(u, v) = (7, 2)$. Recall that $u \leq \max_{w \in W}\{\dim \phi^{-1}(w)\} + 1$ since τ coincides with the ϕ-nefvalue of $(X, \mathcal{H})$ by (1.5.4). It follows that ϕ contracts X to a point, so that $2K_X + 7\mathcal{H} \approx \mathcal{O}_X$. Hence there exists an ample line bundle A on X such that $K_X \approx -7A$, $\mathcal{H} \approx 2A$ and this implies that $(X, \mathcal{H}) \cong (\mathbb{P}^6, \mathcal{O}_{\mathbb{P}^6}(2))$. Then $\mathcal{D} \approx \mathcal{O}_{\mathbb{P}^6}(a)$ for some integer a and therefore $\mathcal{H} \approx K_X + 4\mathcal{D} \approx \mathcal{O}_{\mathbb{P}^6}(2)$ yields the numerical contradiction $9 = 4a$. □

We consider now the case when $n \geq 7$ is odd. The case $n = 5$ will be treated separately.

Theorem 7.7.3. *Let $(X, \mathcal{D})$, $\mathcal{H}$, $\tau(\mathcal{H})$ and $\phi_{\mathcal{H}}$ be as in (7.7.1) with n odd, $n \geq 7$. Then $\tau(\mathcal{H}) \leq n-3$.*

Proof. For simplicity, write $\tau = \tau(\mathcal{H})$, $\phi = \phi_{\mathcal{H}}$. By using Proposition (7.2.2) we know that $\tau \leq n+1$.

Exactly the same argument as in the even case (compare with the proof of (7.7.2)) rules out the cases when either $\tau = n+1$ or $\tau = n$. So we can assume $n-3 < \tau < n$. By the rationality theorem, (1.5.2), we can write $2\tau = u/v$ where u, v are coprime positive integers and $u \leq 2(n+1)$. Note that $v = 1$. Otherwise

$$2(n-3) < u/v \leq u/2 \leq n+1$$

leads to the contradiction $n < 7$. Thus $2\tau = u$ is an integer and

$$2n-5 \leq 2\tau \leq 2n-1.$$

The following case by case analysis proves the theorem. We will use over and over the rationality theorem and (7.5.8.3) without explicitly referring to them.

Case $2\tau = 2n-1$. The ϕ-nefvalue of $2\mathcal{D}$ is

$$\tau(2\mathcal{D}) = \frac{\tau(n-2)}{2(\tau+1)} = \frac{(2n-1)(n-2)}{2(2n+1)}.$$

Then

$$2\tau(2\mathcal{D}) = (2n-1)(n-2)/(2n+1) = a/b,$$

where a, b are coprime integers with $a \leq 2(n+1)$ and either

$$a = (2n-1)(n-2) \text{ if } ((2n-1)(n-2), 2n+1) = 1, \text{ i.e., if } n \not\equiv 2(5)$$

or

$$a = (2n-1)(n-2)/5 \text{ if } ((2n-1)(n-2)/5, (2n+1)/5) = 1, \text{ i.e., if } n \equiv 2(5).$$

Since $n \geq 7$, the first case contradicts the bound $a \leq 2(n+1)$. In the second case,

$$(2n-1)(n-2) \leq 10(n+1)$$

yields $n \leq 8$ and hence $n = 7$, $\tau = 13/2$. We claim that this case does not occur.

Subcase $n = 7$, $\tau = 13/2$. Assume first that the nefvalue morphism $\phi : X \to W$ has lower dimensional image. If $\dim W \geq 1$, let F be a general fiber of ϕ. Note that F is smooth since X has isolated singularities. Let $\mathcal{H}_F$ denote the restriction of $\mathcal{H}$ to F. From $2K_F + 13\mathcal{H}_F \approx \mathcal{O}_F$ we infer that $K_F \approx -13A$ for some ample line bundle A on F. This implies $\dim F \geq 12$ (see (3.3.2)), so it is not possible for dimension reasons.

If $\dim W = 0$, we have $2K_X + 13\mathcal{K} \approx \mathcal{O}_X$. Assume that $\mathcal{K}$ is very ample. Then, since X has isolated singularities, there exists a smooth element $D \in |\mathcal{K}|$. Let $\mathcal{K}_D$ denote the restriction of $\mathcal{K}$ to D. By using the adjunction formula we find $2K_D + 11\mathcal{K}_D \approx \mathcal{O}_D$. Then, as in the case above, $K_D \approx -11A$ for some ample line bundle A on D. This implies $\dim D \geq 10$, so it is not possible for dimension reasons. To prove that $\mathcal{K}$ is very ample it suffices to show that $(X, \mathcal{K})$ has Δ-genus zero (see Fujita [257, I, (4.12)]). Let $p(t) := \chi(t\mathcal{K})$. Since $t\mathcal{K} - K_X \sim (t + (13/2))\mathcal{K}$ is ample for $t > -13/2$, we see by the Kodaira vanishing theorem that $p(t) = h^0(t\mathcal{K})$ for $t > -13/2$ and therefore $p(t) = 0$ for $t = -6, \dots, -1$. Also, $p(0) = h^0(\mathcal{O}_X) = 1$. Let $-c$ be the remaining zero of $p(t)$. By the above we see that

$$p(t) = \frac{(t+1)\cdots(t+6)(t+c)}{6!c}.$$

Then $\mathcal{K}^7/7! = 1/(6!c)$, that is $\mathcal{K}^7 = 7/c$. Furthermore $h^0(\mathcal{K}) = p(1) = 7 + 7/c$. Hence $\Delta(X, \mathcal{K}) = 0$ and we are done.

Thus we can assume that $K_X + (13/2)\mathcal{K}$ is nef and big, i.e., ϕ is a birational morphism which contracts some curve. Let R be an extremal ray contracted by ϕ (see (4.2.14)). Let $\rho : X \to Y$ be the contraction of R and let E be the locus of R. Note that ϕ factors through ρ, $\phi = \alpha \circ \rho$. We review here the argument as in the proof of Theorem (7.2.3).

Let $\mathcal{K}' := \mathcal{K} + \rho^*A$ for some ample line bundle A on Y. We claim that $K_X + (13/2)\mathcal{K}' \approx \rho^*H$ for some ample $\mathbb{Q}$-Cartier divisor H on Y. Take a large integer m such that $m(K_X + (13/2)\mathcal{K}) \approx \phi^*\mathcal{H}$ for some ample line bundle $\mathcal{H}$ on W. Then $K_X + (13/2)\mathcal{K} \approx \rho^*D$ for some nef and big $\mathbb{Q}$-Cartier divisor D on Y (in fact $mD = \alpha^*\mathcal{H}$). Thus

$$K_X + (13/2)\mathcal{K}' = K_X + (13/2)\mathcal{K} + (13/2)\rho^*A \approx \rho^*(D + (13/2)A),$$

so that the claim is proved by taking $H = D + (13/2)A$.

It follows that for some large integer m, $|m(K_X + (13/2)\mathcal{K}')|$ gives the morphism ρ and $K_X + (13/2)\mathcal{K}'$ is nef and big but not ample since $(K_X + (13/2)\mathcal{K}') \cdot R = 0$. Thus $13/2$ is the ρ-nefvalue of $(X, \mathcal{K}')$. Therefore by the rationality theorem we know that

$$13 = u/v \leq u \leq 2(\max_{y \in Y}\{\dim \rho^{-1}(y)\} + 1)$$

where u, v are coprime positive integers. Then we see that ρ has a 6-dimensional fiber and hence the locus E of R is a prime divisor (see the contraction theorem and compare also with the proof of (4.2.18)). We show that this leads to a contradiction. The argument runs parallel to that of (7.2.3). For any integer t consider the exact sequence

$$0 \to t\mathcal{K} - E \to t\mathcal{K} \to t\mathcal{K}_E \to 0.$$

Recalling that $K_X + (13/2)\mathcal{K}$ is trivial on E and $-E$ is ρ-ample we have that

$$t\mathcal{K} - K_X \ \text{ is } \rho\text{-ample for } \ t > -13/2$$

and

$$t\mathcal{K} - K_X - E \ \text{ is } \rho\text{-ample for } \ t \geq -13/2.$$

By using the relative Kodaira vanishing theorem, (2.2.1), we have by the above, for $t > -13/2$, $i > 0$, $\rho_{(i)}(t\mathcal{K}) = \rho_{(i)}(t\mathcal{K} - E) = 0$. Therefore $\rho_{(i)}(t\mathcal{K}_E) \cong H^i(t\mathcal{K}_E) = (0)$ for $i > 0$ and $t > -13/2$. Let $q(t) := \chi(t\mathcal{K}_E)$. Then $q(t) = h^0(t\mathcal{K}_E)$ for $t > -13/2$, so that $q(t) = 0$ for $t = -1, \ldots, -6$ and $q(0) = h^0(\mathcal{O}_E) = 1$. Hence we have $q(t) = (t+1)\cdots(t+6)/6!$. Then in particular $\mathcal{K}_E^6 = 1$ and $h^0(\mathcal{K}_E) = q(1) = 7$. Therefore $(E, \mathcal{K}_E)$ has Δ-genus $\Delta(E, \mathcal{K}_E) = 0$, so we conclude that $(E, \mathcal{K}_E) \cong (\mathbb{P}^{n-1}, \mathcal{O}_{\mathbb{P}^{n-1}}(1))$ by (3.1.2). Thus $K_{X|E} \approx \mathcal{O}_{\mathbb{P}^6}(a)$ for some integer a and from $(K_X + (13/2)\mathcal{K})_E \approx \mathcal{O}_E$ we get the contradiction $a + 13/2 = 0$.

In the following analysis a, b always denote coprime positive integers with $a \leq 2(n+1)$.

Case $2\tau = 2n-2$. The ϕ-nefvalue of $2\mathcal{D}$ is

$$\tau(2\mathcal{D}) = \frac{(n-1)(n-2)}{2n}.$$

Then $2\tau(2\mathcal{D}) = (n-1)(n-2)/n = a/b$. Since $((n-1)(n-2), n) = 1$, we have $(n-1)(n-2) \leq 2(n+1)$, which leads to the contradiction $n \leq 5$.

Case $2\tau = 2n-3$. The ϕ-nefvalue of $2\mathcal{D}$ is

$$\tau(2\mathcal{D}) = \frac{(2n-3)(n-2)}{2(2n-1)}.$$

Then $2\tau(2\mathcal{D}) = (2n-3)(n-2)/(2n-1) = a/b$ and either

$$a = (2n-3)(n-2) \ \text{ if } \ ((2n-3)(n-2), 2n-1) = 1, \ \text{ i.e., if } \ n \not\equiv 2(3),$$

or

$$a = (2n-3)(n-2)/3 \ \text{ if } \ ((2n-3)(n-2), 2n-1) = 3, \ \text{ i.e., if } \ n \equiv 2(3).$$

Since $n \geq 7$, both the cases above contradict the bound $a \leq 2(n+1)$.

Case $2\tau = 2n-4$. The ϕ-nefvalue of $2\mathcal{D}$ is

$$\tau(2\mathcal{D}) = \frac{(n-2)^2}{2(n-1)}.$$

Then $2\tau(2\mathcal{D}) = (n-2)^2/(n-1) = a/b$. Since $((n-2)^2, n-1) = 1$, we have $(n-2)^2 \leq 2n+2$ and hence $n \leq 5$, a contradiction.

Case $2\tau = 2n-5$. The ϕ-nefvalue of $2\mathcal{D}$ is

$$\tau(2\mathcal{D}) = \frac{(2n-5)(n-2)}{2(2n-3)}.$$

Then $2\tau(2\mathcal{D}) = (2n-5)(n-2)/(2n-3) = a/b$. Since $((2n-5)(n-2), 2n-3) = 1$ we have $(2n-5)(n-2) \leq 2(n+1)$ which implies $n \leq 4$, a contradiction. □

Remark 7.7.4. Let $(X, \mathcal{D})$, $\mathcal{H}$ be as in (7.7.1). As noted in (7.6.3), $K_X + (n-3)\mathcal{H}$ is nef if and only if $K_X + (n-3)\mathcal{D}$ is nef.

The following settles the case $n = 5$. It follows by combining [106, Proposition (5.1)] with [261, §3].

Theorem 7.7.5. *Let $(X, \mathcal{H}), \mathcal{D}, \tau(\mathcal{H})$ and $\phi_{\mathcal{H}}$ be as in (7.7.1) with $n = 5$. Then $\tau(\mathcal{H}) \leq n - 3 = 2$ unless either*

1. *$\tau = 5$, $(X, \mathcal{H}) \cong (Q, \mathcal{O}_Q(1))$, Q a hyperquadric in $\mathbb{P}^6$; or*
2. *$\tau = 5$, X is a $\mathbb{P}^4$-bundle over a smooth curve and the restriction, $\mathcal{H}_F$, of $\mathcal{H}$ to any fiber F is $\mathcal{O}_F(1)$; or*
3. *$\tau = 7/2$ and X is a singular 2-Gorenstein Fano 5-fold with $2K_X \approx -7\mathcal{H}$, given by the projective cone over $(\mathbb{P}^4, \mathcal{O}_{\mathbb{P}^4}(2))$. Moreover $\mathcal{M} \cong \mathbb{P}(\mathcal{O}_{\mathbb{P}^4}(3) \oplus \mathcal{O}_{\mathbb{P}^4}(1))$ and $\mathcal{L} \cong \mathcal{O}_{\mathcal{M}}(1)$, $\mathcal{L}^5 = 121$.*

Proof. For simplicity, write $\tau = \tau(\mathcal{H})$, $\phi = \phi_{\mathcal{H}} : X \to W$. By Proposition (7.2.2) we have $\tau \leq n + 1 = 6$. Exactly the same argument as in the proof of (7.7.2) rules out the case $\tau = n + 1 = 6$, while, if $\tau = n = 5$, we find that either $(X, \mathcal{H}) \cong (Q, \mathcal{O}_Q(1))$ with Q a hyperquadric in $\mathbb{P}^6$, or X is a $\mathbb{P}^4$-bundle over a smooth curve and $\mathcal{H}_F \approx \mathcal{O}_F(1)$ for any fiber F. Thus we find the classes (7.7.5.1), (7.7.5.2). Let us assume that $7/2 < \tau < 5$. By the rationality theorem we can write $2\tau = u/v$ where u, v are coprime positive integers and $u \leq 2(n+1) = 12$. Note that $v = 1$, since otherwise we have the contradiction $7 < 2\tau = u/v \leq u/2 \leq 6$. Thus $2\tau = u$ is an integer and hence in particular $2\tau \leq 2n - 1$. Then $7 < 2\tau = u \leq 2n - 1 = 9$, so there are only two possible cases: $u = 8$, $\tau = 4$ or $u = 9$, $\tau = 9/2$.

Recall that by (7.5.8.3) the ϕ-nefvalue of $2\mathcal{D}$ is $\tau(2\mathcal{D}) = \tau(n-2)/(2(\tau+1))$ and by the rationality theorem $2\tau(2\mathcal{D}) = a/b$ where a, b are coprime positive integers with

$$a \leq 2(\max_{w \in W}\{\phi^{-1}(w)\} + 1) \leq 2(n+1) = 12.$$

If $u = 9$, we have $2\tau(2\mathcal{D}) = 27/11$. This contradicts the bound $a \leq 12$.

Thus it remains to consider the case when $u = 8$, $\tau = 4$. In this case $2\tau(2\mathcal{D}) = 12/5$. Then we see that ϕ contracts X to a point and hence $K_X + 4\mathcal{H} \sim \mathcal{O}_X$ or $2K_X + 8\mathcal{H} \approx \mathcal{O}_X$. This is equivalent to $5(2K_X) + 12(2\mathcal{D}) \approx \mathcal{O}_X$. Therefore $2K_X \approx -12A$ for some ample line bundle A on X. Let $p(t) := \chi(tA)$. Since $K_X \sim -6A$, note that $tA - K_X \sim (t+6)A$ is ample for $t > -6$. Then, by the Kodaira vanishing theorem, $h^i(tA) = 0$ for $i > 0$ and $t > -6$, so that $p(t) = h^0(tA)$ for $t > -6$ and therefore $p(t) = 0$ for $t = -1, \dots, -5$ as well as $p(0) = h^0(\mathcal{O}_X) = 1$. It follows that $p(t) = (t+1)\cdots(t+5)/5!$. Hence in particular $A^5 = 1$ and $h^0(A) = p(1) = 6$. So we conclude that $(X, A) \cong (\mathbb{P}^5, \mathcal{O}_{\mathbb{P}^5}(1))$ by (3.1.4). Let

$\mathcal{D} \approx \mathcal{O}_{\mathbb{P}^5}(a)$ for some integer a. From $10K_X + 24\mathcal{D} \approx \mathcal{O}_X$ we find the numerical contradiction $2a = 5$. Thus we can assume $\tau \leq 7/2$. The result now follows from Fujita [261, §3] (see also Beltrametti, Fania, and Sommese [86, (4.2)] and [109, (1.1)] for the case when $\mathcal{L}$ is very ample). □

The following describes the cases when $n \geq 5$ and $K_X + (n-3)\mathcal{K}$ is nef and not big.

Proposition 7.7.6. *Let $(X, \mathcal{K})$ be as in (7.7.1) with $n \geq 5$. Assume that $K_X + (n-3)\mathcal{K}$ is nef. Then $K_X + (n-3)\mathcal{K}$ is big unless either:*

1. *$(X, \mathcal{K})$ is a scroll over a normal 4-fold and the restriction $\mathcal{K}_F$ of $\mathcal{K}$ to a general fiber F is isomorphic to $\mathcal{O}_{\mathbb{P}^{n-4}}(1)$; or*
2. *$(X, \mathcal{K})$ is a quadric fibration over a normal 3-fold; or*
3. *$(X, \mathcal{K})$ is a Del Pezzo fibration over a normal surface; or*
4. *$(X, \mathcal{K})$ is a Mukai fibration over a smooth curve; or*
5. *X is a Fano n-fold with $K_X \approx -(n-3)\mathcal{K}$ if n is even and $2K_X \approx -2(n-3)\mathcal{K}$ if n is odd.*

Proof. Since $\Delta := 2K_X + 2(n-3)\mathcal{K}$ is a nef Cartier divisor and $\Delta - K_X$ is ample, the Kawamata-Shokurov basepoint free theorem applies to give the nefvalue morphism, $\phi = \phi_{\mathcal{K}}$, associated to $|m\Delta|$ for all $m \gg 0$. Note that ϕ has connected fibers and normal image for m large enough.

If $\dim \phi(X) > 0$, let F be a general fiber of ϕ. Note that F is smooth since X has only isolated singularities. Let $\mathcal{K}_F$ be the restriction of $\mathcal{K}$ to F. We have $K_F + (n-3)\mathcal{K}_F \sim \mathcal{O}_F$ which implies that $\dim F \geq n-4$ (see (3.3.2)). Then by (3.1.6), $(F, \mathcal{K}_F)$ is isomorphic to either $(\mathbb{P}^{n-4}, \mathcal{O}_{\mathbb{P}^{n-4}}(1))$ or $(Q, \mathcal{O}_Q(1))$ with Q a smooth hyperquadric in $\mathbb{P}^{n-2}$, if either $\dim F = n-4$ or $\dim F = n-3$. Moreover $(F, \mathcal{K}_F)$ is either a Del Pezzo variety or a Mukai variety if either $\dim F = n-2$ or $\dim F = n-1$. Thus we find the cases (7.7.6.1) — (7.7.6.4).

If $\dim \phi(X) = 0$, then either $K_X \approx -(n-3)\mathcal{K}$ or $2K_X \approx -2(n-3)\mathcal{K}$ according to whether n is even or odd and we are in case (7.7.6.5). □

Remark 7.7.7. Assume that $(X, \mathcal{D})$ is as in Proposition (7.7.6) with $K_X + (n-3)\mathcal{D}$ and hence $K_X + (n-3)\mathcal{K}$ nef but not big. If the nefvalue morphism of $(X, \mathcal{K})$, $\phi_{\mathcal{K}} : X \to W$, has a lower dimensional image, with $\dim W \neq 0$ and $\dim X \geq 8$, then it follows from [114, Theorem A.4] that the first reduction and second reduction morphisms are isomorphisms.

The case $n = 4$ has been studied by Fujita [261]. We state here the results referring to [261] for the proofs. We also refer to [109] for results in the case when $\mathcal{L}$ is very ample on $\mathcal{M}$.

Theorem 7.7.8. *Let $\mathcal{M}$ be a 4-dimensional, smooth, connected variety and $\mathcal{L}$ an ample line bundle on $\mathcal{M}$. Let $(X, \mathcal{D})$ be the second reduction of $(\mathcal{M}, \mathcal{L})$. Let $\tau(\mathcal{K})$ be the nefvalue of $(X, \mathcal{K})$, where $\mathcal{K} \approx K_X + 2\mathcal{D}$. Let $(M^\#, L^\#)$ be the factorial stage of the second reduction map $\varphi : M \to M^\# \to X$ of $(\mathcal{M}, \mathcal{L})$. Then $\tau = \tau(\mathcal{K}) \leq 1$ unless one of the following cases occurs:*

1. *$\tau = 5$, $(X, \mathcal{K}) \cong (\mathbb{P}^4, \mathcal{O}_{\mathbb{P}^4}(1))$;*
2. *$\tau = 3$, $(X, \mathcal{K})$ is a Del Pezzo 4-fold, $K_X \approx -3\mathcal{K}$;*
3. *$\tau = 3$, $(X, \mathcal{K})$ is a quadric fibration over a smooth curve under $\phi_{\mathcal{K}}$;*
4. *$\tau = 3$, $(X, \mathcal{K})$ is a scroll over a normal surface under $\phi_{\mathcal{K}}$;*
5. *$\tau = 2$, X is a quadric in $\mathbb{P}^5$, $\mathcal{K} \approx \mathcal{O}_X(2)$;*
6. *$\tau = 2$, $M^\#$ is a blowup of $\mathbb{P}^4$ along a cubic surface contained in a hyperplane (only a single $(E, L^\#_E) \cong (\mathbb{P}^3, \mathcal{O}_{\mathbb{P}^3}(1))$ with $\mathcal{N}_{\mathbb{P}^3/M^\#} \cong \mathcal{O}_{\mathbb{P}^3}(-2)$ is contracted under $M^\# \to X$);*
7. *$\tau = 2$, $M^\#$ is a quadric fibration over $\mathbb{P}^3$ (only a single $(E, L^\#_E) \cong (\mathbb{P}^3, \mathcal{O}_{\mathbb{P}^3}(1))$ with $\mathcal{N}_{\mathbb{P}^3/M^\#} \cong \mathcal{O}_{\mathbb{P}^3}(-2)$ is contracted under $M^\# \to X$);*
8. *$\tau = 2$, $(X, \mathcal{K})$ is a $(\mathbb{P}^3, \mathcal{O}_{\mathbb{P}^3}(2))$-fibration over a curve;*
9. *$\tau = 5/3$, $(X, \mathcal{K}) \cong (\mathbb{P}^4, \mathcal{O}_{\mathbb{P}^4}(3))$; or*
10. *$\tau = 3/2$, $(X, \mathcal{K})$ is the cone over $(\mathbb{P}^3, \mathcal{O}_{\mathbb{P}^3}(2))$.*

The following consequence of the above results is analogous to Proposition (7.6.9).

Proposition 7.7.9. *Let $\mathcal{M}$ be an n-dimensional, smooth, connected variety and $\mathcal{L}$ an ample line bundle on $\mathcal{M}$. Assume $n \geq 4$. Then $\kappa(K_{\mathcal{M}} + (n-3)\mathcal{L}) \geq 0$ (respectively $= n$) if and only if the second reduction, $(X, \mathcal{D})$, of $(\mathcal{M}, \mathcal{L})$ exists, and either $K_X + (n-3)\mathcal{D}$ or equivalently $K_X + (n-3)\mathcal{K}$ is nef (respectively nef and big).*

Proof. Assume $\kappa(K_{\mathcal{M}} + (n-3)\mathcal{L}) \geq 0$. Hence in particular $\kappa(K_{\mathcal{M}} + (n-2)\mathcal{L}) = n$ so that, by Proposition (7.6.9), the first reduction (M, L) of $(\mathcal{M}, \mathcal{L})$ exists and $K_M + (n-2)L$ is nef and big. Therefore the second reduction $(X, \mathcal{D})$ of $(\mathcal{M}, \mathcal{L})$ exists also (see §7.5). Let $\varphi : M \to X$ be the second reduction map and let $\mathcal{D} := (\varphi_* L)^{**}$. To show that $K_X + (n-3)\mathcal{K}$ is nef, assume otherwise. Then by the previous results of this section, either $n = 6$ and $(X, \mathcal{D}) \cong (\mathbb{P}^6, \mathcal{O}_{\mathbb{P}^6}(2))$, or $n = 5$ and $(X, \mathcal{D})$ is as in (7.7.5), or $n = 4$ and $(X, \mathcal{D})$ is as in (7.7.8). In all these cases an easy check shows that there are no positive integers t such that $t(K_X + (n-3)\mathcal{D})$

is effective. By using Corollary (7.6.2.2), and since $\kappa(K_{\mathcal{M}} + (n-3)\mathcal{L}) \geq 0$, we get a contradiction. We also conclude that $\kappa(K_{\mathcal{M}} + (n-3)\mathcal{L}) = n$ implies the nefness and bigness of $K_X + (n-3)\mathcal{D}$.

If the second reduction, $(X, \mathcal{D})$, exists and $K_X+(n-3)\mathcal{D}$ is nef, then for $m \gg 0$, $m(n-2)(K_X + (n-3)\mathcal{D})$ is spanned. To see this note that if $K_X + (n-3)\mathcal{D}$ is nef, then $K_X + (n-3)\mathcal{K} \cong (n-2)(K_X + (n-3)\mathcal{D})$ is nef. Since $\mathcal{K}$ is an ample line bundle, the conclusion follows from the basepoint free theorem. Similarly if the second reduction, $(X, \mathcal{D})$, exists and $K_X + (n-3)\mathcal{D}$ is nef and big, then for $m \gg 0$, $m(n-2)(K_X + (n-3)\mathcal{D})$ is spanned and $\Gamma(m(K_X + (n-3)\mathcal{D}))$ gives a birational morphism. Thus, by using again Corollary (7.6.2.2), we have $\kappa(K_{\mathcal{M}} + (n-3)\mathcal{L}) \geq 0$ (respectively $\kappa(K_{\mathcal{M}} + (n-3)\mathcal{L}) = n$). □

Remark 7.7.10. Fix the notation as in Proposition (7.7.9). Assume that $\mathcal{L}$ is spanned by its global sections and let V be a smooth 3-fold obtained as transversal intersection of $n-3$ members of $|\mathcal{L}|$. Assume that $\kappa(K_{\mathcal{M}} + (n-3)\mathcal{L}) = n$. Then $\kappa(V) = 3$.

This will follow by descending induction on the dimension if we verify that $\kappa(K_A + (n-4)\mathcal{L}_A) = n-1$ for any smooth $A \in |\mathcal{L}|$. To see this use Lemma (2.5.5). By the assumption, there exist an effective divisor D, an ample divisor H and a positive integer N, such that $N(K_{\mathcal{M}} + (n-3)\mathcal{L}) \approx D + H$. Since $\mathcal{L}$ is ample, it can be assumed that D contains no $A \in |\mathcal{L}|$. By restriction on A we get

$$N(K_A + (n-4)\mathcal{L}_A) \approx N(K_{\mathcal{M}} + (n-3)\mathcal{L})_A \approx D_A + H_A,$$

where D_A is effective and H_A is ample. This implies $\kappa(K_A + (n-4)\mathcal{L}_A) = n-1$.

In particular if $\kappa(\mathcal{M}) \geq 0$, then $\kappa(K_{\mathcal{M}} + (n-3)\mathcal{L}) = n$ and $\kappa(V) = 3$.

Let us conclude this section recalling some recent developments of the adjunction process due to Nakamura and Fania.

Remark 7.7.11. (The third reduction) Let $\mathcal{M}$ be a smooth manifold of dimension $n \geq 4$, and let $\mathcal{L}$ be an ample line bundle on $\mathcal{M}$. Let (M, L), $\pi : \mathcal{M} \to M$, be the first reduction of $(\mathcal{M}, \mathcal{L})$, $L := (\pi_*\mathcal{L})^{**}$. Let $(X, \mathcal{D})$, $\varphi : M \to X$, $\mathcal{D} := (\varphi_* L)^{**}$, be the second reduction of $(\mathcal{M}, \mathcal{L})$. Recall that $K_M + (n-2)L \approx \varphi^*\mathcal{K}$, where $\mathcal{K}$ is an ample Cartier divisor on X such that $\mathcal{K} \approx K_X + (n-2)\mathcal{D}$ and $2\mathcal{D}$ is a Cartier divisor (see (7.5.8)).

A natural question is that of trying to push further the adjunction process. From the results of this section we may assume that $K_X + (n-3)\mathcal{K}$ is nef and big. Since

$$(n-2)(K_X + (n-3)\mathcal{D}) \approx K_X + (n-3)\mathcal{K}$$

(see (7.6.3)) we conclude that for $m \gg 0$, m even, the complete linear system

$$|m(n-2)(K_X + (n-3)\mathcal{D})| = |m(K_X + (n-3)\mathcal{K})|$$

defines a birational morphism with connected fibers, ψ. Let $Z := \psi(X)$ and $\Delta := (\psi_* \mathcal{D})^{**}$. Then we can define (Z, Δ), $\psi : X \to Z$, as the *third reduction* of $(\mathcal{M}, \mathcal{L})$. Note that $K_X + (n-3)\mathcal{D} \approx \psi^* \mathcal{P}$, for some ample $(n-2)$-Cartier divisor, $\mathcal{P}$, on Z. The natural question here is to try to understand the pairs (Z, Δ), or $(Z, \mathcal{P})$, and the structure of the morphism $\psi : X \to Z$. This turns out to be very hard. The difficulties come essentially from the facts that X has isolated 2-factorial singularities, small contractions of extremal rays (of flipping type), through which ψ factors, occur, and $\mathcal{D}$ is not necessarily ample or even a line bundle.

In this direction some partial results are contained in Fania [221], where the pairs $(\mathcal{M}, \mathcal{L})$, of dimension $n \geq 5$, with $K_{\mathcal{M}} + (n-3)\mathcal{L}$ nef and $K_{\mathcal{M}} + (n-4)\mathcal{L}$ nonnef are studied. Since $K_{\mathcal{M}} + (n-3)\mathcal{L}$ is nef the linear system $|m(K_{\mathcal{M}} + (n-3)\mathcal{L})|$, for $m \gg 0$, defines a morphism, f. From Mori's cone theorem (see also Lemma (4.2.14)) we know that there exists an extremal rational curve, C, such that $(K_{\mathcal{M}} + (n-4)\mathcal{L}) \cdot C < 0$. Let $\rho := \mathrm{cont}_R : X \to Y$. In [221] the structure of the morphism ρ is studied, in both the cases when ρ is of fiber type or birational. This leads to the knowledge of the structure of f and to the classification of smooth polarized pairs $(\mathcal{M}, \mathcal{L})$, of dimension $n \geq 5$, $\mathcal{L}$ ample on $\mathcal{M}$, such that $K_{\mathcal{M}} + (n-3)\mathcal{L}$ is nef and $K_{\mathcal{M}} + (n-4)\mathcal{L}$ is nonnef. We refer to Fania [221] for the complete list and full details. In Nakamura [471, 473] a detailed study of the structure of the third reduction morphism $\psi : X \to Z$ and the singularities of Z is worked out under the further assumption that $\mathcal{L}$ is very ample. Further refined results are obtained in [471, §4] assuming the extra condition that the unnormalized spectral value $u(\mathcal{M}, \mathcal{L})$ satisfies $u(\mathcal{M}, \mathcal{L}) > 2\lceil (n-1)/3 \rceil$. Note that under this condition it must be $n \geq 5$ and both the first and the second reduction morphisms are isomorphisms by Theorem (7.6.5). For the proofs and full details we refer to Nakamura [471, 473].

We also refer to Andreatta and Wiśniewski [28] for general results about the structure of contractions of extremal rays whose supporting function is an adjoint line bundle.

7.8 The three dimensional case

Let $(\mathcal{M}, \mathcal{L})$ be a smooth threefold polarized with an ample line bundle $\mathcal{L}$ and let (M, L), $\pi : \mathcal{M} \to M$ be the first reduction of $(\mathcal{M}, \mathcal{L})$ as in (7.3.3). Let $\tau(L)$ be the nefvalue of (M, L). By Theorems (7.2.4), (7.3.2) and (7.5.3) we may assume that $\tau(L) \leq 1$ and that $K_M + L$ is nef and big. Thus the second reduction, $(X, \mathcal{D})$, $\varphi : M \to X$, $\mathcal{D} := (\varphi_* L)^{**}$, $\mathcal{K} \approx K_X + \mathcal{D}$, exists.

The following result is due to Fujita. The result is announced in [263] and it can be proved by the same methods as in [261], by using Mori theory (we also refer to [86, §3], for results in the case when $\mathcal{L}$ is very ample on $\mathcal{M}$). Note that the values of u, τ in the theorem below follow from the main result of [263] by using Lemma (7.1.6) and Proposition (7.6.4) and recalling that the Kodaira energy $\kappa\epsilon(\mathcal{M}, \mathcal{L})$ as in [263] coincides with $-u(\mathcal{M}, \mathcal{L})$ (see (7.1.2)).

Theorem 7.8.1. *Let $(\mathcal{M}, \mathcal{L})$ be a smooth threefold, polarized with an ample line bundle $\mathcal{L}$. Let (M, L), $(X, \mathcal{D})$, $\mathcal{K} \approx K_X + \mathcal{D}$, be the first and the second reduction of $(\mathcal{M}, \mathcal{L})$. Let $\tau = \tau(\mathcal{K})$ be the nefvalue and let $\phi_{\mathcal{K}}$ be the nefvalue morphism of $(X, \mathcal{K})$. Let $u = u(\mathcal{M}, \mathcal{L})$ be the unnormalized spectral value of $(\mathcal{M}, \mathcal{L})$. Then*

i) *X has either hypersurface singularities of the type $\{x^2 + y^2 + z^2 + u^k = 0\}$, or quotient singularities isomorphic to the vertex of the cone over the Veronese surface $(\mathbb{P}^2, \mathcal{O}_{\mathbb{P}^2}(2))$. In particular $\mathcal{D}$ is invertible except at quotient singularities.*

ii) *Either $u \le 1/2$, or one of the following occurs:*
   - (a) *$u = 4/5$, $\tau = 4$, $(X, \mathcal{D}) \cong (\mathbb{P}^3, \mathcal{O}_{\mathbb{P}^3}(5))$;*
   - (b) *$u = 3/4$, $\tau = 3$, $(X, \mathcal{D}) \cong (Q, \mathcal{O}_Q(4))$, Q a hyperquadric in $\mathbb{P}^4$;*
   - (c) *$u = 3/4$, $\tau = 3$, $(X, \mathcal{K})$ is a scroll over a smooth curve under $\phi_{\mathcal{K}}$;*
   - (d) *$u = 5/7$, $\tau = 5/2$, $(X, \mathcal{K})$ is a cone over $(\mathbb{P}^2, \mathcal{O}_{\mathbb{P}^2}(2))$;*
   - (e) *$u = 2/3$, $\tau = 2$, $(X, \mathcal{K})$ is a Del Pezzo threefold, i.e., $K_X \approx -2\mathcal{K}$;*
   - (f) *$u = 2/3$, $\tau = 2$, $(X, \mathcal{K})$ is a hyperquadric fibration over a smooth curve under $\phi_{\mathcal{K}}$;*
   - (g) *$u = 2/3$, $\tau = 2$, $(X, \mathcal{K})$ is a scroll over a normal surface under $\phi_{\mathcal{K}}$.*

In the following cases, $(X, \mathcal{K})$ may need to be further blown down to another model $(X', \mathcal{K}')$. This new pair has no worse singularities than $(X, \mathcal{K})$.

1. *$u = 3/5$, $\tau \ge 3/2$, $(X', \mathcal{K}') \cong (Q, \mathcal{O}_Q(2))$, Q a hyperquadric in $\mathbb{P}^4$;*

2. *$u = 3/5$, $\tau \ge 3/2$, X' has exactly one quotient singularity, and the blown up, $\sigma : X'' \to X'$, at this point is isomorphic to the blowup of $\mathbb{P}^3$ along a smooth plane cubic, C. Moreover $\sigma^*\mathcal{K}' \cong 3H - E_C$ where H is the pullback of $\mathcal{O}(1)$ of $\mathbb{P}^3$ and E_C is the exceptional divisor over C;*

3. *$u = 3/5$, $\tau \ge 3/2$, X' has exactly two quotient singularities, and the blowing up, $\sigma : X'' \to X'$ at these points is a smooth member of $|2\xi_{\mathcal{E}}|$ on $\mathbb{P}(\mathcal{E})$, where $\mathcal{E} = \mathcal{O}_{\mathbb{P}^2}(2) \oplus \mathcal{O}_{\mathbb{P}^2} \oplus \mathcal{O}_{\mathbb{P}^2}$ and $\xi_{\mathcal{E}}$ is the tautological line bundle on $\mathbb{P}(\mathcal{E})$. Moreover $\sigma^*\mathcal{K}'$ is the restriction of $\xi_{\mathcal{E}}$ to X'';*

4. *$u = 3/5$, $\tau \ge 3/2$, X' is a $\mathbb{P}^2$ -fibration over a curve and $\mathcal{K}_F \cong \mathcal{O}_{\mathbb{P}^2}(2)$ for a general fiber $F \cong \mathbb{P}^2$;*

5. *$u = 4/7$, $\tau \ge 4/3$, $(X', \mathcal{K}') \cong (\mathbb{P}^3, \mathcal{O}_{\mathbb{P}^3}(3))$;*

6. *$u = 5/9$, $\tau \ge 5/4$, $(X', \mathcal{B})$ is a cone over $(\mathbb{P}^2, \mathcal{O}_{\mathbb{P}^2}(2))$ for some line bundle $\mathcal{B}$ such that $\mathcal{K}' \cong 2\mathcal{B}$.*

7.9 Applications

In this section we give some applications of the above results. The first is a coarse classification of threefolds, $\mathcal{M}$, carrying an ample line bundle, $\mathcal{L}$, such that $|\mathcal{L}|$ contains a smooth surface, S, of given Kodaira dimension. The second shows that for a smooth 3-fold, V, which is obtained as transversal intersection of $n-3$ hyperplane sections of a smooth n-fold, $\mathcal{M}$, there is a simple birational morphism which maps V onto a minimal model, W, of V. The third is a vanishing theorem which has interest by its own and also has some nice consequences. The last application is a new proof of a well-known result due to Bădescu, which will be used later in Chapter 11.

7.9.1 A coarse classification of 3-folds of special type

Let $\mathcal{M}$, $\mathcal{L}$, S be as above. By the results of §§ 7.2, 7.3, 7.5 we have the following list of possible cases.

1. $(\mathcal{M}, \mathcal{L}) \cong (\mathbb{P}^3, \mathcal{O}_{\mathbb{P}^3}(1))$;
2. $\mathcal{M}$ is a $\mathbb{P}^2$-bundle, $p : \mathcal{M} \to C$, over a smooth curve C and $K_{\mathcal{M}} + 3\mathcal{L} \approx p^*H$ for some ample line bundle H on C;
3. $(\mathcal{M}, \mathcal{L}) \cong (Q, \mathcal{O}_Q(1))$, Q a hyperquadric in $\mathbb{P}^4$;
4. $(\mathcal{M}, \mathcal{L})$ is a Del Pezzo threefold, i.e., $K_{\mathcal{M}} \approx -2\mathcal{L}$;
5. $(\mathcal{M}, \mathcal{L})$ is a quadric fibration, $p : \mathcal{M} \to C$, over a smooth curve C;
6. $(\mathcal{M}, \mathcal{L})$ is a scroll, and in fact a $\mathbb{P}^1$-bundle, $p : \mathcal{M} \to Y$, over a smooth surface Y;

   There exists the first reduction (M, L) of $(\mathcal{M}, \mathcal{L})$, $\pi : \mathcal{M} \to M$, and either:

7. $(M, L) \cong (Q, \mathcal{O}_Q(2))$, Q a hyperquadric in $\mathbb{P}^4$;
8. $(M, L) \cong (\mathbb{P}^3, \mathcal{O}_{\mathbb{P}^3}(3))$;
9. there exists a surjective morphism, $p : M \to C$, onto a smooth curve C and $2K_M + 3L \approx p^*H$ for some ample line bundle H on C;
10. (M, L) is a Mukai threefold, i.e., $K_M \approx -L$;
11. (M, L) is a Del Pezzo fibration, $p : M \to C$, over a smooth curve C;
12. (M, L) is a quadric fibration, $p : M \to Y$, over a normal surface Y;

13. $K_M + L$ is nef and big.

By looking over the list above we can describe $(\mathcal{M}, \mathcal{L})$ (or the first reduction (M, L)) according to the Kodaira dimension, $\kappa(S)$, of S. We summarize the results in Table (7.3).

Table 7.3. Classification of threefolds according to the Kodaira dimension of an ample divisor

Case	*smooth* $S \in \lvert\mathcal{L}\rvert$	*Description of* $(\mathcal{M}, \mathcal{L})$ *or* (M, L)	τ
1)		$(\mathcal{M}, \mathcal{L}) \cong (\mathbb{P}^3, \mathcal{O}_{\mathbb{P}^3}(1))$	4
2)		$(\mathcal{M}, \mathcal{L})$ a scroll (and thus a $\mathbb{P}^2$-bundle) over $\mathbb{P}^1$	3
3)		$(\mathcal{M}, \mathcal{L}) \cong (Q, \mathcal{O}_Q(1))$, Q a hyperquadric in $\mathbb{P}^4$	3
4)	$\kappa(S) = -\infty$	$(\mathcal{M}, \mathcal{L})$ a Del Pezzo 3-fold, $K_{\mathcal{M}} \approx -2\mathcal{L}$	2
5)	$q(S) = 0$	$(\mathcal{M}, \mathcal{L})$ a quadric fibration over $\mathbb{P}^1$	2
6)		$(\mathcal{M}, \mathcal{L}) \cong (Q, \mathcal{O}_Q(2))$, Q a hyperquadric in $\mathbb{P}^4$	3/2
7)		$(\mathcal{M}, \mathcal{L}) \cong (\mathbb{P}^3, \mathcal{O}_{\mathbb{P}^3}(3))$	4/3
8)		there exists a surjective morphism, $p : M \to \mathbb{P}^1$, $2K_M + 3L \approx p^*H$, where H is an ample line bundle on $\mathbb{P}^1$	3/2
9)		$(\mathcal{M}, \mathcal{L})$ a scroll (and thus a $\mathbb{P}^2$-bundle) over a genus $q(S)$ curve	3
10)	$\kappa(S) = -\infty$	$(\mathcal{M}, \mathcal{L})$ a quadric fibration over a genus $q(S)$ curve	2
11)	$q(S) > 0$	there exists a surjective morphism, $p : M \to C$, with C a positive genus curve, $2K_M + 3L \approx p^*H$, H an ample line bundle on C	3/2
12)	$\kappa(S) = 0$	(M, L) a Mukai 3-fold, $K_M \approx -L$	1
13)	$\kappa(S) = 1$	(M, L) a Del Pezzo fibration over a smooth curve	1
14)	$\kappa(S) = 2$	(M, L) a quadric fibration over a normal surface	1
15)		$K_M + L$ nef and big	1

Let us give a few comments on the table. The "trivial" scroll case over a surface is omitted. Furthermore τ denotes the nefvalue of either $(\mathcal{M}, \mathcal{L})$ or (M, L). Note that the case 6) of the previous list always occurs as "trivial" case for each value of $\kappa(S)$. In this case S is a meromorphic section of $p : \mathcal{M} \to Y$, but S cannot be isomorphic to Y by (5.1.5). Also, in this case, (Y, L_Y), $L_Y := (p_{S*}L_S)^{**}$, where L_S denotes the restriction of L to S, is in fact a reduction of (S, L_S) and p_S is the reduction map (see [104, (0.6.2)] and also Theorem (11.1.2)). In cases 8), 13), 14) write $A = \pi(S)$, where $\pi : \mathcal{M} \to M$ denotes the first reduction map. Then A is a smooth surface in $|L|$. In case 8), let $L' = K_M + 2L$ and let L'_A be the restriction of L' to A. Since $2K_M + 3L$ is nef and big, L' is ample, and (A, L'_A) is a conic fibration under the restriction map p_A since $K_A + L'_A \approx (2K_M + 3L)_A \approx p_A^*(H)$. In case 13) we have $\kappa(S) = 1$. Indeed $K_M + L \approx p^*H$ for some ample line bundle on the curve C, so that $K_A \approx p^*H$ is nef and not big. Note also that A maps onto the curve C since A meets each fiber of p, A being ample. Then $\kappa(S) = \kappa(K_A) = \dim C = 1$. In case 14), $K_X + 2L \approx p^*H$ for some ample line bundle H on the surface Y, so that K_A is nef and big, i.e., $\kappa(S) = 2$.

7.9.2 A minimal model for 3-fold sections

The following result is the 3-dimensional analogous of (7.6.10). We also refer to the second author's survey paper [577] for discussions on the 2-dimensional case.

Proposition 7.9.1. *Let $\mathcal{M}$ be a smooth n-fold, $n \geq 4$, and let $\mathcal{L}$ be a very ample line bundle on $\mathcal{M}$. Let V be a smooth 3-fold obtained as transversal intersection of $n-3$ elements of $|\mathcal{L}|$. Let $\mathcal{L}_V$ be the restriction of $\mathcal{L}$ to V. Assume that the first reduction, (M, L), $\pi : \mathcal{M} \to M$, and the second reduction, $(X, \mathcal{D})$, $\varphi : M \to X$, of $(\mathcal{M}, \mathcal{L})$ exist. Let $\psi = \varphi \circ \pi$ and let $W := \psi_V(V)$, where ψ_V denotes the restriction of ψ to V. Then ψ_V is the composition of the first and the second reduction mapping of $(V, \mathcal{L}_V)$. Hence in particular W is 2-Gorenstein with terminal singularities.*

Furthermore either $n = 6$ and $(X, \mathcal{D}) \cong (\mathbb{P}^6, \mathcal{O}_{\mathbb{P}^6}(2))$, $n = 4, 5$ and $(X, \mathcal{D})$ is described as in (7.7.5), (7.7.8), or K_W is nef, i.e., W is a minimal model of V.

Proof. The fact that ψ_V is the composition of the first and the second reduction mapping for $(V, \mathcal{L}_V)$ is a standard consequence of the structure Theorems (7.3.2) and (7.5.3) for π and φ. The fact that W is 2-Gorenstein with terminal singularities then follows from Proposition (12.2.6).

Let $\mathcal{K} \approx K_X + (n-2)\mathcal{D}$, $\mathcal{D} = (\psi_*\mathcal{L})^{**}$, as in (7.5.8). Since K_X and $\mathcal{D}$ are 2-Cartier divisors, the restriction $(K_X + (n-3)\mathcal{D})_W$ of $K_X + (n-3)\mathcal{D}$ to W is a well-defined 2-Cartier divisor on W. From Theorems (7.3.2), (7.5.3) we know that ψ_V is an isomorphism outside of a subset, $Z \subset W$, with $\operatorname{cod}_W Z \geq 2$. Let $W_0 = W - Z$. Then we have an isomorphism of line bundles, $K_{W_0} \cong (K_X + (n-3)\mathcal{D})_{W_0}$ and $2K_W$ and $2(K_X + (n-3)\mathcal{D})_W$ are line bundles. Let $i : W_0 \hookrightarrow W$ be the open inclusion. Then

$$2K_W \cong i_*(2K_{W_0}) \cong i_*(2(K_X + (n-3)\mathcal{D})_{W_0}) \cong 2(K_X + (n-3)\mathcal{D})_W. \qquad (7.8)$$

Assuming that $(X, \mathcal{D})$ is not as in one of the cases listed in the statement, we can assume that $K_X + (n-3)\mathcal{K}$ is nef. Hence $K_X + (n-3)\mathcal{D}$ is nef and therefore, by (7.8), K_W is nef. □

7.9.3 A vanishing theorem

The following result was proved first in the case when $\mathcal{L}$ is very ample by Beltrametti, Fania, and Sommese [86, (0.2.8)]. That works in the general case with the structure Theorem (7.5.3) in place of the structure theorem of [86].

Lemma 7.9.2. *Let $\mathcal{M}$ be a smooth connected n-fold polarized by an ample line bundle, $\mathcal{L}$. Let (M, L), $\pi : \mathcal{M} \to M$, and $(X, \mathfrak{D})$, $\varphi : M \to X$, be the first and the second reduction of $(\mathcal{M}, \mathcal{L})$. Let $\mathcal{K} \approx K_X + (n-2)\mathfrak{D}$ be as in (7.5.8). Assume that $K_X + (n-3)\mathcal{K}$ is nef and big. Then for all positive integers t and $i > 0$ $h^i(t(K_X + (n-3)\mathfrak{D})) = 0$.*

Proof. Recall that $K_X + (n-3)\mathcal{K} \approx (n-2)(K_X + (n-3)\mathfrak{D})$ and note that

$$t(K_X + (n-3)\mathfrak{D}) - K_X = (t - (n-2))(K_X + (n-3)\mathfrak{D}) + (n-3)\mathcal{K}.$$

Thus $t(K_X + (n-3)\mathfrak{D}) - K_X$ is nef and big for $t \geq n-2$, so that in this case the result follows from the Kawamata-Viehweg vanishing theorem. Therefore we can assume from now on in the proof that $1 \leq t \leq n-3$.

From Theorem (7.5.3.5) it follows that φ factors as $\varphi = \rho \circ q$, where $\rho : W \to X$ is the blowing up of the curves occurring in (7.5.3.5a). Note that ρ is an isomorphism in a neighborhood of $\mathrm{Sing}(X)$. Let Z_1 be the union of such curves and let $D := \rho^{-1}(Z_1)$. Let $H := \rho^*\mathfrak{D} - D$. Then $H = (q_*L)^{**}$ is ample on W. Indeed L is ample on M and $q : M \to W$ is an isomorphism outside of a finite set of points. Since ρ is an isomorphism in a neighborhood of the set where $t(K_X + (n-3)\mathfrak{D})$ is not a line bundle, we have the projection formula

$$\rho_{(i)}(\rho^*(t(K_X + (n-3)\mathfrak{D}))) \cong \rho_{(i)}(\mathcal{O}_W) \otimes (t(K_X + (n-3)\mathfrak{D})).$$

The latter sheaf is the zero sheaf if $i > 0$ and is isomorphic to $t(K_X + (n-3)\mathfrak{D})$ if $i = 0$. Thus by the Leray spectral sequence we have for $i > 0$

$$h^i(t(K_X + (n-3)\mathfrak{D})) = h^i(\rho^*(t(K_X + (n-3)\mathfrak{D}))).$$

Note that $K_W \approx \rho^*K_X + (n-2)D$. Therefore it suffices to show that, for $i > 0$ and $t = 1, \ldots, n-3$,

$$h^i(\rho^*(t(K_X + (n-3)\mathfrak{D}))) = h^i(t(K_W + (n-3)H - D)) = 0. \tag{7.9}$$

For $0 \leq s \leq t$, let

$$\mathcal{F}_s := t(K_W + (n-3)H) - sD.$$

Since $\mathcal{F}_s$ is a line bundle in a neighborhood of D, we have the exact sequence

$$0 \to \mathcal{F}_s \to \mathcal{F}_{s-1} \to \mathcal{F}_{s-1|D} \to 0. \tag{7.10}$$

Note also that, since $K_W + (n-2)H \approx \rho^*\mathcal{K}$, one has

$$\mathcal{F}_{s-1|D} = t(K_W + (n-2)H)_D - tH_D - (s-1)D_D \approx -tH_D - (s-1)D_D. \quad (7.11)$$

Let F ($\cong \mathbb{P}^{n-2}$) be any fiber of the restriction $\sigma : D \to Z_1$ of ρ to D. From (7.11), and since $\mathcal{N}_{F/D} \cong \mathcal{O}_{\mathbb{P}^{n-2}}(-1)$, we have $\mathcal{F}_{s-1|F} \approx \mathcal{O}_{\mathbb{P}^{n-2}}(s-t-1)$. Therefore we see that the i-th direct image sheaves $\sigma_{(i)}(\mathcal{F}_{s-1|D})$ are zero for $i \geq 0$ and $1 \leq s \leq t$. The Leray spectral sequence gives, for $i \geq 0$,

$$h^i(\mathcal{F}_{s-1|D}) \leq \sum_{j=0}^{i} h^{i-j}(\sigma_{(j)}(\mathcal{F}_{s-1|D})).$$

Hence we conclude that $h^i(\mathcal{F}_{s-1|D}) = 0$ for $i \geq 0$ and $1 \leq s \leq t$. Then by using (7.10) we see that for $i \geq 0$,

$$h^i(t(K_W + (n-3)H - D)) = h^i(\mathcal{F}_t) = \ldots = h^i(\mathcal{F}_0) = h^i(t(K_W + (n-3)H)).$$

Therefore, recalling (7.9), we are reduced to showing that $h^i(t(K_W+(n-3)H)) = 0$ for $i > 0$. Hence it is enough to show that $t(K_W + (n-3)H) - K_W = (t-1)K_W + t(n-3)H$ is nef and big. To see this write

$$(t-1)K_W + t(n-3)H = (t-1)(K_W + (n-2)H) + (t(n-3) - (t-1)(n-2))H.$$

Note that $K_W + (n-2)H$ is nef and $t(n-3) - (t-1)(n-2) > 0$ is equivalent to $t < n-2$, which is our present assumption. □

The lemma above has the following consequence (compare with the second author's results [576] and [577, Corollary II-A]).

Corollary 7.9.3. *Let $\mathcal{M}$ be a smooth connected n-fold polarized by an ample line bundle, $\mathcal{L}$. Let (M, L) be the first reduction of $(\mathcal{M}, \mathcal{L})$ and let $(X, \mathcal{D})$, $\varphi : M \to X$, $\mathcal{K} \approx K_X + (n-2)\mathcal{D}$, $\mathcal{D} = (\varphi_* L)^{**}$, be the second reduction of $(\mathcal{M}, \mathcal{L})$. Then there exist degree n polynomials $p(z)$, $q(z) \in \mathbb{Q}[z]$ such that, for a positive integer t, $p(t) = h^0(2t(K_X + (n-3)\mathcal{D}))$, and $q(t) = h^0((2t+1)(K_X + (n-3)\mathcal{D}))$. Furthermore $p(z)$, $q(z)$ have the same leading term.*

Proof. Write $\mathcal{F} := K_X + (n-3)\mathcal{D}$ and $H := 2(K_X + (n-3)\mathcal{D})$. Then $\mathcal{F}$ is a 2-Cartier divisor and H is a line bundle. Let $p(z) \in \mathbb{Q}[z]$ be the Hilbert polynomial of H. Then by Lemma (7.9.2) we have for positive integers, t, $p(t) = \chi(tH) = h^0(t(2K_X + 2(n-3)\mathcal{D}))$. Furthermore, for positive integers t, $q(t) := \chi(\mathcal{F} + tH) = h^0((2t+1)(K_X + (n-3)\mathcal{D}))$. □

Remark 7.9.4. Notation as in the above lemma. Lemma (7.9.2) can be also applied to get effectiveness results, e.g., for $(n+1)K_X + n(n-3)\mathcal{D}$ and $(n+1)K_{\mathcal{M}} + n(n-3)\mathcal{L}$. We refer for this to our paper with Fania [86, (3.1), (4.1), (4.2)]. We also refer to our paper with Biancofiore [85, (2.2)] and our paper [111] for effectiveness results of appropriate adjoint bundles.

7.9.4 A result of Bădescu

The following result is due to Bădescu [42, 43] (see also the second author's paper [576]). The alternate proof we give is based on adjunction theory results. Here, as usual, if $\mathbb{F}_r$ is a Hirzebruch surface, E, f denotes a section of minimal self-intersection and a fiber of $\mathbb{F}_r$, respectively.

Theorem 7.9.5. *Let $\mathcal{M}$ be a smooth, projective threefold, and $\mathcal{L}$ an ample line bundle on $\mathcal{M}$. Assume that there is a smooth $A \in |\mathcal{L}|$ such that A is a $\mathbb{P}^1$-bundle, $p : A \to C$, over a smooth curve, C.*

1. *If $g(C) > 0$, then $\mathcal{M}$ is a $\mathbb{P}^2$-bundle $\psi : \mathcal{M} \to C$ over C such that the restriction $\psi_A = p$ and $K_{\mathcal{M}} + 3\mathcal{L} \approx \psi^* H$ for some ample line bundle H on C. Hence in particular $\mathcal{L}_F \cong \mathcal{O}_{\mathbb{P}^2}(1)$ for each fiber $F = \mathbb{P}^2$ of ψ.*
2. *If $g(C) = 0$, then p extends to a $\mathbb{P}^2$-bundle, $\psi : \mathcal{M} \to C$ as in 1) above, unless either:*
   - (a) *$A = \mathbb{F}_0$, $\mathcal{L}_{\mathbb{F}_0} \approx \mathcal{O}_{\mathbb{F}_0}(1, t)$, $t \geq 2$, with $\mathcal{L}_f \approx \mathcal{O}_f(t)$ where f is a fiber of $p : \mathbb{F}_0 \to C$; or*
   - (b) *$A = \mathbb{F}_0$, $\mathcal{L}_{\mathbb{F}_0} \approx \mathcal{O}_{\mathbb{F}_0}(1, 1)$, $(\mathcal{M}, \mathcal{L}) \cong (Q, \mathcal{O}_Q(1))$, Q a hyperquadric in $\mathbb{P}^4$, or*
   - (c) *$A = \mathbb{F}_0$, $(\mathcal{M}, \mathcal{L}) \cong (\mathbb{P}^3, \mathcal{O}_{\mathbb{P}^3}(2))$.*

Proof. Note that $K_{\mathcal{M}}$ is nef under the assumption on A. By the results of §§7.2, 7.3, 7.5 we see that one of the following holds:

1. $(\mathcal{M}, \mathcal{L}) \cong (\mathbb{P}^3, \mathcal{O}_{\mathbb{P}^3}(1))$;
2. $\mathcal{M}$ is a $\mathbb{P}^2$-bundle, $\psi : \mathcal{M} \to Y$, over a smooth curve Y and $K_{\mathcal{M}} + 3\mathcal{L} \cong \psi^* H$ for some ample line bundle H on Y;
3. $(\mathcal{M}, \mathcal{L}) \cong (Q, \mathcal{O}_Q(1))$, Q a hyperquadric in $\mathbb{P}^4$;
4. $(\mathcal{M}, \mathcal{L})$ is a Del Pezzo threefold, i.e., $K_{\mathcal{M}} \cong -2\mathcal{L}$;
5. $(\mathcal{M}, \mathcal{L})$ is a quadric fibration, $\psi : \mathcal{M} \to Y$, over a smooth curve Y;
6. $(\mathcal{M}, \mathcal{L})$ is a scroll, and in fact a $\mathbb{P}^1$-bundle, $\psi : \mathcal{M} \to Y$, over a smooth surface Y; or
7. there exists the first reduction, (M, L), $\pi : \mathcal{M} \to M$, of $(\mathcal{M}, \mathcal{L})$ with $\mathcal{M}$ not isomorphic to M.

   If $\mathcal{M}$ is not as in one of the cases above then $\mathcal{M}$ is its own reduction and either:

8. $(\mathcal{M}, \mathcal{L}) \cong (\mathbb{P}^3, \mathcal{O}_{\mathbb{P}^3}(3))$;

9. $(\mathcal{M}, \mathcal{L}) \cong (Q, \mathcal{O}_Q(2))$, Q a hyperquadric in $\mathbb{P}^4$;

10. there exists a surjective morphism with connected fibers, $\psi : \mathcal{M} \to Y$, onto a smooth curve Y and $2K_{\mathcal{M}} + 3\mathcal{L} \approx \psi^* H$ for some ample line bundle H on Y; or

11. $K_{\mathcal{M}} + \mathcal{L}$ is nef.

We will go on by a case by case analysis, following the list above.

First note that, since A is a $\mathbb{P}^1$-bundle, cases 1), 8), 9), 11) can be easily ruled out. Indeed in case 1) we would have $A \cong \mathbb{P}^2$. In case 8), A would be a cubic surface in $\mathbb{P}^3$, so that $K_A^2 = 3$, which is not possible since $K_A^2 = 8(1 - q(A))$. In case 9), $K_A \approx \mathcal{O}_A(-1)$, so that $K_A^2 = 4$, the same contradiction as above. Finally case 11) is excluded since in that case K_A would be nef.

Thus we have to take care of cases 2)—7) and 10).

Case 2). If $C \cong Y$ then $p : A \to C$ extends and we are done. So we can assume $C \not\cong Y$. Since $K_{\mathcal{M}} + 3\mathcal{L} \approx \psi^* H$ for some ample line bundle H on Y, we have $\mathcal{L}_F \approx \mathcal{O}_{\mathbb{P}^2}(1)$ for each fiber $F = \mathbb{P}^2$ of ψ and then the restriction, $\psi_A : A \to Y$, is a $\mathbb{P}^1$-bundle. In this case each fiber, $\mathbb{P}^1$, of ψ_A goes onto the curve C under p. Therefore $C \cong \mathbb{P}^1$, so that $A \cong \mathbb{F}_r$, for some $r \geq 0$. Hence in particular $Y \cong \mathbb{P}^1$. Since A has two different $\mathbb{P}^1$-bundle projections onto $\mathbb{P}^1$, it thus follows that $A \cong \mathbb{F}_0$.

Note that $\mathcal{L}_A \not\cong \mathcal{O}_{\mathbb{F}_0}(1, 1)$. Indeed, if it was, then

$$(K_{\mathcal{M}} + 3\mathcal{L})_A \approx K_A + 2\mathcal{L}_A \approx \mathcal{O}_A$$

and therefore $K_{\mathcal{M}} + 3\mathcal{L} \approx \mathcal{O}_{\mathcal{M}}$ by the Lefschetz theorem (see (2.3.1)). This implies that $(\mathcal{M}, \mathcal{L}) \cong (Q, \mathcal{O}_Q(1))$, Q a hyperquadric in $\mathbb{P}^4$, by the Kobayashi-Ochiai theorem. This contradicts the fact that $\mathcal{M}$ is a $\mathbb{P}^2$-bundle. Thus $\mathcal{L}$ is $\mathcal{O}(1)$ on the fibers of $\psi_A : A \to Y$ and $\mathcal{L}$ is $\mathcal{O}(t)$, $t \geq 2$, on the fibers of $p : A \to C$. We have case a) of (7.9.5.2).

Case 3). It can happen and leads to the case b) of (7.9.5.2). Note that the $\mathbb{P}^1$-bundle $p : A \to C$ does not extend to a $\mathbb{P}^2$-bundle $\psi : \mathcal{M} \to C$ since $\mathrm{Pic}(\mathcal{M}) \cong \mathbb{Z}$.

Case 4). From $K_{\mathcal{M}} \approx -2\mathcal{L}$, we get $K_A \approx -\mathcal{L}_A$. Hence in particular $q(A) = 0$ and therefore $A \cong \mathbb{F}_r$, $r \geq 0$. Since $-K_A$ is ample it is an easy numerical check to show that either $A \cong \mathbb{F}_0$ or $A \cong \mathbb{F}_1$.

Let $A \cong \mathbb{F}_0$. Then $K_A \approx (K_{\mathcal{M}} + \mathcal{L})_A \approx -2(E + f)$. Since the restriction map $\mathrm{Pic}(\mathcal{M}) \to \mathrm{Pic}(A)$ has torsion free cokernel by the Lefschetz theorem, it thus follows that $K_{\mathcal{M}} + \mathcal{L} \approx -2\mathcal{H}$ for some $\mathcal{H} \in \mathrm{Pic}(\mathcal{M})$. Therefore $\mathcal{L} \approx 2\mathcal{H}$ so that $\mathcal{H}$ is ample and $-K_{\mathcal{M}} \approx 4\mathcal{H}$. Thus we conclude that $(\mathcal{M}, \mathcal{L}) \cong (\mathbb{P}^3, \mathcal{O}_{\mathbb{P}^3}(2))$ by the Kobayashi-Ochiai result. This leads to the case c) of (7.9.5.2). Note that $p : A \to C$ does not extend to a $\mathbb{P}^2$-bundle $\psi : \mathcal{M} \to C$ since $\mathrm{Pic}(\mathcal{M}) \cong \mathbb{Z}$.

Let $A \cong \mathbb{F}_1$. Then $\mathcal{L}^3 = \mathcal{L}_A^2 = K_{\mathbb{F}_1}^2 = (2E+3f)^2 = 8$. Thus $\mathcal{M}$ is a Del Pezzo threefold of degree 8 and hence $(\mathcal{M}, \mathcal{L}) \cong (\mathbb{P}^3, \mathcal{O}_{\mathbb{P}^3}(2))$ by Fujita [257, (8.11), p. 72]. This leads to the contradiction $A \cong \mathbb{F}_1 \in |\mathcal{O}_{\mathbb{P}^3}(2)|$.

Case 5). We claim that the restriction $\psi_A : A \to Y$ is a $\mathbb{P}^1$-bundle. If $C \cong Y$, this is clear. Thus, assume $C \not\cong Y$. In this case each fiber, $\mathbb{P}^1$, of $p : A \to C$ goes onto Y under ψ. Therefore $Y \cong \mathbb{P}^1$. Since $K_{\mathcal{M}} + 2\mathcal{L} \approx \psi^* H$ for some ample line bundle H on Y, we have $\mathcal{L}_Q \cong \mathcal{O}_Q(1)$ for each fiber Q, Q a quadric in $\mathbb{P}^3$, of ψ and hence the restriction, $\psi_A : A \to Y$, is a conic fibration. Let # be the number of singular fibers of ψ_A. Then

$$e(A) = e(\mathbb{P}^1)e(f) + \# = 4 + \#, \tag{7.12}$$

where $f \cong \mathbb{P}^1$ is a general fiber of $\psi_A : A \to Y$. We also have that each general fiber $f \cong \mathbb{P}^1$ of ψ_A goes onto C under p and therefore $C \cong \mathbb{P}^1$. Then $e(A) = e(\mathbb{P}^1)e(\mathbb{P}^1) = 4$, so that (7.12) gives $\# = 0$. Thus we conclude that $\psi_A : A \to Y$ is a $\mathbb{P}^1$-bundle. By the Lefschetz theorem we have an isomorphism $H^2(\mathcal{M}, \mathbb{Z}) \cong H^2(A, \mathbb{Z}) \cong \mathbb{Z} \oplus \mathbb{Z}$. We also have a commutative diagram

$$\begin{array}{ccc} H^2(\mathcal{M}, \mathbb{Z}) & \stackrel{\alpha}{\to} & H^2(A, \mathbb{Z}) \\ \sigma \downarrow & & \downarrow \sigma' \\ H^2(F, \mathbb{Z}) & \stackrel{\beta}{\to} & H^2(f, \mathbb{Z}) \end{array}$$

where α is an isomorphism, F, f denote a general fiber of ψ and ψ_A respectively and σ' is onto since ψ_A is a $\mathbb{P}^1$-bundle. Since $\mathcal{L}_F \cong \mathcal{O}_Q(1)$ we see that the image of $H^2(\mathcal{M}, \mathbb{Z})$ in $H^2(F, \mathbb{Z})$ is $\mathbb{Z}[\mathcal{L}_F]$. Since the image of $\mathcal{L}_F$ in $H^2(f, \mathbb{Z})$ is $\mathcal{L}_f$ and $\mathcal{L}_f \cong \mathcal{O}_{\mathbb{P}^1}(2)$ we conclude that the image of $H^2(\mathcal{M}, \mathbb{Z})$ under the composition $\beta \circ \sigma$ is $2H^2(f, \mathbb{Z})$. Hence in particular $\beta \circ \sigma$ is not onto. This contradicts the fact that $\sigma' \circ \alpha$ is onto. Thus case 5) does not occur.

Case 6). Since $(\mathcal{M}, \mathcal{L})$ is a scroll, $\psi : \mathcal{M} \to Y$, over a surface Y, general results from adjunction theory apply to say that the restriction map $\psi_A : A \to Y$ is the reduction map for $(A, \mathcal{L}_A)$ (see [104, (0.6.2)] and (11.1.2)). As shown by Proposition (5.1.5), ψ_A is not an isomorphism. Since the only $\mathbb{P}^1$-bundle containing a smooth rational curve ℓ with $\ell^2 = -1$ is $\mathbb{F}_1$, we conclude that $A \cong \mathbb{F}_1$. Hence Y is rational and from $K_Y^2 \geq K_A^2 + 1 = 9$ we infer that $Y \cong \mathbb{P}^2$. Then in particular $\operatorname{Pic}(\mathcal{M}) \cong \operatorname{Pic}(A) \cong \mathbb{Z} \oplus \mathbb{Z}$. Let $H := \mathcal{O}_{\mathbb{P}^1}(1)$. Then $p^* H \approx \mathcal{H}_A$ where $\mathcal{H}_A$ denotes the restriction to A of some line bundle $\mathcal{H}$ on $\mathcal{M}$. Furthermore, for all $t \geq 1$,

$$H^1(p^*H - t\mathcal{L}_A) \cong H^1(\mathcal{H}_A - t\mathcal{L}_A) \cong H^1(K_A + (t\mathcal{L}_A - \mathcal{H}_A))$$

and $t\mathcal{L}_A - \mathcal{H}_A = (t-1)\mathcal{L}_A + (\mathcal{L}_A - \mathcal{H}_A)$ is nef and big. To see this, let $\mathcal{L}_A \sim aE + bf$ and note that $\mathcal{L}_A - \mathcal{H}_A \sim aE + (b-1)f$ with $b - 1 \geq a$ by the ampleness of $\mathcal{L}_A$. The condition $b - 1 \geq a$ implies that $\mathcal{L}_A - \mathcal{H}_A$ is nef and big and hence $t\mathcal{L}_A - \mathcal{H}_A$ is also. Thus the Kawamata-Viehweg vanishing theorem gives

$$h^1(p^*H - t\mathcal{L}_A) = 0 \ \text{ for } \ t \geq 1. \tag{7.13}$$

In view of (7.13) and recalling that $p^*H \approx \mathcal{H}_A$, the extension theorem, (5.2.1), applies to say that $p : A \to \mathbb{P}^1$ extends to a morphism $\psi' : \mathcal{M} \to \mathbb{P}^1$. Look at the exact sequence

$$0 \to \mathcal{O}_{\mathcal{M}} \to \mathcal{L} \to \mathcal{L}_{\mathbb{F}_1} \to 0.$$

Since any ample line bundle on $\mathbb{F}_1$ is in fact very ample and $h^1(\mathcal{O}_{\mathcal{M}}) = 0$, we conclude that $\mathcal{L}$ is spanned on $\mathcal{M}$. Therefore $\mathcal{E} := \psi_*\mathcal{L}$ is an ample and spanned rank 2 vector bundle on $\mathbb{P}^2$ and since $\psi_A : \mathbb{F}_1 \to \mathbb{P}^2$ is a blowing up at a single point we have that $c_2(\mathcal{E}) = 1$. To see this, note that sections of $\mathcal{L}$ correspond to sections of $\mathcal{E}$. Furthermore a general section of $\mathcal{E}$ vanishes on $c_2(\mathcal{E})$ points and a section of $\mathcal{E}$ vanishes at a point, $y \in \mathbb{P}^2$, if and only if the corresponding section of $\mathcal{L}$ vanishes on the fiber $\psi^{-1}(y)$. Therefore $c_2(\mathcal{E})$ coincides with the number of positive dimensional fibers of ψ_A. Thus by the main result of Lanteri and Sommese [389] (see also (11.1.3)) we conclude that $\mathcal{E} \cong \mathcal{O}_{\mathbb{P}^2}(1) \oplus \mathcal{O}_{\mathbb{P}^2}(1)$. Then $\mathcal{M} \cong \mathbb{P}^1 \times \mathbb{P}^2$, $\mathcal{L} \approx \mathcal{O}_{\mathbb{P}^1 \times \mathbb{P}^2}(1, 1)$ and, by construction, the extension $\psi' : \mathcal{M} \to \mathbb{P}^1$ is the projection $\mathcal{M} = \mathbb{P}^1 \times \mathbb{P}^2 \to \mathbb{P}^1$. Then we proved that $p : A \to C$ extends to a $\mathbb{P}^2$-bundle $\psi' : \mathcal{M} \to \mathbb{P}^1$ with $\mathcal{L}_{\mathbb{P}^2} \cong \mathcal{O}_{\mathbb{P}^2}(1)$.

Case 7). Let (M, L) be the first reduction of $(\mathcal{M}, \mathcal{L})$ and let $\pi : \mathcal{M} \to M$ be the reduction morphism. The image $A' := \pi(A)$ is a smooth element in $|L|$ and the restriction $\pi_A : A \to A'$ of π to A is the reduction morphism for $(A, \mathcal{L}_A)$. Since π is not an isomorphism, A contains a -1 curve at least. Therefore the same argument as in the case above shows that $A \cong \mathbb{F}_1$ and $A' \cong \mathbb{P}^2$. Since $\mathbb{P}^2$ is ample in M it thus follows from (5.4.10) that $M \cong \mathbb{P}^3$. Let ℓ' be a line in M, ℓ' not contained in $\mathbb{P}^2$, and passing through a point x' blown up under π. Let ℓ be the proper transform of ℓ'. Then we get

$$\mathcal{L} \cdot \ell \leq (\pi^*L - \pi^{-1}(x')) \cdot \ell = L \cdot \ell' - 1 = 0,$$

which contradicts the ampleness of $\mathcal{L}$.

Case 10). The general fiber F of $\psi : \mathcal{M} \to Y$ is $\mathbb{P}^2$ and $\mathcal{L}_F \cong \mathcal{O}_{\mathbb{P}^2}(a)$ (see case 4) of Theorem (7.3.4)). Let $\mathcal{H} := K_{\mathcal{M}} + 2\mathcal{L}$. Since $2K_{\mathcal{M}} + 3\mathcal{L} \approx \psi^*H$ for some ample line bundle H on Y, we see that $\mathcal{H}$ is ample. Note that $(\mathcal{M}, \mathcal{H})$ is a scroll under ψ since $K_{\mathcal{M}} + 3\mathcal{H} \approx \psi^*(2H)$ and $(A, \mathcal{H}_A)$ is a conic fibration under the restriction ψ_A since $K_A + \mathcal{H}_A \approx (2K_{\mathcal{M}} + 3\mathcal{L})_A \approx \psi_A^*(H)$. The same argument as in case 5) shows that the restriction $\psi_A : A \to Y$ is a $\mathbb{P}^1$-bundle and leads to a contradiction. □

Chapter 8

Background for classical adjunction theory

In this chapter we collect many miscellaneous results we need later on in the book to develop the classical adjunction theory, i.e., the study of the adjunction process for pairs, $(\mathcal{M}, \mathcal{L})$, with $\mathcal{L}$ a very ample or ample and spanned line bundle on a manifold, $\mathcal{M}$, (see also the introductions of the next two chapters).

Instead of listing the contents of the single sections, for which we refer the reader to the table of contents, we only note here that the first 4 sections and §8.6 mainly contain basic numerical results. One of those, Theorem (8.4.5), suggests the interesting and probably hard question to classify $\mathbb{P}^k$-degenerate pairs $(\mathcal{M}, \mathcal{L})$, with $\mathcal{M}$ an n-dimensional variety, $n \geq 3$, and $\mathcal{L}$ a very ample line bundle on $\mathcal{M}$, that is pairs $(\mathcal{M}, \mathcal{L})$ which contain a linear $\mathbb{P}^k$, in the sense that $\mathcal{L}^k \cdot \mathbb{P}^k = 1$, and such that $(\pi^*\mathcal{L} - \pi^{-1}(\mathbb{P}^k))^n = 0$, where $\pi : Z \to \mathcal{M}$ is the blowing up of $\mathcal{M}$ along the $\mathbb{P}^k$.

In §§8.7, 8.9, 8.10 we prove some elementary classification results for varieties with small invariants in the form we need in the sequel.

In §8.8 thespannedness of $K_{\mathcal{M}} + (\dim \mathcal{M})\mathcal{L}$ in presence of singularities is studied.

Finally in §8.5 we define and we give the first properties of k-very ampleness, a strong and natural notion of k-th order embedding. This concept is closely related to the study of the Hilbert scheme of 0-cycles. Here one main result is Theorem (8.5.1), the k-very ample version of the usual Reider's criterion. See also §10.7 for related results.

8.1 Numerical implications of nonnegative Kodaira dimension

The following is a generalization of Livorni and Sommese [415, (0.6)] of a result of Griffiths and Harris [292, (2.23)]. The result of Griffiths and Harris used smoothness and the assumption $h^0(K_X) > 0$, rather than just that X has nonnegative Kodaira dimension in the weak sense that $h^0(tK_X) > 0$ for some $t \geq 1$. We follow the argument of [415, (0.6)].

Theorem 8.1.1. *Let $\varphi : X \to \mathbb{P}^N$ be a finite generically one-to-one morphism where X is an n-dimensional irreducible normal projective variety of nonnegative Kodaira dimension in the sense that $h^0(tK_X) > 0$ for some $t \geq 1$. Assume that $\varphi(X)$ is nondegenerate in $\mathbb{P}^N$, i.e., $\varphi(X)$ is not contained in a hyperplane. Let d be the degree of $\varphi(X)$ in $\mathbb{P}^N$. Then $d \geq n(N-n)+2$, with equality only if $tK_X \approx \mathcal{O}_X$ for any $t > 0$ such that $h^0(tK_X) > 0$. In this case $\varphi(X)$ has codimension ≥ 2 singularities and $g(\varphi^*\mathcal{O}_{\mathbb{P}^N}(1))$ is the maximum allowed by Castelnuovo's bound for the genus of a nondegenerate irreducible and reduced degree d curve in $\mathbb{P}^{N-n+1}$.*

Proof. Let $\overline{C}$ be a smooth curve obtained as the pullback of the curve, C, gotten by transversal intersection of $\varphi(X)$ with a general $\mathbb{P}^{N-n+1} \subset \mathbb{P}^N$. Let $g(C)$ be the arithmetic genus of C. Note that $g(\overline{C}) \leq g(C)$, with equality only if C is smooth, and hence equal to $\overline{C}$.

Let $L := \varphi^*\mathcal{O}_{\mathbb{P}^N}(1)$ and let $L_{\overline{C}}$ be the restriction of L to $\overline{C}$. Then $d = \deg(L_{\overline{C}})$. Note that L is ample and spanned. By Lemma (1.7.6) we have the adjunction formula, $K_{\overline{C}} \approx (K_X + (n-1)L)_{\overline{C}}$. Therefore

$$2g(\overline{C}) - 2 = \deg(K_{X|\overline{C}}) + (n-1)d. \tag{8.1}$$

Note that, since $\kappa(X) \geq 0$, tK_X has a section, s, for some $t > 0$. Since L is spanned and big it follows that, for a general $\mathbb{P}^{N-n+1}$, $\overline{C}$ does meet the divisor where the section s vanishes (unless $s^{-1}(0)$ is empty) and hence

$$\deg(K_{X|\overline{C}}) = \frac{\deg(tK_{X|\overline{C}})}{t} \geq 0.$$

Thus from (8.1) we obtain

$$d \leq \frac{2g(\overline{C}) - 2}{n-1}. \tag{8.2}$$

By Castelnuovo's bound (1.3) we have

$$g(C) \leq \left[\frac{d-2}{N-n}\right]\left(d - N + n - 1 - \left(\left[\frac{d-2}{N-n}\right] - 1\right)\frac{N-n}{2}\right). \tag{8.3}$$

Assume that $d < n(N-n) + 2$. Then

$$\left[\frac{d-2}{N-n}\right] \leq n - 1.$$

Therefore by (8.3) and by

$$d - N + n - 1 = 2\left(\frac{d-2}{N-n} - 1\right)\frac{N-n}{2} + 1$$

it follows that

$$g(C) \leq (n-1)\left(1 + \frac{N-n}{2}\left(\frac{d-2}{N-n} - 1 + \frac{d-2}{N-n} - \left[\frac{d-2}{N-n}\right]\right)\right).$$

Since

$$0 \leq \frac{d-2}{N-n} - \left[\frac{d-2}{N-n}\right] \leq 1$$

we find

$$g(C) \leq (n-1)\left(1 + \frac{d-2}{2}\right) = \frac{(n-1)d}{2}.$$

By using (8.2) this leads to the contradiction $2g(C) \leq (n-1)d \leq 2g(\overline{C}) - 2$. Thus we conclude that $d \geq n(N-n) + 2$.

If we have equality then (8.3) becomes

$$g(C) \leq n\left(d - N + n - 1 - (n-1)\frac{N-n}{2}\right),$$

or

$$2g(C) \leq n(d - N + n). \tag{8.4}$$

Assume $tK_X \not\approx \mathcal{O}_X$ for all $t > 0$. We claim that in (8.2) the strict inequality holds. Indeed, assume otherwise. Then by ampleness of L either the section, s, of tK_X is nowhere vanishing or $\overline{C}$ would have to meet the divisor where s vanishes. By the generality of the $\mathbb{P}^{N-n+1}$ pulling back to $\overline{C}$ under φ, we can assume without loss of generality that if $\overline{C}$ meets $s^{-1}(0)$, then it meets $\mathrm{reg}(X) \cap s^{-1}(0)$. In the latter case we have a strict inequality in (8.2) and we are done. In the first case we have $tK_{\mathrm{reg}(X)} \approx \mathcal{O}_{\mathrm{reg}(X)}$ and hence, by the fact that K_X is reflexive on a normal variety X, we conclude that $tK_X \approx \mathcal{O}_X$, which contradicts our present assumption. Thus we conclude that if $tK_X \not\approx \mathcal{O}_X$ for all $t > 0$, then $d < (2g(\overline{C}) - 2)/(n-1)$. Combining this with (8.4) we find the contradiction

$$\begin{aligned} 2g(C) &\leq (n-1)d + d + n(n-N) \\ &< 2g(\overline{C}) - 2 + d + n(n-N) = 2g(\overline{C}). \end{aligned}$$

Thus we conclude that $d = n(N-n) + 2$ implies $tK_X \approx \mathcal{O}_X$ for the any $t > 0$ such that $h^0(tK_X) > 0$.

To see that $\varphi(X)$ has codimension ≥ 2 singularities it is enough to show that $g(C) = g(\overline{C})$. By combining (8.4) with the equality $d = n(N-n) + 2$ we get

$$2g(C) \leq (n-1)d + d + n(n-N) = 2g(\overline{C}).$$

Therefore $g(C) = g(\overline{C})$ since $g(C) \geq g(\overline{C})$. An easy calculation shows the assertion about the curve section of $\varphi(X)$ having the maximum genus allowed by Castelnuovo's bound. □

Remark 8.1.2. The condition that the curve section of $\varphi(X)$ has the maximum genus allowed by Castelnuovo's bound has some very strong consequences. See Corollary (8.6.2).

Corollary 8.1.3. *Let $\varphi : X \to \mathbb{P}^N$ be a finite generically one-to-one morphism where X is an n-dimensional irreducible normal projective variety with $L := \varphi^*\mathcal{O}_{\mathbb{P}^N}(1)$. Assume that $h^0(t(K_X + mL)) > 0$ for some positive integers t, $m < n$, or more generally that the intersection, M, of $m < n$ general elements of $|L|$ is of nonnegative Kodaira dimension in the sense that $h^0(tK_M) > 0$ for some integer $t > 0$. Assume that $\varphi(X)$ is nondegenerate in $\mathbb{P}^N$, i.e., $\varphi(X)$ is not contained in a hyperplane. Then $\deg \varphi(X) \geq (n-m)(N-n)+2$. In the case of equality $\varphi(X)$ has codimension ≥ 2 singularities.*

Proof. Let $\{H_i \in |L| \big| 1 \leq i \leq m\}$ be general elements, and let $M := H_1 \cap \ldots \cap H_m$. By Theorem (1.7.1) we can assume without loss of generality that M is normal. By Lemma (1.7.6) we conclude that $(K_X + mL)_M \cong K_M$. Therefore since the H_i are general, we conclude that $h^0(t(K_X+mL)) > 0$ implies $h^0(tK_M) > 0$. Applying Theorem (8.1.1) to the restriction map $\varphi_M : M \to \mathbb{P}^{N-m}$ and using the fact that $\deg \varphi_M(M) = \deg \varphi(X)$, we obtain the desired inequality. If we have equality, we conclude that the singularities of $\varphi(M)$ have codimension ≥ 2 in $\varphi(M)$ which implies that the singularities of $\varphi(X)$ have codimension ≥ 2 in X. □

Corollary 8.1.4. *Let $\varphi : X \to \mathbb{P}^N$ be a finite generically one-to-one morphism where X is an n-dimensional irreducible normal projective variety. Assume that $\varphi(X)$ is nondegenerate in $\mathbb{P}^N$. If* $\operatorname{Irr}(X)$ *is nonempty and* $\dim \operatorname{Irr}(X) = m$ *then* $\deg \varphi(X) \geq (n-m-1)(N-n)+2$. *In the case of equality $\varphi(X)$ has codimension ≥ 2 singularities.*

Proof. Choose m general elements, $H_1, \ldots, H_m$, of $|L|$ and let $M := H_1 \cap \ldots \cap H_m$. We will be done by the last corollary if we know that $h^0(K_M + L_M) > 0$, where L_M denotes the restriction of L to M. By Theorem (1.7.1) we can assume without loss of generality that M is normal and $\operatorname{Irr}(M)$ is nonempty of dimension 0. Therefore we are done by Corollary (2.2.8). □

8.2 The double point formula for surfaces

The following is a special case of the double point formula (13.1.5).

Lemma 8.2.1. *Let L be a very ample line bundle on a smooth connected projective surface, S. Let $d = L^2$. Assume that $\Gamma(L)$ embeds S in $\mathbb{P}^N$, $N \geq 4$. Then*

$$d(d-5) - 10(g(L)-1) + 12\chi(\mathcal{O}_S) \geq 2K_S^2$$

with equality if $N = 4$.

Proof. If $N = 4$, the result follows from the exact sequence of tangent bundles

$$0 \to \mathcal{T}_S \to i^*\mathcal{T}_{\mathbb{P}^4} \to \mathcal{N}_{S/\mathbb{P}^4} \to 0$$

where $i : S \hookrightarrow \mathbb{P}^4$, as in Hartshorne [318, p. 434].

If $N \geq 5$, look at the generic projection, S_0, of S into a $\mathbb{P}^4$. The surface S_0 has a finite number, #, of nodes as singular points. Let S' be the minimal desingularization of S_0. Then there exists a connected complex manifold $\mathcal{U}$ that contains S' as a submanifold and such that $S' \to \mathbb{P}^4$ extends to a holomorphic immersion $\theta : \mathcal{U} \to \mathbb{P}^4$, see, e.g., [569, Lemma 2.2]. Hence we have the exact sequence

$$0 \to \mathcal{T}_{S'} \to u^*\mathcal{T}_{\mathcal{U}} \to \mathcal{N}_{S'/\mathcal{U}} \to 0$$

where u denotes the inclusion of S' in $\mathcal{U}$ and $\mathcal{T}_{\mathcal{U}} \cong \theta^*\mathcal{T}_{\mathbb{P}^4}$. The same argument as in [318, p. 434] works, noting that $c_2(\mathcal{N}_{S'/\mathcal{U}}) = (S'^2)_{\mathcal{U}} = (S^2)_{\mathbb{P}^N} - 2\#$. □

8.3 Smooth double covers of irreducible quadric surfaces

Let S be a smooth connected surface and Q an irreducible quadric in $\mathbb{P}^3$. Let $p : S \to Q$ be a finite double cover of Q. The ramification set is the fixed point set of the $\mathbb{Z}_2$ involution of S obtained by interchanging sheets. Since the fixed point set of a finite (or more generally compact) group of automorphisms is smooth, we see that the ramification set of p is the union of smooth curves and a finite set, with the image of the finite set under p singular.

If Q is smooth, then p is branched along a smooth curve, B, belonging to $|\mathcal{O}_{\mathbb{P}^1 \times \mathbb{P}^1}(a, b)|$ for some even integers a, b.

Assume now that Q is a quadric cone in $\mathbb{P}^3$ with vertex x as its singular set. Then the branch locus of p is the union $\{x\} \cup B$ where B is a smooth curve on Q not containing the vertex. Furthermore there exists a commutative diagram

$$\begin{array}{ccc} S & \xrightarrow{p} & Q \\ \alpha \uparrow & & \uparrow \beta \\ S^\wedge & \xrightarrow{p^\wedge} & \mathbb{F}_2 \end{array}$$

where α is the blowing up of S at $p^{-1}(x)$ (note that $p^{-1}(x)$ is a single point since p is branched at x), β is the contraction of the unique irreducible curve, E, on $\mathbb{F}_2$ with $E^2 = -2$ to the vertex x of Q and $p^\wedge$ is a double cover branched along $E \cup \beta^{-1}(B)$.

Let $\beta^{-1}(B) = aE + bf$ in $\mathbb{F}_2$, f fiber of the ruling. Since $x \notin B$ we have $\beta^{-1}(B) \cdot E = 0$ and therefore $b = 2a$. Thus the branch locus of $p^\wedge$ is the union $E \cup a(E + 2f)$ and hence it is a curve, $B^\wedge$, in the linear system $|(a+1)E + 2af|$. In particular $a + 1$ is even.

Lemma 8.3.1. *Notations as above. Let S be a smooth connected surface, Q an irreducible quadric in $\mathbb{P}^3$. Let $p : S \to Q$ be a finite double cover of Q.*

1. *Assume $Q \cong \mathbb{P}^1 \times \mathbb{P}^1$ and let $B \in |\mathcal{O}_{\mathbb{P}^1 \times \mathbb{P}^1}(a, b)|$ be the branch locus of p, with a and b even integers. Then $K_S^2 = (a - 4)(b - 4)$.*

2. *Assume Q is a cone in $\mathbb{P}^3$ and let $p^\wedge : S^\wedge \to \mathbb{F}_2$ be the finite double cover associated to p. Let $B^\wedge \in |(a+1)E + 2af|$ be the branch locus of $p^\wedge$, with $a + 1$ an even integer. Then $K_S^2 = (a - 4)^2$.*

Proof. Let $Q \cong \mathbb{P}^1 \times \mathbb{P}^1$. We have

$$K_S \approx p^*\left(K_Q + \frac{B}{2}\right) \approx p^*\mathcal{O}_{\mathbb{P}^1 \times \mathbb{P}^1}\left(\frac{a}{2} - 2, \frac{b}{2} - 2\right).$$

Therefore

$$K_S^2 = 4\left(\frac{a}{2} - 2\right)\left(\frac{b}{2} - 2\right) = (a - 4)(b - 4).$$

Let Q be a quadric cone in $\mathbb{P}^3$ with a single point, x, as its singular set. We have $K_{\mathbb{F}_2} \approx -2E - 4f$ and

$$K_{S^\wedge} \approx p^{\wedge *}\left(K_{\mathbb{F}_2} + \frac{B^\wedge}{2}\right) \approx p^{\wedge *}\left(\frac{a-3}{2}E + (a-4)f\right).$$

Hence

$$\begin{aligned} K_{S^\wedge}^2 &= 2\left(\frac{a-3}{2}E + (a-4)f\right)^2 = -(a-3)^2 + 2(a-3)(a-4) \\ &= (a-3)(a-5). \end{aligned}$$

Since $K_{S^\wedge}^2 = K_S^2 - 1$ we find $K_S^2 = (a-3)(a-5) + 1 = (a-4)^2$. □

8.4 Surfaces with one dimensional projection from a line

The Theorem (8.4.2) below slightly improves the second author's result [574, (4.2)]. To prove it we need a special case of the following general lemma due to Matsumura [428, (2.2), p. 38] (see also Sommese [571, (0.7)]).

Lemma 8.4.1. *Let D be a smooth divisor on a connected complex manifold, X, of dimension n. Let*

$$0 \to K_X \to K_X \otimes \mathcal{O}_X(D) \stackrel{\mathcal{R}_D}{\to} K_D \to 0$$

be the usual exact sequence, where $\mathcal{R}_D$ denotes the residue mapping. Then the diagram

$$\begin{array}{ccc} H^p(K_D) & \stackrel{\delta}{\longrightarrow} & H^{p+1}(K_X) \\ & \nwarrow_{r_D} & \uparrow \theta_D \\ & & H^p(\wedge^{n-1}\mathcal{T}_X^*) \end{array}$$

commutes, where δ is the connecting homomorphism of the long exact cohomology sequence associated to the sequence above, r_D is the restriction map, and θ_D is obtained by wedging with the first Chern class of D.

Furthermore if D is ample and X is projective, then θ_D is an isomorphism for $p = 0$. Thus in this case $H^0(K_D)$ splits as direct sum

$$H^0(K_D) \cong \mathrm{Im}(r_D) \oplus \mathrm{Im}(\mathcal{R}_D).$$

Proof. We refer to Matsumura's article quoted above for a Čech cohomology proof of the first part of the statement, or to [571, (0.7)] for a differential geometric proof of the statement.

Note that if D is ample, X is projective, and $p = 0$, then θ_D is an isomorphism by the hard Lefschetz theorem (see Griffiths and Harris [293, p. 122]). Then we have a morphism $\eta := r_D \circ \theta_D^{-1} : H^1(K_X) \to H^0(K_D)$ compatible with the diagram above and which gives the desired splitting. □

Theorem 8.4.2. *Let S be a smooth projective surface and let C be a smooth curve which is an ample divisor in S. Let $L := \mathcal{O}_S(C)$ and let $N = h^0(L) - 1$. Assume further that there is an integer $k > 0$ with a point x so that any holomorphic 1 form on C that vanishes on x to order k, vanishes to order $k+1$. Then there exists a locally closed algebraic subset, $\mathcal{S} \subset |L|$, of curves with this property and such that* $\dim \mathcal{S} \geq N - k - 2$.

Furthermore if $q(S) = 0$ then there exists $\mathcal{S}$ as above with $\dim \mathcal{S} \geq N - k - 1$.

Proof. Let $\mathcal{P}$ be the set of all curves $C' \in |L|$ with C' tangent to C at x to the $(k+1)$-st order. Then $\mathcal{P}$ is a linear $\mathbb{P}^r \subset |L|$ with $r \geq N - k - 2$ and since $C \in \mathcal{P}$ is smooth there exists by Bertini's theorem, (1.7.9), a nonempty Zariski open set, $\mathcal{S}$, of $\mathcal{P}$ such that the curves $C' \in \mathcal{S}$ are smooth.

We can choose a trivialization e of L, i.e., a local section, e, of L not vanishing at x, and local coordinates (z, w) in some neighborhood U of x so that the defining equations of C and C' are given on U by

$$h(z, w) = we \quad \text{and} \quad h'(z, w) = (w - z^{k+2} g(z))e$$

respectively for some holomorphic function $g(z)$. Let t_1, t_2 be local coordinates of $U \cap C$ and $U \cap C'$ respectively that are zero at x. Then under the embeddings $U \cap C \hookrightarrow S$ and $U \cap C' \hookrightarrow S$, we have the maps $t_1 \mapsto (t_1, 0)$ and $t_2 \mapsto (t_2, t_2^{k+2} g(t_2))$ respectively. Let $f(z, w) dz \wedge dw \otimes e$ be a section of $K_S + L$ in U. Therefore the image of this section under the residue maps, $\mathcal{R}_C$, $\mathcal{R}_{C'}$, associated to the above explicit defining equations $h(z, w)$, $h'(z, w)$ for C, C' are

$$f(z, w) dz \wedge dw \otimes e \mapsto f(t_1, 0) dt_1 \quad \text{on } C \tag{8.5}$$

and

$$f(z, w) dz \wedge dw \otimes e \mapsto f(t_2, t_2^{k+2} g(t_2)) dt_2 \quad \text{on } C'. \tag{8.6}$$

Furthermore an element

$$a(z, w) dz + b(z, w) dw \in \Gamma(U, \mathcal{T}^*_{S|U})$$

goes under restriction to $a(t_1, 0) dt_1$ on C and to

$$a(t_2, t_2^{k+2} g(t_2)) dt_2 + b(t_2, t_2^{k+2} g(t_2))((k+2) t_2^{k+1} g(t_2) + t_2^{k+2} g'(t_2)) dt_2$$

on C'. Up to terms of order $k+1$ or greater the above becomes

$$a(t_1, 0) dt_1 \quad \text{on} \quad C \quad \text{and} \quad a(t_2, 0) dt_2 + \mathrm{O}(t_2^{k+1}) dt_2 \quad \text{on} \quad C'. \tag{8.7}$$

Let now $\omega \in \Gamma(K_{C'})$ vanish to the k-th order. By Lemma (8.4.1) we have a decomposition $\omega = \mathcal{R}_{C'}(\omega_1) + r_{C'}(\omega_2)$ with $\omega_1 \in \Gamma(K_S + L)$ and $\omega_2 \in \Gamma(\mathcal{T}^*_S)$. By (8.5), (8.6) we see that $\mathcal{R}_C(\omega_1)$ and $\mathcal{R}_{C'}(\omega_1)$ agree up to terms of order $k+1$. By (8.7) we see that $r_C(\omega_2)$ and $r_{C'}(\omega_2)$ agree up to terms of order k. Then $\mathcal{R}_C(\omega_1) + r_C(\omega_2)$ and $\mathcal{R}_{C'}(\omega_1) + r_{C'}(\omega_2)$ agree up to terms of order k. Thus, by the assumption on ω, we conclude that $\mathcal{R}_C(\omega_1) + r_C(\omega_2)$ vanishes to the k-th order and hence by hypothesis to the $(k+1)$-st order. By using again (8.5), (8.6) and (8.7) we conclude that $\mathcal{R}_{C'}(\omega_1) + r_{C'}(\omega_2)$ vanishes at x to the $(k+1)$-st order. This proves the first part of the statement.

Assume now $q(S) = 0$. Let $\mathcal{S}$ be the set of all smooth $C' \in |L|$ with C' tangent to C at x to the k-th order. Then $\dim \mathcal{S} \geq N - k - 1$. Under the present assumption $H^0(\mathcal{T}^*_S) \cong H^1(K_S) = (0)$, relation (8.6) holds true with $k+1$ instead of $k+2$ and (8.7) is trivial. Let $\omega \in \Gamma(K_{C'})$ vanish on x to the k-th order. By Lemma (8.4.1), $\omega = \mathcal{R}_{C'}(\omega_1)$ with $\omega_1 \in \Gamma(K_S + L)$. By (8.5) and (8.6) we see that $\mathcal{R}_C(\omega_1)$ and $\mathcal{R}_{C'}(\omega_1)$ agree up to terms of order k. Thus by the assumption on ω, we conclude that $\mathcal{R}_C(\omega_1)$ vanishes to the k-th order and hence by hypothesis to the $(k+1)$-st order. By using again (8.5), (8.6) we conclude that $\omega = \mathcal{R}_{C'}(\omega_1)$ vanishes on x to the $(k+1)$-st order. □

Remark 8.4.3. Note that with the assumptions and notation as in (8.4.2) one has $h^0((k+1)x) \geq 2$. To see this, look at the long exact cohomology sequence associated to the exact sequence

$$0 \to K_C \otimes \mathfrak{m}_x^{k+1} \to K_C \otimes \mathfrak{m}_x^k \to \mathbb{C}_x \to 0,$$

where $\mathfrak{m}_x$ denotes the ideal sheaf of x in C. The assumption on C implies that

$$h^0(K_C - (k+1)x) = h^0(K_C - kx)$$

and therefore

$$h^1(K_C - (k+1)x) = 1 + h^1(K_C - kx),$$

or, by Serre duality,

$$h^0((k+1)x) = 1 + h^0(kx) \geq 2.$$

Thus there exists a rational function on C of degree $d \leq k+1$ with a pole at x of order d. This gives a morphism, $C \to \mathbb{P}^1$, of degree d.

Conversely, the same argument as above shows that if $h^0((k+1)x) \geq 2$, then there exists a $k' \leq k$ such that if $\omega \in h^0(K_C)$ vanishes on x to order k' it vanishes to order $k'+1$.

In particular, note that a smooth curve C of genus $g \geq 2$ is hyperelliptic if and only if C satisfies the condition of (8.4.2) with $k = 1$.

Note also that the bound given in (8.4.2) is sharp. This follows from the Geiser involution example described in (10.2.4).

Let S be a smooth connected surface and L a very ample line bundle on S. We say that an irreducible curve, ℓ, is a *line*, relative to L, if $L \cdot \ell = 1$. Then ℓ is a smooth $\mathbb{P}^1$. In this section we prove a few general facts on surfaces S, containing a line, we need in the sequel (see §10.6).

Look at the embedding $S \hookrightarrow \mathbb{P}^N$ given by $\Gamma(L)$, $N = h^0(L) - 1$. By Lemma (6.6.1) we have that $\Gamma(L - \ell)$ spans $L - \ell$ on S. Consider the exact sequence

$$0 \to L - \ell \to L \to L_\ell \cong \mathcal{O}_{\mathbb{P}^1}(1) \to 0.$$

Since L is spanned, the image of $\Gamma(L)$ in $\Gamma(L_\ell)$ spans $\Gamma(L_\ell)$. Since L_ℓ is spanned exactly by $2 = h^0(L_\ell)$ sections we see that $\Gamma(L)$ surjects on $\Gamma(L_\ell)$. Therefore

$$h^0(L - \ell) = h^0(L) - 2. \tag{8.8}$$

Then we have a commutative diagram

$$\begin{array}{ccc} S & \hookrightarrow & \mathbb{P}^N \\ \pi \downarrow & & \downarrow \rho \\ C & \overset{i}{\hookrightarrow} & \mathbb{P}^{N-2} \end{array}$$

where ρ is the projection of $\mathbb{P}^N$ to $\mathbb{P}^{N-2}$ from ℓ and π, the restriction of ρ to S, is the morphism defined by $\Gamma(L - \ell)$. From now on we assume that $L - \ell$ is not big, that is

$$(L - \ell)^2 = 0. \tag{8.9}$$

Here we make the further blanket assumption that $(S, L) \not\cong (\mathbb{P}^2, \mathcal{O}_{\mathbb{P}^2}(1))$. Note that the image of S under ρ is a curve, C, because of the assumptions (8.9) and $(S, L) \not\cong (\mathbb{P}^2, \mathcal{O}_{\mathbb{P}^2}(1))$.

Note that $L \not\cong \mathcal{O}_S(\ell)$. Otherwise we would have $L^2 = 1$, which implies that $(S, L) \cong (\mathbb{P}^2, \mathcal{O}_{\mathbb{P}^2}(1))$.

Claim 8.4.4. Notation as above. The image $C := \pi(S)$ is a smooth $\mathbb{P}^1$ embedded in $\mathbb{P}^{N-2}$ as a curve of degree $N-2$ (i.e., $i^*\mathcal{O}_{\mathbb{P}^{N-2}}(1) \cong \mathcal{O}_{\mathbb{P}^1}(N-2)$).

Proof. Note that $(L-\ell)\cdot\ell > 0$. Otherwise $\ell^2 = L\cdot\ell = 1$, so that by the Hodge index theorem either $(L-\ell)^2 < 0$, which is not the case since $L-\ell$ is nef, or $L \sim \ell$. In this case we would have $L^2 = L\cdot\ell = 1$, implying $(S, L) \cong (\mathbb{P}^2, \mathcal{O}_{\mathbb{P}^2}(1))$. Thus $\pi(\ell)$ is not a point, hence $\pi(\ell) = C$, so $C \cong \mathbb{P}^1$.

Let $i : C \hookrightarrow \mathbb{P}^{N-2}$ be the embedding of C in $\mathbb{P}^{N-2}$ and let $i^*\mathcal{O}_{\mathbb{P}^{N-2}}(1) \cong \mathcal{O}_{\mathbb{P}^1}(t)$ for some t. Then $L-\ell \cong \pi^*\mathcal{O}_{\mathbb{P}^1}(t)$ and therefore $h^0(L-\ell) = t+1$. From equation (8.8) we find $t = N-2$. □

Let Y be the cone which projects $C \cong \mathbb{P}^1$ from ℓ in $\mathbb{P}^N$. Then S is a divisor on Y. Let $\sigma : \mathcal{X} \to \mathbb{P}^N$ be the blowing up of $\mathbb{P}^N$ along ℓ, let X, $S^\wedge$ be the proper transforms of Y, S under σ, respectively. Then $S^\wedge$ is a Cartier divisor on X, $p : \mathcal{X} \to \mathbb{P}^{N-2}$ is a $\mathbb{P}^2$-bundle, where p is the composition $p = \sigma \circ \rho$, $X = p^{-1}(C)$ and the restriction, $\varphi : X \to C$, of p to X is a $\mathbb{P}^2$-bundle over $C \cong \mathbb{P}^1$. Let $\pi^\wedge : S^\wedge \to C$ be the restriction of φ to $S^\wedge$. Note also that $S^\wedge \cong S$ under σ. We have a commutative diagram

$$\begin{array}{ccccccccc} S^\wedge & \hookrightarrow & X & \hookrightarrow & \mathcal{X} & \xrightarrow{\sigma} & \mathbb{P}^N & \hookleftarrow & S \\ & \pi^\wedge \searrow & \varphi\downarrow & & \downarrow p & & \downarrow \rho & & \downarrow \pi \\ & & C & \hookrightarrow & \mathbb{P}^{N-2} & \cong & \mathbb{P}^{N-2} & \hookleftarrow & C \end{array} \tag{8.10}$$

We can prove now the main result of this section.

Theorem 8.4.5. *Let S be a smooth connected surface and L a very ample line bundle on S such that $(S, L) \not\cong (\mathbb{P}^2, \mathcal{O}_{\mathbb{P}^2}(1))$. Let $h^0(L) = N+1$. Assume that there exists on S a line, ℓ, relative to L such that $(L-\ell)^2 = 0$. Then, up to isomorphism, S is embedded as a Cartier divisor in*

$$X := \mathbb{P}(\mathcal{O}_{\mathbb{P}^1} \oplus \mathcal{O}_{\mathbb{P}^1} \oplus \mathcal{O}_{\mathbb{P}^1}(N-2)).$$

Furthermore, let ξ be the tautological bundle of X and let $F \cong \mathbb{P}^2$ be a fiber of $\varphi : X \to \mathbb{P}^1$ as in (8.10). Then $S = a\xi + F$ in Pic(X), *for some positive integer a, and $L \approx \xi_S$, the restriction of ξ to S.*

Proof. Let us go back to the diagram (8.10). Recalling that $\mathcal{X}$ is the blowing up of $\mathbb{P}^N$ along $\ell \cong \mathbb{P}^1$ we see that $p : \mathcal{X} \to \mathbb{P}^{N-2}$ is the compactification as a $\mathbb{P}^2$-bundle of

$$\mathbb{P}^N - \mathbb{P}^1 \cong \mathcal{O}_{\mathbb{P}^{N-2}}(1) \oplus \mathcal{O}_{\mathbb{P}^{N-2}}(1).$$

Let $V := \mathcal{O}_{\mathbb{P}^{N-2}}(1) \oplus \mathcal{O}_{\mathbb{P}^{N-2}}(1)$. Then

$$\begin{aligned} \mathcal{X} \cong \mathbb{P}(\mathcal{O}_{\mathbb{P}^{N-2}} \oplus V^*) &= \mathbb{P}(\mathcal{O}_{\mathbb{P}^{N-2}}(-1) \oplus \mathcal{O}_{\mathbb{P}^{N-2}}(-1) \oplus \mathcal{O}_{\mathbb{P}^{N-2}}) \\ &\cong \mathbb{P}(\mathcal{O}_{\mathbb{P}^{N-2}} \oplus \mathcal{O}_{\mathbb{P}^{N-2}} \oplus \mathcal{O}_{\mathbb{P}^{N-2}}(1)). \end{aligned}$$

Since X is the restriction of $p : \mathcal{X} \to \mathbb{P}^{N-2}$ to $C \hookrightarrow \mathbb{P}^{N-2}$ and $C \cong \mathbb{P}^1$ is embedded in $\mathbb{P}^{N-2}$ as a degree $N-2$ curve (see Claim (8.4.4)) we have

$$X = \mathbb{P}(\mathcal{O}_{\mathbb{P}^1}(2-N) \oplus \mathcal{O}_{\mathbb{P}^1}(2-N) \oplus \mathcal{O}_{\mathbb{P}^1}) \cong \mathbb{P}(\mathcal{O}_{\mathbb{P}^1} \oplus \mathcal{O}_{\mathbb{P}^1} \oplus \mathcal{O}_{\mathbb{P}^1}(N-2)). \quad (8.11)$$

Let

$$\mathcal{E} := \mathcal{O}_{\mathbb{P}^1} \oplus \mathcal{O}_{\mathbb{P}^1} \oplus \mathcal{O}_{\mathbb{P}^1}(N-2).$$

The general relation $\xi^3 = (\varphi^* c_1(\mathcal{E})) \cdot \xi^2$ gives

$$\xi^3 = (\varphi^* \mathcal{O}_{\mathbb{P}^1}(N-2)) \cdot \xi^2 = (N-2)F \cdot \xi^2.$$

Since $F \cdot \xi^2 = 1$, we find

$$\xi^3 = N-2. \quad (8.12)$$

Let $\alpha : X \to \mathbb{P}^N$ be the composition of the blowing up $\sigma : \mathcal{X} \to \mathbb{P}^N$ with the inclusion $X \hookrightarrow \mathcal{X}$ and let $\beta : S^\wedge \to \mathbb{P}^N$ be the restriction of α to $S^\wedge$. Then $\beta(S^\wedge) \cong S$.

Let $\mathcal{L} := \alpha^* \mathcal{O}_{\mathbb{P}^N}(1) = a\xi + bF$ for some integers a, b. Note that $\mathcal{L}$ is spanned, so that $a \geq 0, b \geq 0$ by (3.2.4.1), but it is not ample since α has positive dimensional fibers. We also have $\mathcal{L}^2 \cdot F = 1$. By noting that $F^3 = \xi \cdot F^2 = 0$ we find $1 = \mathcal{L}^2 \cdot F = a^2$ whence $a = 1$. Therefore $b = 0$, since otherwise $\mathcal{L}$ would be ample by (3.2.4.2). Thus $\mathcal{L} := \alpha^* \mathcal{O}_{\mathbb{P}^N}(1) = \xi$. Therefore

$$\xi_{S^\wedge} \cong \beta^*(\mathcal{O}_{\mathbb{P}^N}(1)_{|S}) \cong \beta^* L.$$

Note that the image of X under $\alpha : X \to \mathbb{P}^N$ is the cone, Y, which projects the curve $C \cong \mathbb{P}^1$ from ℓ in $\mathbb{P}^N$. Also, the restriction, α, of $\sigma : \mathcal{X} \to \mathbb{P}^N$ to X is the blowing up of Y along ℓ. Let E be the exceptional divisor of $\alpha : X \to Y = \alpha(X)$. From (8.11) we see that $E = \mathbb{P}(W)$ where $W = \mathcal{O}_{\mathbb{P}^1}(2-N) \oplus \mathcal{O}_{\mathbb{P}^1}(2-N)$. Hence

$$E \cong \mathbb{P}(\mathcal{O}_{\mathbb{P}^1} \oplus \mathcal{O}_{\mathbb{P}^1}) \cong \mathbb{P}^1 \times \mathbb{P}^1.$$

Write $E = a\xi + bF$ in $\mathrm{Pic}(X)$. Since $E \cap F = \mathbb{P}^1$ we see that $\xi \cdot E \cdot F = 1$. Now,

$$1 = \xi \cdot E \cdot F = a\xi^2 \cdot F + b\xi \cdot F^2 = a$$

so that $E = \xi + bF$. Since $|\xi|$ collapses E to ℓ we have $\xi^2 \cdot E = 0$. Recall equation (8.12) and compute

$$0 = \xi^2 \cdot E = \xi^3 + b\xi^2 \cdot F = N - 2 + b.$$

Thus $E = \xi + (2 - N)F$. Let $S^\wedge = a\xi + bF$ in $\mathrm{Pic}(X)$. Note that $E \cap S^\wedge \cong \mathbb{P}^1$ since $\beta(E \cap S^\wedge) = \ell \cong \mathbb{P}^1$ and $S^\wedge$, S are isomorphic under β. Hence we have

$$S^\wedge \cdot E \cdot \xi = \xi_{S^\wedge} \cdot E_{S^\wedge} = \beta^* L \cdot E_{S^\wedge} = L \cdot \ell = 1.$$

Then compute

$$\begin{aligned} 1 = S^\wedge \cdot E \cdot \xi &= (a\xi + bF) \cdot (\xi + (2 - N)F) \cdot \xi = \\ &= a\xi^3 + (b + (2 - N)a)\xi^2 \cdot F = a(N - 2) + b + a(2 - N) = b. \end{aligned}$$

Thus $S = a\xi + F$ in $\mathrm{Pic}(X)$. Note also that $a > 0$ since $a = S \cdot \xi \cdot F > 0$. □

Corollary 8.4.6. *Let S be a smooth connected surface and L a very ample line bundle on S. Assume that there exists on S a line, ℓ, relative to L such that $(L - \ell)^2 = 0$. Then $K_S + L$ is very ample unless either*

1. $(S, L) \cong (\mathbb{P}^2, \mathcal{O}_{\mathbb{P}^2}(1))$ *or* $(S, L) \cong (Q, \mathcal{O}_Q(1))$, *$Q$ a quadric in $\mathbb{P}^3$; or*
2. *(S, L) is a scroll over $\mathbb{P}^1$; or*
3. *(S, L) is a conic fibration over $\mathbb{P}^1$.*

Proof. Let $h^0(L) = N + 1$. We can assume $N \geq 4$. Otherwise S is embedded by $\Gamma(L)$ in $\mathbb{P}^3$ as a surface of degree $d = L^2$. Then either $K_S + L$ is very ample or $d = 1, 2$ and we are as in (8.4.6.1).

From Theorem (8.4.5) we know that, up to isomorphism, S is contained as a Cartier divisor in $X = \mathbb{P}(\mathcal{E})$, with

$$\mathcal{E} = \mathcal{O}_{\mathbb{P}^1} \oplus \mathcal{O}_{\mathbb{P}^1} \oplus \mathcal{O}_{\mathbb{P}^1}(N - 2).$$

Let $\varphi : X \to \mathbb{P}^1$ be the bundle projection and let ξ, F denote the tautological bundle of X and a fiber of φ respectively. Let $\pi : S \to \mathbb{P}^1$ be the restriction of φ to S. The canonical bundle formula yields

$$K_X + 3\xi \cong \varphi^*(K_{\mathbb{P}^1} + \det \mathcal{E}) \cong \varphi^* \mathcal{O}_{\mathbb{P}^1}(N - 4),$$

i.e., $K_X \approx (N - 4)F - 3\xi$. By (8.4.5) we also know that $S = a\xi + F$ in $\mathrm{Pic}(X)$, for some positive integer a and $L = \xi_S$. Therefore by the adjunction formula

$$K_S \approx (K_X + S)_S \approx ((a - 3)\xi + (N - 3)F)_S.$$

Let $a = 1$. Then $K_S + 2L \approx (N-3)F_S \approx \pi^*\mathcal{O}_{\mathbb{P}^1}(N-3)$, so that (S, L) is a scroll over $\mathbb{P}^1$ as in (8.4.6.2).

Let $a = 2$. Then $K_S + L \approx (N-3)F_S \approx \pi^*\mathcal{O}_{\mathbb{P}^1}(N-3)$, so that (S, L) is a conic fibration over $\mathbb{P}^1$ as in (8.4.6.3).

Thus we can assume $a \geq 3$. We have $K_S + L \approx ((a-2)\xi + (N-3)F)_S$ and hence K_S+L is ample and spanned, by Lemma (3.2.4), since $a-2 \geq 1, N-3 \geq 1$.

Let us show that $\lambda\xi + \mu F$ is in fact very ample on X as soon as $\lambda \geq 1$, $\mu \geq 1$, so we are done. Use the notation as in diagram (8.10). Let $\alpha : X \to \mathbb{P}^N$ be the composition of $\sigma : \mathcal{X} \to \mathbb{P}^N$ with the inclusion of X in $\mathcal{X}$. Recall that $\alpha^*\mathcal{O}_{\mathbb{P}^N}(1) \cong \xi$ (see the proof of (8.11)). Consider the composition map $X \xrightarrow{\theta} \mathbb{P}^N \times \mathbb{P}^1 \xrightarrow{\psi} \mathbb{P}^r$ where $\theta = (\alpha, \varphi)$ is the product map and ψ is the morphism associated to the line bundle $\mathcal{M} := q_1^*\mathcal{O}_{\mathbb{P}^N}(\lambda) \otimes q_2^*\mathcal{O}_{\mathbb{P}^1}(\mu)$, where $q_1 : \mathbb{P}^N \times \mathbb{P}^1 \to \mathbb{P}^N$, $q_2 : \mathbb{P}^N \times \mathbb{P}^1 \to \mathbb{P}^1$ are the projections on the two factors. Note that ψ is an embedding (the Segre embedding) since $\mathcal{M}$ is very ample. Note also that the composition $\theta \circ \psi$ is the morphism associated to $\lambda\xi + \mu F$. Indeed,

$$\begin{aligned}(\theta \circ \psi)^*\mathcal{O}_{\mathbb{P}^r}(1) &= \theta^*(q_1^*\mathcal{O}_{\mathbb{P}^N}(\lambda) \otimes q_2^*\mathcal{O}_{\mathbb{P}^1}(\mu)) = \\ &= \alpha^*\mathcal{O}_{\mathbb{P}^N}(\lambda) \otimes \varphi^*\mathcal{O}_{\mathbb{P}^1}(\mu) = \lambda\xi + \mu F.\end{aligned}$$

Therefore to show that $\lambda\xi + \mu F$ is very ample is equivalent to show that θ is an embedding.

Let x, y be two distinct points of X and assume $\theta(x) = \theta(y)$, i.e., $\alpha(x) = \alpha(y)$ and $\varphi(x) = \varphi(y)$. Since $\varphi(x) = \varphi(y)$ we have that x, y belong to the same fiber, $F = \mathbb{P}^2$, of φ. Then $\alpha(x) = \alpha(y)$ implies $x = y$ since α is an isomorphism on $F = \mathbb{P}^2$.

Let x be a point of X and *le* τ_x be a nonzero element of $\mathcal{T}_{X,x}$. If $d\theta(\tau_x) = (d\alpha(\tau_x), d\varphi(\tau_x)) = 0$, we have $d\varphi(\tau_x) = 0$, this implying that τ_x is tangent to $\varphi^{-1}(\varphi(x)) = \mathbb{P}^2$ at x, i.e., $\tau_x \in \mathcal{T}_{\mathbb{P}^2,x}$. Therefore $d\alpha(\tau_x) \neq 0$ since α is an isomorphism on $\varphi^{-1}(\varphi(x)) = \mathbb{P}^2$. This contradicts $d\theta(\tau_x) = 0$. □

8.5 k-very ampleness

Let X be an algebraic variety defined over the complex field $\mathbb{C}$. We denote the Hilbert scheme of 0-dimensional subschemes $(\mathfrak{X}, \mathcal{O}_{\mathfrak{X}})$ of X with $\text{length}(\mathcal{O}_{\mathfrak{X}}) = r$ by $X^{[r]}$. Since we are working in characteristic zero, $\text{length}(\mathcal{O}_{\mathfrak{X}}) = h^0(\mathcal{O}_{\mathfrak{X}})$.

We say that a line bundle L on X is *k-very ample* if the restriction map $\Gamma(L) \to \Gamma(\mathcal{O}_{\mathfrak{X}}(L))$ is onto for any $\mathfrak{X} \in X^{[k+1]}$. Note that L is 0-very ample if and only if L is spanned by its global sections and L is 1-very ample if and only if L is very ample. Note also that for smooth surfaces with $k \leq 2$, L being k-very ample is equivalent to L being *k-spanned* in the sense of Beltrametti, Francia, and Sommese

[88], i.e., $\Gamma(L)$ surjects on $\Gamma(\mathcal{O}_{\mathfrak{X}}(L))$ for any *curvilinear* 0-cycle $\mathfrak{X} \in X^{[k+1]}$, i.e., any 0-dimensional subscheme, $\mathfrak{X} \subset X$, such that length$(\mathcal{O}_{\mathfrak{X}}) = k+1$ and $\mathfrak{X} \subset C$ for some smooth curve, C, on X ([88, see (0.4), (3.1)]).

We refer to [88, 102, 103, 152, 108, 191, 431, 476] for further results on k-very ampleness.

The following theorem of ours [103, (2.1)], see also Kotschick [361], generalizes the well-known Reider criterion [522] for the adjoint bundle to be spanned (case $k = 0$) or very ample (case $k = 1$).

Theorem 8.5.1. *Let L be a nef line bundle on a smooth surface S. Let $L^2 \geq 4k+5$. Then either $K_S + L$ is k-very ample or there exists an effective divisor, D, such that $L - 2D$ is $\mathbb{Q}$-effective, D contains some 0-dimensional subscheme $(\mathfrak{X}, \mathcal{O}_{\mathfrak{X}})$ of S where the k-very ampleness fails and*

$$L \cdot D - k - 1 \leq D^2 < L \cdot D/2 < k+1.$$

The use of this result is illustrated by the following result from Sommese and Van de Ven [586, Prop. (0.5.1)] (cases $k = 0, 1$) we need in the sequel (compare also with Beltrametti, Francia, and Sommese [88, (2.6)], for a general k).

Proposition 8.5.2. *Let S be a smooth, connected, projective surface with $-K_S$ ample. Then for any positive integer t with $t + K_S^2 \geq 3$, the line bundle $-tK_S$ is spanned by global sections, and for any positive integer t with $t + K_S^2 \geq 4$, the line bundle $-tK_S$ is very ample.*

Proof. Note that $t + K_S^2 \geq 3$ is equivalent to the condition $(t+1)^2 K_S^2 \geq 5$ and $t + K_S^2 \geq 4$ is equivalent to the condition $(t+1)^2 K_S^2 \geq 10$.

Let $L := -(t+1)K_S$. Then $K_S + L = -tK_S$ is spanned unless there exists an effective divisor D such that

$$L \cdot D - 1 \leq D^2 < L \cdot D/2 < 1.$$

Since $L \cdot D > 0$, this gives the contradiction

$$1 \geq L \cdot D = -(t+1)K_S \cdot D \geq t+1 \geq 2.$$

Similarly, $K_S + L = -tK_S$ is very ample unless there exists an effective divisor D such that

$$L \cdot D - 2 \leq D^2 < L \cdot D/2 < 2.$$

Then either $L \cdot D = 3$, $D^2 = 1$ or $L \cdot D = 2$, $D^2 = 0$ or $L \cdot D = 1$, $D^2 = 0, -1$. The third case leads to the numerical contradiction $-(t+1)K_S \cdot D = 1$. In the second case, $-(t+1)K_S \cdot D = 2$ implies $t = 1$, $-K_S \cdot D = 1$. Since $D^2 = 0$, the genus formula gives a parity contradiction. In the first case we have the numerical contradiction $10 \leq L^2 = L^2 D^2 \leq (L \cdot D)^2 = 9$. □

For general k, k-very ampleness gives a natural and strong notion of k-th order embedding, expressed in Theorem (8.5.3) below (see also Tikhomirov [595] and Tikhomirov and Troshina [596], where the result is used to study the Hilbert scheme of points on a surface).

Let $(\mathfrak{X}, \mathcal{O}_{\mathfrak{X}})$ be a 0-dimensional subscheme of a surface S belonging to the Hilbert scheme $S^{[r]}$, $r \geq 1$. Let $\mathcal{L}$ be a line bundle on S and suppose that the restriction map $\Gamma(\mathcal{L}) \to \Gamma(\mathcal{O}_{\mathfrak{X}}(\mathcal{L}))$ is onto. Then we can define the associated rational map

$$\varphi_{r-1} : S^{[r]} \to \mathrm{Grass}(\Gamma(\mathcal{L}), r)$$

at $(\mathfrak{X}, \mathcal{O}_{\mathfrak{X}})$, where $\mathrm{Grass}(\Gamma(\mathcal{L}), r)$ denotes the Grassmannian of all r-dimensional quotients of $\Gamma(\mathcal{L})$, by sending $(\mathfrak{X}, \mathcal{O}_{\mathfrak{X}})$ into the quotient $\Gamma(\mathcal{L}) \to \Gamma(\mathcal{O}_{\mathfrak{X}}(\mathcal{L}))$.

Note that if $\Gamma(\mathcal{L}) \to \Gamma(\mathcal{O}_{\mathfrak{X}}(\mathcal{L}))$ is onto for all length r zero cycles, then φ_{r-1} is a globally defined morphism. Further, if $r = 1$, this is the usual morphism to $\mathbb{P}(\Gamma(\mathcal{L}))$ associated to $\Gamma(\mathcal{L})$.

Note also that if a line bundle $\mathcal{L}$ is k-very ample the maps φ_j, for $j \leq k$, are morphisms and for $k = 1$, i.e., $\mathcal{L}$ very ample, φ_0 is an embedding.

Though we will not use it we call attention to the following result of Catanese and Göttsche [152] which improves a result [103, (3.1)] that we previously obtained. It is true in the more general case of an n-dimensional complete algebraic variety defined over any algebraically closed field.

Theorem 8.5.3. *Let S be a smooth connected surface and let $\mathcal{L}$ be a line bundle on S. Let $\varphi_{r-1} : S^{[r]} \to \mathrm{Grass}(\Gamma(\mathcal{L}), r)$ be the rational map associated to $\mathcal{L}$, defined as above. Then $\mathcal{L}$ is k-very ample if and only if φ_j is an embedding for $j \leq k - 1$.*

Finally we would like to discuss the relation of k-very ampleness to gonality of curves. A smooth projective curve, C, is said to be *r-gonal* if there is a degree r map of C onto $\mathbb{P}^1$, but no map of lower degree of C onto $\mathbb{P}^1$. We showed in [102] that if L is a k-very ample line bundle on a smooth projective curve, C, and if $h^1(L) \neq 0$ then K_C is k-very ample. We also noted that in this case, since the restriction map $\Gamma(K_C) \to \Gamma(\mathcal{O}_{\mathfrak{X}}(K_C))$ is surjective for any 0-cycle $\mathfrak{X}$ with $\mathrm{length}(\mathcal{O}_{\mathfrak{X}}) \leq k + 1$, it follows $h^1(K_C - \mathfrak{X}) = 1$, and hence by Serre duality that $h^0(\mathcal{O}_{\mathfrak{X}}) = 1$ for all 0-cycles with $\mathrm{length}(\mathcal{O}_{\mathfrak{X}}) \leq k + 1$. Thus the curve is r-gonal for some $r \geq k + 2$. Since $g(C) \geq 2r - 3$ by Meis's theorem we obtain the useful inequality, $g(C) \geq 2k + 1$. As noted by Mella and Palleschi [431], this argument is immediately reversible: C is r-gonal if and only if K_C is $(r - 2)$-very ample.

The refined Clifford inequality of Coppens and Martens [169] has the following consequence (see Ballico and Sommese [65, Theorem 2.3]).

Theorem 8.5.4. *Let L be a k-very ample line bundle of degree d on a smooth projective curve of genus g. If $h^1(L) \geq 2$, then $d - k + 2 \geq 2h^0(L)$.*

Brill-Noether theory can be applied much more extensively than the application of Meis's theorem mentioned above. For example, there is the following result [65, Theorem 2.5].

Theorem 8.5.5. *Let L be a k-very ample line bundle on a smooth curve C of genus g. If $k \geq 1$, $h^1(L) > 0$, and $K_C \not\approx L$, then $d := \deg L \geq k + g - h^1(L) + (k + 1)/h^1(L)$.*

8.5.6. $\mathbb{P}^1$-bundles over an elliptic curve with invariant $e = -1$. Let L be a very ample line bundle on a smooth connected projective surface, S. Assume that (S, L) admits as a reduction, (Y, L_Y), $\pi : S \to Y$, a $\mathbb{P}^1$-bundle, $p : Y \to R$, over an elliptic curve, R, of invariant $e = -1$. This special situation occurs often in the theory. Here we prove some results which hold true in this special case.

Let C be any smooth curve on Y. Let B be the set of points blown up under $\pi : S \to Y$ and let # be the number of points of $B \cap C$. Let $C^\wedge$ be the proper transform of C under π. Note that $C^\wedge \cong C$ since C is smooth. Then we have

$$L_Y \cdot C \geq L \cdot C^\wedge + \#. \tag{8.13}$$

This easily follows from $L \cdot C^\wedge = (\pi^* L_Y - \pi^{-1}(B)) \cdot C^\wedge \leq L_Y \cdot C - \#$. In particular, if C is an elliptic curve, one has $L \cdot C^\wedge \geq 3$ and hence

$$L_Y \cdot C \geq 3 + \# \quad \text{(the elliptic case).} \tag{8.14}$$

Lemma 8.5.7. *Let Y be a $\mathbb{P}^1$-bundle, $p : Y \to R$, over a smooth elliptic curve R, of invariant $e = -1$. Let E be a section of minimal self-intersection $E^2 = 1$. Then for any given point $y \in Y$ there exists a smooth elliptic curve C on Y, C algebraically equivalent to E and such that $y \in C$.*

Proof. Since $\deg(\mathcal{N}_{E/Y}) = 1$, it follows that $h^1(\mathcal{N}_{E/Y}) = h^0(K_E(-1)) = 0$ and hence by the Riemann-Roch theorem, $h^0(\mathcal{N}_{E/Y}) = 1$. Therefore by Theorem (4.1.9) there exists a 1-dimensional family of deformations of E. Let $\mathcal{H}$ be the irreducible 1-dimensional component of $\mathrm{Hilb}(Y, E)$ containing E. Let

$$\mathcal{A} := \{(h, y) \in \mathcal{H} \times Y \mid y \in h\} \subset \mathcal{H} \times Y$$

be the universal family over $\mathcal{H}$ and let $\pi_Y : \mathcal{H} \times Y \to Y$, $\pi_{\mathcal{H}} : \mathcal{H} \times Y \to \mathcal{H}$ be the projections on the two factors. For every fiber, F, of the restriction, π', of $\pi_{\mathcal{H}}$ to $\mathcal{A}$ we have $E^2 = E \cdot F = 1$, so that F is irreducible since E is ample. Then $\mathcal{A}$ is irreducible since F and $\mathcal{H}$ are both irreducible. Moreover $\pi_Y(\mathcal{A}) = Y$. Indeed clearly $\pi_Y(\mathcal{A}) \subset Y$ and $\dim \pi_Y(\mathcal{A}) = 2$ since there exists a 1-dimensional family of deformations of E. Thus, for every $y \in Y$, there exists a fiber F of π' such that $\pi_Y(F) \ni y$. Let $C := \pi_Y(F)$. Then C is algebraically equivalent to E and $C \cdot f = 1$ for every fiber, f, of $p : Y \to R$. Therefore, since C is irreducible, C is isomorphic to R under p, so that C is a smooth elliptic curve which is a section of p and contains y. □

We use the following k-very ampleness criterion only in the case $k = 1$. The result is a consequence of (8.5.1). For the proof we refer to our paper [108, (2.2)].

Proposition 8.5.8. *Let Y be a $\mathbb{P}^1$-bundle, $p : Y \to R$, over a smooth elliptic curve R of invariant $e = -1$. Let E, f denote a section of minimal self-intersection $E^2 = 1$ and a fiber of the ruling. Let $L \sim aE + bf$ be a line bundle on Y. Then L is k-very ample if and only if $a \geq k$, $a + b \geq k + 2$, and $a + 2b \geq k + 2$.*

Lemma 8.5.9. *Let S be a smooth connected surface and L a very ample line bundle on S. Assume that (S, L) admits as a reduction, (Y, L_Y), $\pi : S \to Y$, a $\mathbb{P}^1$-bundle $p : Y \to R$ over an elliptic curve R of invariant $e = -1$. Let E, f denote a section of minimal self-intersection $E^2 = 1$ and a fiber of the ruling. Let $L_Y \sim aE + bf$. We have*

1. *L_Y is very ample;*
2. *if π is not an isomorphism, then $a + b \geq 4$.*

Proof. One has $-K_Y \sim 2E - f$ and it is a well-known fact that $2E - f$ is numerically equivalent to a smooth effective elliptic curve, C (see, e.g., [99, (0.15)]). From (8.14) we know that $L_Y \cdot E \geq 3$ as well as $L_Y \cdot C \geq 3$. This leads to $a + b \geq 3$ and $a + 2b \geq 3$ respectively. Moreover $L_Y \cdot f = a \geq 1$ since L_Y is ample. Thus we conclude by (8.5.8) that L_Y is very ample.

Let $y \in Y$ be a point blown up under $\pi : S \to Y$. From (8.5.7) we know that there exists a smooth elliptic curve, $C \sim E$, passing through y. Therefore (8.14) applies to give $L_Y \cdot C = L_Y \cdot E \geq 4$, that is $a + b \geq 4$. □

8.6 Surfaces with Castelnuovo curves as hyperplane sections

In this section we prove two theorems which lead to considerable simplification in the development of classical adjunction theory. The first result is mostly due to Roth [526, (3.61)]. The equality for $p_g(S)$ goes back to Castelnuovo, see [526, (5.24)]. We do not know an explicit reference for the second result. In what follows, Castel(a, b) denotes the maximum genus allowed by the Castelnuovo bound for a smooth curve, C, of degree a in a projective space $\mathbb{P}^b$, with C nondegenerate, i.e., not contained in any hyperplane (see (1.3)).

Theorem 8.6.1. *Let $\varphi : S \to \mathbb{P}^N$ be a finite generically one-to-one map of an irreducible normal projective surface, S, into $\mathbb{P}^N$, such that $\varphi(S)$ is not contained in any hyperplane. Let $L := \varphi^*\mathcal{O}_{\mathbb{P}^N}(1)$, and $d := \deg\varphi(S) = L^2$. Assume that $g(L) = \text{Castel}(d, N-1)$, i.e., that the genus of a general $C \in |L|$ equals the Castelnuovo bound for a nondegenerate degree d curve in $\mathbb{P}^{N-1}$. Then $h^1(tL) = 0$ for $t \geq 0$, L is very ample, and S is projectively normal. In particular $h^0(L) = N+1$ and $h^0(L_C) = N$. Moreover:*

$$p_g(S) = h^2(\mathcal{O}_S) = \sum_{t=1}^{\left[\frac{2g(L)-2}{d}\right]} h^1(tL_C).$$

Proof. Let $V := \varphi^*H^0(\mathbb{P}^N, \mathcal{O}_{\mathbb{P}^N}(1))$ denote the space of sections of L obtained by pullback under φ. Since V spans L, a general element of $|V|$ is a smooth curve, $C \subset \text{reg}(S)$. Since C is a general element of $|V|$, and since φ is generically one-to-one, it follows that φ_C is generically one-to-one. Note that the restriction map φ_C is an embedding, since otherwise it would follow that $h^1(\mathcal{O}_{\varphi(C)}) > h^1(\mathcal{O}_C) = g(L)$. This would give a contradiction since, by (1.4.9), $h^1(\mathcal{O}_{\varphi(C)}) \leq \text{Castel}(d, N-1)$ with $g(L) = \text{Castel}(d, N-1)$ by hypothesis.

We now will show that $h^1(tL) = 0$ for $t \geq 0$. Letting L_C be the restriction of L to C, we have a surjection

$$\Gamma(L) \to \Gamma(L_C) \to 0. \tag{8.15}$$

Indeed, if not, then $|V|$ embeds C in $\mathbb{P}^{N-1}$ as a curve of degree $d := L^2$ and $\Gamma(L_C)$ embeds C in $\mathbb{P}^r$ with $r \geq N$. Let C' be the image of C in $\mathbb{P}^r$. Hence in particular

$$g(C') \leq \text{Castel}(d, r) < \text{Castel}(d, N-1) = g(C).$$

This is absurd since $g(C') = g(C)$.

The above argument also shows that $h^0(L) = N+1$ and $h^0(L_C) = N$.

Thus we have a commutative diagram, for any integer $t \geq 0$,

$$\begin{array}{ccc} \Gamma(L)^{\otimes t} & \xrightarrow{\alpha_t} & \Gamma(L_C)^{\otimes t} \\ \beta'_t \downarrow & & \downarrow \beta_t \\ \Gamma(tL) & \xrightarrow{\alpha'_t} & \Gamma(tL_C) \end{array}$$

where α_t is surjective by (8.15) and β_t is surjective for $t \geq 1$ since by (1.4.9) any Castelnuovo curve is projectively normal. Therefore we conclude that α'_t is surjective for $t \geq 0$. From Lemma (4.1.5) it follows that $h^1(tL) = 0$ for $t \geq 0$.

Since as noted above, $h^0(L) = N+1$, proving the projective normality of $\varphi(S)$ comes down to showing that β'_t is surjective for $t \geq 1$. Since the kernel of α'_t is $\Gamma((t-1)L)$ and since α_t and β_t are surjective for all $t \geq 1$ we are reduced to the fact that β'_1 is a surjection.

Since $h^1(tL) = 0$ for all $t \geq 0$ and since $h^2(tL) = 0$ for $t \gg 0$ we see that (see also the second author's paper [575, (2.2)])

$$p_g(S) = \sum_{t\geq 1} h^1(tL_C).$$

Since $h^1(tL_C) = 0$ by the Kodaira vanishing theorem when $td > 2g(C) - 2$, we are done.

Since tL is very ample for some $t \geq 1$ and since we have shown that $\Gamma(L)^{\otimes t} \to \Gamma(tL)$ is onto for all $t \geq 1$, we conclude that L is very ample. □

Corollary 8.6.2. *Let $\varphi : S \to \mathbb{P}^N$ be a finite generically one-to-one map of an irreducible normal projective surface, S, into $\mathbb{P}^N$, such that $\varphi(S)$ is not contained in any hyperplane. Let $L := \varphi^*\mathcal{O}_{\mathbb{P}^N}(1)$, and $d := \deg \varphi(S) = L^2$. Assume that $g(L) = \mathrm{Castel}(d, N-1)$, i.e., that the genus of a general $C \in |L|$ equals the Castelnuovo bound for a nondegenerate degree d curve in $\mathbb{P}^{N-1}$. Assume also that $\kappa(S) \geq 0$ and $d = 2N - 2$. Then $K_S \cong \mathcal{O}_S$ and $h^1(\mathcal{O}_S) = 0$.*

Proof. By Theorem (8.6.1) we know that $h^1(\mathcal{O}_S) = 0$ and that $h^2(\mathcal{O}_S) \geq h^1(L_C)$. By Theorem (8.1.1) we know that $tK_S \cong \mathcal{O}_S$ for some $t \geq 1$ and $K_S \cong \mathcal{O}_S$ if $h^2(\mathcal{O}_S) \geq 1$. Therefore it suffices to show that $h^1(L_C) \geq 1$. Since $d = 2g(L)-2 = 2N-2$ it suffices to show that $h^0(L_C) \geq N$. This will follow from $h^0(L) \geq N+1$. This is equivalent to the assertion that $\varphi(S)$ is not contained in any hyperplane. □

Theorem 8.6.3. *Let $\varphi : S \to \mathbb{P}^N$ be an embedding of a normal irreducible projective surface, S, into $\mathbb{P}^N$. Assume that $\varphi(S)$ is not contained in any hyperplane. Let $L := \varphi^*\mathcal{O}_{\mathbb{P}^N}(1)$, and $d := \deg \varphi(S) = L^2$. Assume that $g(L) = \mathrm{Castel}(d, N-1)$, i.e., that the genus of a hyperplane section of $\varphi(S)$ equals the Castelnuovo bound for the genus of an irreducible degree d curve in the hyperplane. Then $\Gamma(K_S)$ spans $K_{\mathrm{reg}(S)}$ if and only if $h^2(\mathcal{O}_S) > 0$, if and only if $h^1(tL_C) > 0$ for some smooth $C \in |L|$ and $t > 0$.*

Proof. By Theorem (8.6.1), $h^0(L) = N + 1$ and $h^0(L_C) = N$.

If $\Gamma(K_S)$ spans $K_{\mathrm{reg}(S)}$ then $h^0(K_S) > 0$. If $h^0(K_S) > 0$ then for a general $C \in |L|$, the map $\Gamma(K_S) \to \Gamma(K_{S|C})$ has a nontrivial image. Since $K_{S|C} \cong K_C - L_C$, we have by Serre duality that $h^1(L_C) = h^0(K_{S|C}) > 0$. Thus it only remains to show that if $h^1(tL_C) > 0$ for some $t > 0$ then $\Gamma(K_S)$ spans $K_{\mathrm{reg}(S)}$. Since $h^1(tL_C) = h^0(K_C - tL_C) > 0$ for some $t > 0$ implies that $h^1(L_C) = h^0(K_C - L_C) > 0$, we can assume without loss of generality that $h^1(L_C) > 0$. Since $h^1(L) = 0$ by Theorem (8.6.1), $h^1(L_C) > 0$ for one smooth $C \in |L|$ implies that $h^1(L_C) > 0$ for all smooth $C \in |L|$. Indeed letting $C' \in |L|$ be smooth we have $h^1(L_{C'}) = h^2(\mathcal{O}_S) - h^2(L) = h^1(L_C)$.

Let $x \in \mathrm{reg}(S)$. Choose a general $C \in |L|$ that contains x. Since C is general it misses the finite set $\mathrm{Sing}(S)$, and thus $C \subset \mathrm{reg}(S)$. Moreover, by (1.7.8), C is smooth. Consider the exact sequence

$$0 \to K_S - L \to K_S \to K_{S|C} \to 0.$$

Since $h^1(L) = 0$ by Theorem (8.6.1), the restriction map $\Gamma(K_S) \to \Gamma(K_{S|C})$ is onto. Therefore we are reduced to showing that $K_{S|C}$ is spanned. We assume otherwise and we show that this leads to a contradiction.

Since $h^0(K_C - L_C) = h^1(L_C) > 0$, the basepoints of $K_{S|C} \approx K_C - L_C$ are isolated. Let p be a basepoint of $K_C - L_C$. Then $h^0(K_C - L_C - p) = h^0(K_C - L_C)$, or, by Serre duality,

$$h^1(L_C + p) = h^1(L_C). \tag{8.16}$$

Consider the exact sequence

$$0 \to L_C \to L_C \otimes \mathbb{O}_C(p) \to (L_C \otimes \mathbb{O}_C(p))_p \to 0.$$

By (8.16) we have a surjection $\Gamma(L_C + p) \to \mathbb{C} \to 0$. This means that $L_C + p$ is spanned at p. Since $|L_C|$ gives an embedding, we conclude that $|L_C + p|$ embeds C in $\mathbb{P}^N$ as a curve of degree $d + 1$. Let C' be the image of this embedding of C in $\mathbb{P}^N$. Hence

$$\mathrm{Castel}(d, N-1) = g(C) = g(C') \le \mathrm{Castel}(d+1, N). \tag{8.17}$$

Let $\ell := N$. Therefore

$$\mathrm{Castel}(d, N-1) = \sigma_2 \left(d - \ell + 1 - (\sigma_2 - 1)\frac{\ell - 2}{2} \right), \quad \sigma_2 = \left[\frac{d-2}{\ell - 2} \right], \tag{8.18}$$

and

$$\mathrm{Castel}(d+1, N) = \sigma_1 \left(d - \ell + 1 - (\sigma_1 - 1)\frac{\ell - 1}{2} \right), \quad \sigma_1 = \left[\frac{d-1}{\ell - 1} \right]. \tag{8.19}$$

We claim that $\sigma_1 < \sigma_2$. To see this write

$$d - 2 = \sigma_2(\ell - 2) + \varepsilon_2, \tag{8.20}$$

where $\varepsilon_2 \le \ell - 3$, so that

$$d - 1 = \sigma_2(\ell - 2) + \varepsilon_2 + 1 = \sigma_2(\ell - 1) + \varepsilon_2 + 1 - \sigma_2,$$

or

$$\sigma_1 = \sigma_2 + \left[\frac{\varepsilon_2 + 1 - \sigma_2}{\ell - 1} \right].$$

Since $\varepsilon_2+1-\sigma_2 \leq \ell-2$ we have $(\varepsilon_2+1-\sigma_2)/(\ell-1) \leq 0$ and hence $\sigma_1 \leq \sigma_2$.

Now, assume $\sigma_1 = \sigma_2$. In this case (8.17), (8.18), (8.19) yield $\sigma_2(\sigma_2-1)/2 \leq 0$. Therefore either $\sigma_2 = 0$ or $\sigma_2 = 1$. If $\sigma_2 = 0$ we find $g(C) = 0$. In this case $h^1(L_C) = 0$, which contradicts our assumption. If $\sigma_2 = 1$, then $g(C) = d-\ell+1$, or $h^0(L_C) = d - g(C) + 1$ and hence by the Riemann-Roch theorem we conclude that $h^1(L_C) = 0$, which again contradicts our hypothesis. Thus we conclude that $\sigma_1 < \sigma_2$.

From (8.17), (8.18), (8.19) we get

$$\sigma_2\left((d-\ell+1)-(\sigma_2-1)\frac{\ell-2}{2}\right) \leq \sigma_1\left((d-\ell+1)-(\sigma_1-1)\frac{\ell-1}{2}\right).$$

Therefore

$$\begin{aligned}(\sigma_2-\sigma_1)(d-\ell+1) &\leq \sigma_2(\sigma_2-1)\frac{\ell-2}{2} - \sigma_1(\sigma_1-1)\frac{\ell-1}{2}\\ &= (\sigma_2^2-\sigma_1^2)\frac{\ell-2}{2} - (\sigma_2-\sigma_1)\frac{\ell-2}{2} - \frac{\sigma_1(\sigma_1-1)}{2},\end{aligned}$$

or, since $\sigma_2 - \sigma_1 > 0$,

$$d-\ell+1 \leq (\ell-2)\left(\frac{\sigma_2+\sigma_1-1}{2}\right) - \frac{\sigma_1(\sigma_1-1)}{2(\sigma_2-\sigma_1)}. \tag{8.21}$$

Writing

$$d-1-\varepsilon_1 = \sigma_1(\ell-1),\ \varepsilon_1 \leq \ell-2, \tag{8.22}$$

and recalling (8.20), the inequality (8.21) yields

$$\begin{aligned}d-\ell+1 &\leq \frac{(\ell-2)\sigma_2}{2} + \frac{(\ell-1)\sigma_1}{2} - \frac{\sigma_1}{2} - \frac{\ell-2}{2} - \frac{\sigma_1(\sigma_1-1)}{2(\sigma_2-\sigma_1)}\\ &= \frac{d-2}{2} + \frac{d-1}{2} - \frac{\varepsilon_2}{2} - \frac{\varepsilon_1}{2} - \frac{\sigma_1}{2} - \frac{\ell-2}{2} - \frac{\sigma_1(\sigma_1-1)}{2(\sigma_2-\sigma_1)},\end{aligned}$$

or

$$\ell \geq \varepsilon_1+\varepsilon_2+3+\sigma_1+\frac{\sigma_1(\sigma_1-1)}{\sigma_2-\sigma_1}. \tag{8.23}$$

Note that by combining (8.20) and (8.22) and recalling that $\sigma_2 > \sigma_1$ we get

$$\ell-2 \leq (\sigma_2-\sigma_1)(\ell-2) = \sigma_1+\varepsilon_1-\varepsilon_2-1$$

or $\ell \leq \sigma_1+\varepsilon_1-\varepsilon_2+1$, which contradicts (8.23). □

8.7 Polarized varieties (X, L) with sectional genus $g(L) = h^1(\mathcal{O}_X)$

Let X be an irreducible normal variety and L an ample and spanned line bundle on X. Though the results of this section follow from the result (7.2.10), the following less abstract and more direct proof is important. In this section the case when $q(X) := h^1(\mathcal{O}_X) = g(L) \neq 0$ will be treated. The case when $q(X) = g(L) = 0$ will be treated separately in Theorem (8.9.1).

Theorem 8.7.1. *Let X be an irreducible normal n-dimensional projective variety and let L be an ample and spanned line bundle on X. Then $g(L) \geq q(X)$. If in addition X has at worst Cohen-Macaulay singularities and $q(X) \geq 1$, then $g(L) = q(X)$ if and only if $(X, L) \cong (\mathbb{P}(\mathcal{E}), \xi_{\mathcal{E}})$ for an ample and spanned vector bundle $\mathcal{E}$ over a smooth curve R of genus $g(R) = q(X)$. In particular $g(R) = q(X) > 0$ implies that X is smooth.*

Proof. We first prove the case when $n = 2$. In this case the singularities of X are finite and we will use the vanishing theorem, (2.2.4), without further comment. Let C be a smooth curve $C \in |L|$. Then $q(X) \leq g(L)$. This immediately follows from the exact sequence

$$0 \to -L \to \mathcal{O}_X \to \mathcal{O}_C \to 0$$

since $h^1(-L) = h^1(K_X + L) = 0$ and $h^1(\mathcal{O}_C) = g(C) = g(L)$.

Let C be a smooth curve in $|L|$ and consider the long exact cohomology sequence of

$$0 \to K_X \to K_X + L \to K_C \to 0.$$

Since $g(L) = h^0(K_C) = h^1(\mathcal{O}_X)$ and $h^1(K_X+L) = 0$ we see that $\Gamma(K_C) \cong H^1(\mathcal{O}_X)$. Therefore $\Gamma(K_X) \cong \Gamma(K_X + L)$ and the restriction map $\Gamma(K_X + L) \to \Gamma(K_C)$ is the zero map.

We claim that $h^0(K_X + L) = 0$. If not, take a section $s \in \Gamma(K_X + L)$ not identically zero. Since L is spanned we can assume that C is not contained in $s^{-1}(0)$. Thus C meets $s^{-1}(0)$ in at most a finite set of points. Then the restriction of s to C has only a finite set of zeroes. This contradicts the fact that $\Gamma(K_X+L) \to \Gamma(K_C)$ is the zero map. Therefore X has at worst rational singularities by (2.2.8). In particular $h^i(K_X) = h^i(K_{\overline{X}})$ for $i \geq 1$ and any desingularization $p : \overline{X} \to X$. Moreover the Albanese map exists (see §2.4).

Thus we conclude that $p_g(X) = h^0(K_X) = 0$. Let $\alpha : X \to \mathrm{Alb}(X)$ be the Albanese map. Since $p_g(X) = 0$ it follows that $\dim \alpha(X) < 2$, since otherwise we would have holomorphic 2-forms on X. Therefore, since $q(X) > 0$, the image $\alpha(X)$ is a curve, R. From the general theory of Albanese mapping (see §2.4) we know that R is a smooth curve of genus $g(R) = q(X)$ and $\alpha : X \to R$ has connected fibers.

Let $\alpha_C : C \to R$ be the restriction of α to C. We claim that α_C is an isomorphism. Indeed the Hurwitz formula yields

$$2g(L) - 2 = (\deg \alpha_C)(2g(R) - 2) + \rho,$$

where ρ is the degree of the ramification divisor of α_C. Note that $g(L) = g(R)$ since $g(R) = q(X)$. Then, if $g(L) \geq 2$, we see that $\rho = 0$ and $\deg \alpha_C = 1$, so α_C is an isomorphism. If $g(L) = 1$, then $\rho = 0$. Let $(\alpha_C)_* : H_1(C, \mathbb{Z}) \to H_1(R, \mathbb{Z})$ be the induced map on the integral homology. Since $\rho = 0$, $\deg \alpha_C$ is equal to the index of $(\alpha_C)_* H_1(C, \mathbb{Z})$ as subgroup of $H_1(R, \mathbb{Z})$. Therefore if we show that $(\alpha_C)_*$ is surjective, we will have $\deg \alpha_C = 1$, which will imply that α_C is an isomorphism. Assume otherwise and take the fiber product, $X' := X \times_R C$, of X and C over R. Note that the induced map $\alpha' : H_1(X, \mathbb{Z}) \to H_1(R, \mathbb{Z})$ is surjective since $\alpha : X \to R$ has connected fibers. Thus the pullback, C', of C to X' is connected. Since C is ample and the map $X' \to X$ is finite we conclude that C' is ample. Since we are assuming that $\deg \alpha_C > 1$, we conclude C' is disconnected. This implies that $h^1(-C') > 0$ which contradicts the usual vanishing theorem.

Thus we conclude that $L \cdot f = C \cdot f = 1$ for each fiber f of $\alpha : X \to R$. This implies that $f \cong \mathbb{P}^1$ and hence α is a $\mathbb{P}^1$-bundle by (3.1.4).

Now we will prove the general case by induction. Assume that $n \geq 3$ and that we have proven the theorem when $\dim X < n$. Choose a general $A \in |L|$. By (1.7.1) we can assume that A is normal. Since $h^1(-L) = 0$ by (2.2.7), we conclude from

$$0 \to -L \to \mathcal{O}_X \to \mathcal{O}_A \to 0$$

that $h^1(\mathcal{O}_X) \leq h^1(\mathcal{O}_A)$. By our induction hypothesis we know that $h^1(\mathcal{O}_A) \leq g(L_A)$. Since

$$2g(L) - 2 = (K_X + (n-1)L) \cdot L^{n-1} = (K_A + (n-2)L_A) \cdot L_A^{n-2} = 2g(L_A) - 2,$$

we conclude that $h^1(\mathcal{O}_X) \leq h^1(\mathcal{O}_A) \leq g(L_A) = g(L)$.

Therefore we can assume without loss of generality that A is in addition Cohen-Macaulay and we have equality, $q(A) = g(L_A)$. To see that the divisor $A \in |L|$ is Cohen-Macaulay note that it is a Cartier divisor on a Cohen-Macaulay variety.

By the induction hypothesis, A is a smooth $\mathbb{P}^{n-2}$-bundle, $p : A \to R$, over a smooth curve R of genus $q \geq 1$. Thus the singularities of X are finite and we will use the vanishing theorem, (2.2.4), without further comment. We claim that $h^0(K_X + L) = 0$. If not then since $A \in |L|$ is general, the restriction of a nontrivial element of $\Gamma(K_X + L)$ to A will give rise to a nontrivial element of K_A. Since A is a $\mathbb{P}^{n-2}$-bundle and $n \geq 3$, $h^0(K_A) = 0$. Thus $h^0(K_X + L) = 0$. Therefore X has at worst rational singularities by (2.2.8). Thus by (5.2.3) the map p extends to a morphism, $\overline{p} : X \to R$. All fibers of $\overline{p}$ are equal dimensional. Moreover given a general fiber F of $\overline{p}$, $L^{n-1} \cdot F = L_A^{n-2} \cdot \mathbb{P}^{n-2} = 1$. Now use (3.2.1). □

The following simple result will be needed in §8.10.

Corollary 8.7.2. *Let $\varphi : X \to \mathbb{P}^N$ be a finite map of an irreducible normal projective variety, X, to $\mathbb{P}^N$. Let $L := \varphi^*\mathcal{O}_{\mathbb{P}^N}(1)$. Assume that $\varphi(X)$ is nondegenerate, i.e., $\varphi(X)$ is contained in no linear $\mathbb{P}^{N-1} \subset \mathbb{P}^N$, and that φ gives an embedding off a finite set. If $n := \dim X \geq 2$, $q := h^1(\mathcal{O}_X) > 0$, and q equals the sectional genus, $g(L)$, of X, then $d := \deg \varphi(X) \geq 5$.*

Proof. Assume that $d \leq 4$. By taking the intersection of $\varphi(X)$ with a general $\mathbb{P}^{N-n+2}$, it follows from (1.7.1) and (8.7.1) that we can assume without loss of generality that $\varphi(X)$ is an irreducible normal surface.

First note that $\varphi(X)$ cannot be a hypersurface of $\mathbb{P}^3$. Indeed if $\varphi(X)$ was a hypersurface of $\mathbb{P}^3$ it is Cohen-Macaulay and $h^1(\mathcal{O}_{\varphi(X)}) = 0$. Since by the above X has a finite singular set and φ gives an embedding in the complement of a finite set, $\varphi(X)$ has at most finite singular set, it is normal, and thus by Zariski's main theorem φ is an isomorphism. Therefore we conclude that $h^1(\mathcal{O}_X) = h^1(\mathcal{O}_{\varphi(X)}) = 0$ in contradiction to $q > 0$.

Therefore $N \geq 4$. Let C' be the pullback of a general hyperplane section, C, of $\varphi(X) \subset \mathbb{P}^N$. The curve C is a nondegenerate smooth curve of genus $q \geq 1$ in $\mathbb{P}^{N-1}$ with $N - 1 \geq 3$. Castelnuovo's inequality (1.4.9) implies that $g(L) = g(C') = g(C) \leq 1$. By using the hypothesis $q(X) = g(L) \geq 1$ we conclude that $q(X) = g(L) = 1$. But in this case the upper bound allowed by (1.4.9) is taken on and (8.6.1) implies the contradiction that $q(X) = 0$. □

8.8 Spannedness of $K_X + (\dim X)L$ for ample and spanned L

We need some useful results proved by the second author with Andreatta [24] and with Lanteri and Palleschi [382].

Recall that given an irreducible normal variety, X, the Grauert-Riemenschneider canonical sheaf, $\overline{K}_X$, is defined to be $h_* K_{\overline{X}}$ where $h : \overline{X} \to X$ is a desingularization of X. Recall also that $\overline{K}_X$ is independent of the desingularization chosen and that there is a canonical injection $0 \to \overline{K}_X \to K_X$ whose cokernel is supported on the singular set of X.

Given an integer $k \geq 0$, we say that a section, s, of a line bundle, L, on a variety, X, *vanishes to the k-th order* at $x \in X$ if $s \in \Gamma(L \otimes \mathcal{O}_X/\mathfrak{m}_x^k)$ with $\mathfrak{m}_x$ the maximal ideal sheaf at x.

Theorem 8.8.1. *Let L be an ample and spanned line bundle on a normal irreducible projective variety X of dimension n. Given an integer $k \geq 1$ and a smooth point $x \in X$ assume that the vector subspace $V \subset \Gamma(L)$ of sections that vanish to the k-th order at x defines a linear system $|V|$ with a finite base locus. Then $\Gamma(\overline{K}_X + tL)$ spans $\overline{K}_X + tL$ at x if either $kt > n$ or $kt = n$ and $L^n \geq k^n + 1$.*

Proof. Note that for $n = 1$ the result is straightforward. Thus we can assume without loss of generality that $n \geq 2$.

Let $p : X' \to X$ denote the blowup of X at x, and let $E := p^{-1}(x)$. By hypothesis all the $A \in |V|$ vanish to the k-th order at x. Then $L' := p^*L - kE$ is nef. To see this assume to the contrary that there is an irreducible curve $C \subset X'$ such that $L' \cdot C < 0$. Note that given any $A \in |L - x|$, $A' := p^*A - kE$ is effective and $A' \in |L'|$, and thus $L' \cdot C < 0$ implies that $C \subset A'$. This implies that the image of C in X is contained in the base locus of $|L - x|$. Since it is assumed that the base locus is finite, we conclude that $C \subset E$. This gives the contradiction $L' \cdot C = -kE \cdot C > 0$. Note that the same argument shows that $p^*L - k'E$ is nef for $k' \leq k$. Moreover if $k' < k$, then $p^*L - k'E$ is big since

$$\begin{aligned}(p^*L - k'E)^n &= (p^*L)^n - k'^n = (p^*L)^n - k^n + k^n - k'^n \\ &> (p^*L)^n - k^n = (p^*L - kE)^n \geq 0.\end{aligned}$$

In particular $tp^*L - t'E$ is nef and big if $kt > t' \geq 0$ or $kt = t' \geq 0$ and $L^n \geq k^n + 1$.

Let $p : X' \to X$ be as above, and let $h : \overline{X} \to X'$ be a desingularization of X' such that the restriction $h_{h^{-1}(\mathrm{reg}(X'))}$ is an isomorphism. Since h is an isomorphism in a neighborhood of E, we identify E and $h^{-1}(E)$. Let $g = p \circ h : \overline{X} \to X$.

If $tp^*L - t'E$ is nef and big, $tg^*L - t'E$ is nef and big. Thus in this case

$$h^1(K_{\overline{X}} + tg^*L - t'E) = 0$$

and we have a surjection

$$\Gamma(K_{\overline{X}} + tg^*L - (t' - 1)E) \to \Gamma(K_E + (tg^*L - t'E)_E) \to 0 \quad \text{for} \quad kt > t' \geq 0.$$

Since

$$K_E + (tg^*L - t'E)_E \cong \mathbb{O}_{\mathbb{P}^{n-1}}(t' - n)$$

is spanned for $t' \geq n$ we see that

$$\Gamma(K_{\overline{X}} + tg^*L - (n-1)E) \text{ spans } K_{\overline{X}} + tg^*L - (n-1)E$$

over E if either $kt > n$ or $kt = n$ and $L^n \geq k^n + 1$.

Choosing a smooth neighborhood U of x we have

$$g^*(K_U + tL) \cong K_{g^{-1}(U)} + tg^*L - (n-1)E.$$

Since $g_*K_{\overline{X}} \cong \overline{K}_X$ we see that

$$g_*(K_{\overline{X}} + tg^*L - (n-1)E) \cong g_*K_{\overline{X}} + tL \cong \overline{K}_X + tL.$$

Thus by the last paragraph we see that the image of $\Gamma(g_*(K_{\overline{X}} + tg^*L - (n-1)E))$ in $\Gamma(\overline{K}_X + tL)$ spans $\overline{K}_X + tL$ at x. □

Corollary 8.8.2. *Let L be an ample and spanned line bundle on an irreducible normal n-dimensional projective variety, X. Then $\Gamma(\overline{K}_X + tL)$ spans $\overline{K}_{\mathrm{reg}(X)} + tL_{\mathrm{reg}(X)}$ for $t \geq n$ except if $t = n$ and $(X, L) \cong (\mathbb{P}^n, \mathcal{O}_{\mathbb{P}^n}(1))$.*

Proof. We can assume that $(X, L) \not\cong (\mathbb{P}^n, \mathcal{O}_{\mathbb{P}^n}(1))$, and therefore by (3.1.5) that $L^n \geq 2$.

Let $x \in \mathrm{reg}(X)$. Since $|L - x|$ has a finite base locus we obtain our result from (8.8.1). □

The next result is a special case of the second author's results [581, §3]. The argument for spannedness at rational singularities is based on the argument of Bombieri [135] (see also Sakai [534]).

Lemma 8.8.3. *Let L be an ample and spanned line bundle on an irreducible normal Gorenstein projective variety, X, of dimension $n \geq 2$. If $L^n \geq 3$ and* $\dim \mathrm{Irr}(X) \leq 0$, *then $\Gamma(K_X + (n-1)L)$ spans $K_X + (n-1)L$ on* $\mathrm{Sing}(X)$.

Proof. First we prove the result for surfaces. At irrational points the result follows from (2.2.8).

Let $x \in X$ be a rational singular point, and let $p : \overline{X} \to X$ be a minimal partial desingularization of X at x, i.e., the restriction $p_{p^{-1}(X-x)}$ is an isomorphism, $p^{-1}(x) \subset \mathrm{reg}(\overline{X})$, and if $E \subset \overline{X}$ is a smooth rational curve satisfying $E^2 = -1$, then $\dim p(E) = 1$. Let $\mathfrak{m}_x$ denote the ideal sheaf of x. Since x is a rational singularity we know from Artin [35] (see §1.2) that $p^*\mathfrak{m}_x$ is invertible and defines a divisor, $Z := -p^*\mathfrak{m}_x$, called the fundamental cycle of x, with the properties that $Z^2 = -2$ and $Z \cdot C \leq 0$ for all irreducible curves C such that $p(C) = x$.

Note that $p^*L - Z$ is nef and big. Let C be an irreducible curve on $\overline{X}$. If $(p^*L - Z) \cdot C < 0$ then $C \subset D$ for all effective $D \in |p^*L - Z|$. Since every $D' \in |L - x|$ gives rise to an effective $D \in |p^*L - Z|$ we conclude that $p(C) \subset \mathrm{Bs}|L - x|$. If $C \subset p^{-1}(x)$ we have that $(p^*L - Z) \cdot C = -Z \cdot C \geq 0$. If $p(C)$ is a point other than x, $Z \cdot C = 0$ and thus $(p^*L - Z) \cdot C = 0$. This shows nefness of $p^*L - Z$. Since $(p^*L - Z)^2 = L^2 - 2 > 0$ we have that $p^*L - Z$ is big. Thus $h^1(K_{\overline{X}} + p^*L - Z) = 0$. Since $K_{\overline{X}} \cong p^*K_X$ we have

$$p_*(K_{\overline{X}} + p^*L - Z) \cong K_X \otimes L \otimes \mathfrak{m}_x$$

and

$$p_{(1)}(K_{\overline{X}} + p^*L - Z) \cong (K_X + L) \otimes p_{(1)}(-Z).$$

Since $p_{(1)}(-Z) = 0$ we have from the Leray spectral sequence for p and $K_{\overline{X}} + p^*L - Z$ that $h^1(K_X \otimes L \otimes \mathfrak{m}_x) = 0$. We therefore have that $K_X + L$ is spanned at $x \in X$.

Now assume that the result is true if $2 \leq \dim X < n$ and assume that $\dim X = n$. Note that the divisors $A \in |L - x|$ are Gorenstein and singular at x. By (1.7.2) we can choose an $A \in |L - x|$ which is normal and irreducible with $\dim \mathrm{Irr}(A) \leq 0$. Thus by the induction hypothesis $\Gamma(K_A + (n-2)L_A)$ spans $(K_A + (n-2)L_A)_x$. Since $\dim \mathrm{Irr}(X) \leq 0$ and since $n \geq 3$ we have that $h^1(K_X + (n-2)L) = 0$ by (2.2.4). Thus using the exact sequence

$$0 \to K_X + (n-2)L \to K_X + (n-1)L \to K_A + (n-2)L_A \to 0$$

we see that $\Gamma(K_X + (n-1)L)$ spans $(K_X + (n-1)L)_x$. □

Lemma 8.8.4. *Let L be an ample and spanned line bundle on an irreducible normal Gorenstein surface, S. For $t \geq 2$, $K_S + tL$ is spanned by global sections except if $t = 2$ and $(S, L) \cong (\mathbb{P}^2, \mathcal{O}_{\mathbb{P}^2}(1))$.*

Proof. By (8.8.2) we can assume that $L^2 \geq 2$ and that $\Gamma(K_S + tL)$ spans $(K_S + tL)_{\mathrm{reg}(S)}$. Thus we can assume that $(tL)^2 \geq 8 \geq 3$. By (8.8.3) we are done. □

Theorem 8.8.5. *Let L be an ample and spanned line bundle on a normal irreducible n-dimensional Gorenstein projective variety, X. Assume that* $\dim \mathrm{Irr}(X) \leq 0$. *Then $K_X + tL$ is spanned for $t \geq n$ except if $(X, L) \cong (\mathbb{P}^n, \mathcal{O}_{\mathbb{P}^n}(1))$.*

Proof. By (3.1.5) we can assume that $L^n \geq 2$. We can assume without loss of generality that $n \geq 2$ since the result is trivial for $n = 1$. Also we can assume that $t = n$ since the result for $t > n$ is a consequence of the result for $t = n$.

We prove the theorem by induction on $\dim X$. First the theorem is true for $\dim X = 2$ by (8.8.2) and (8.8.3). Assume as an induction hypothesis that the result has been proven for $2 \leq \dim X < n$. Given an $x \in X$ we can by (1.7.2) find an $A \in |L - x|$ such that A is irreducible and normal with $\dim \mathrm{Irr}(A) \leq 0$. Since X is Gorenstein it follows that A is also Gorenstein. By the induction hypothesis $K_A + (n-1)L_A$ is spanned by global sections. Using the exact sequence

$$0 \to K_X + (n-1)L \to K_X + nL \to K_A + (n-1)L_A \to 0$$

we see that the sections that span $K_A + (n-1)L_A$ lift to give global sections of $K_X + nL$ that span X at x. □

For the analogue of this for embeddings, which follows along the same lines as the above argument, see [382].

Corollary 8.8.6. *Let L be an ample and spanned line bundle on a normal irreducible n-dimensional Gorenstein variety, X. Assume that* $\dim \mathrm{Irr}(X) \leq 0$. *Then $K_X + nL$ is spanned and big except if:*

1. *$(X, L) \cong (\mathbb{P}^n, \mathcal{O}_{\mathbb{P}^n}(1))$ or $(X, L) \cong (\mathbb{P}^1, \mathcal{O}_{\mathbb{P}^1}(2))$;*

2. $X \subset \mathbb{P}^{n+1}$ *is a quadric with* $L \cong \mathcal{O}_{\mathbb{P}^{n+1}}(1)_X$*;*

3. $(X, L) \cong (\mathbb{P}(\mathcal{E}), \xi_{\mathcal{E}})$ *for an ample and spanned vector bundle,* $\mathcal{E}$*, on a curve of genus* $h^1(\mathcal{O}_X) = g(L)$*.*

Proof. The case when $n = 1$ is easy and so we assume that $n \geq 2$. By Theorem (8.8.5) we can assume that $K_X + nL$ is spanned. Let $\phi : X \to \mathbb{P}_{\mathbb{C}}$ be the morphism associated to $|K_X + nL|$. Let $\phi = s \circ \pi$ be the Remmert-Stein factorization of ϕ, i.e., $\pi : X \to Y$ is a map with connected fibers onto a normal projective variety, Y, and $s : Y \to \mathbb{P}_{\mathbb{C}}$ is a finite map. Assuming that $K_X + nL$ is not big, we have that π has positive dimensional fibers.

Let F be a general fiber of π of dimension $f \leq n$. Since F is general and defined in a neighborhood of F by the pullbacks of $\dim Y$ coordinate functions on Y, we have that F is Gorenstein and by an application of (1.7.1) that it has rational singularities. We also have $K_F + nL_F \cong \mathcal{O}_F$ since the normal bundle of F is trivial. From (3.3.2) we see that $n \leq f + 1$, i.e., $0 \leq \dim Y \leq 1$. By (3.1.6) we have in the former case that we are in case 2), and in the latter case we have $(F, L_F) \cong (\mathbb{P}^{n-1}, \mathcal{O}_{\mathbb{P}^{n-1}}(1))$. Using (3.2.1) we are done. □

We close this section with a special result in the special case of surfaces.

Lemma 8.8.7. *Let* L *be an ample line bundle on an irreducible normal Gorenstein surface,* S*. For* $t \geq 2$ *and* $t^2L^2 \geq 5$*,* $\Gamma(K_S+tL)$ *spans* K_S+tL *at* $x \in \mathrm{Irr}(S) \cup \mathrm{reg}(S)$*. If in addition* $K_S + 2L$ *is big then it is ample.*

Proof. At irrational points the result follows from (2.2.8). Therefore we have to show that $\Gamma(K_S+tL)$ spans K_S+tL at an arbitrary point $x \in \mathrm{reg}(S)$. Let $p : \overline{S} \to S$ be a desingularization of S with the restriction $p_{p^{-1}(\mathrm{reg}(S))}$ is a biholomorphism. Let $x' = p^{-1}(x)$. Consider the exact sequence

$$0 \to p_*K_{\overline{S}} + tL \to K_S + tL \to \mathcal{S} \to 0,$$

where $\mathcal{S}$ is a skyscraper sheaf supported on $\mathrm{Irr}(S)$. Since

$$p_*K_{\overline{S}} + tL \cong p_*(K_{\overline{S}} + tp^*L)$$

it suffices to show that there is a global section of $K_{\overline{S}} + tp^*L$ that does not vanish at x'. By using Reider's criterion we see that if such a section does not exist, there is an effective nontrivial divisor $D \subset \overline{S}$ such that $x' \in D$ and either:

1. $tp^*L \cdot D = 0$ and $D^2 = -1$; or

2. $tp^*L \cdot D = 1$ and $D^2 = 0$.

If case 1) occurred, then it would imply that $L \cdot p(D) = 0$. Since L is ample this implies that $p(D)$ is a point. Since $x' \in D$, this implies that $p(D) = x$, which is impossible since $p_{p^{-1}(\mathrm{reg}(S))}$ is a biholomorphism. Therefore we can assume without loss of generality that we are in case 2). But $t \leq tp^*L \cdot D = 1$ is absurd because $t \geq 2$.

Now assume that $K_S + 2L$ is big. Since $K_S + 2L$ is spanned off a finite set it is straightforward to check that some finite power of $K_S + 2L$ is spanned, e.g., use (1.1.2). From this it follows that if $K_S + 2L$ is not ample there exists an irreducible curve $C \subset S$ such that $(K_S + 2L) \cdot C = 0$. Assume that the above desingularization is minimal in the sense that for any smooth rational curve, $E \subset \overline{S}$, satisfying $E^2 = -1$, $\dim p(E) = 1$. Then $K_{\overline{S}} \cong p^*K_S - \Delta$ where $\Delta \subset p^{-1}(\mathrm{Sing}(S))$ is an effective divisor with support equal to $p^{-1}(\mathrm{Irr}(S))$ (see, e.g., Sakai [529]). Let $\overline{C}$ denote the proper transform of C under p. Since $p^*(K_S + 2L)$ is big and $p^*(K_S + 2L) \cdot \overline{C} = 0$, we have by the Hodge index theorem that $\overline{C}^2 < 0$. We also have

$$K_{\overline{S}} \cdot \overline{C} = (p^*K_S - \Delta) \cdot \overline{C} = (-2p^*L) \cdot \overline{C} - \Delta \cdot \overline{C} \leq -2L \cdot C \leq -2.$$

This implies the absurdity that $g(\overline{C}) \leq -1$. □

8.9 Polarized varieties (X, L) with sectional genus $g(L) \leq 1$

Let L be an ample and spanned line bundle on an irreducible normal Gorenstein projective variety, X. In this section we give the classification of polarized pairs (X, L) with sectional genus $g(L) \leq 1$. The results in this section are special cases of Fujita's results, (3.1.1), (3.1.2), and (3.1.3), and also follow in greater generality from Proposition (7.2.2). Because of the importance of this result in classical adjunction theory, we give an alternate less abstract proof.

Theorem 8.9.1. *Let X be an irreducible normal n-dimensional Gorenstein projective variety with $n \geq 2$, and let L be an ample line bundle on X with $g(L) = 0$. Assume that either $n = 2$ and there is an irreducible curve $C \in |L|$, or that $n \geq 3$ and L is ample and spanned. Then either*

1. $(X, L) \cong (\mathbb{P}^n, \mathcal{O}_{\mathbb{P}^n}(1))$ *or* $(\mathbb{P}^2, \mathcal{O}_{\mathbb{P}^2}(2))$*; or*

2. $X \subset \mathbb{P}^{n+1}$ *is a quadric and* $L \cong \mathcal{O}_{\mathbb{P}^{n+1}}(1)_X$*; or*

3. $(X, L) \cong (\mathbb{P}(\mathcal{E}), \xi_{\mathcal{E}})$ *for an ample and spanned vector bundle, $\mathcal{E}$, on $\mathbb{P}^1$.*

Proof. First assume that $n = 2$. In this case the singularities of X are finite and we will use the vanishing theorem, (2.2.4), without further comment.

L is spanned by global sections. To see this note that since C corresponds to a section of L vanishing only on C, it suffices to show that L_C is spanned by restrictions of global sections of L. Since $h^1(\mathcal{O}_C) = g(L) = 0$, C is isomorphic to a smooth $\mathbb{P}^1$ by Lemma (1.4.1). Therefore L_C is spanned on C, and if we show that $h^1(\mathcal{O}_X) = 0$ we will have a surjection $H^0(L) \to H^0(L_C)$ which will show the spannedness. Consider the sequence

$$0 \to -L \to \mathcal{O}_X \to \mathcal{O}_C \to 0.$$

Since $h^1(-L) = 0$ and $h^1(\mathcal{O}_C) = g(L) = 0$, we conclude that $h^1(\mathcal{O}_X) = 0$.

Note also that $h^2(\mathcal{O}_X) = 0$. This follows from Serre duality, $h^0(K_X) = h^2(\mathcal{O}_X)$, and the fact that by the genus formula $K_X \cdot L = -L^2 - 2 < 0$.

By (8.8.6) it can be assumed that $K_X + 2L$ is spanned and big. By (8.8.7) we have that $K_X + 2L$ is also ample. If $(K_X + 2L)^2 = 1$ then we have an isomorphism $(X, K_X + 2L) \cong (\mathbb{P}^2, \mathcal{O}_{\mathbb{P}^2}(1))$, i.e., $(X, L) \cong (\mathbb{P}^2, \mathcal{O}_{\mathbb{P}^2}(2))$. Thus we can assume without loss of generality that $(K_X + 2L)^2 \geq 2$. Let $d := L^2$. Since $K_X + 2L$ is ample and spanned, $g(K_X + 2L) \geq 0$, i.e.,

$$(K_X + K_X + 2L) \cdot (K_X + 2L) \geq -2.$$

This gives $(K_X + L) \cdot (K_X + 2L) \geq -1$ or $(K_X + L)^2 \geq 1$. By the Hodge index theorem this gives

$$d \leq d(K_X + L)^2 \leq ((K_X + L) \cdot L)^2 = 4,$$

i.e., $d \leq 4$. The Hodge index theorem also gives

$$2d \leq d(K_X + 2L)^2 \leq (d - 2)^2.$$

Since $1 \leq d \leq 4$ this is absurd.

We will now use induction on $\dim X$ to prove the result. Assume that it has been shown for $2 \leq \dim X < n$. We must show that it is true for $\dim X = n$. Choose a general $A \in |L|$. By (1.7.2) we know that (A, L_A) satisfies all the conditions of the theorem. Thus by the induction hypothesis A has rational singularities. Therefore by Elkik's theorem [213] we conclude that $\dim \operatorname{Irr}(X) \leq 0$. Thus by (8.8.6) it can be assumed that $K_X + nL$ is spanned and big. In particular $(K_X + nL)_A \cong K_A + (\dim A)L_A$ is spanned and big, which contradicts the induction hypothesis unless $n = 3$ and $(A, L_A) \cong (\mathbb{P}^2, \mathcal{O}_{\mathbb{P}^2}(2))$. In this case consider $2K_X + 5L$. We have the exact sequence

$$0 \to 2K_X + 4L \to 2K_X + 5L \to \mathcal{O}_A \to 0.$$

Since $K_X + 3L$ is spanned and big, we conclude that $h^1(2K_X + 4L) = 0$ and thus that there exists a nontrivial section of $2K_X + 5L$. Since

$$(2K_X + 5L) \cdot L^2 = \mathcal{O}_A \cdot L_A = 0$$

we conclude that $2K_X + 5L \cong \mathcal{O}_X$. Thus there exists a line bundle H on X such that $-K_X \cong 5H$ and $L \cong 2H$. Since $L \cong 2H$ we conclude that H is ample, but $-K_X \cong 5H$ is absurd by (3.3.2). □

Remark 8.9.2. Note that one consequence of the above is that if L is an ample divisor on an irreducible normal Gorenstein projective surface, S, with $L^2 = 1$, $h^0(L) > 0$, and $g(L) = 0$, then $(S, L) \cong (\mathbb{P}^2, \mathcal{O}_{\mathbb{P}^2}(1))$. Indeed $L^2 = 1$ implies that $C \in |L|$ is irreducible. Therefore we can use Theorem (8.9.1). We can proceed by computing degrees or by noting that by the proof of Theorem (8.9.1) it follows that L is spanned and $h^0(L) = 1 + h^0(\mathbb{P}^1, \mathcal{O}_{\mathbb{P}^1}(1)) = 3$. Thus $|L|$ gives a finite-to-one, generically one-to-one map of S onto $\mathbb{P}^2$. By Zariski's main theorem it follows that $S \cong \mathbb{P}^2$.

Theorem 8.9.3. *Let X be an irreducible normal n-dimensional Gorenstein projective variety with* $\dim \operatorname{Irr}(X) \leq 0$ *and with* $n \geq 2$. *Let L be an ample and spanned line bundle on X with $g(L) = 1$. Then either*

1. $h^1(\mathcal{O}_X) = 1$ *and* $(X, L) \cong (\mathbb{P}(\mathcal{E}), \xi_{\mathcal{E}})$ *for an ample and spanned vector bundle, $\mathcal{E}$, on a smooth curve of genus* 1; *or*
2. $h^1(\mathcal{O}_X) = 0$ *and* $K_X + (n-1)L \cong \mathcal{O}_X$.

Proof. Let $q := h^1(\mathcal{O}_X)$. By (8.7.1) we know that $1 = g(L) \geq q \geq 0$ with $q = g(L)$ only if we are in case 1). Thus we can assume without loss of generality that $h^1(\mathcal{O}_X) = 0$. By the usual induction argument we can find a nontrivial section of $K_X + (n-1)L$ that restricts to a nontrivial section, s, of $(K_X + (n-1)L)_C \cong \mathcal{O}_C$ for the smooth transverse intersection, C, of $n-1$ general elements of $|L|$. Since $(K_X + (n-1)L) \cdot L^{n-1} = 0$, we conclude that s is nowhere zero, i.e., $K_X + (n-1)L \cong \mathcal{O}_X$. □

8.10 Classification of varieties up to degree 4

The classification of algebraic sets up to degree 4 is classically well-known, see Weil [616], Swinnerton-Dyer [592], and Hartshorne's survey article [317]. In this section we give a crude classification of these sets that is sufficient for the results proved in the sequel.

Proposition 8.10.1. *Let L be an ample and spanned line bundle on an irreducible normal n-dimensional Gorenstein projective variety, X, with* $\dim \operatorname{Irr}(X) \leq 0$ *and* $n \geq 2$. *Assume that $d := L^n \leq 4$ and that $|L|$ gives rise to a morphism which is an embedding off a finite set.*

1. *If* $d = 1$ *then* $g(L) = 0$, $(X, L) \cong (\mathbb{P}^n, \mathcal{O}_{\mathbb{P}^n}(1))$.

2. *If* $d = 2$ *then* $g(L) = 0$, $X \subset \mathbb{P}^{n+1}$, *and* $L \cong \mathcal{O}_{\mathbb{P}^{n+1}}(1)_X$.

3. *If* $d = 3$ *then either:*

   (a) $X \subset \mathbb{P}^{n+1}$, $g(L) = 1$, *and* $L \cong \mathcal{O}_{\mathbb{P}^{n+1}}(1)_X$; *or*

   (b) $(X, L) \cong (\mathbb{P}(\mathcal{E}), \xi_{\mathcal{E}})$, *where* $\mathcal{E}$ *is an ample and spanned vector bundle on* $\mathbb{P}^1$ *with* $c_1(\mathcal{E}) = 3$, $n \leq 3$, *and* $g(L) = 0$.

4. *If* $d = 4$ *then either:*

   (a) $X \subset \mathbb{P}^{n+1}$, $g(L) = 3$, *and* $L \cong \mathcal{O}_{\mathbb{P}^{n+1}}(1)_X$; *or*

   (b) $(X, L) \cong (\mathbb{P}(\mathcal{E}), \xi_{\mathcal{E}})$, *where* $\mathcal{E}$ *is an ample and spanned vector bundle on* $\mathbb{P}^1$ *with* $c_1(\mathcal{E}) = 4$, $n \leq 4$, *and* $g(L) = 0$; *or*

   (c) $(X, L) \cong (\mathbb{P}^2, \mathcal{O}_{\mathbb{P}^2}(2))$ *and* $g(L) = 0$; *or*

   (d) $X \subset \mathbb{P}^{n+2}$ *is the complete intersection of* 2 *quadric hypersurfaces of* $\mathbb{P}^{n+2}$, $g(L) = 0$, *and* $K_X + (n-1)L \cong \mathcal{O}_X$.

In particular L is very ample in the above cases.

Proof. Note that $r := h^0(L) - n - 1 \geq 0$ with equality only if $d = 1$ and $(X, L) \cong (\mathbb{P}^n, \mathcal{O}_{\mathbb{P}^n}(1))$. Let $\varphi : X \to \mathbb{P}^{n+r}$ denote the morphism associated to $|L|$.

If $r = 1$ then since $\varphi(X)$ is a Cartier divisor of a manifold, it follows that $\varphi(X)$ is Cohen-Macaulay. Since φ embeds off a finite set, it follows that $\varphi(X)$ has at most a codimension 2 singular set, and, being Cohen-Macaulay, is normal. Thus by Zariski's main theorem it follows that φ is an embedding.

If $r \geq 2$ then by applying Castelnuovo's bound (1.4.9) to the pullback of a general linear $\mathbb{P}^{r+1}$ we see that $g(L) \leq 1$ with equality only if $(d, r) = (4, 2)$. If $g(L) = 0$ then by (8.9.1) we get all the $g(L) = 0$ cases. Using (3.2.4) we see that L is very ample in these cases.

Therefore we can assume without loss of generality that $(d, r, g(L)) = (4, 2, 1)$. By (8.6.1) we conclude that $h^1(\mathcal{O}_X) = 0$ and thus $K_X + (n-1)L \cong \mathcal{O}_X$ by (8.9.3). By (8.6.1) we also conclude that L is very ample, $h^1(tL) = 0$ for all $t \geq 1$, and that $\Gamma(L)^{\otimes t} \to \Gamma(tL)$ is surjective for all $t \geq 1$. By a simple induction down to a smooth curve section, by using the exact sequence

$$0 \to tL \to (t+1)L \to (t+1)L_A \to 0,$$

for a smooth $A \in |L|$ and $t = 0, 1$, we conclude that

$$h^0(2L) = 8 + \sum_{i=2}^{n}(i+3) = \frac{n^2 + 7n + 8}{2}.$$

We claim that X is the intersection of 2 quadric hypersurfaces. Note that

$$h^0(\mathcal{O}_{\mathbb{P}^{n+2}}(2)) = \frac{(n+3)(n+4)}{2}.$$

Let $\mathcal{J}_X$ be the ideal sheaf of X in $\mathbb{P}^{n+2}$. Thus

$$h^0(\mathcal{O}_{\mathbb{P}^{n+2}}(2) \otimes \mathcal{J}_X) \geq h^0(\mathcal{O}_{\mathbb{P}^{n+2}}(2)) - h^0(2L) \geq 2.$$

This means that there are two distinct quadrics, Q_1, Q_2, in $\mathbb{P}^{n+2}$ containing X. Since X is nondegenerate in $\mathbb{P}^{n+2}$, the quadrics are both irreducible, since otherwise X would be contained in a hyperplane of $\mathbb{P}^{n+2}$. Therefore $Q_1 \cap Q_2$ is of codimension 2 in $\mathbb{P}^{n+2}$. Thus $Q_1 \cap Q_2$ is a Cartier divisor in Q_1. Since Q_1 is Cohen-Macaulay we conclude that $Q_1 \cap Q_2$ is Cohen-Macaulay. Since the intersection $Q_1 \cap Q_2$ is of degree 4 and contains X, it must equal X up to algebraic sets of lower dimension. The intersection $Q_1 \cap Q_2$ is clearly connected, e.g., from the exact sequence

$$0 \to \mathcal{O}_{Q_1}(-(Q_1 \cap Q_2)) \to \mathcal{O}_{Q_1} \to \mathcal{O}_{Q_1 \cap Q_2} \to 0$$

we find $h^0(\mathcal{O}_{Q_1 \cap Q_2}) = 1$. Since connected Cohen-Macaulay algebraic sets are pure dimensional and reduced if generically reduced, we conclude that X is the intersection $Q_1 \cap Q_2$. □

Corollary 8.10.2. *Let L be an ample and spanned line bundle on an irreducible normal n-dimensional Gorenstein projective variety, X, with* $\dim \operatorname{Irr}(X) \leq 0$ *and $n \geq 2$. Assume that $L^n \leq 4$ and that $|L|$ gives rise to a morphism $\varphi : X \to \mathbb{P}^{n+r}$ which is an embedding off a finite set. If $g(L) \neq h^1(\mathcal{O}_X)$, then $K_X + (n-1)L$ is spanned by global sections.*

Proof. This is a simple check of the list in (8.10.1). □

Chapter 9

The adjunction mapping

From the classical survey [148] of Castelnuovo and Enriques, it is clear that the adjunction process of replacing a linear system on a smooth surface with its adjoint bundle was a major algebraic geometric tool at the turn of the century. Here we see applications to the birational classification of surfaces with hyperplane sections of genus less than or equal to 3. In Roth's paper [526] we see the technique used to carry out the birational classification of surfaces of low genus and degree. It was well understood that "often" when you project from a point on a surface, the surface is transformed into a birational surface in a lower dimensional space with the point of projection replaced with an exceptional line and with the degree reduced by 1. When trying to classify surfaces projectively, it was natural to try to classify the surfaces which were not the projections of other surfaces in this way. Classically, e.g., Semple and Roth [552, Chap. I, §4.52], such surfaces were called supernormal surfaces. The first reduction captures all the essentials of the supernormal surface, and has the important virtue of existing except in a few special completely understood cases.

Reading the paper of Roth, it was clear to the second author that the classical theory of projective classification could be redone biregularly if it was known that given a very ample line bundle, L, on a smooth projective surface, S, the spannedness and very ampleness of $K_S + L$ were completely understood. In the original version of the second author's paper [571], the theory of the first reduction, and the basic structure of the adjunction mapping were presented. Let d and $g(L)$ be the degree and the sectional genus of (S, L) respectively. The optimal spannedness result was conjectured, and proved by the classical technique of restriction to smooth curves in a wide variety of cases, e.g., if $h^1(\mathcal{O}_S) = 0$, or if $d \geq g(L)+1$, or if $g(L)$ is prime, etc. Ideas from Sakai's work [530] on the logarithmic canonical maps was one of the sources of inspiration for the second author's paper [571]. In [604] Van de Ven gave the first complete proof of the spannedness. This proof was based on Bombieri's technique [135] of using the Ramanujam vanishing theorem [510] to show the spannedness of pluricanonical maps. In [571] the second author finished classifying surfaces, whose projection from a line was lower dimensional, and as a consequence gave a second completely different proof of the existence theorem, and showed in the appendix of [571] and in [574] how to extend the results to higher dimensions.

The theory of the first reduction was developed in more generality by the second author [575, 576, 577].

The second author showed in [581] that for an ample and spanned line bundle, L, on a normal Gorenstein surface, S, satisfying very minimal conditions, $\Gamma(K_S+L)$ spans $K_S + L$ on $\mathrm{Sing}(S)$, and in fact the associated adjunction mapping embeds $\mathrm{Sing}(S)$. In [582], Sommese proved the fundamental spannedness theorem for $K_X+(n-1)L$ where L is a very ample line bundle on normal projective Gorenstein n-folds, X. In [24], Andreatta and Sommese proved a sharpened version of the fundamental spannedness of $K_X + (n-1)L$ that included the case of ample and spanned line bundles, L, on smooth n-folds, X. The spannedness result is an immediate consequence of the Reider criterion if $n = 2$, but for $n \geq 3$ is nontrivial since, given a point $x \in X$, the general $A \in |L|$ that contains x may be singular. The new technique added in [24], to deal with points where all sections are singular, was sharpened further by Lanteri, Palleschi and the second author in [382]. As a consequence they obtained optimal results under very general circumstances for the bundles $K_X + tL$ for $t \geq n$.

In §9.1 and §9.2 we will establish the basic spannedness theorem for $K_X + tL$ where $t \geq n-1$, and where L is an ample and spanned line bundle on an n-dimensional irreducible Gorenstein projective variety, X, with $\mathrm{cod}_X\mathrm{Sing}(X) \geq 3$ and $\dim \mathrm{Irr}(X) \leq 0$. We follow [581, 582, 24, 382], but with a new trick that allows us to bypass the second author's residue argument [582]. For more results on the spannedness of $K_X + (n-1)L$ with X singular we refer to the discussion after Theorem (9.2.1) and the quoted papers.

The precise theory that we develop in the following chapters requires smoothness of X plus for the most part very ampleness of L. Because of its foundational importance we have done the extra work to present a sharp form of the basic spannedness theorem for $K_X + (n-1)L$. For readers only interested in the case of very ample L on a smooth n-fold, §9.1 can be skipped.

9.1 Spannedness of adjoint bundles at singular points

We prove first the following general result.

Lemma 9.1.1. *Let L be an ample and spanned line bundle on an n-dimensional normal irreducible Gorenstein projective variety, X. If $L^n \leq 2$ then either $K_X + (n-1)L$ is spanned, or*

1. *$(X, L) \cong (\mathbb{P}^n, \mathcal{O}_{\mathbb{P}^n}(1))$; or*
2. *X is a quadric hypersurface in $\mathbb{P}^{n+1}$ and $L \cong \mathcal{O}_{\mathbb{P}^{n+1}}(1)_X$.*

Proof. If $L^n = 1$ then it follows from (3.1.4) that $(X, L) \cong (\mathbb{P}^n, \mathcal{O}_{\mathbb{P}^n}(1))$. Thus we can assume without loss of generality that $L^n = 2$.

Since L is spanned we can choose an $(n+1)$-dimensional linear subspace, $V \subset \Gamma(L)$, which spans L at all points. The morphism, $p : X \to \mathbb{P}^n$, given by V, expresses X as a two sheeted branched cover of $\mathbb{P}^n$. Thus $K_X \cong p^*\mathcal{O}_{\mathbb{P}^n}(b - n - 1)$ where the branch locus $B \in |\mathcal{O}_{\mathbb{P}^n}(2b)|$. Thus $K_X + (n-1)L \cong p^*\mathcal{O}_{\mathbb{P}^n}(b-2)$. This is spanned except in the case $b = 1$. In this case $K_X \cong -nL$, and (X, L) is as in case 2) by (3.1.6). □

The following lemma will allow us to avoid the residue arguments of [582]. Given a line bundle, L, on a projective variety, X, and a positive dimensional linear subspace $V \subset \Gamma(L)$, we let $|V| \subset |L|$ denote the corresponding linear subsystem of $|L|$. For a $x \in X$ we let $|V - x|$ denote the linear subsystem of $|V|$ consisting of the divisors $D \in |V|$ that contain x.

Recall that given a desingularization, $p : \overline{X} \to X$, the Grauert-Riemenschneider canonical sheaf, $\overline{K}_X := p_* K_{\overline{X}}$, of a normal projective variety is independent of the desingularization.

Lemma 9.1.2. *Let L be an ample line bundle on an irreducible normal projective surface, X. Assume that there is a vector subspace $V \subset H^0(X, L)$ that spans L at all points of X. Assume that X is not smooth and $L^2 \geq 5$. If $\mathrm{Sing}(X) \subset \mathrm{Bs}|V - x|$ for some $x \in \mathrm{reg}(X)$, then $\Gamma(\overline{K}_X + L)$ spans $\overline{K}_X + L$ at x.*

Proof. Let $p : \overline{X} \to X$ be a desingularization of X with the restriction $p_{p^{-1}(\mathrm{reg}(X))}$ a biholomorphism. Let $x' = p^{-1}(x)$. Consider the exact sequence

$$0 \to p_* K_{\overline{X}} + L \to K_X + L \to \mathcal{S} \to 0,$$

where $\mathcal{S}$ is a skyscraper sheaf supported on $\mathrm{Irr}(X)$. Since $p_* K_{\overline{X}} + L \cong p_*(K_{\overline{X}} + p^*L)$ it suffices to show that there is a global section of $K_{\overline{X}} + p^*L$ that does not vanish at x'. By using Reider's criterion (8.5.1) we see that if such a section does not exist, there is an effective nontrivial divisor $D \subset \overline{X}$ such that $x' \in D$ and either

1. $p^*L \cdot D = 0$ and $D^2 = -1$; or

2. $p^*L \cdot D = 1$ and $D^2 = 0$.

If case 1) occurs, then it would imply that $L \cdot p(D) = 0$. Since L is ample this implies that $p(D)$ is a point. Since $x' \in D$, this implies that $p(D) = x$, which is impossible since $p_{p^{-1}(\mathrm{reg}(X))}$ is a biholomorphism. Therefore we can assume without loss of generality that we are in case 2).

Write $D := A + B$ where $B \subset p^{-1}(\mathrm{Sing}(X))$ and $A \cap p^{-1}(\mathrm{reg}(X))$ is a dense open set of A. Note that $B \cdot p^*L = 0$ and therefore that $A \cdot p^*L = 1$. Therefore

$p(A) \cdot L = 1$ and we conclude that A is a smooth $\mathbb{P}^1$ mapped biholomorphically onto its image $A' := p(A)$.

We claim that A is disjoint from $p^{-1}(\mathrm{Sing}(X))$. Since L is ample and spanned, the base locus of $|L - x|$ is finite. Thus a general $\mathfrak{D} \in |L - x|$ does not contain A'. If A met $p^{-1}(\mathrm{Sing}(X))$ then A' would contain at least one singular point of X. Thus, since $\mathrm{Sing}(X) \subset \mathrm{Bs}|L - x|$, $\mathfrak{D} \cdot p(A) \geq 2$. This contradicts $L \cdot A' = 1$. Thus A is disjoint from B and $A' \subset \mathrm{reg}(X)$.

Since B goes to a finite set under p we know that $B^2 < 0$ by the Hodge index theorem unless B is empty. In any case $B^2 \leq 0$. Since $A' \subset \mathrm{reg}(X)$, $A^2 = A'^2$. Since by the last paragraph $A \cdot B = 0$, we conclude that $A^2 \geq 0$. By the Hodge index theorem we have $5A^2 \leq L^2 A'^2 \leq (L \cdot A')^2 = 1$. Thus we conclude that $A'^2 \leq 0$, and thus that $A'^2 = A^2 = 0$. By Lemma (4.1.10) we have the contradiction that X is a smooth $\mathbb{P}^1$-bundle. □

Lemma 9.1.3. *Let X be a smooth connected projective surface and L an ample and spanned line bundle on X. Assume that $L^2 \geq 5$. Then $K_X + L$ is spanned unless (X, L) is a scroll, and therefore a $\mathbb{P}^1$-bundle, over a smooth curve.*

Proof. From (8.5.1) we know that either $K_X + L$ is spanned or there exists an effective divisor D on X such that $L \cdot D - 1 \leq D^2 < L \cdot D/2 < 1$. Then $L \cdot D = 1$ and $D^2 = 0$. Since L is ample and spanned we have that $D \cong \mathbb{P}^1$. Therefore Lemma (4.1.10) implies that X is birationally a ruled surface and D is a fiber of a ruling of X. Since $L \cdot D = 1$ we see that X is in fact a $\mathbb{P}^1$-bundle, $p : X \to C$, over a smooth curve C. Moreover $(K_X + 2L) \cdot f = 0$ for any fiber, f, of p, so that $K_X + 2L \approx p^*H$ for some line bundle H on C. From the proof of Corollary (8.8.6) we see that either H is ample, and hence (X, L) is a scroll over C, or $X \cong \mathbb{P}^1 \times \mathbb{P}^1$ and $L \cong \mathcal{O}_{\mathbb{P}^1 \times \mathbb{P}^1}(1, 1)$, which is not the case since $L^2 \geq 5$. □

9.2 The adjunction mapping

In this section we prove the fundamental existence theorem for the adjunction mapping. For readers only interested in the case of a very ample line bundle, L, on a smooth n-dimensional projective manifold, X, the key fact (1.7.9) allows a significant simplification of the proof.

Theorem 9.2.1. *Let L be an ample and spanned line bundle on an irreducible n-dimensional normal Cohen-Macaulay projective variety, X. Assume that $n \geq 2$, $\mathrm{cod}_X \mathrm{Sing}(X) \geq 3$, $\dim \mathrm{Irr}(X) \leq 0$, and that $L^n \geq 5$. Then $\Gamma(K_X + (n-1)L)$ spans $(K_X + (n-1)L)_{\mathrm{reg}(X)}$, except if $g(L) = h^1(\mathcal{O}_X)$.*

Proof. First note that $g(L) = h^1(\mathcal{O}_X)$ implies that $K_X + (n-1)L$ has no sections. If $g(L) \geq 1$ this follows from (8.7.1). If $g(L) = 0$ then we have $(K_X + (n-1)L) \cdot L^{n-1} = -2 < 0$. Thus we can assume without loss of generality that $g(L) \neq h^1(\mathcal{O}_X)$. If $n = 2$ the result is true by Lemma (9.1.3).

Note that (1.7.2) implies that for $n > 2$, general $A \in |L - x|$, $x \in \text{reg}(X)$, satisfy all the hypotheses of the theorem. Note that for such an A it follows from the sequence

$$0 \to K_X + (n-2)L \to K_X + (n-1)L \to K_A + (n-2)L_A \to 0$$

that $\Gamma(K_X+(n-1)L)$ spans $(K_X+(n-1)L)_x$ if and only if $\Gamma(K_A+(n-2)L_A)$ spans $(K_A + (n-2)L_A)_x$. For $n > 2$ we also have $h^1(\mathcal{O}_X) = h^1(\mathcal{O}_A)$ and $g(L) = g(L_A)$. Thus the spannedness condition on X is equivalent to the condition on A.

For $n \geq 3$ we will prove the result by induction. First we must prove the case when $\dim X = 3$. If $n = 3$ then choose a point $x \in \text{reg}(X)$. If all elements $A \in |L - x|$ are singular at x then global sections of $K_X + 2L$ span $K_X + 2L$ at x by Theorem (8.8.1). If the general element $A \in |L - x|$ is smooth everywhere then $\Gamma(K_A + L_A)$ spans $(K_A + L_A)_x$ by (9.1.3) and $\Gamma(K_X + 2L)$ spans $(K_X + 2L)_x$. If the general element $A \in |L - x|$ is not everywhere smooth then since $\text{cod}_X\text{Sing}(X) \geq 3$ it follows that $\text{Sing}(A) \subset \text{Bs}|L - x|$. Moreover by (1.7.2) we have that A is irreducible and normal, and we can use (9.1.2) in place of (9.1.3).

As an induction hypothesis we assume that the result is true when $3 \leq \dim X < n$. Assume now that $n \geq 4$.

Choose a point $x \in \text{reg}(X)$. If all elements $A \in |L - x|$ are singular at x then global sections of $K_X + (n-1)L$ span $K_X + (n-1)L$ at x by Theorem (8.8.1). Thus we can assume that the general element $A \in |L - x|$ is smooth at x. By (1.7.2), (A, L_A) satisfies the conditions of the theorem. By the induction hypothesis $\Gamma(K_A + (n-2)L_A)$ spans $(K_A + (n-2)L_A)_x$. By the discussion in the second paragraph of the proof we are done. □

Recall that for an ample and spanned line bundle, L, on an irreducible projective variety, X, a line $\ell \subset X$ is a rational curve, satisfying $L \cdot \ell = 1$. Since L is spanned such a curve is smooth. By [24], the above proof for spannedness works when the codimension of the singular locus is ≥ 2, except at $x \in \text{reg}(X)$ where the set of points on lines containing x contains a codimension 2 component of $\text{Sing}(X)$. We do not know if such bad points can actually exist, but note that in the theorem of Lipman and the second author [408] there is a codimension 3 condition that can also be relaxed to the assumption that such bad points do not exist.

Corollary 9.2.2. *Let L be an ample and spanned line bundle on an irreducible n-dimensional Gorenstein projective variety, X. Assume that* $\dim \text{Irr}(X) \leq 0$ *and* $\text{cod}_X\text{Sing}(X) \geq 3$. *Assume that $n \geq 2$, and that either $L^n \geq 5$ or $|L|$ embeds X on the complement of a finite set. Then $K_X + (n-1)L$ is spanned except if either:*

1. $(X, L) \cong (\mathbb{P}^n, \mathcal{O}_{\mathbb{P}^n}(1))$; *or*

2. $(X, L) \cong (\mathbb{P}^2, \mathcal{O}_{\mathbb{P}^2}(2))$; *or*

3. $X \subset \mathbb{P}^{n+1}$ *is a quadric hypersurface and* $L \cong \mathcal{O}_{\mathbb{P}^{n+1}}(1)_X$; *or*

4. (X, L) *is a scroll over a curve.*

Proof. First assume that $L^n \geq 5$. By (8.7.1), (8.9.1), and Theorem (9.2.1) we are reduced to proving the corollary at singularities $x \in \mathrm{Sing}(X)$. This follows from (8.8.3).

Now assume that $L^n \leq 4$. By (8.10.1) we are done in this case. □

Corollary 9.2.3. *Let L be an ample and spanned line bundle on an irreducible n-dimensional Gorenstein projective variety, X. Assume that* $\dim \mathrm{Irr}(X) \leq 0$ *and* $\mathrm{cod}_X \mathrm{Sing}(X) \geq 3$. *Assume that* $n \geq 2$, *and that either* $L^n \geq 5$ *or* $|L|$ *embeds X on the complement of a finite set. Then the following are equivalent statements.*

1. $K_X + (n-1)L$ *is spanned.*

2. $g(L) \neq h^1(\mathcal{O}_X)$.

3. $K_X + (n-1)L$ *is nef.*

4. $h^0(K_X + (n-1)L) \neq 0$.

Proof. Use (9.2.2) combined with (8.7.1) and (8.9.1) to see the equivalence of the first two statements. The equivalence of the first and last two statements follows by noting that the exceptions, (X, L), listed in Corollary (9.2.2) to $K_X + (n-1)L$ being spanned have $K_X + (n-1)L$ nonnef and $h^0(K_X + (n-1)L) = 0$. □

Let L be an ample line bundle on a normal irreducible projective variety, X, of dimension $n \geq 2$. If $h^0(K_X + (n-1)L) \geq 1$, the rational mapping $\phi : X \to \mathbb{P}_{\mathbb{C}}$ associated to $H^0(K_X+(n-1)L)$ is called the *adjunction mapping*. Theorem (9.2.1) shows that this map is actually a morphism under modest conditions.

The Example (2.2.10) shows that the most we can expect if $\dim \mathrm{Irr}(X) > 0$ is spannedness on $X - \mathrm{Irr}(X)$.

Lemma 9.2.4. *Let L be a nef and big line bundle on an irreducible normal* $\mathbb{Q}$-*Gorenstein Cohen-Macaulay projective variety, X, of dimension n. Let* $a > 0$ *and* $b > 0$ *be integers such that* $aK_X + bL$ *is invertible and spanned by global sections. Let* ϕ *be the morphism associated to* $aK_X + bL$. *Either* $\dim \phi(X) = n$ *or* $\dim \phi(X) \leq n - b/a + 1$. *In particular if* $K_X + (n-1)L$ *is invertible and spanned by global sections, then either* $\dim \phi(X) = n$ *or* $\dim \phi(X) \leq 2$.

Proof. Any connected component, F, of a general fiber of ϕ is Cohen-Macaulay, normal, and $\mathbb{Q}$-Gorenstein, e.g., by (1.7.1). By (2.1.2) the restriction, L_F, of L to F is nef and big. Note that $aK_F + bL_F \cong \mathcal{O}_F$ and apply (3.3.2) to (F, L_F). □

Problem 9.2.5. *Find all pairs (X, L) consisting of an ample and spanned line bundle, L, on a connected n-dimensional projective manifold, X, such that $K_X + (n-1)L$ is not spanned, but $h^0(K_X + (n-1)L) > 0$.*

Using the results in this chapter it is easy to see that we need only look at the problem when $n \geq 2$, $L^n = 3, 4$, and $|L|$ does not give an embedding off a finite set. Note that such examples exist. For example take S a smooth projective surface with $-K_S$ ample and $K_S^2 = 1$. Then, as we have seen in (8.5.2), $L := -2K_S$ is spanned, but $-K_S$ cannot be spanned since it is ample with $h^0(-K_S) = 2$.

Chapter 10

Classical adjunction theory of surfaces

Let S be a smooth surface and let L be a very ample line bundle on S. If $h^1(\mathcal{O}_S) = 0$, then the second author showed in [571] that the adjunction mapping was an embedding on the first reduction except in some completely classified cases. Moreover Van de Ven [604] proved a general result on when $|K_S + L|$ gives an embedding for surfaces without lines, with $L^2 \geq 10$, and with $h^0(L) \geq 7$.

Ein [200] made some improvements of the embedding result for the adjunction mapping in the cases not covered by the results of [571, 604]. There quickly began to be many papers using and developing adjunction theory. We note especially a few of the early papers dealing with the first reduction, e.g., the classification (especially in degree ≤ 8) by Okonek [485, 486, 487, 488, 489], the work of low degree and low genus classification by Ionescu, [331, 333, 334], the work of Livorni [409, 410, 411, 412] on classification of surfaces with sectional genus ≤ 6, the results of Lanteri and Palleschi [375] on the very useful stable part of adjunction theory for merely ample divisors on surfaces (see also Sommese [581]), the applications to dual varieties by Ein [203, 204].

Optimal results on the very ampleness of adjunction bundles on surfaces were found by Serrano [554] and Sommese and Van de Ven [586]. The method of Serrano was based on Miyaoka's development of the technique of composition series used by Kodaira in his work on pluricanonical mappings (see Sakai [534] for further development of this technique). The method used by Sommese and Van de Ven is based on Reider's results [522] on very ampleness of adjoint bundles. The results of Sommese and Van de Ven [586] are sharper than the other results, e.g., they include the complete result on the existence of the 4 exceptions to the very ampleness of the adjoint bundle, and they cover the case of dimension ≥ 2.

The main references for this chapter are Sommese [571] and Sommese and Van de Ven [586]. We follow the second author's development [571] of the general theory of the adjunction mapping, and the classification of the pairs, (S, L), with L very ample on a smooth projective surface S, $h^1(\mathcal{O}_S) = 0$, and $K_S + L$ not very ample. This argument applies to cases with $\deg S \leq 8$, which cannot be handled by any other method.

Recently there has been a developing interest in adjunction theory in characteristic $p > 0$. The key papers here are due to Andreatta and Ballico [11, 12, 13], Ballico, Chiantini, and Monti [60], and Nakashima [476].

10.1 When the adjunction mapping has lower dimensional image

We need the concept of a reduction, which is defined more generally in Chapter 7. Let L be an ample line bundle on a smooth connected projective surface, S. We say that a pair, (S', L') and a surjective morphism $\pi : S \to S'$ is a *reduction* if:

1. S' is a smooth projective surface, and π expresses S as S' with a finite set B blown up; and
2. L' is an ample line bundle such that $L \cong \pi^* L' - \pi^{-1}(B)$.

Note that given a map $\pi : S \to S'$ satisfying 1), the bundle $L' := (\pi_* L)^{**}$ satisfies 2), and $K_S + L \cong \pi^*(K_{S'} + L')$.

Lemma 10.1.1. *Let L be an ample line bundle on a smooth connected projective surface, S. Let $r : S \to Y$ be a morphism with connected fibers onto a normal variety Y such that $K_S + L \cong r^* H$ for an ample line bundle H on Y.*

1. *If $\dim r(S) = 0$, then $L \cong -K_S$.*
2. *If $\dim r(S) = 1$, then there is a not necessarily unique factorization, $r = p \circ \pi$, where $\pi : S \to S'$ is a reduction and $p : S' \to Y$ is a $\mathbb{P}^1$-bundle with $L'_f \cong \mathcal{O}_{\mathbb{P}^1}(2)$ for any fiber f of p. In particular $q(S) = q(Y)$.*
3. *If $\dim r(S) = 2$, then (Y, L_Y) with $L_Y := (r_* L)^{**}$ and r is a reduction, and in this case (Y, L_Y) and the morphism $r : S \to Y$ is called the first reduction.*

Proof. Since case 1) is trivial, we assume without loss of generality that $\dim r(S) \geq 1$.

First assume that $\dim r(S) = 1$. Since $(K_S + L)_f \cong \mathcal{O}_f$ for a general fiber, f of π, we see that $(f, L_f) \cong (\mathbb{P}^1, \mathcal{O}_{\mathbb{P}^1}(2))$. Note that a singular fiber cannot be irreducible by Corollary (4.1.3) applied to a neighborhood of the fiber. Note also that since Y is normal fibers of r are Cartier divisors. Thus if there is a singular fiber, f', it follows from $L \cdot f' = L \cdot f = 2$ that either $f = 2\ell$ for an irreducible curve ℓ or $f' = A + A'$ where A and A' are irreducible. In the former case $K_S \cdot \ell = K_S \cdot f/2 = -1$ and $\ell^2 = f^2/4 = 0$ which contradicts parity. In the latter case $A \cdot (A + A') = A \cdot f = 0$ and similarly $A' \cdot (A + A') = 0$. Since all fibers are connected we see that $A^2 = -A \cdot A' < 0$ and $A'^2 = A^2 < 0$. Since $K_S \cdot (A + A') = -2$ we see that either $K_S \cdot A < 0$ or $K_S \cdot A' < 0$. By renaming if necessary we can assume that $K_S \cdot A < 0$. Thus $2g(A) - 2 = K_S \cdot A + A^2 < 0$. This implies by Lemma (1.4.1) that $A \cong \mathbb{P}^1$ and $K_S \cdot A + A^2 = -2$. Thus $A^2 = -1$, $K_S \cdot A = -1$, and therefore we conclude $K_S \cdot A' = -1$. From this we conclude that $A' \cong \mathbb{P}^1$. Note also that $A \cdot A' = 1$. Let $\pi : S \to S'$ be the morphism obtained by blowing down one irreducible component for each reducible fiber of r. Let $L' = (\pi_* L)^{**}$. Since S' is smooth all Weil divisors are Cartier, and thus L' being a reflexive rank 1 coherent sheaf is invertible. Then L' is ample since L' is ample outside of a finite set of points. Since $L \cdot \ell = 1$ for any positive dimensional fiber, ℓ, of the map π, we see that $L \cong \pi^* L' - \pi^{-1}(B)$ where B is the finite set of S' blown up by π. Thus (S', L') is a (not necessarily uniquely determined) reduction of (S, L). By construction all fibers of $S' \to Y$ are smooth $\mathbb{P}^1$'s, and thus S' is a $\mathbb{P}^1$-bundle, $p : S \to Y$, over Y.

Now assume that $\dim r(S) = 2$. If r is an isomorphism there is nothing to prove. Therefore let A be an irreducible component of a positive dimensional fiber of r. Since $r^* H$ is nef and big and $A \cdot r^* H = 0$ we conclude that $A^2 < 0$. Since $0 = (K_S + L) \cdot A > K_S \cdot A$ we conclude by Lemma (1.4.1) that $A \cong \mathbb{P}^1$, $A^2 = -1$, and $K_S \cdot A = -1$. Therefore $L \cdot A = 1$. We claim that positive dimensional fibers of r are irreducible. Assume otherwise that there are two distinct components A, A' of a positive dimensional fiber. Since fibers are connected we can assume that $A \cdot A' > 0$. We have $A^2 = A'^2 = -1$. Since $(A + A') \cdot r^* H = 0$ we conclude as before that $(A + A')^2 < 0$. This gives the absurdity that $-1 \geq (A + A')^2 = -2 + 2A \cdot A' \geq 0$. Thus fibers are irreducible. Since smooth rational curves with self-intersection -1 can be smoothly contracted we see that there is a map $\pi : S \to S'$ with S' the smooth compact complex surface obtained by smoothly contracting all the positive dimensional fibers of r. The map r factors as $\sigma \circ \pi$ with $\sigma : S' \to Y$ a one-to-one morphism from a smooth surface onto a normal surface. By Zariski's main theorem we conclude that σ is an isomorphism. The rest of the argument is the same as in the case $\dim r(S) = 1$. □

Lemma 10.1.2. *Let L be an ample line bundle on a smooth connected projective surface S. Let $q := h^1(\mathcal{O}_S)$, $p_g := h^2(\mathcal{O}_S)$, and $g := g(L)$. Then $h^0(K_S + L) = p_g - q + g$. If $p_g > 0$ and $h^0(L) > 0$ then $g - q \geq h^0(L) - 1$. Assume further that $K_S + L$ is spanned by global sections, and that letting $\phi : S \to \mathbb{P}_{\mathbb{C}}$ be the morphism associated to $|K_S + L|$, we have $\dim \phi(S) = 2$. Let $\phi = s \circ r$ be the Remmert-Stein factorization with $r : S \to Y$ a morphism with connected fibers onto a normal surface Y, and $s : Y \to \mathbb{P}_{\mathbb{C}}$ has finite fibers. In the following note that $K_S^2 - L^2 = K_Y^2 - L_Y^2$, where $L_Y := (r_* L)^{**}$.*

1. $(K_S + L)^2 \geq g + q - 2$, *i.e.*, $K_S^2 - L^2 + 3g \geq q + 2$. *Moreover equality happens only if* $(Y, K_Y + L_Y)$ *is either a scroll over a smooth curve of genus* q, *or* $(\mathbb{P}^2, \mathcal{O}_{\mathbb{P}^2}(a))$ *with* $a = 1, 2$, *or* $(\mathbb{P}^1 \times \mathbb{P}^1, \mathcal{O}_{\mathbb{P}^1 \times \mathbb{P}^1}(1, 1))$.

2. *If the Kodaira dimension of* S *is greater than or equal to zero, then* $(K_S + L)^2 \geq 2(p_g - q + g - 2)$, *i.e.*, $K_S^2 - L^2 + 2g + 2q \geq 2p_g$. *If equality happens and* $|K_S + L|$ *gives rise to a generically one-to-one map, then* Y *is a* $K3$ *surface, i.e.*, $K_Y \cong \mathcal{O}_Y$ *and* $h^1(\mathcal{O}_Y) = 0$.

Proof. Using the vanishing of $h^i(K_S + L) = 0$ for $i > 0$ and Serre duality, we see that $h^0(K_S + L) = \chi(K_S + L) = \chi(-L)$. Using the Riemann-Roch theorem we get

$$\chi(-L) = \frac{L^2 + K_S \cdot L}{2} + \chi(\mathcal{O}_S) = p_g + g - q.$$

If $h^0(K_S) > 0$ and $h^0(L) > 0$ then $h^0(K_S + L) \geq h^0(K_S) + h^0(L) - 1$, by Lemma (1.1.6), and hence $g - q \geq h^0(L) - 1$.

In what follows not that $(K_S + L) \cdot r^* H = (K_Y + L_Y) \cdot H$ for any Cartier divisor H on Y, e.g., $(K_S + L)^2 = (K_Y + L_Y)^2$.

Pass to the first reduction (Y, L_Y) of (S, L). Since $K_Y + L_Y$ is ample and spanned, it follows from Theorem (8.9.1) and Theorem (8.7.1) that $g(K_Y + L_Y) \geq q$ with equality only if $(Y, K_Y + L_Y)$ is as described in case 1). Since $g(K_Y + L_Y) \geq q$ is equivalent to

$$(K_Y + L_Y) \cdot (K_Y + K_Y + L_Y) = 2g(K_Y + L_Y) - 2 \geq 2q - 2,$$

we obtain case 1). Indeed note that

$$(K_Y + L_Y) \cdot (K_Y + K_Y + L_Y) = 2(K_Y + L_Y)^2 - 2g + 2 = 2(K_S + L)^2 - 2g + 2.$$

If the map $s : Y \to \mathbb{P}^{p_g + g - q - 1}$ is generically one-to-one then Theorem (8.1.1) gives the inequality in 2). Corollary (8.6.2) implies the equality statement of case 2). If s is generically t-to-one for some $t \geq 2$ then we have $(K_Y + L_Y)^2 = t \deg s(Y) \geq 2(p_g + g - q - 2)$. □

Let S be a smooth surface and L a very ample line bundle on S. In view of (9.2.2) we may assume that $K_S + L$ is spanned. Let ϕ be the morphism associated to $|K_S + L|$. In the remaining part of this section we will consider the case when $\dim \phi(S) < 2$. The aim is to classify the cases when s is not an embedding. Sommese [571, Proposition (2.1.1)] showed that when $\dim \phi(S) = 1$, then s is an embedding except possibly if $d = 7, 8$, $g(L) = 3$, and $q(S) = 1$. Ionescu [333, Proposition 3.9] showed $d = 7$ did not happen, but $d = 8$ does. The use of the double point inequality, Lemma (8.2.1), simplifies the original arguments.

Theorem 10.1.3. *Let L be a very ample line bundle on a smooth connected projective surface, S. Furthermore assume that $K_S + L$ is spanned, and let $\phi = s \circ r$ be the Remmert-Stein factorization of the morphism, ϕ, given by $\Gamma(K_S + L)$. If* $\dim \phi(S) = 1$, *then s is an embedding unless $d = 8$, $q(S) = 1$, $g(L) = 3$. In this case S is a $\mathbb{P}^1$-bundle over an elliptic curve of invariant $e = -1$ under r and $L \sim 2E + f$, where E, f denote a section of minimal self-intersection $E^2 = 1$ and a fiber of r.*

Proof. By Lemma (10.1.1), $r : S \to Y$ is a conic fibration over a smooth curve Y of genus $g(Y) = q(S)$ with $K_S + L \approx r^*H$ for some ample line bundle H on Y. Note that $H = s^*\mathcal{O}_{\mathbb{P}_\mathbb{C}}(1)$. If $q(S) = q(Y) = 0$, the line bundle H is very ample and therefore s is an embedding. So we can assume $q(S) > 0$. Let C be a smooth element of $|L|$. Since $L \cdot f = 2$ for any fiber, f, of r, the restriction r_C, of r to C is a two-to-one cover $r_C : C \to Y$, so that $r^*H \cdot L = \deg(r_C^* H) = 2\deg H$. Therefore

$$2g(L) - 2 = (K_S + L) \cdot L = r^*H \cdot L = 2\deg H,$$

which gives

$$\deg H = g(L) - 1. \tag{10.1}$$

Let (S', L') and $\pi : S \to S'$ be a reduction as is obtained in Lemma (10.1.1) by blowing down one irreducible component for each reducible fiber of r. Since S' is a $\mathbb{P}^1$-bundle, $p : S \to Y$, over Y, we find from

$$0 = (K_S + L)^2 = (K_{S'} + L')^2 = K_{S'}^2 + 4g(L) - 4 - L'^2$$

and $K_{S'}^2 = 8(1 - q(S))$ that $4g(L) + 4 - 8q(S) - L'^2 = 0$. By combining this with equation (10.1) we get

$$\deg H = g(L) - 1 = \frac{L'^2}{4} + 2q(S) - 2. \tag{10.2}$$

Hence L'^2 is divisible by 4. Let $L'^2 \geq 9$. Then equation (10.2) gives $\deg H > 2q(S) = 2g(Y)$, so that H is very ample. Therefore s is an embedding. Thus we can assume that $L'^2 \leq 8$.

Since $\deg H \geq 1$ we conclude that $g(L) \geq 2$, and since $q(S) > 0$ we conclude that $h^0(L) \geq 5$. Therefore by Castelnuovo's bound for the genus of a curve section we see that $L'^2 \geq L^2 := d \geq 5$. Since L'^2 is divisible by 4 and $L'^2 \geq d \geq 5$, we must have $L'^2 = 8$, so equation (10.2) yields

$$\deg H = g(L) - 1 = 2q(S). \tag{10.3}$$

Note that π is the blowing up of S' at $L'^2 - d = 8 - d$ points. Then

$$K_S^2 = K_{S'}^2 - (8 - d) = d - 8q(S). \tag{10.4}$$

Hence Lemma (8.2.1) applies to give

$$d(d-5) - 10g(L) + 22 - 12q(S) \geq 2(d - 8q(S))$$

or $d(d-7) - 10g(L) + 22 \geq -4q(S)$. By using equation (10.3) we find

$$d(d-7) + 20 \geq 8g(L). \tag{10.5}$$

If $d \leq 7$, the above inequality gives $g(L) \leq 2$. Since $q(S) > 0$ this contradicts equation (10.3).

Thus we conclude that $d = d' = 8$. Hence $(S, L) \cong (S', L')$. From equation (10.5) we get $g(L) \leq 3$ and from equation (10.3) we conclude that $q(S) = 1$, $g(L) = 3$.

By the above, we know that S is a $\mathbb{P}^1$-bundle, $r : S \to Y$, over a smooth elliptic curve, Y, of invariant e. Let E, f denote a section of minimal self-intersection $E^2 = -e$ and a fiber of r. Since (S, L) is a conic fibration under r we have $L \sim 2E + bf$. From $d = L^2 = 8$ we get $b = e + 2$. Note that $L \cdot E = 2 - e \geq 3$ since L is very ample and E is an elliptic curve. Thus $e \leq -1$. Since $e \geq -1$ by Theorem (3.2.11), we conclude $e = -1$ and $b = 1$. Note also that $L \sim 2E + f$ is in fact very ample by (8.5.8). □

Remark 10.1.4. Let S be a Del Pezzo surface as in (10.1.3). Then $d = L^2 \leq 9$ with equality if and only if $(S, L) \cong (\mathbb{P}^2, \mathcal{O}_{\mathbb{P}^2}(3))$.

Let $d = 8$. Since $L \approx -K_S$, we have $K_S^2 = 8(1 - q(S)) = 8$. Therefore S is a Hirzebruch surface $\mathbb{F}_r$, $r \geq 0$. Then $-K_S \sim 2E + (2 + r)f$ where E, f is a basis for the second integral homology $H_2(\mathbb{F}_r, \mathbb{Z})$. Since $-K_S$ is ample we have $2 + r \geq 2r + 1$ (see Theorem (3.2.11)). Then either $r = 0$ and $L \sim 2E + 2f$ or $r = 1$ and $L \sim 2E + 3f$.

10.2 Surfaces with sectional genus $g(L) \leq 3$

Let S be a smooth connected surface and L a very ample line bundle on S. In this section we give the classification of (S, L) with $g(L) \leq 3$. This classification will be needed in the study of the very ampleness of the adjoint bundle on the first reduction, and leads to two of the four exceptions to the very ampleness of the adjoint bundle. Let $d := L^2$.

We start with a classification of first reductions when the adjoint bundle $K_S + L$ is nef and big and of degree 1 or 2.

To start we need a lemma of Lanteri, Palleschi, and Sommese [382].

Lemma 10.2.1. *Let H be an ample and spanned line bundle on a smooth n-dimensional projective manifold, X. Let $p : \overline{X} \to X$ be the blowup of X at a* *single point $x \in X$. Then $H' := p^*H - p^{-1}(x)$ is nef, and if $(X, H) \not\cong (\mathbb{P}^n, \mathcal{O}_{\mathbb{P}^n}(1))$ then H' is also big.*

Proof. To show nefness we must show that given an irreducible curve C on $\overline{X}$, it follows that $(p^*H - p^{-1}(x)) \cdot C \geq 0$. If $p(C) = x$ we have $H' \cdot C = -p^{-1}(x) \cdot C > 0$. Since H is ample and spanned the base locus of $|H - x|$ is finite. Thus if $p(C) \neq x$ we conclude that there is a divisor $D \in |H|$ which contains x and which does not contain $p(C)$. Thus $D' := p^*D - p^{-1}(x)$ is an effective divisor that does not contain C. This shows that $H' \cdot C = D' \cdot C \geq 0$.

If H' is not big we have $0 = H'^n = H^n - 1$. Since $H^n = 1$ the map given by $|H|$ is therefore finite-to-one, generically one-to-one onto a degree 1 subspace of $\mathbb{P}_{\mathbb{C}}$. Thus the image is $\mathbb{P}^n$ and by Zariski's main theorem we have an isomorphism. □

Proposition 10.2.2. *Let Y be a smooth connected projective surface with $-K_Y$ ample and $K_Y^2 = 2$. Let $p : S \to Y$ be the blowing up at any point $y \in Y$. If we set $p^{-1}(y) = E$, then $p^*(-2K_Y) - E$ is very ample.*

Proof. Let $\mathcal{L} := -3p^*K_Y - 2E$. Write $\mathcal{L} = -p^*K_Y + 2(-p^*K_Y - E)$. Note that $-K_Y$ is ample and spanned by (8.5.2) since $K_Y^2 = 2$. Therefore from Lemma (10.2.1) we know that $-K_S \cong -p^*K_Y - E$ is nef, and it is also big since $(-p^*K_Y - E)^2 = 2 - 1 = 1$. Thus $\mathcal{L}$ is nef and big.

Since $\mathcal{L}^2 = 14$, we can use Theorem (8.5.1), with $k = 1$, to show that $p^*(-2K_Y) - E \approx K_S + \mathcal{L}$ is very ample. Note that $\mathcal{L} \cdot D > 0$ for any effective curve D. To see this note that we can assume that D is irreducible and reduced. Therefore if $D \neq E$, then $\mathcal{L} \cdot D \geq -p^*K_Y \cdot D > 0$. If $D = E$, then $\mathcal{L} \cdot D = 2(-p^*K_Y - E) \cdot D = 2$. Thus to show that $p^*(-2K_Y) - E \approx K_S + \mathcal{L}$ is very ample we have only to use Reider's criterion (8.5.1) to exclude the following possibilities for an effective divisor D on S.

1. $\mathcal{L} \cdot D = 1$, $D^2 = 0, -1$;
2. $\mathcal{L} \cdot D = 2$, $D^2 = 0$.

Assume that $\mathcal{L} \cdot D = 1$. Then

$$\mathcal{L} \cdot D = p^*(-K_Y) \cdot D + 2(-p^*K_Y - E) \cdot D = 1.$$

Since $-p^*K_Y$ and $-(p^*K_Y + E)$ are nef and since $K_S \cong p^*K_Y + E$ we conclude that $K_S \cdot D = (p^*K_Y + E) \cdot D = 0$. Thus we have the contradiction to parity unless $K_S \cdot D = D^2 = 0$. But since $-K_S$ is nef and big, the Hodge index theorem gives the contradiction that $D^2 < 0$.

Let D be as in 2). Since $\mathcal{L} \cdot D = 2$ we see that either $p^*(-K_Y) \cdot D = 0$, $(-p^*K_Y - E) \cdot D = 1$ or $p^*(-K_Y) \cdot D = 2$, $(-p^*K_Y - E) \cdot D = 0$. In the first case, we get the parity contradiction $(K_S + D) \cdot D = -1$. In the second case since $-p^*K_Y - E$ is nef and big, we conclude by the Hodge index theorem that either $D^2 < 0$ or $D \sim 0$. This contradicts the assumption that $D^2 = 0$ and D is effective. □

Theorem 10.2.3. *Let L be a very ample line bundle on a smooth connected projective surface, S. Assume that $K_S + L$ is spanned and big and let (Y, L_Y) be the first reduction of (S, L). If $(K_S + L)^2 = 1$, then $(Y, L_Y) \cong (\mathbb{P}^2, \mathcal{O}_{\mathbb{P}^2}(4))$. If $(K_S + L)^2 = 2$ then (Y, L_Y) is one of the following:*

1. *$(Y, L_Y) \cong (\mathbb{P}^1 \times \mathbb{P}^1, \mathcal{O}_{\mathbb{P}^1 \times \mathbb{P}^1}(3, 3))$, and $K_Y + L_Y$ is very ample;*
2. *$-K_Y$ is ample with $K_Y^2 = 2$, $L_Y \approx -2K_Y$, and $|K_Y + L_Y|$ expresses Y as a 2-sheeted branched cover of $\mathbb{P}^2$. In this case, $g(L) = 3$, and if $(S, L) \ncong (Y, L_Y)$, then S is obtained from Y by blowing up a single point.*

Proof. Recall that $K_Y + L_Y$ is ample. Note also that $K_Y + L_Y$ is spanned since $K_S + L$ is spanned. If $(K_Y + L_Y)^2 = 1$, then Y is isomorphic to $\mathbb{P}^2$ under the morphism given by $\Gamma(K_Y + L_Y)$. Let $L_Y \cong \mathcal{O}_{\mathbb{P}^2}(a)$. From $K_Y + L_Y \cong \mathcal{O}_{\mathbb{P}^2}(a - 3) \cong \mathcal{O}_{\mathbb{P}^2}(1)$ we find $L_Y \cong \mathcal{O}_{\mathbb{P}^2}(4)$.

Thus we can assume $(K_Y + L_Y)^2 = 2$. Then either $h^0(K_Y + L_Y) = 3$ or $h^0(K_Y + L_Y) = 4$. If $h^0(K_Y + L_Y) = 4, \Gamma(K_Y + L_Y)$ gives a degree one finite morphism, $s : Y \to Q$, with Q a quadric in $\mathbb{P}^3$. Note that Q is normal since Y is irreducible. Hence s is an isomorphism by the Zariski main theorem. Therefore $K_Y + L_Y$ is very ample and we are in case 1).

Thus we can assume that $h^0(K_Y + L_Y) = 3$, so that $\Gamma(K_Y + L_Y)$ gives a two-to-one cover, $s : Y \to \mathbb{P}^2$. Let $B \in |\mathcal{O}_{\mathbb{P}^2}(2b)|$ be the branch locus of s, and let b be a positive integer. From $K_Y \cong s^*\mathcal{O}_{\mathbb{P}^2}(b - 3)$, $K_Y + L_Y \cong s^*\mathcal{O}_{\mathbb{P}^2}(1)$ we see that $L_Y \cong s^*\mathcal{O}_{\mathbb{P}^2}(4 - b)$. Then $1 \leq b \leq 3$.

If $b = 1$, $K_Y \cong s^*\mathcal{O}_{\mathbb{P}^2}(-2)$ and $s^*\mathcal{O}_{\mathbb{P}^2}(1)$ is ample since s is a finite map. It thus follows that $Y \cong \mathbb{P}^1 \times \mathbb{P}^1$, $L_Y \cong s^*\mathcal{O}_{\mathbb{P}^2}(3)$ and hence $K_Y + L_Y$ is very ample since it is ample. If $b = 3$, then $L_Y \cong s^*\mathcal{O}_{\mathbb{P}^2}(1)$, so $L^2 \leq L_Y^2 = 2$ and $K_Y \cong \mathcal{O}_Y$. This contradicts Proposition (8.10.1).

Let $b = 2$. Then $K_Y \cong s^*\mathcal{O}_{\mathbb{P}^2}(-1)$, $L_Y \cong s^*\mathcal{O}_{\mathbb{P}^2}(2)$. Therefore Y is a Del Pezzo surface with $K_Y^2 = 2$ and $L_Y \approx -2K_Y$. Note that L_Y is very ample by (8.5.2). To conclude the proof we have to show that, if $(S, L) \ncong (Y, L_Y)$, then (S, L) is obtained from the reduction, (Y, L_Y), by blowing up a single point. The Riemann-Roch theorem gives

$$h^0(L_Y) = h^0(-2K_Y) = 7. \tag{10.6}$$

Assume that $r : S \to Y$ is the blowing up at two distinct points $x, y \in Y$. Let $\mathfrak{m}_x$, $\mathfrak{m}_y$ be the ideal sheaves defining x and y in Y. By using the Leray spectral sequence, we have

$$h^0(L) = h^0(r^*L_Y - r^{-1}(x) - r^{-1}(y)) = h^0(L_Y \otimes \mathfrak{m}_x \otimes \mathfrak{m}_y). \tag{10.7}$$

Since $L_Y \approx -2K_Y$ is very ample, the exact sequence

$$0 \to L_Y \otimes \mathfrak{m}_x \otimes \mathfrak{m}_y \to L_Y \to \mathbb{C}_x \oplus \mathbb{C}_y \to 0$$

gives $h^0(L_Y \otimes \mathfrak{m}_x \otimes \mathfrak{m}_y) = h^0(L_Y) - 2$. Thus by combining equations (10.6) and (10.7) we get $h^0(L) = 5$. Note that $K_S^2 = K_Y^2 - 2 = 0$, $L^2 = L_Y^2 - 2 = 6$, $g(L) = g(L_Y) = 3$ and $\chi(\mathcal{O}_S) = 1$. Then Lemma (8.2.1) applies to give a numerical contradiction.

If $r : S \to Y$ is the blowing up at more than two points, say at # points $y_1, \ldots, y_\#$, $\# \geq 3$, we have

$$h^0(L) = h^0(L_Y \otimes \mathfrak{m}_{y_1} \otimes \cdots \otimes \mathfrak{m}_{y_\#}) \leq h^0(L_Y) - 2 = 5$$

as well as $d := L^2 = 8 - \#$, $K_S^2 = 2 - \#$. If $h^0(L) = 5$, Lemma (8.2.1) applies again to give a numerical contradiction. If $h^0(L) = 4$, then $K_S \cong \mathcal{O}_S(d - 4)$. Since S is rational, $d \leq 3$, which implies that $g(L) \leq 1$. This is impossible since $2g(L) - 2 = (K_Y + L_Y) \cdot L_Y \geq 1$ by the ampleness of $K_Y + L_Y$.

Now use Proposition (10.2.2) to deal with the remaining case of one point blown up. □

The surfaces (S, L), (Y, L_Y) in (10.2.3.2) can be described as follows.

Example 10.2.4. (The Geiser involution [586, (1.1)]) Let S be a smooth connected surface with $-K_S$ ample and $K_S^2 = 2$. Then by (8.5.2) the line bundle $L := -2K_S$ is very ample. We have $L^2 = 8$, $g(L) = 3$, $h^0(L) = 7$ and (S, L) is its own reduction since $K_S + L$ is ample. This bundle is also spanned by (9.2.3), but it is not very ample. In fact, by Riemann-Roch and Kodaira vanishing, $h^0(K_S + L) = 3$, so the adjunction map expresses S as a two-to-one cover $p : S \to \mathbb{P}^2$. Let $B \in |\mathcal{O}_{\mathbb{P}^2}(2b)|$ be the branch locus of p. Since p is two-to-one, B is smooth and in fact it is a quartic curve in $\mathbb{P}^2$. To see this, note that $K_S \approx p^*\mathcal{O}_{\mathbb{P}^2}(b - 3)$. Since $-K_S$ is ample we have $b \leq 2$. From $2 = K_S^2 = 2\mathcal{O}_{\mathbb{P}^2}(b - 3)^2$ we infer that $b = 2$. Hence $L \approx -2K_S \approx p^*\mathcal{O}_{\mathbb{P}^2}(2)$. The codimension 1 set of smooth $C \in |L|$ which are inverse images of lines in $\mathbb{P}^2$ under p are hyperelliptic curves.

Conversely, given $p : S \to \mathbb{P}^2$, a double cover branched along a smooth quartic, then $K_S \approx p^*\mathcal{O}_{\mathbb{P}^2}(-1)$, and we can conclude that $-K_S$ is ample with $K_S^2 = 2$. The map p is classically called the Geiser involution, e.g., Semple and Roth [552, p. 148].

Example 10.2.5. ([586, (1.2)]) Let Y be as in (10.2.4) and let $r : (S, E) \to (Y, y)$ be the blowing up at a point $y \in Y$. By Proposition (10.2.2), $L = r^*(-2K_Y) - E$ is very ample. Clearly $(Y, -2K_Y)$ is the reduction of (S, L).

Theorem 10.2.6. *Let L be an ample and spanned line bundle on a smooth connected projective surface, S.*

1. *If $g(L) = q(S) > 0$, (S, L) is as described in* (8.7.1)*;*
2. *if $g(L) = 1$ and $q(S) = 0$, (S, L) is a Del Pezzo surface, i.e., $L \cong -K_S$.*

Proof. If $g(L) = q(S) > 0$ we conclude by (8.7.1).

To show 2), let C be a smooth element in $|L|$. By assumption $g(L) = 1$ and $q(S) = 0$. Use the exact sequence

$$0 \to K_S \to K_S + L \to K_C \cong \mathcal{O}_C \to 0.$$

Note that $\Gamma(K_S + L) \to \Gamma(\mathcal{O}_C) \cong \mathbb{C}$ is onto since $h^1(K_S) = q(S) = 0$. Therefore we can choose a section $s \in \Gamma(K_S + L)$ such that C meets transversally the zero set, $s^{-1}(0)$, of s and such that the restriction, s_C, of s to C, is nowhere zero. Then $s^{-1}(0)$ misses all C. Since C is ample we conclude that s is nowhere zero. This implies that $K_S + L \cong \mathcal{O}_S$ or $L \cong -K_S$. Then (S, L) is a Del Pezzo surface as in (10.2.6.2). □

The following is the main result of this section. In the theorem below ϕ is the morphism given by $\Gamma(K_S + L)$ and $\phi = s \circ r$ is the Remmert-Stein factorization of ϕ.

Theorem 10.2.7. *Let L be a very ample line bundle on a smooth connected projective surface, S. Assume that $q(S) < g(L) \leq 3$. Then we have*

1. *$g(L) = 1$, (S, L) is a Del Pezzo surface, i.e., $K_S \approx -L$; or*
2. *$g(L) = 2$ and (S, L) is a conic fibration over $\mathbb{P}^1$ under ϕ; or*
3. *$g(L) = 3$ and either*
   (a) *(S, L) is a conic fibration over $\mathbb{P}^1$ under ϕ; or*
   (b) *S is a $\mathbb{P}^1$-bundle over an elliptic curve of invariant $e = -1$ under r, $L \sim 2E + f$, where E and f denote a section of minimal self-intersection $E^2 = 1$ and a fiber of r respectively; or*
   (c) *(S, L) has $(\mathbb{P}^2, \mathcal{O}_{\mathbb{P}^2}(4))$, $r : S \to \mathbb{P}^2$, as the reduction; or*
   (d) *$S \subset \mathbb{P}^3$ and $L \cong \mathcal{O}_{\mathbb{P}^3}(1)_S$, with $\deg S = 4$; or*
   (e) *the reduction (Y, L_Y), $r : S \to Y$, of (S, L) is a Del Pezzo surface with $K_Y^2 = 2$, $L_Y \approx -2K_Y$.*

If $K_S + L$ is spanned and big, then $K_Y + L_Y$ is very ample except in the last case which is described in more detail in (10.2.4) *and* (10.2.5).

Proof. In view of the results of §8.10 and Theorem (10.2.6) we are reduced to the cases

$$d := L^2 \geq 5 \text{ and } g(L) \geq 2. \tag{10.8}$$

Note that the assumption $d \geq 5$ implies that S is ruled. Indeed, by the genus formula, we have $K_S \cdot L = 2g(L) - 2 - d \leq 4 - d < 0$.

Since $q(S) < g(L)$ it follows by (9.2.3) that the adjoint bundle $K_S + L$ is spanned. Let $\phi : S \to \mathbb{P}_{\mathbb{C}}$ be the morphism given by $\Gamma(K_S + L)$ and let $\phi = s \circ r$ be the Remmert-Stein factorization of ϕ, where $r : S \to Y$ has connected fibers and s is a finite-to-one morphism. Note that the pair (S, L) in (10.2.7.3b) below is the same as the $\mathbb{P}^1$-bundle in (10.1.3).

Let $g(L) = 2$. Then $q(S) \leq 1$. The genus formula gives $(K_S + L) \cdot L = 2$. Note that $h^0(K_S + L) \geq 2$ since otherwise $\dim \phi(S) = 0$, i.e., $K_S \approx -L$, implying that $g(L) = 1$. Let C be a smooth curve in $|L|$. Look at the long exact cohomology sequence of

$$0 \to K_S \to K_S + L \to K_C \to 0.$$

From $p_g(S) = h^0(K_S) = 0$, $h^0(K_C) = 2$, $h^0(K_S + L) \geq 2$ and $h^1(K_S + L) = 0$, we infer that $h^0(K_S + L) = 2$ and $h^1(\mathbb{O}_S) = q(S) = 0$. Then S is a rational surface and $\phi(S) \cong \mathbb{P}^1$, $\phi : S \to \mathbb{P}^1$. From (10.1.3) we know that s is an embedding, i.e., ϕ has connected fibers. Therefore (S, L) is a conic fibration as in (10.2.7.2).

Let $g(L) = 3$. Then $q(S) \leq 2$. We have

$$h^0(L) \geq 5. \tag{10.9}$$

Indeed, clearly $h^0(L) \geq 3$. If $h^0(L) = 3$ we would have $S \cong \mathbb{P}^2$ and $g(L) = 0 \neq 3$. If $h^0(L) = 4$, then S is a surface of degree d embedded by $\Gamma(L)$ in $\mathbb{P}^3$ and the genus formula would imply $d = 4$. We also have $\dim \phi(S) \geq 1$, since otherwise $g(L) = 1$.

Assume $\dim \phi(S) = 1$. Then (S, L) is a conic fibration under r and one has

$$0 = (K_S + L)^2 = K_S^2 + 4g(L) - 4 - d = K_S^2 + 8 - d. \tag{10.10}$$

Therefore $K_S^2 \leq 8(1 - q(S))$ gives $d \leq 16 - 8q(S)$ and hence $q(S) \leq 1$.

If $q(S) = 0$, then $r(S) \cong \mathbb{P}^1$, s is an embedding by (10.1.3) and (S, L) is a conic fibration over $\mathbb{P}^1$ under ϕ. We are in (10.2.7.3a).

Assume $q(S) = 1$. If $h^0(L) = 5$, Lemma (8.2.1) yields

$$d(d - 5) - 20 = 2K_S^2. \tag{10.11}$$

By combining equation (10.10) and equation (10.11) we find $d^2 - 7d - 4 = 0$ which has no integer solutions. Therefore $h^0(L) \geq 6$ and hence $h^0(L_C) \geq 5$. Note that $h^1(L_C) = h^0(K_C - L_C) = 0$ since $\deg(K_C - L_C) = 4 - d < 0$ by the assumption $d \geq 5$. Then the Riemann-Roch theorem gives $h^0(L_C) = d - g(L) + 1 = d - 2 \geq 5$, i.e., $d \geq 7$. Since $K_S^2 \leq 8(1 - q(S)) = 0$ we see from equation (10.10) that $d \leq 8$ and either $d = 7$, $K_S^2 = -1$ or $d = 8$, $K_S^2 = 0$.

First consider the case when $d = 7$, $K_S^2 = -1$. Since (S, L) is a conic fibration over a smooth curve under r, there exists a finite number of reducible fibers, f, of r of the form $f = f_1 + f_2$ with $L \cdot f_i = 1$, $f_i^2 = -1$, $f_1 \cdot f_2 = 1$, $i = 1, 2$. Let $\pi : S \to S'$ be the morphism obtained by blowing down one irreducible component for each reducible fiber of r. Let $L' := (\pi_* L)^{**}$. Then L' is ample since L' is ample outside of a finite set of points and (S', L') is a (not uniquely determined) reduction of (S, L). Since $q(S) = 1$, (S', L') is a $\mathbb{P}^1$-bundle, $p : S' \to Y$, over a smooth elliptic curve $Y = r(S)$ of invariant e. Let γ be the number of points blown up under π. From

$$-1 = K_S^2 = K_{S'}^2 - \gamma = 8(1 - q(S)) - \gamma = -\gamma$$

we find $\gamma = 1$. Let E, f denote a section of minimal self-intersection $E^2 = -e$ and a fiber of p. Let $L' \sim aE + bf$. Note that $L' \cdot f = 2$, so $a = 2$, and $L'^2 = -4e + 4b = L^2 + 1 = 8$. This gives $b = e + 2$. From (8.14) we get $L' \cdot E \geq 3$, implying $b - 2e = 2 - e \geq 3$. Thus we conclude that $e = -1$. Therefore Lemma (8.5.9) applies to give $a + b = 4 + e \geq 4$, which contradicts $e = -1$. This shows that the case $d = 7$, $K_S^2 = -1$ does not occur.

Let us go back to the case $d = 8$, $K_S^2 = 0$. In this case $K_S^2 = 8(1 - q(S)) = 0$, so that S is in fact a $\mathbb{P}^1$-bundle over an elliptic curve of invariant e under r. Let E, f denote a curve of minimal self-intersection $E^2 = -e$ and a fiber of r. Since (S, L) is a conic fibration under r we have $L \sim 2E + bf$. From $d = L^2 = 8$ we get $b = e + 2$. Note that $L \cdot E = -2e + b = 2 - e \geq 3$ since L is very ample and E is an elliptic curve. Thus $e = -1$ and $b = 1$. Note also that $L \sim 2E + f$ is in fact very ample by (8.5.9). We are in (10.2.7.3b).

Assume now $\dim \phi(S) = 2$. Let $L_Y = (r_* L)^{**}$. Then (Y, L_Y), $r : S \to Y$, is the reduction of (S, L). Using the exact sequence,

$$0 \to K_S \to K_S + L \to K_C \to 0,$$

$h^0(K_S) = 0$, and $h^0(K_C) = 3$ we have $h^0(K_S + L) \leq 3$ and in fact $h^0(K_S + L) = 3$. Thus $\phi(S) \cong \mathbb{P}^2$. Note that $h^1(L_C) = h^0(K_C - L_C) = 0$ since $\deg(K_C - L_C) = 4 - d < 0$ and equation (10.9) implies $h^0(L_C) \geq 4$. Then $h^0(L_C) = d - g(L) + 1 = d - 2 \geq 4$ gives $d \geq 6$. By the Hodge index theorem and the genus formula we have $(K_S + L)^2 \cdot L^2 \leq 16$, so that $d \geq 6$ implies $(K_S + L)^2 = 1, 2$. Here we use Theorem (10.2.3). □

Lemma 10.2.8. *Let L be a very ample line bundle on a smooth connected projective surface, S. If $L^2 \leq 6$ and $h^0(L) \geq 5$ then $g(L) \leq 4$ with equality only if $L^2 = 6$ and S is a K3 surface, i.e., only if $K_S \cong \mathcal{O}_S$ and $q(S) = 0$.*

Proof. By Castelnuovo's bound (1.4.9), $g(L) \leq 3$ unless $L^2 = 6$ and $h^0(L) = 5$, in which case the upper bound $g(L) = 4$ in Castelnuovo's bound is taken on and $q(S) = 0$ by Theorem (8.6.1). Since $g(L) \neq q(S)$ we conclude from Corollary (9.2.3) that $K_S + L$ is spanned. Thus by Lemma (10.1.2.1) we see that $K_S^2 \geq -4$. Combining this with $K_S^2 = 6\chi(\mathcal{O}_S) - 12$ which follows from the double point formula, (8.2.1), we conclude that $p_g(S) \geq 1$. Since $K_S \cdot L = 0$ we conclude that $K_S \cong \mathcal{O}_S$. □

10.3 Very ampleness of the adjoint bundle

The rest of this chapter will be devoted to proving the next result; see Serrano [554], Sommese and Van de Ven [586], and also the later paper by Ionescu [336].

Let S be a smooth connected surface and L a very ample line bundle on S. From the results of §9.2 we can assume that $K_S + L$ is spanned and the morphism, $\phi : S \to \mathbb{P}_{\mathbb{C}}$, associated to $\Gamma(K_S + L)$ has a 2-dimensional image. Let $\phi = s \circ r$ be the Remmert-Stein factorization of ϕ, where $r : S \to Y$ has connected fibers and s is a finite-to-one morphism. Let $L_Y = (r_* L)^{**}$. Then (Y, L_Y) is the reduction of (S, L).

Theorem 10.3.1. *With the above notation, $K_Y + L_Y$ is very ample with the exceptions of degrees* 9, 8, 7, *and* 9 *given in*

1. *$S \cong Y$, S is a Del Pezzo surface, $L \approx -3K_S$, $K_S^2 = 1$ as described in* (10.4.3).
2. *$S \cong Y$, S is a Del Pezzo surface, $L \approx -2K_S$, $K_S^2 = 2$ as described in* (10.2.4).
3. *(Y, L_Y) is as in* 2), *$r : S \to Y$ is the blowing up at a point, as described in* (10.2.5).
4. *$S \cong Y$, S is a $\mathbb{P}^1$-bundle over a smooth elliptic curve of invariant $e = -1$, $L \sim 3E$, E a section of minimal self-intersection as described in* (10.4.4).

10.4 Very ampleness of the adjoint bundle for degree $d \geq 9$

Let S be a smooth connected surface and L a very ample line bundle on S. From the results of §9.2 we can assume that $K_S + L$ is spanned and the morphism, $\phi :$ $S \to \mathbb{P}_{\mathbb{C}}$, associated to $\Gamma(K_S + L)$ has a 2-dimensional image. Let $\phi = s \circ r$ be the Remmert-Stein factorization of ϕ, where $r : S \to Y$ has connected fibers and s is a finite-to-one morphism. Let $L_Y = (r_*L)^{**}$. Then (Y, L_Y) is the reduction of (S, L). Note that $K_Y + L_Y$ is spanned since $K_S + L$ is spanned and $K_S + L \approx r^*(K_Y + L_Y)$. Recall also that $K_Y + L_Y$ is ample. In this section we study the very ampleness of $K_Y + L_Y$ under the assumption that $d_Y := L_Y^2 \geq 9$.

Theorem 10.4.1. *Let L be a very ample line bundle on a smooth connected projective surface, S. Assume that $K_S + L$ is spanned and big. Let (Y, L_Y) be the reduction of (S, L). Furthermore assume that $d_Y := L_Y^2 \geq 9$. Then $K_Y + L_Y$ is very ample unless $(S, L) \cong (Y, L_Y)$, $L^2 = L_Y^2 = 9$ and either*

1. *S is a Del Pezzo surface with $L \approx -3K_S$; or*
2. *S is a $\mathbb{P}^1$-bundle, $p : S \to C$, over an elliptic curve C of invariant $e = -1$ and $L \sim 3E$, where E denotes a section with self-intersection $E^2 = 1$.*

Proof. Since L_Y is ample and $L_Y^2 \geq 9$ we use Reider's criterion (8.5.1) for $k = 1$. If $K_Y + L_Y$ is not very ample, there exists an effective divisor D on Y such that

$$L_Y \cdot D - 2 \leq D^2 < L_Y \cdot D/2 < 2.$$

Then one of the following cases occurs:

i) $L_Y \cdot D = 1$, $D^2 = 0$ or -1;

ii) $L_Y \cdot D = 2$, $D^2 = 0$; or

iii) $L_Y \cdot D = 3$, $L_Y \sim 3D$ and $D^2 = 1$.

We will consider these cases separately. Let B be the finite set of points blown up under r and let $\mathscr{E} = r^{-1}(B)$ be the exceptional locus of r. Let $D^\wedge$ be the proper transform of D on S. Let # denote the number of points of $D \cap B$. Then we have

$$L \cdot D^\wedge = r^*L_Y \cdot D^\wedge - \mathscr{E} \cdot D^\wedge \leq L_Y \cdot D - \#. \tag{10.12}$$

Hence $L_Y \cdot D \geq L \cdot D^\wedge$ with equality if and only if $D \cap B = \emptyset$.

In case i), since $L_Y \cdot D = 1$, we have $L_Y \cdot D = L \cdot D^\wedge = 1, \# = 0$. So we conclude that D is a smooth rational curve. If $D^2 = 0$, then $K_Y \cdot D = -2$ and $(K_Y + L_Y) \cdot D = -1$, contradicting the fact that $K_Y + L_Y$ is spanned. If $D^2 = -1$, then $K_Y \cdot D = -1$ and $(K_Y + L_Y) \cdot D = 0$, contradicting the ampleness of $K_Y + L_Y$.

Consider case ii). Here D is either irreducible and reduced or $D = D_1 + D_2$ where D_1, D_2 are irreducible, reduced, possibly coinciding, components. Moreover $2 = L_Y \cdot D \geq L \cdot D^\wedge$. Hence $\# = 0$ or 1 by (10.12), so that $D \cong D^\wedge$. If D is irreducible, then $D, D^\wedge$ are smooth rational curves. Therefore $K_Y \cdot D = -2$ and $(K_Y + L_Y) \cdot D = 0$, contradicting again the ampleness of $K_Y + L_Y$. Assume $D = D_1 + D_2$. Then we must have $L_Y \cdot D_j = L \cdot D_j^\wedge = 1$, where $D_j^\wedge$ denotes the proper transform of D_j, $j = 1, 2$. Then both D_1 and D_2 are linear $\mathbb{P}^1$'s such that $D_1 \cap B = D_2 \cap B = \emptyset$. If $D_1 = D_2$, then $D^2 = D_j^2 = 0$, $K_Y \cdot D_j = -2$, and $(K_Y + L_Y) \cdot D_j = -1$ for $j = 1, 2$, which contradicts the ampleness of $K_Y + L_Y$. Let $D_1 \neq D_2$. Since $D_j^\wedge$ is a line relative to L and $r^*(D_j) \cong D_j^\wedge$, $j = 1, 2$, we have $0 \leq D_1 \cdot D_2 = D_1^\wedge \cdot D_2^\wedge \leq 1$. If $D_1^2 = 0$ or $D_2^2 = 0$ we find the same contradiction as above. If $D_j^2 > 0$ for $j = 1$ or 2, the Hodge index theorem leads to the contradiction $9 \leq (D_j^2)(L_Y^2) \leq (L_Y \cdot D_j)^2 = 1$. Therefore we can assume that $D_j^2 < 0$ for $j = 1, 2$. From

$$0 = D^2 = D_1^2 + 2D_1 \cdot D_2 + D_2^2 \leq D_1^2 + 2 + D_2^2$$

we see that $D_1^2 = D_2^2 = -1$, $D_1 \cdot D_2 = 1$. Then $K_Y \cdot D_j = -1$ and $(K_Y + L_Y) \cdot D_j = 0$, $j = 1, 2$, which contradicts ampleness.

In case iii), since $D^2 = 1$ and $L_Y \sim 3D$, we see that D is ample, irreducible, and reduced. From (10.12) we have that either $\# = 0$, $L \cdot D^\wedge = 3$; $\# = 1$, $L \cdot D^\wedge = 1, 2$; or $\# = 2$, $L \cdot D^\wedge = 1$.

Let $\# = 1$, $L \cdot D^\wedge = 1, 2$. Then $D^\wedge \cong D \cong \mathbb{P}^1$. Since D is ample, $(Y, D) \cong (\mathbb{P}^2, \mathcal{O}_{\mathbb{P}^2}(1))$ by Remark (8.9.2). Thus $(Y, L_Y) \cong (\mathbb{P}^2, \mathcal{O}_{\mathbb{P}^2}(3))$, which gives the contradiction $K_Y + L_Y \cong \mathcal{O}_Y$.

Let $\# = 2$. Then $L \cdot D^\wedge = 1$, so that $D^\wedge$ is a linear $\mathbb{P}^1$. Note that $D^\wedge$ meets each exceptional line, ℓ, of $\mathcal{E}$ in a point at most, since if two lines meet in two points they are the same. Therefore there are exactly two distinct exceptional lines, ℓ_1, ℓ_2, such that $\ell_1 \cdot D^\wedge = \ell_2 \cdot D^\wedge = 1$. Hence $D^\wedge \cong D$ so $D \cong \mathbb{P}^1$. As in the case above we have the contradiction $(Y, L_Y) \cong (\mathbb{P}^2, \mathcal{O}_{\mathbb{P}^2}(3))$.

Let $\# = 0$, $L \cdot D^\wedge = 3$. Then $D^\wedge$ is a degree three curve, so that $g(D^\wedge) \leq 1$. Since $D^\wedge \cong D$, either $g(D) = 0$ or $g(D) = 1$. Note that D is irreducible and reduced since D is ample and $D^2 = 1$. We claim that $g(D) = 1$. Indeed otherwise Theorem (8.9.1) applies to say that either $(Y, D) \cong (\mathbb{P}^2, \mathcal{O}_{\mathbb{P}^2}(a))$, $a = 1, 2$, or $Y \cong \mathbb{F}_e$, $e \geq 0$, $D \sim E + bf$, $b \geq e + 1$, where E, f denote a section of minimal self-intersection $E^2 = -e$ and a fiber of $\mathbb{F}_e$. If $(Y, D) \cong (\mathbb{P}^2, \mathcal{O}_{\mathbb{P}^2}(1))$ we have the contradiction $K_Y + L_Y \cong \mathcal{O}_{\mathbb{P}^2}$. The case $(Y, D) \cong (\mathbb{P}^2, \mathcal{O}_{\mathbb{P}^2}(2))$ is not possible since $D^2 = 1$. If $Y \cong \mathbb{F}_e$ we get the contradiction $1 = D^2 = 2b - e \geq e + 2 \geq 2$. Thus we conclude that $g(D) = 1$, so D is an elliptic curve and, by the genus formula,

$$(K_Y + D) \cdot D = 0. \tag{10.13}$$

Using the exact sequence

$$0 \to K_Y \to K_Y + D \to K_D \to 0,$$

$h^0(K_D) = 1$, and $h^1(K_Y + D) = 0$ we see that

$$q(S) \le 1 \tag{10.14}$$

and $q(S) = 0$ implies $h^0(K_Y + D) > 0$.

Let $q(S) = 0$, $h^0(K_Y + D) > 0$. Since D is ample, we see from equation (10.13) that $K_Y \approx -D$. Therefore $L_Y \approx -3K_Y$, $K_Y^2 = 1$ and we are in case (10.4.1.1) after proving the following

Claim 10.4.2. With the notation as above, $(S, L) \cong (Y, L_Y)$.

Proof. Note that $h^i(-K_Y) = 0$ for $i > 0$ so that $h^0(-K_Y) = 2$ by the Riemann-Roch theorem. All members $\Gamma \in |-K_Y|$ are irreducible, reduced curves since $\Gamma \cdot (-K_Y) = 1$ and any two distinct curves, Γ_1, Γ_2 in $|-K_Y|$ intersect in a single point, $x = \Gamma_1 \cap \Gamma_2$, which is smooth for Γ_1, Γ_2. Then we have a pencil of elliptic curves having x as basepoint. Assume that $S \ncong Y$ and let $y_i \in B = \{y_1, \cdots, y_\gamma\}$ be a point blown up under r. Then there exists a (possibly singular) elliptic curve $\Gamma \in |-K_Y|$ such that $y_i \in \Gamma$. Let μ_i be the multiplicity of Γ at y_i, $\mu_i \ge 1$. Let $\ell_j = r^{-1}(y_j)$, $j = 1, \cdots, \gamma$. Let $\Gamma^\wedge$ be the proper transform of Γ under r. Therefore, for some nonnegative integers λ_j, $j = 1, \cdots, \gamma$, $j \neq i$, we have

$$\begin{aligned} 1 \le L \cdot \Gamma^\wedge &= (r^* L_Y - \ell_i - \sum_{j \neq i} \ell_j) \cdot (r^*\Gamma - \mu_i \ell_i - \sum_{j \neq i} \lambda_j \ell_j) = \\ &= L_Y \cdot \Gamma - \mu_i - \sum_{j \neq i} \lambda_j. \end{aligned}$$

Since $L_Y \cdot \Gamma = 3K_Y^2 = 3$, either

$$\begin{aligned} \mu_i &= 2 \text{ and } \lambda_j = 0 \text{ for each } j = 1, \cdots, \gamma,\ \ j \neq i, \text{ or} \\ \mu_i &= 1, \lambda_j \le 1 \text{ for each } j \text{ and } \lambda_j = 1 \text{ for one index } j \text{ at most.} \end{aligned}$$

In the first case $L \cdot \Gamma^\wedge = 1$, so that $\Gamma^\wedge$ is a linear $\mathbb{P}^1$ and $\ell_i \cdot \Gamma^\wedge = 2$. Then we would have two distinct lines, $\Gamma^\wedge$ and ℓ_i, meeting at two points, which is not possible. In the second case $L \cdot \Gamma^\wedge \le 2$ so that $\Gamma^\wedge$ is a rational curve. Recall that Γ is irreducible and reduced, and $\Gamma^\wedge \cong \Gamma$. This contradicts $g(\Gamma) = 1$. □

Thus, recalling equation (10.14), we can assume without loss of generality that $q(S) = 1$. Since $K_Y \cdot D < 0$ by equation (10.13), we know that Y is a ruled surface, $p : Y \to C$, over a smooth curve C of genus $q(S) = 1$. We claim that D is smooth. To see this, assume to the contrary that $\nu : \mathbb{P}^1 \to D$ is the normalization of D. Note that D maps surjectively onto C under the restriction, p_D, of p to D, since D is ample. Then we have a surjective morphism $p_D \circ \nu : \mathbb{P}^1 \to C$. This is not possible, so D is smooth. Look at the commutative diagram

$$\begin{array}{ccc} H_1(D, \mathbb{Z}) & \xrightarrow{\alpha} & H_1(Y, \mathbb{Z}) \\ p_{D*} & \searrow & \downarrow p_* \\ & & H_1(C, \mathbb{Z}) \end{array}$$

Note that α is surjective since D is ample and p_* is surjective since p has connected fibers. Then p_{D*} is surjective too. By using the Hurwitz formula we know that $p_D : D \to C$ is an unbranched covering. Therefore $\deg p_D$ is the index of $p_{D*}(H_1(D, \mathbb{Z}))$ as subgroup of $H_1(C, \mathbb{Z})$. Since p_{D*} is surjective we conclude that $\deg p_D = 1$. Thus $p_D : D \to C$ is in fact an isomorphism. Hence $D \cdot f = 1$ for every fiber f of p. This means that $p : Y \to C$ is a $\mathbb{P}^1$-bundle. Let E be a section of minimal self-intersection, $E^2 = -e$, on Y. Then $D \sim E + bf$, for some integer b. From $D \cdot E > 0$ we find $b > e$ and hence $1 = D^2 = -e + 2b \geq e + 2$. Therefore $e \leq -1$. Since $e \geq -g(C) = -1$, we get $e = -1, b = 0$ and $D \sim E$. Thus $L_Y \sim 3E$ and we are in case (10.4.1.2) after proving that $(S, L) \cong (Y, L_Y)$.

To see this, assume that there is a point, $y \in Y$, blown up under $r : S \to Y$. From Lemma (8.5.7) we know that there exists a smooth elliptic curve Γ on Y, Γ algebraically equivalent to E, such that Γ contains y. Let $\Gamma^\wedge$ be the proper transform of Γ under r. Then we have $L \cdot \Gamma^\wedge = (r^* L_Y - \mathscr{E}) \cdot \Gamma^\wedge < L_Y \cdot \Gamma = L_Y \cdot E = 3$. Since Γ is smooth we have $\Gamma^\wedge \cong \Gamma$ and hence $g(\Gamma^\wedge) = 1$. Then the inequality $L \cdot \Gamma^\wedge < 3$ contradicts the very ampleness of L. □

The surfaces in (10.4.1.1), (10.4.1.2) can be described as follows.

Example 10.4.3. (The Bertini involution [586, (1.3)]) Let S be a smooth surface with $-K_S$ ample and $K_S^2 = 1$. By (8.5.2), the line bundle $L = -3K_S$ is very ample. We have $L^2 = 9$, $g(L) = 4$, $h^0(L) = 7$ and (S, L) is its own reduction. By (8.5.2), the ample bundle $K_S + L = -2K_S$ is also spanned but it is not very ample. In fact, $h^0(K_S + L) = 4$, and since $(K_S + L)^2 = 4$, the surface S would have trivial canonical bundle if $K_S + L$ were very ample. It can be proved (see Sommese [571, Proposition (3.1)]) that the adjunction map expresses S as a double cover $p : S \to Q$, where $Q \subset \mathbb{P}^3$ is a quadric cone. The branch locus of p consists of the vertex, x, and a smooth curve, B, which is the transversal intersection of Q with a cubic surface. Conversely, let Q again be a quadric cone in $\mathbb{P}^3$ with vertex x and B the transversal intersection of Q with some cubic surface. Let $\pi : (\mathbb{F}_2, E) \to (Q, x)$ be the blowing up of Q at the vertex x and let f be a fiber of the Hirzebruch surface $\mathbb{F}_2$. By abuse of notation we denote the inverse image $\pi^{-1}(B)$ by B. Then $B = 3E + 6f$ in homology, and $\mathcal{O}_{\mathbb{F}_2}(B + E) \cong 2\mathscr{L}$ with $\mathscr{L} \cong \mathcal{O}_{\mathbb{F}_2}(2E + 3f)$. Consequently there is a double cover $p^\wedge : S^\wedge \to \mathbb{F}_2$ with branch locus $B \cup E$. Then, since $2p^{\wedge -1}(E) \approx p^{\wedge *}(E)$, $p^{\wedge -1}(E)$ is a -1 curve which can be blown down. Let $\sigma : S^\wedge \to S$ be the contraction of $p^{\wedge -1}(E)$. The map $p^\wedge$ descends to a (finite) double cover $p : S \to Q$ with branch locus $B \cup \{x\}$. Note that $K_{S^\wedge} \approx p^{\wedge *}(K_{\mathbb{F}_2} + \mathscr{L}) \approx p^{\wedge *}(-f)$ and hence $\sigma^* K_S \approx p^{\wedge *}(-f) - p^{\wedge -1}(E)$. On the other hand, since $\mathcal{O}_Q(1)$ pulls back to $E + 2f$ on $\mathbb{F}_2$,

$$\sigma^* p^* \mathcal{O}_Q(1) \approx p^{\wedge *}(E + 2f) \approx 2(p^{\wedge *} f + p^{\wedge -1}(E)).$$

Therefore $-2\sigma^* K_S \approx \sigma^* p^* \mathcal{O}_Q(1)$. It thus follows that $-2K_S \approx p^*\mathcal{O}_Q(1)$ and we see that $-2K_S$ is ample with $K_S^2 = 1$. The involution $p : S \to Q$ is classically called the Bertini involution, e.g., Semple and Roth ([552, p. 149]).

Example 10.4.4. ([586, (1.4)]) Let C be a smooth elliptic curve and $\mathcal{M}$ a line bundle on C of degree 1. Consider a non-split extension $0 \to \mathcal{O}_C \to V \to \mathcal{M} \to 0$ and let $S = \mathbb{P}(V)$. This is (up to isomorphism) the unique $\mathbb{P}^1$-bundle over C with invariant $e = -1$ (see Hartshorne [318, Chap V, 2]). Let $\xi = \mathcal{O}_{\mathbb{P}(V)}(1)$ be the tautological line bundle on S. If $p : S \to C$ is the bundle projection then $p_*\xi = V$. Note that $K_S \approx -2\xi + p^*\mathcal{M}$. It is known that $L = 3\xi$ is very ample (see also (8.5.8)). We have $L^2 = 9$, $g(L) = 4$, $h^0(L) = 6$. Furthermore $K_S + L$ is ample (see Theorem (3.2.11)) and it is spanned by (9.1.3). However this bundle is not very ample since $K_S + L$ restricted to smooth $C \in |\xi|$ has degree 2. In fact $(K_S + L)^2 = 3$ and the adjunction map is a three-to-one map onto $\mathbb{P}^2$.

10.5 Very ampleness of the adjoint bundle when $h^1(\mathcal{O}_S) > 0$

Let L be a very ample line bundle on a smooth connected projective surface, S. Let $d := L^2$ be the degree of (S, L). We assume that $K_S + L$ is spanned and the morphism, $\phi : S \to \mathbb{P}_\mathbb{C}$, associated to $\Gamma(K_S + L)$ has a 2-dimensional image. Let $\phi = s \circ r$ be the Remmert-Stein factorization of ϕ, where $r : S \to Y$ has connected fibers and s is a finite-to-one morphism. Let $L_Y = (r_*L)^{**}$ and $d_Y = L_Y^2$. In this section we study the very ampleness of $K_Y + L_Y$ in the case when $q(S) > 0$. Let $g(L)$ be the sectional genus of L. Under the present assumptions, $\dim \phi(S) = 2$, $q(S) > 0$, and using Theorem (10.2.6), Theorem (10.2.7), Lemma (10.2.8) and §10.4, we can assume that we have

$$g(L) \geq 4 \quad \text{and} \quad 8 \geq d_Y \geq d \geq 7. \tag{10.15}$$

Note also that since $q(S) > 0$, we conclude from Theorem (10.2.3) that $(K_S+L)^2 \geq 3$ or equivalently

$$K_S^2 + 4g(L) - d \geq 7. \tag{10.16}$$

Theorem 10.5.1. *Let L be a very ample line bundle on a smooth connected projective surface, S. Assume that $K_S + L$ is spanned and big. Furthermore assume that $d := L^2 \leq 8$ and $q(S) > 0$. Then $K_S + L$ is very ample. Hence (S, L) is its own reduction.*

Proof. We go on by a case by case analysis, according to the possible values of d. We will often use Castelnuovo's bound (1.4.9) without explicitly referring to it. Note that since $q(S) > 0$ we can assume without loss of generality that $h^0(L) \geq 5$.

Let $d = 7$. We have $h^0(L) = 5$ since $h^0(L) \geq 6$ implies $g(L) \leq 3$, which contradicts (10.15). Therefore Lemma (8.2.1) applies to give

$$12 - 5g(L) + 6\chi(\mathcal{O}_S) = K_S^2. \tag{10.17}$$

We also have $g(L) \leq 6$ since $h^0(L) = 5$. By combining equations (10.16) and (10.17) we find

$$6\chi(\mathcal{O}_S) \geq g(L) + 2. \tag{10.18}$$

Then $\chi(\mathcal{O}_S) \geq 1$, which implies $p_g(S) > 0$ since $q(S) \geq 1$.

If $g(L) = 4$, we have $d > 2g(L)-2$ and hence $K_S \cdot L < 0$ by the genus formula. This contradicts $p_g(S) > 0$, so we conclude that $g(L) \geq 5$. From equation (10.18) it thus follows $\chi(\mathcal{O}_S) \geq 2$, implying $p_g(S) \geq 2$.

If $g(L) = 5$, then the genus formula yields $K_S \cdot L = 1$. Thus by using Lemma (1.4.2) we conclude that $h^0(K_{S|C}) \leq 1$, where C is a smooth element of $|L|$. Since $(K_S - L) \cdot L < 0$ we have $h^0(K_S) \leq h^0(K_{S|C}) \leq 1$. This contradicts $p_g(S) \geq 2$.

If $g(L) = 6$, then the maximum for the Castelnuovo bound is taken on which implies the contradiction that $q(S) = 0$ by Theorem (8.6.1).

Let $d = 8$. We have $h^0(L) \leq 6$ since $h^0(L) \geq 7$ implies $g(L) \leq 3$. Furthermore $g(L) \leq 5$ if $h^0(L) = 6$ and $g(L) \leq 9$ if $h^0(L) = 5$. Recall that, if $h^0(L) = 5$, either $g(L) = 9$ or $g(L) \leq 7$ by the Gruson-Peskine bound, (1.4). Since by Theorem (8.6.1), $q(S) = 0$ for curves with genus equal to the maximum allowed by the Castelnuovo bound, we have the following possibilities

$$\begin{array}{lclcl} h^0(L) & = & 6 & \text{and} & g(L) = 4; \text{ or} \\ h^0(L) & = & 5 & \text{and} & g(L) = 4, 5, 6, 7. \end{array} \tag{10.19}$$

Let $g(L) = 4$. Then $K_S \cdot L = -2$. Let C be a smooth curve in $|L|$. Note that $h^1(L_C) = h^0(K_C - L_C) = 0$ since

$$\deg(K_C - L_C) = 2g(L) - 2 - d < 0.$$

We also have $h^2(L) = h^0(K_S - L) = 0$ since $(K_S - L) \cdot L < 0$. Therefore from the exact sequence

$$0 \to K_S - L \to K_S \to K_{S|C} \to 0$$

we infer that $p_g(S) = h^2(\mathcal{O}_S) = 0$. Look at the long exact cohomology sequence of

$$0 \to K_S \to K_S + L \to K_C \to 0.$$

Since $h^0(K_S) = 0$, $h^0(K_C) = g(L) = 4$, $h^0(K_S + L) \geq 3$, $h^1(K_S + L) = 0$, and $q(S) > 0$ we see that $h^0(K_S + L) = 3$ and $q(S) = 1$. Thus $\chi(\mathcal{O}_S) = 0$, so that Lemma (8.2.1) gives $K_S^2 \leq -3$. This contradicts the inequality $K_S^2 \geq -1$ which comes from inequality (10.16).

Let $g(L) = 5$. From (10.16) we have $K_S^2 \geq -5$. Lemma (8.2.1) gives $K_S^2 \leq -8 + 6\chi(\mathcal{O}_S)$. Thus $6\chi(\mathcal{O}_S) - 8 \geq -5$, that is $\chi(\mathcal{O}_S) \geq 1$. Hence $p_g(S) \geq 1$, so K_S

is effective. Since $K_S \cdot L = 2g(L) - 2 - d = 0$ we obtain $K_S \cong \mathcal{O}_S$ and therefore $K_S + L$ is very ample.

From now on we can assume $g(L) \geq 6$, so $h^0(L) = 5$ by (10.19).

Let $g(L) = 6$. Lemma (8.2.1) gives

$$K_S^2 = -13 + 6\chi(\mathcal{O}_S). \tag{10.20}$$

Moreover $K_S^2 \geq -9$ by (10.16). Therefore $6\chi(\mathcal{O}_S) \geq 4$, so that $\chi(\mathcal{O}_S) \geq 1$ and hence $p_g(S) \geq 1$. Then Corollary (7.4.5) applies to give

$$2 = K_S \cdot L \geq -K_S^2. \tag{10.21}$$

From the Hodge inequality $dK_S^2 \leq (L \cdot K_S)^2 = 4$ we infer that $K_S^2 \leq 0$. Then by equation (10.20) we have $\chi(\mathcal{O}_S) \leq 2$ and either $\chi(\mathcal{O}_S) = 1$, $K_S^2 = -7$ or $\chi(\mathcal{O}_S) = 2$, $K_S^2 = -1$. The first case contradicts inequality (10.21). If $\chi(\mathcal{O}_S) = 2$, $K_S^2 = -1$, one has $p_g(S) \geq 2$. Since K_S is effective and $L \cdot K_S = 2$ we conclude that $K_S \cong \mathcal{O}_S(\gamma)$ where either γ is a smooth conic or $\gamma = A + B$ where $L \cdot A = L \cdot B = 1$, so that A, B are linear $\mathbb{P}^1$'s. If γ is a smooth conic, then $\gamma^2 \geq 0$ since $h^0(\gamma) = h^0(K_S) \geq 2$. This contradicts $\gamma^2 = K_S^2 = -1$. If $\gamma = A + B$, either $h^0(A) \geq 2$ or $h^0(B) \geq 2$. Therefore there exists on S a pencil of rational curves. This contradicts the fact that $p_g(S)$ is positive.

Let $g(L) = 7$. Lemma (8.2.1) yields

$$K_S^2 = -18 + 6\chi(\mathcal{O}_S). \tag{10.22}$$

Moreover $K_S^2 \geq -13$ by (10.16). Hence $6\chi(\mathcal{O}_S) \geq 5$, so that $\chi(\mathcal{O}_S) \geq 1$ and hence $p_g(S) \geq 1$. Then Corollary (7.4.5) applies to give

$$4 = K_S \cdot L \geq -K_S^2. \tag{10.23}$$

By combining equations (10.22) and (10.23) we get $\chi(\mathcal{O}_S) \geq 3$. From the Hodge inequality $(K_S^2)(L^2) \leq (L \cdot K_S)^2 = 16$ we find $K_S^2 \leq 2$, so that $\chi(\mathcal{O}_S) \leq 3$, and in fact $\chi(\mathcal{O}_S) = 3$, $K_S^2 = 0$, by equation (10.22). Note that Castelnuovo's bound (1.3) yields $h^0(L_C) \leq 4$ since otherwise $g(L) \leq 5$. Therefore $h^0(L_C) = 4$ since $h^0(L) = 5$. Note also that

$$h^1(L_C) = h^0(L_C) - d + g(L) - 1 = 2$$

and $h^2(L) = h^0(K_S - L) = 0$ since $(K_S - L) \cdot L < 0$. Thus from the long cohomology exact sequence of

$$0 \to \mathcal{O}_S \to L \to L_C \to 0$$

we conclude that there is a surjective map $H^1(L_C) \to H^2(\mathcal{O}_S)$. This implies that

$$h^2(\mathcal{O}_S) = p_g(S) = \chi(\mathcal{O}_S) + q(S) - 1 = q(S) + 2 \leq 2,$$

which contradicts the assumption $q(S) > 0$. □

10.6 Very ampleness of the adjoint bundle when $h^1(\mathcal{O}_S) = 0$

Let S be a smooth connected surface and L a very ample line bundle on S. As in §§10.4, 10.5, we assume that $K_S + L$ is spanned and the morphism, $\phi : S \to \mathbb{P}_{\mathbb{C}}$, associated to $\Gamma(K_S+L)$ has a 2-dimensional image. Let $\phi = s \circ r$ be the Remmert-Stein factorization of ϕ, where $r : S \to Y$ has connected fibers and s is a finite-to-one morphism. Let $L_Y = (r_*L)^{**}$. Recall that $K_Y + L_Y$ is ample. Note also that $K_Y + L_Y$ is spanned since $K_S + L$ is spanned and $\Gamma(K_Y + L_Y)$ defines the finite morphism $s : Y \to \mathbb{P}_{\mathbb{C}}$. Let $d := L^2$.

In this section we study the very ampleness of $K_Y + L_Y$ in the case when $q(S) = 0$. Most of the section deals with the case that there exists a smooth hyperelliptic curve $C \in |L|$.

We need some preliminary results.

Proposition 10.6.1. *Let S be a smooth connected surface and L a very ample line bundle on S. Assume that there is a smooth hyperelliptic curve $C \in |L|$. Let L_C be the restriction of L to C, let $g(L)$ be the sectional genus of (S, L) and let $d := L^2$. Then*

1. $h^1(tL_C) = 0$ *for* $t > 0$;

2. $d \geq g(L) + 3$;

3. $h^2(\mathcal{O}_S) = 0$.

Proof. To show (10.6.1.1), assume $h^1(L_C) = h^0(K_C - L_C) > 0$. Then we can choose a not identically zero section $\omega \in \Gamma(K_C - L_C)$. Note that $\omega \otimes \Gamma(L_C)$ gives rise to sections of $\Gamma(K_C)$ which embed C outside of $\omega^{-1}(0)$. This contradicts the fact that $\Gamma(K_C)$ gives a two-to-one morphism because C is hyperelliptic. Hence $h^1(L_C) = 0$. It thus follows that $h^1(tL_C) = 0$ for $t > 0$. Indeed, since $h^0(L_C) > 0$, if $h^1(tL_C) = h^0(K_C - tL_C) > 0$, we would have $h^0(K_C - L_C) = h^1(L_C) > 0$.

To show (10.6.1.2), note that $h^0(L_C) \geq 4$ since there are no hyperelliptic curves in $\mathbb{P}^2$. Then by (10.6.1.1) and the Riemann-Roch theorem we get $h^0(L_C) = d - g(L) + 1$. Therefore $d \geq g(L) + 3$.

To show (10.6.1.3), use Lemma (4.1.5). □

Note that we also have, under the assumptions as in (10.6.1) and $q(S) = 0$,

$$h^0(K_S + L) = g(L). \tag{10.24}$$

This immediately follows from (10.6.1.3) and the long exact cohomology sequence of

$$0 \to K_S \to K_S + L \to K_C \to 0.$$

Proposition 10.6.2. *Let S be a smooth connected surface and L a very ample line bundle on S. Assume that K_S+L is spanned and big. Let (Y, L_Y) be the reduction of (S, L). Further assume that $q(S)=0$ and there exists a smooth hyperelliptic curve in $|L|$. Then L_Y is very ample.*

Proof. Let L_C be the restriction of L to C. If C is hyperelliptic, then $h^1(L_C)=0$ by Proposition (10.6.1). Since $q(S)=0$ we then conclude that $h^1(L)=0$. Thus the result follows from Corollary (7.4.2). □

We can prove now the main result of this section.

Theorem 10.6.3. *Let L be a very ample line bundle on a smooth connected projective surface, S. Assume that K_S+L is spanned and big, and let (Y, L_Y) be the reduction of (S, L). Further assume that $d:=L^2\leq 8$, $q(S)=0$ and there exists a smooth hyperelliptic curve, C, in $|L|$. Then $L_Y^2=8$, $(K_S+L)^2=2$ and (S, L) is as in* (10.2.3.2).

Proof. Let us show that $d_Y:=L_Y^2\leq 8$. Assume otherwise. Hence $(S, L)\ncong(Y, L_Y)$. Note that for a smooth hyperelliptic $C\in|L_Y|$, $K_C\cong(K_Y+L_Y)_C$ is not very ample. Thus since smooth elements of $|L|$ give rise to smooth elements of $|L_Y|$, we conclude that K_Y+L_Y is not very ample. Using this and the fact that L_Y is very ample by Proposition (10.6.2), Theorem (10.4.1) applies to say that (Y, L_Y) is either as in (10.4.1.1) or as in (10.4.1.2). In both the cases (Y, L_Y) is not a reduction of a given pair (S, L), that is $(Y, L_Y)\cong(S, L)$. This leads to a contradiction.

By Theorem (10.2.6) and Theorem (10.2.7) we can assume without loss of generality that $g(L)\geq 4$. Note that when $(K_S+L)^2=2$, Theorem (10.2.3) gives the example with $L_Y^2=8$ from (10.2.3.2) listed in the theorem. By using the classification in Theorem (10.2.3) we can assume without loss of generality that $(K_S+L)^2\geq 3$. We are thus reduced to showing that under these assumptions there are no L with a smooth hyperelliptic curve C in $|L|$. Since $d_Y\leq 8$ we can use Proposition (10.6.2) to assume without loss of generality that $S=Y$ and $L=L_Y$. Then we can also assume that K_S+L is ample.

Note that since there are no hyperelliptic curves in $\mathbb{P}^2$ we have that $h^0(L)\geq 5$. Assume first $d\leq 6$. Thus using Lemma (10.2.8) and the fact that, by (10.24), $g(L)\geq 4$ we see that $K_S\cong\mathcal{O}_S$ which implies that K_S+L is very ample, and in particular that there are no hyperelliptic curves $C\in|L|$. Thus we can assume without loss of generality that $7\leq d\leq 8$. Since $q(S)=0$, we also have $\chi(\mathcal{O}_S)=1$ by Proposition (10.6.1).

Let $d=7$. As noted above we have that (S, L) is its own reduction and K_S+L is ample. From (10.6.1.2) we get $g(L)\leq 4$, and therefore that $g(L)=4$. Since $h^1(L_C)=0$ by (10.6.1.1), we have $h^0(L_C)=d-g(L)+1=7-4+1=4$, so we conclude that $h^0(L)=5$. From Lemma (8.2.1) we find $K_S^2=-2$. To compute $g(K_S+L)$, note that $2g(K_S+L)-2=(2K_S+L)\cdot(K_S+L)=0$. Thus by

Theorem (10.2.6) we conclude that $K_S + L \cong -K_S$. This implies the contradiction that $K_S^2 > 0$.

Let $d = 8$. By (10.6.1.2) we have $8 = d \geq g(L) + 3$, so that $4 \leq g(L) \leq 5$.

Let $g(L) = 4$ first. We claim that $K_S^2 \geq 0$. To see this assume that $K_S^2 < 0$. Computing $g(K_S + L)$ we obtain, $2g(K_S + L) - 2 = (2K_S + L) \cdot (K_S + L) \leq 0$, i.e., $g(K_S + L) \leq 1$. If $g(K_S + L) = 1$ we conclude as in the case $d = 7$ the contradiction that $K_S^2 > 0$. If $g(K_S + L) = 0$ we conclude by Theorem (8.9.1) the same contradiction that $K_S^2 > 0$. By the Hodge index theorem we have $8K_S^2 = dK_S^2 \leq (K_S \cdot L)^2 = 4$. Thus we conclude that $K_S^2 = 0$. Since there is a smooth hyperelliptic curve, $C \in |L|$ by hypothesis, Theorem (8.4.2) applies, with $k = 1$, to say that there exists a subset, $\mathcal{S}$, of $|L|$ consisting of smooth hyperelliptic curves and such that $\dim \mathcal{S} \geq h^0(L) - 1 - k - 2 \geq 1$. Recall that, by (10.6.1.1), $h^0(L) = h^0(L_C) + 1 = d - g(L) + 2 = 6$. Since $q(S) = 0$, the restriction of the adjoint mapping, ϕ, to each of such smooth hyperelliptic curves, $\mathcal{C}$, in $\mathcal{S}$ is the canonical map given by $\Gamma(K_{\mathcal{C}})$. Thus there exists a 1-dimensional family (at least) of curves $\mathcal{C}$ on S such that the restriction, $\phi_{\mathcal{C}}$, of ϕ to $\mathcal{C}$ is a two-to-one morphism. This shows that ϕ is a morphism of degree 2 at least. Therefore the finite morphism $s : S \to \mathbb{P}_{\mathbb{C}}$ given by $\Gamma(K_S + L)$ is of degree two at least. Since $h^0(K_S + L) = 4$ by equation (10.24), and $(K_S + L)^2 = 4$, we see that $s : S \to \mathbb{P}^3$ is in fact a degree two morphism onto a normal quadric Q in $\mathbb{P}^3$. Assume that Q is a singular quadric. Since $K_S^2 = 0$, (8.3.1.2) applies to give a numerical contradiction. Thus $Q \cong \mathbb{P}^1 \times \mathbb{P}^1$. Let $B \subset Q$ be the branch locus of s. Then $B \in |\mathcal{O}_{\mathbb{P}^1 \times \mathbb{P}^1}(a, b)|$ where a, b are even positive integers. Since $K_S^2 = 0$, either $a = 4$ or $b = 4$ by (8.3.1.1). Let $a = 4$. Then

$$K_S \cong s^* \left(K_{\mathbb{P}^1 \times \mathbb{P}^1} + \mathcal{O}_{\mathbb{P}^1 \times \mathbb{P}^1} \left(\frac{a}{2}, \frac{b}{2} \right) \right) \cong s^* \mathcal{O}_{\mathbb{P}^1 \times \mathbb{P}^1} \left(0, \frac{b-4}{2} \right),$$

so that

$$L \cong s^* \left(\mathcal{O}_{\mathbb{P}^1 \times \mathbb{P}^1}(1, 1) + \mathcal{O}_{\mathbb{P}^1 \times \mathbb{P}^1} \left(0, \frac{4-b}{2} \right) \right) \cong s^* \mathcal{O}_{\mathbb{P}^1 \times \mathbb{P}^1} \left(1, \frac{6-b}{2} \right).$$

The genus formula yields

$$6 = (K_S + L) \cdot L = 2\mathcal{O}_{\mathbb{P}^1 \times \mathbb{P}^1}(1, 1) \cdot \mathcal{O}_{\mathbb{P}^1 \times \mathbb{P}^1} \left(1, \frac{6-b}{2} \right) = 6 - b + 2 = 8 - b.$$

Therefore $b = 2$ and $L \cong s^* \mathcal{O}_{\mathbb{P}^1 \times \mathbb{P}^1}(1, 2)$. Let $H = \mathcal{O}_{\mathbb{P}^1 \times \mathbb{P}^1}(1, 2)$. Since

$$s_* \mathcal{O}_S \cong \mathcal{O}_{\mathbb{P}^1 \times \mathbb{P}^1} \oplus \mathcal{O}_{\mathbb{P}^1 \times \mathbb{P}^1} \left(-\frac{a}{2}, -\frac{b}{2} \right) \cong \mathcal{O}_{\mathbb{P}^1 \times \mathbb{P}^1} \oplus \mathcal{O}_{\mathbb{P}^1 \times \mathbb{P}^1}(-2, -1),$$

by using the projection formula we have

$$h^0(L) = h^0(s^* H) = h^0(s_* \mathcal{O}_S \otimes H) = h^0(H) + h^0(\mathcal{O}_{\mathbb{P}^1 \times \mathbb{P}^1}(-1, 1)) = h^0(H).$$

The equality $h^0(L) = h^0(H)$ implies that each section of $\Gamma(L)$ is the pullback of a section of $\Gamma(H)$. Hence we have the contradiction that $|L|$ does not embed the fibers of s.

It remains to consider the case when $d = 8$ and $g(L) = 5$. In this case

$$h^0(L) = 1 + h^0(L_C) = 1 + 8 - 5 + 1 = 5.$$

Lemma (8.2.1) applies again to give $K_S^2 = -2$. Hence $(K_S + L)^2 = 6$. Once again, the same argument above, as in the case $d = 7$, shows that the adjunction mapping is of degree two at least. Since $h^0(K_S + L) = 5$ by equation (10.24), we conclude that $\Gamma(K_S + L)$ defines a two-to-one finite cover $\phi : S \to \mathbb{P}^4$, whose image is a cubic surface in $\mathbb{P}^4$.

Let $\phi = s \circ r$ be the Remmert-Stein factorization of ϕ. Then we can assume that r is a biholomorphism and $\phi = s$. Let $\alpha : S \to Z$ be the map that ϕ induces from S to the normalization, Z, of $\phi(S)$. Let B be the codimension 1 divisor of the ramification locus in S of α. Let E_x for $x \in B$ be the subspace of $C \in |L|$ with $x \in C$ and such that the differential $d\alpha_x$ is zero on the Zariski tangent space of C at x. Now E_x is of codimension 2 for a generic x, and therefore $\cup_{x \in B} E_x$ is at most of codimension 1. Therefore a general $C' \in |L|$ is transverse to B, misses the singular points of B, and $d\alpha$ is nonzero on all nonzero $\tau \in \mathcal{T}_{C',x}$ for all $x \in C'$. Thus C' is not hyperelliptic and $\alpha_{C'}$ is an embedding; otherwise, if C' were hyperelliptic, $d\alpha$ would be zero on $\mathcal{T}_{C',x}$ at the hyperelliptic branch point of $\alpha_{C'}$. Now α is given by quotienting S by an involution σ that extends the hyperelliptic involution on any smooth hyperelliptic element of $|L|$. Thus $\sigma(C') \in |L|$, since $\sigma(C) = C$ for a hyperelliptic $C \in |L|$, implying $\mathcal{O}_S(\sigma(C)) \cong L$, or $\sigma^* L \approx L$. Therefore, since $\alpha^{-1}(\alpha(C')) = C' + \sigma(C')$ and $C' \cdot \sigma(C') = C' \cdot B$, and since this is true for all $C \in |L|$ near C', $\phi = s \circ r$, it follows that $\mathcal{O}_S(B)_{C'} \approx L_{C'}$ for an open set of $C' \in |L|$. Thus $\mathcal{O}_S(B) \approx L$ by Proposition (2.3.6), i.e., $B \in |L|$.

Now either there is or is not an $x \in S$ such that $d\alpha_x$ is identically zero on $\mathcal{T}_{S,x}$. If there is no such point, then B is the whole branch locus, and any smooth hyperelliptic $C_0 \in |L|$ must meet B in at least the $2g(C) + 2$ hyperelliptic branch points. Thus $L^2 = C_0 \cdot B \geq 2g(L) + 2$ in homology. If there is such a point x, a smooth $C_0 \in |L|$ can be found with $x \in C_0$ and C_0 transverse to B, containing only regular points of B, and missing all the finitely many other points $y \in S$ such that $d\alpha_y$ is identically zero on $\mathcal{T}_{S,y}$. Now since $d\alpha_x$ is zero on $\mathcal{T}_{C_0,x}$, it follows that C_0 is hyperelliptic. This implies that $L^2 = C_0 \cdot L = C_0 \cdot B \geq 2g(L) + 1$. Thus $L^2 \geq 2g(L) + 1 = 11$, contradicting $L^2 = 8$. □

Let (S, L), (Y, L_Y), $r : S \to Y$, be as at the beginning of this section. We will now study the very ampleness of $K_Y + L_Y$ under the assumption that $q(S) = 0$ and there are no smooth hyperelliptic curves in $|L|$.

The following result completes the proof of Theorem (10.3.1).

Theorem 10.6.4. *Let L be a very ample line bundle on a smooth connected projective surface, S. Let $d := L^2$. Assume that $K_S + L$ is spanned and big, and let (Y, L_Y) be the reduction of (S, L). Furthermore assume that $d \leq 8$, $q(S) = 0$, and there are no hyperelliptic curves in $|L|$. Then $K_Y + L_Y$ is very ample.*

Proof. Using Theorem (10.2.6), Theorem (10.2.7), Theorem (10.4.1), and Lemma (10.2.8) we can assume without loss of generality that $7 \leq d \leq d_Y \leq 8$, $g(L) \geq 4$. Using the classification given in Theorem (10.2.3) we can also assume without loss of generality $(K_S + L)^2 \geq 3$.

Let $r : S \to Y$ be the reduction morphism. Take two distinct points $y_1, y_2 \in Y$ and let $x_1, x_2 \in S$ be two points such that $r(x_i) = y_i$, $i = 1, 2$.

Assume that there exists a smooth curve $C \in |L|$ passing through x_1, x_2. Look at the exact sequence

$$0 \to K_S \to K_S + L \to K_C \to 0.$$

Since C is a smooth nonhyperelliptic curve of genus $g(C) \geq 4$, the canonical map given by $\Gamma(K_C)$ is an embedding (see Hartshorne [318, IV, 5.2]). Since $q(S) = 0$ we see from the exact sequence above that $\Gamma(K_S + L)$ surjects onto $\Gamma(K_C)$. Therefore the restriction to C of the adjoint mapping, $\phi : S \to \mathbb{P}_{\mathbb{C}}$, coincides with the morphism given by $\Gamma(K_C)$. Thus ϕ is an embedding on C. Since ϕ factors through s, $\phi = s \circ r$, it follows that $s(y_1) \neq s(y_2)$, i.e., $\Gamma(K_Y + L_Y)$ separates y_1, y_2.

If there are no smooth curves $C \in |L|$ passing through x_1, x_2, then Theorem (1.7.9) applies to say that there is a line, ℓ, of S relative to L containing x_1, x_2, $\ell \cong \mathbb{P}^1$. We know that $L - \ell$ is spanned on S (compare with §8.4). We can also assume that $L - \ell$ is nef and big. Otherwise from Corollary (8.4.6) we know that $K_S + L$ is very ample. Look at the exact sequence

$$0 \to K_S + L - \ell \to K_S + L \to (K_S + L)_\ell \to 0.$$

Let $K_S \cdot \ell = k$. Since $L - \ell$ is nef and big, we have a surjective map $\Gamma(K_S + L) \to \Gamma(\mathcal{O}_{\mathbb{P}^1}(k+1))$. Therefore the restriction to ℓ of the adjunction mapping ϕ coincides with the map given by $\Gamma(\mathcal{O}_{\mathbb{P}^1}(k+1))$. Since $K_S + L$ is nef we have $(K_S + L) \cdot \ell \geq 0$, i.e., $k = K_S \cdot \ell \geq -1$.

If $k \geq 0$, $\mathcal{O}_{\mathbb{P}^1}(k+1)$ is very ample, and thus ϕ is an embedding on ℓ. Since ϕ factors through s, it follows that $s(y_1) \neq s(y_2)$, i.e., $\Gamma(K_Y + L_Y)$ separates y_1, y_2.

If $k = -1$, then $\ell^2 = -1$, so that ℓ contracts to a point under r. This contradicts the fact that $r(\ell)$ contains y_1, y_2.

To conclude that $K_Y + L_Y$ is very ample we have to show that $\Gamma(K_Y + L_Y)$ separates infinitely near points (y, τ_y), where τ_y is a nonzero element of $\mathcal{T}_{Y,y}$ for $y \in Y$. Let x be a point on S such that $r(x) = y$ and let τ' be a nonzero element of $\mathcal{T}_{S,x}$ such that $dr_x(\tau') = \tau_y$, where $dr_x : \mathcal{T}_{S,x} \to \mathcal{T}_{Y,y}$ denotes the differential map. By Theorem (1.7.9) either a smooth curve $C \in |L|$ can be found with $x \in C$ and $\tau' \in \mathcal{T}_{C,x}$; or a line $\ell \subset S$ relative to L can be found with $x \in \ell$ and $\tau' \in \mathcal{T}_{\ell,x}$. Note that such a line ℓ cannot be contracted to a point under r, since otherwise $dr_x(\tau')$ would be the zero element in $\mathcal{T}_{Y,y}$. Now an argument exactly like the above shows that s is an embedding, i.e., $K_Y + L_Y$ is very ample. □

10.7 Preservation of *k*-very ampleness under adjunction

It is very natural to study the preservation of k-very ampleness under adjunction, i.e., to ask about the generalization of the results in this chapter to the case $k \geq 2$.

Question 10.7.1. *Let L be a k-very ample line bundle on a smooth connected projective surface, S. When is the adjoint bundle, $K_S + L$, k-very ample?*

The problem divides naturally into two cases, the "general" case when $L^2 \geq 4k+5$ and the "special" case when $L^2 \leq 4k+4$. This breakup stems from the Reider type existence theorem, (8.5.1), for the case $L^2 \geq 4k + 5$. For very ample line bundles this division also occurred; the case $L^2 \leq 8$ required detailed classification (see §10.5 and §10.6). Note that the low degree classification ultimately rests on the results on the adjunction mapping in the case $h^1(\mathcal{O}_S) = 0$ (see §10.6). In the general case $L^2 \geq 4k + 5$ the problem is settled by Beltrametti and Sommese [108] (see also Andreatta and Palleschi [23] for the case $k = 2$). In the special case $L^2 \leq 4k + 4$ the problem is solved in Ballico and Sommese [65], where the polarized pairs (S, L) of degree $L^2 \leq 4k + 4$ are classified. The results of both [108] and [65] rely on strong results for k-very ample line bundles on curves (see [65, §1]).

Before we describe the answer to the above question, let us survey the main papers in the rapidly growing literature on k-very ampleness. We refer to Andreatta [9], Beltrametti and Sommese [102, §5], and Lanteri [366] for the classification of polarized pairs (S, L) with L k-very ample and of sectional genus $g(L) \leq 8$. We refer to Ballico and Beltrametti [57, 58, 59] for results on k-very ampleness of the adjoint line bundle in dimension ≥ 3. For further results on k-spannedness and k-very ampleness we refer to Ballico [52, 53, 54], our paper [101], and our paper [88] with Francia. We refer to the paper [431] of Mella and Palleschi for a study of k-very ampleness on elliptic surfaces. Finally we refer to our paper [107] for the study of the k-jet ampleness, a stronger notion of higher order embedding.

Let L be a k-very ample line bundle on a smooth connected projective surface, S. We always assume that $k \geq 2$ in the rest of this section. The approach in our paper [108] is to use the notion of a k-reduction which generalizes the notion of reduction in the case of very ample line bundles. The first observation is that the k-very ampleness of L implies that for any effective divisor $C \subset S$, we have

$L \cdot C \geq k$ with equality only if C is biholomorphic to $\mathbb{P}^1$. Thus there are no exceptional lines, $\ell \subset S$, with $\ell^2 = -1$ if $k \geq 2$. The analogue of exceptional lines is played by smooth rational curves, $\ell \subset S$, with $L \cdot \ell = k$ and $\ell^2 = -1$.

A *k-reduction* is a pair, (S', L'), consisting of an ample line bundle, L', on a smooth projective surface, S', such that there is a morphism, $\pi : S \to S'$, expressing S as the blowup of S' at a finite set $B \subset S'$ with $L \cong \pi^* L' - k\pi^{-1}(B)$. The strategy of [108] is to show that for a k-very ample line bundle, L, $aK_S + L$ is very ample for $a \leq k - 1$ and $kK_S + L$ is spanned and big except for a short list of simple degenerate surfaces, e.g., $\mathbb{P}^1$-bundles with L of degree at most $2k$ on fibers of the bundle morphism, surfaces with $L \cong -kK_S$, etc. If $kK_S + L$ is nef and big then there exists a unique k-reduction, (S', L'), with $\pi : S \to S'$, such that $kK_S + L \cong \pi^*(kK_{S'} + L')$ with $kK_{S'} + L'$ ample. Using (8.5.1), Beltrametti and Sommese [108] showed that a very complete analogue of all the results in this chapter hold for k-very ample L with $L^2 \geq 4k + 5$, i.e., except for a few simple degenerate surfaces, $K_{S'} + L'$ is k-very ample. This result had been obtained in the case $k = 2$ by Andreatta and Palleschi [23].

Ballico and Sommese [65] classify k-very ample line bundles L on a smooth projective surface S satisfying $L^2 \leq 4k + 4$. Two representative results [65, Theorem (5.4), Theorem (6.1)] of their paper are the following.

Theorem 10.7.2. *Let L be a k-very ample line bundle on a smooth connected projective surface, S. Assume that $k \geq 2$ and that $d := L^2 \leq \max\{11, 4k + 2\}$. Then (S, L) is one of the following:*

1. *$(S, L) \cong (\mathbb{P}^2, \mathcal{O}_{\mathbb{P}^2}(a))$ with $2 \leq k \leq 3$ for $k \leq a \leq 3$ and with $a = k = 4$;*
2. *S is a $K3$ surface with $d = 4k,\ 4k + 2$;*
3. *$k = 2$ and $(S, L) \cong (\mathbb{P}^1 \times \mathbb{P}^1, \mathcal{O}_{\mathbb{P}^1 \times \mathbb{P}^1}(2, 2))$; or*
4. *S is a Del Pezzo surface and $L \cong -2K_S$ with $k = 2 = K_S^2$.*

Theorem 10.7.3. *Let L be a k-very ample line bundle on a smooth, connected, projective surface, S, with $d := L^2 \leq 4k + 4$.*

1. *If $\kappa(S) = -\infty$ then $k \leq 8$;*
2. *If $\kappa(S) \geq 0$ and $k \geq 5$, then either S is a $K3$ surface with $d = 4k,\ 4k + 2,\ 4k + 4$; or $d = 4k + 4$ and S is an Enriques surface, i.e., $2K_S \cong \mathcal{O}_S$, $q(S) = 0$, and $K_S \ncong \mathcal{O}_S$.*

As a corollary they contain an almost complete answer on the k-very ampleness of $K_S + L$ for $L^2 \leq 4k + 4$. The unanswered question in their result is whether given a k-very ample line bundle of degree, $L^2 = 4k + 4$, on an Enriques surface, S, it follows that $K_S + L$ is k-very ample. From (8.5.1) it follows that $K_S + L$ is $(k - 1)$-very ample in this case.

Chapter 11

Classical adjunction theory in dimension $\geq$ 3

Our main debt in this chapter is to the paper [586] of Sommese and Van de Ven, which we follow closely.

In §11.1 we prove some results on scrolls we need in the sequel and we also include the proof of a result of Lanteri and Sommese [389], Theorem (11.1.3), which characterizes the pairs, $(X, \mathscr{E})$, where $\mathscr{E}$ is an ample and spanned vector bundle on a normal projective surface, X, with $c_2(\mathscr{E}) = 1$.

In §11.2 we study the adjunction mapping in the case when it has a lower dimensional image. In §11.3 we study the very ampleness of the adjoint bundle $K_M + (n-1)L$ on the first reduction, (M, L), of an n-dimensional smooth variety $\mathcal{M}$ polarized by a very ample line bundle $\mathscr{L}$. Following [586], we prove the main result of this chapter, which states that $K_M + (n-1)L$ is very ample if $n \geq 3$ (see §10.3 for the case $n = 2$).

In §11.5 we give an outline of the proof of a result from Andreatta, Ballico, and the second author [27] and Ein and Lazarsfeld [207] on the projective normality of adjoint bundles.

As one application of the results of §11.2 and §11.3 we give in §11.4 the classification of projective manifolds with hyperelliptic curve sections. As a consequence of this classification we derive an analogue of (11.1.3) above in the case when $c_2(\mathscr{E}) = 2$. To illustrate how adjunction theory is used in projective classification, we give in §11.6 a rough classification of projective manifolds with curve sections of genus $g = 3, 4$. This is based on the more detailed classification of Livorni [409, 410, 411, 412] and Ionescu [333, 334].

We end with a discussion of the adjunction processes of Fano and Morin and as a consequence we derive some strong nefness results for adjoint bundles, that were derived in a different way in Chapter 7.

11.1 Some results on scrolls

Scrolls come up naturally in adjunction theory. In this section we collect a few results we will need throughout this chapter. We will return to more general versions of some of these results, e.g., Proposition (14.1.3), later in this book.

Theorem 11.1.1. *Let L be an ample and spanned line bundle on an n-dimensional connected projective manifold, X. Let (X, L) be a scroll over an m-dimensional normal projective variety, Y, i.e., there exists a morphism, $p : X \to Y$, with connected fibers of X onto Y such that the general fiber has dimension $f = n - m$, $K_X + (f+1)L \cong p^*H$ for some ample line bundle H on Y. If $n > m \leq 2$ then Y is smooth, p has equidimensional fibers and $(X, L) \cong (\mathbb{P}(\mathcal{E}), \xi_{\mathcal{E}})$, with $\mathcal{E} := p_*L$ a locally free sheaf of rank $f+1$.*

Proof. First let us show that Y is smooth and the fibers are all equal dimensional. If $\dim Y = 1$ then the smoothness of Y follows immediately from the normality of Y. Moreover if $\dim Y = 1$ then all fibers of p are at least $(n-1)$-dimensional, and hence exactly $(n-1)$-dimensional. Therefore we can assume without loss of generality that $\dim Y = 2$. Note that $n-3$ general sections $A_1, \ldots, A_{n-3} \in |L|$ intersect transversally in a smooth 3-fold, M. Note also that (M, L_M) is a scroll with respect to the restriction map $p_M : M \to Y$ over Y, and that if all fibers of p_M are 1-dimensional, then all fibers of p are $(n-2)$-dimensional. Thus we can assume without loss of generality that $n = 3$. Let $S \in |L|$ be a smooth divisor. Note that $K_S + L_S \cong (K_X + 2L)_S \cong p_S^* H$. Thus $K_S + L$ is nef and big. Consider the exact sequence

$$0 \to K_X + N(K_X + 2L) + L \to (N+1)(K_X + 2L) \to (N+1)(K_S + L_S) \to 0.$$

Using the Kodaira vanishing theorem we see that there is a surjection

$$\Gamma((N+1)(K_X + 2L)) \to \Gamma((N+1)(K_S + L_S)) \to 0 \text{ for all } N \geq 0.$$

Since the map p is the map given by $|(N+1)(K_X + 2L)|$ for large N we conclude that p_S is the map given by $|(N+1)(K_S + L_S)|$ for large N. In particular p_S is the first reduction mapping and Y is therefore smooth. To see that all fibers of p are 1-dimensional, assume to the contrary that there is a fiber, F, of p which is not of dimension 1. Then $\dim F = 2$. Choose a general $S \in |L|$. Note that $S \cap F$ is a 1-dimensional fiber of p_S. Since p_S is the first reduction morphism, this implies that $S \cap F$ is a smooth curve $C \cong \mathbb{P}^1$ with $L \cdot C = 1$. Since F has a smooth ample divisor, C, the singular set of F is at most finite. Thus F is irreducible and generically reduced, and $F' := F_{\text{red}}$, i.e., F with its reduced structure, is normal. Using the classification of normal surfaces with $\mathbb{P}^1$ as an ample divisor (see (8.9.1)) and using the fact that on F', $C \cdot C = L \cdot C = 1$, we see that $F' \cong \mathbb{P}^2$.

Since $(K_X + 2L)_{F'} \cong K_{X|F'} + \mathcal{O}_{\mathbb{P}^2}(2)$, we see that $K_{X|F'} \cong \mathcal{O}_{\mathbb{P}^2}(-2)$, and therefore that $\mathcal{N}_{F'/X} \cong \mathcal{O}_{\mathbb{P}^2}(-1)$. Thus there exists a morphism $g : X \to Z$ from X onto a normal analytic space, Z, with $z := g(F')$ a point, and $g_{X-g^{-1}(z)}$ an isomorphism. By the rigidity lemma, (4.1.13), we can factor p as $h \circ g$ where $h : Z \to Y$ is a surjective map. We have the absurdity that the fiber of h over $p(F')$ is a point but the general fiber of h is 1-dimensional. This implies that all the fibers of p are equal dimensional.

Since $K_F + (f+1)L_F \cong \mathcal{O}_F$ for a general fiber F of p, of dimension f, we conclude from (3.1.6) that $(F, L_F) \cong (\mathbb{P}^f, \mathcal{O}_{\mathbb{P}^f}(1))$, and therefore that $L^f \cdot F = 1$. Since Y is smooth and all the fibers of p are equal dimensional, we see that p is flat, and therefore that $L^f \cdot F' = 1$ for any fiber F' of p. By (3.2.1) we conclude that p is a $\mathbb{P}^f$-bundle over Y with all fibers isomorphic to $(\mathbb{P}^f, \mathcal{O}_{\mathbb{P}^f}(1))$. Thus $\mathcal{E} := p_*L$ is a rank $f+1$ vector bundle. By construction $(X, L) \cong (\mathbb{P}(\mathcal{E}), \xi_{\mathcal{E}})$. □

Theorem 11.1.2. *Let L be an ample and spanned line bundle on a 3-dimensional connected projective manifold, X. Assume that (X, L) is a scroll, $p : X \to Y$, over a normal surface Y, i.e., there exists a morphism, $p : X \to Y$, with connected fibers of X onto Y, $K_X + 2L \cong p^*H$ for some ample line bundle H on Y. Let $S \in |L|$ be a smooth ample divisor on X. Let $\mathcal{E} := p_*L$. Then:*

1. *(S, L_S) has $(Y, \det\mathcal{E})$ as its first reduction. Moreover the reduction map is given by the restriction, p_S, of p to S and the number of points blown up by p_S is equal to $c_2(\mathcal{E})$. In particular $L^3 = (\det\mathcal{E})^2 - c_2(\mathcal{E})$.*

2. *If $A \in |L|$ and $B \in |L|$ are smooth and meet transversally, then the smooth curve $C := A \cap B$ possesses a $c_2(\mathcal{E})$-to-one map of C to $\mathbb{P}^1$.*

3. *If L is ample and spanned (respectively very ample) then* $\det\mathcal{E}$ *is ample and spanned (respectively very ample).*

Proof. The proof of Theorem (11.1.1) shows that (S, L_S) has $(Y, H - K_Y)$ as its first reduction, and that $X \cong \mathbb{P}(\mathcal{E})$, where $\mathcal{E} = p_*L$. By the canonical bundle formula for $\mathbb{P}(\mathcal{E})$ we have that $K_X \cong p^*(K_Y + \det\mathcal{E}) - 2L$. Hence $p^*H \cong K_X + 2L \cong p^*(K_Y + \det\mathcal{E})$. Thus $H - K_Y \cong \det\mathcal{E}$, and therefore (S, L_S) has $(Y, \det\mathcal{E})$ as its first reduction. Choose a smooth $S \in |L|$ corresponding to a section s of $\mathcal{E}$ under the correspondence between sections of $L = \xi_{\mathcal{E}}$ and $\mathcal{E}$. The positive dimensional fibers of p_S are fibers of p, and thus the positive dimensional fibers of p_S correspond to zeroes of s. Since the fibers of p_S are smooth, the zeroes of s have multiplicity 1. Thus $c_2(\mathcal{E})$ equals the number of points blown up by p_S. Since $L^3 = \xi_{\mathcal{E}}^3 = (\det\mathcal{E})^2 - c_2(\mathcal{E})$, item 1) is proven.

Let A and B be as in item 2). Choose sections s and t of $\mathcal{E}$ giving rise to A and B respectively. Since (A, L_A) has $(Y, \det\mathcal{E})$ as its first reduction, we conclude that the smooth curve $A \cap B \in |L_A|$ maps isomorphically under p to a smooth curve

$C' \in |\det \mathcal{E}|$. For $x \in C'$, s and t give rise to linearly dependent sections of $\mathcal{E}_{C',x}$. Note that there is no point of C' where both s and t are zero, since if there was such a point we would have the contradiction that $C = A \cap B$ would contain a fiber of p. Thus s, t span a line subbundle $\mathcal{L}$ of $\mathcal{E}_{C'}$. The degree of $\mathcal{L}$ is the number of zeroes of the section of $\mathcal{L}$ induced by, e.g., s. By construction this is $c_2(\mathcal{E})$. Thus the sections of $\mathcal{L}$ induced by s, t give a $c_2(\mathcal{E})$-to-one map of C' to $\mathbb{P}^1$. This proves item 2).

To prove item 3) we must show that $\det \mathcal{E}$ is spanned if L is ample and spanned and that $\det \mathcal{E}$ is very ample if L is very ample. Choose a point $y \in Y$ and a smooth $S \in |L|$ such that S does not contain $p^{-1}(y)$. Then p_S is an isomorphism on $p_S^{-1}(U)$ for some neighborhood U of y. Thus $(\det \mathcal{E})_U \cong (p_* L_S)_U$, and we see that $\det \mathcal{E}$ is spanned by global sections at y. If L is very ample, the global sections of $\det \mathcal{E}$ give an embedding in a neighborhood of y. Moreover given two distinct points, y_1, y_2 of Y we can choose a smooth $S \in |L|$ such that S does not contain $p^{-1}(y_i)$ for $i = 1, 2$. Therefore the same reasoning as for one point shows that if L is very ample, then the map associated to $|\det \mathcal{E}|$ separates the points y_i, $i = 1, 2$. □

The following result is proved by Lanteri and Sommese in [389]. We refer the reader there and also to Wiśniewski [619] for more results in the same direction. See also Theorem (11.4.5).

Theorem 11.1.3. *Let $\mathcal{E}$ be an ample and spanned vector bundle on a normal projective surface, X. If $c_2(\mathcal{E}) = 1$, then $(X, \mathcal{E}) \cong (\mathbb{P}^2, \mathcal{O}_{\mathbb{P}^2}(1) \oplus \mathcal{O}_{\mathbb{P}^2}(1))$.*

Proof. Let $r := \operatorname{rank}(\mathcal{E})$. Since $c_2(\mathcal{E}) = 1$ we see that $r \geq 2$.

Assume that $r = 2$. Then by Theorem (11.1.2) we know that $g(\det \mathcal{E}) = 0$. Thus $(X, \det \mathcal{E})$ is on the list in Proposition (3.1.2). (Theorem (8.9.1) can be used if the reader is only interested in the case when $\det(\mathcal{E})$ is very ample and X is smooth.) Since $\det \mathcal{E} \cdot C = \deg \mathcal{E}_C \geq r = 2$ for any effective curve C on X, we see that the only possibility is $(\mathbb{P}^2, \mathcal{O}_{\mathbb{P}^2}(2))$. Consider the restriction $\mathcal{E}_\ell$ for any line $\ell \subset \mathbb{P}^2$. Since $\mathcal{E}_\ell \cong \mathcal{O}_{\mathbb{P}^1}(a) \oplus \mathcal{O}_{\mathbb{P}^1}(b)$ with $a > 0$, $b > 0$ and $2 = c_1(\mathcal{E}) \cdot \ell = a + b$, we conclude that $a = b = 1$ and $(\mathcal{E} \otimes \mathcal{O}_{\mathbb{P}^2}(-1))_\ell$ is the trivial bundle. Let $\mathcal{F} := \mathcal{E} \otimes \mathcal{O}_{\mathbb{P}^2}(-1)$. Since $c_1(\mathcal{F}) = 0$ and $c_2(\mathcal{F}) = 0$, we conclude from the Riemann-Roch theorem (see Fulton [267, 15.22]) that $\chi(\mathcal{F}) = 2$. Note that

$$h^2(\mathcal{F}) = h^0(K_{\mathbb{P}^2} \otimes \mathcal{F}^*) = h^0(\mathcal{E}^* \otimes \mathcal{O}_{\mathbb{P}^2}(-2)).$$

Since $\mathcal{E}^* \otimes \mathcal{O}_{\mathbb{P}^2}(-2)$ has an ample dual bundle, we see that $h^0(\mathcal{E}^* \otimes \mathcal{O}_{\mathbb{P}^2}(-2)) = 0$ and hence $h^2(\mathcal{F}) = 0$. Thus $h^0(\mathcal{F}) \geq \chi(\mathcal{F}) = 2$. Note that no section of $\mathcal{F}$ can vanish anywhere. Indeed by restricting to lines of $\mathbb{P}^2$ we see that if a section vanished at a point, $z \in \mathbb{P}^2$, then it would vanish on any line through z, and hence on all of $\mathbb{P}^2$. Thus the 2 linearly independent sections span $\mathcal{F}$. This shows $\mathcal{E} \cong \mathcal{O}_{\mathbb{P}^2}(1) \oplus \mathcal{O}_{\mathbb{P}^2}(1)$.

Now assume that $r > 2$. Then we can find $r-2$ general sections which span a trivial rank $r-2$ subbundle, T, of $\mathcal{E}$ which gives an exact sequence

$$0 \to T \to \mathcal{E} \to \mathcal{E}/T \to 0.$$

Note that $\mathcal{E}/T$ is an ample and spanned rank 2 vector bundle which satisfies $c_2(\mathcal{E}/T) = 1$. By the last paragraph, $\mathcal{E}/T \cong \mathcal{O}_{\mathbb{P}^2}(1) \oplus \mathcal{O}_{\mathbb{P}^2}(1)$. Since $h^1(\mathcal{O}_{\mathbb{P}^2}(-1)) = 0$, then $h^1((\mathcal{E}/T)^*) = 0$, and hence the exact sequence splits. This gives $\mathcal{E} \cong T \oplus \mathcal{E}/T$, which contradicts the ampleness of $\mathcal{E}$. □

11.2 The adjunction mapping with a lower dimensional image

Let $\mathcal{M}$ be a smooth projective variety of dimension $n \geq 3$ and let $\mathcal{L}$ be a very ample line bundle on $\mathcal{M}$. By using the results of §9.2 we can assume that except for some simple, explicit examples, the adjoint bundle, $K_{\mathcal{M}} + (n-1)\mathcal{L}$, is spanned. Let $\phi : \mathcal{M} \to \mathbb{P}_{\mathbb{C}}$ be the morphism associated to $\Gamma(K_{\mathcal{M}} + (n-1)\mathcal{L})$, and let $\phi = s \circ r$ be the Remmert-Stein factorization of ϕ, where $r : \mathcal{M} \to Y$ has connected fibers, Y is normal, and s is a finite-to-one morphism. Note that either $\dim \phi(\mathcal{M}) \leq 2$ or $\dim \phi(\mathcal{M}) = n$, since otherwise a general fiber, F, of ϕ would satisfy the condition $-K_F \approx t\mathcal{L}_F$ with $t > \dim F + 1$, and this is not possible for dimension reasons (see again Lemma (3.3.2)). In the last part of this section we will consider the case when $\phi(\mathcal{M}) < n$. The aim is to show that s is always an embedding in this case (see Proposition (11.2.4) below). The case when $\dim \phi(\mathcal{M}) = n$ will be treated in §11.3.

Throughout this section we will use over and over Lemma (3.3.2), as well as the Kobayashi-Ochiai result, (3.1.6), without explicitly referring to them.

Remark-Example 11.2.1. Let $\mathcal{M}$ be a connected projective manifold of dimension $n \geq 3$, and let $\mathcal{L}$ be a very ample line bundle on $\mathcal{M}$. Note that $K_{\mathcal{M}} + (n-1)\mathcal{L}$ is spanned if $K_{\mathcal{M}} + (n-1)\mathcal{L}$ is nef by Corollary (9.2.3). The analogous "$n-2$" result, i.e., that $K_M + (n-2)L$ nef implies $K_M + (n-2)L$ spanned where (M, L) denotes the first reduction of $(\mathcal{M}, \mathcal{L})$, is not true as shown by the following example from Lanteri, Palleschi, and Sommese [384].

Fujita has constructed a Del Pezzo 3-fold $(\mathcal{M}, H)$, $K_{\mathcal{M}} \approx -2H$, with H ample and of degree $H^3 = 1$ (see Fujita [257, Chap. I, §8]). Precisely, $\mathcal{M}$ is a given hypersurface in the weighted projective space $\mathbb{P}(3, 2, 1, 1)$. In Lanteri, Palleschi, and Sommese [384] it is proved that mH is very ample for $m \geq 3$ and letting $\mathcal{L} := 3H$, $|K_{\mathcal{M}} + \mathcal{L}| = |H|$ has exactly one base point and defines a rational map of $\mathcal{M}$ onto $\mathbb{P}^2$. On the other hand, $K_{\mathcal{M}} + \mathcal{L} = H$ is ample, and hence nef. Note that $(\mathcal{M}, \mathcal{L})$ coincides with its own first reduction. In fact, since $\mathcal{L}$ is divisible by 3, there are no $\mathbb{P}^{n-1}$ in $\mathcal{M}$ such that $\mathcal{L}_{\mathbb{P}^{n-1}} \cong \mathcal{O}_{\mathbb{P}^{n-1}}(1)$.

Note that $2(K_{\mathcal{M}} + \mathcal{L}) = 2H$ is spanned as proved in (13.2.5).

Remark 11.2.2. It should be noted that the varieties, $\mathcal{M}$, listed in (9.2.2), are exactly those Gorenstein projective varieties that contain a family of lines of dimension at least $2n-3$, where $n = \dim \mathcal{M}$. This is a special case of a classical result due to Segre [550], which doesn't put restrictions on the singularities (see also Severi [560] and Rogora [523]).

To prove Proposition (11.2.4) we need the following general nonextension result of Sommese and Van de Ven [586] which will also be used in §11.3.

Theorem 11.2.3. *Let $\mathcal{M}$ be a smooth, connected, projective 3-fold and let $\mathcal{L}$ be a very ample line bundle on $\mathcal{M}$. Then there are no smooth surfaces $S \in |\mathcal{L}|$ with $(S, \mathcal{L}_S)$ as in* (10.3.1.1), (10.3.1.2), (10.3.1.3), (10.3.1.4).

Proof. Let $(S, \mathcal{L}_S)$ be as in (10.3.1.1). Then $-\mathcal{L}_S \approx 3K_S \approx 3K_{\mathcal{M}|S} + 3\mathcal{L}_S$ and hence $-3K_{\mathcal{M}|S} \approx 4\mathcal{L}_S$. From the Lefschetz theorem, (2.3.1), we know that $\mathrm{Pic}(\mathcal{M})$ injects in $\mathrm{Pic}(S)$, and therefore we conclude that $-3K_{\mathcal{M}} \approx 4\mathcal{L}$. Therefore there exists an ample line bundle A on $\mathcal{M}$ such that $-K_{\mathcal{M}} \approx 4A$, $\mathcal{L} \approx 3A$. Thus $(\mathcal{M}, A) \cong (\mathbb{P}^3, \mathcal{O}_{\mathbb{P}^3}(1))$ by the Kobayashi-Ochiai theorem. Hence in particular $\mathcal{L}^3 = 27A^3 = 27$. This contradicts $\mathcal{L}^3 = \mathcal{L}_S^2 = 9$.

Let $(S, \mathcal{L}_S)$ be as in (10.3.1.2). Then $-\mathcal{L}_S \approx 2K_S \approx 2K_{\mathcal{M}|S} + 2\mathcal{L}_S$, and hence $-2K_{\mathcal{M}|S} \approx 3\mathcal{L}_S$. By the Lefschetz theorem, (2.3.1), it thus follows that $-2K_{\mathcal{M}} \approx 3\mathcal{L}$. Therefore $-K_{\mathcal{M}} \approx 3A$, $\mathcal{L} \approx 2A$ for some ample line bundle A on $\mathcal{M}$. By the Kobayashi-Ochiai theorem, we conclude that $(\mathcal{M}, A) \cong (Q, \mathcal{O}_Q(1))$, with Q a quadric in $\mathbb{P}^4$. Hence $\mathcal{L}^3 = 8A^3 = 16$. This contradicts $\mathcal{L}^3 = \mathcal{L}_S^2 = 8$.

Let $(S, \mathcal{L}_S)$ be as in (10.3.1.4). By Bădescu's result, (7.9.5), we know that the $\mathbb{P}^1$-bundle, $p : S \to C$, extends to a $\mathbb{P}^2$-bundle, $\mathcal{M} \to C$ and $\mathcal{L}_{\mathbb{P}^2} \approx \mathcal{O}_{\mathbb{P}^2}(1)$ for each fiber $\mathbb{P}^2$. In particular, $\mathcal{L} \cdot f = 1$ for each fiber, f, of p. This contradicts $\mathcal{L}_S \approx 3E$, where E is a section of minimal self-intersection, $E^2 = 1$, which gives $\mathcal{L} \cdot f = 3$.

Let $(S, \mathcal{L}_S)$ be as in (10.3.1.3). Then $(S, \mathcal{L}_S)$ is the blowing up, $\pi : S \to S'$, of its own reduction, $(S', \mathcal{L}')$, with $\mathcal{L}' \approx -2K_{S'}$, $\mathcal{L}'^2 = 8$, $K_{S'}^2 = 2$, i.e., $(S', \mathcal{L}')$ is as in (10.3.1.2). Note that $\mathcal{L}_S^2 = 7$, $K_S^2 = 1$. We can assume that $K_{\mathcal{M}} + 2\mathcal{L}$ is spanned. Indeed otherwise we know by Corollary (9.2.2) that either $(\mathcal{M}, \mathcal{L}) \cong (\mathbb{P}^3, \mathcal{O}_{\mathbb{P}^3}(1))$, $(Q, \mathcal{O}_Q(1))$, where Q is a quadric in $\mathbb{P}^4$, or $\mathcal{M}$ is a $\mathbb{P}^2$-bundle over a smooth curve, $p : \mathcal{M} \to C$, with $\mathcal{L}_{\mathbb{P}^2} \cong \mathcal{O}_{\mathbb{P}^2}(1)$ for each fiber $\mathbb{P}^2$. The first two cases contradict the assumption that $\mathcal{L}^3 = \mathcal{L}_S^2 = 7$. In the third case, S would be a $\mathbb{P}^1$-bundle over C under the restriction of p to S. Therefore $K_S^2 = 8(1 - q(S))$, which contradicts $K_S^2 = 1$.

Since $K_{\mathcal{M}} + 2\mathcal{L}$ is spanned the adjunction map, ϕ, of $(\mathcal{M}, \mathcal{L})$ is a morphism. Look at the exact sequence

$$0 \to K_{\mathcal{M}} + \mathcal{L} \to K_{\mathcal{M}} + 2\mathcal{L} \to K_S + \mathcal{L}_S \to 0.$$

Since $h^1(K_{\mathcal{M}} + \mathcal{L}) = 0$, the restriction, ϕ_S, of ϕ to S is the adjoint map, φ, for $(S, \mathcal{L}_S)$ associated to $\Gamma(K_S + \mathcal{L}_S)$. Since there exists a reduction, $\pi : S \to S'$ (recall that π is the "connected part" of the morphism φ), we see that the restriction $(K_{\mathcal{M}} + 2\mathcal{L})_S \approx K_S + \mathcal{L}_S$ is big. Hence $\dim \varphi(S) = \dim S$, and so in particular $\dim \phi(\mathcal{M}) \geq 2$.

Let $\dim \phi(\mathcal{M}) = 3$, i.e., $K_{\mathcal{M}} + 2\mathcal{L}$ is spanned and big. Then there exists the first reduction (M, L), $r : \mathcal{M} \to M$, of $(\mathcal{M}, \mathcal{L})$, and $(S', \mathcal{L}')$ is an ample divisor in (M, L). Therefore the same argument as above (case $(S, \mathcal{L}_S)$ as in (10.3.1.2)) applies to S' in M to give a contradiction.

Let $\dim \phi(\mathcal{M}) = 2$. Then $(\mathcal{M}, \mathcal{L})$ is a scroll, and indeed a $\mathbb{P}^1$-bundle, $p : \mathcal{M} \to Y$, over a smooth surface, Y (see (11.1.1)). Let p_S be the restriction of p to S, and let $\mathcal{L}_Y := (p_{S*}\mathcal{L}_S)^{**}$. From Theorem (11.1.2) we know that $(Y, \mathcal{L}_Y)$ is the reduction of $(S, \mathcal{L}_S)$, i.e., $(S', \mathcal{L}') = (Y, \mathcal{L}_Y)$, $\pi = p_S$. Consider the ample and spanned rank 2 vector bundle $\mathcal{E} := p_*\mathcal{L}$ on Y. By Theorem (11.1.2) we have that $c_2(\mathcal{E}) = 1$. Therefore by Theorem (11.1.3) we conclude that $Y \cong \mathbb{P}^2$. This contradicts $K_{S'}^2 = 2$. □

Let us return to the case when the adjunction map, ϕ, has a lower dimensional image.

Theorem 11.2.4. *Let $\mathcal{M}$ be a smooth, connected, projective variety of dimension $n \geq 3$, and let $\mathcal{L}$ be a very ample line bundle on $\mathcal{M}$. Assume that $K_{\mathcal{M}} + (n-1)\mathcal{L}$ is spanned and let $\phi = s \circ r$ be the Remmert-Stein factorization of the morphism, ϕ, given by $\Gamma(K_{\mathcal{M}} + (n-1)\mathcal{L})$. Assume that ϕ has a lower dimensional image. Then* $\dim \phi(\mathcal{M}) \leq 2$ *and s is an embedding.*

Proof. By (9.2.4), $\dim \phi(\mathcal{M}) \leq 2$. If $\dim \phi(\mathcal{M}) = 0$ the statement is clear. Therefore we can assume that $\dim \phi(\mathcal{M}) = 1$ or 2. We will only prove the result for the case $n = 3$. For $n \geq 4$ the result follows since, as in the proof of (11.2.3), one sees that the restriction of ϕ to a 3-fold section, X, is the adjunction map associated to $\Gamma(K_X + 2\mathcal{L}_X)$, where $\mathcal{L}_X$ denotes the restriction of $\mathcal{L}$ to X. Let S be a smooth surface in $|\mathcal{L}|$ and let $\mathcal{L}_S$ be the restriction of $\mathcal{L}$ to S. As in the proof of (11.2.3) one sees that the restriction, ϕ_S, of ϕ to S is the adjunction map, φ, for $(S, \mathcal{L}_S)$ associated to $\Gamma(K_S + \mathcal{L}_S)$. Let r_S be the restriction of r to S. Since $\dim \phi(\mathcal{M}) \leq 2$, we know from (5.1.3) that $r(\mathcal{M}) = r_S(S) := Y$. Hence the same finite morphism s occurs in the Remmert-Stein factorizations of both ϕ and φ, i.e., $\phi = s \circ r$, $\varphi = s \circ r_S$.

Let $\dim \phi(\mathcal{M}) = 1$. Then $(\mathcal{M}, \mathcal{L})$ is a quadric fibration under r. Assume that s is not an embedding. Then by (10.1.3) we know that S is a $\mathbb{P}^1$-bundle over an elliptic curve of invariant $e = -1$ under r_S, and $\mathcal{L}_S \sim 2E + f$, where E, f denote the section of minimal self-intersection $E^2 = 1$, and a fiber of r_S. Bădescu's result (7.9.5) applies to say that S cannot be an ample divisor on $\mathcal{M}$ since $\mathcal{L}_S \cdot f = 2$. This shows that s is an embedding if $\dim \phi(\mathcal{M}) = 1$.

Let $\dim \phi(\mathcal{M}) = 2$. Then $(\mathcal{M}, \mathcal{L})$ is a scroll over $Y := r(\mathcal{M})$ under r. Let $\mathcal{L}_Y := (r_{S*}\mathcal{L}_S)^{**}$. From (11.1.2.1) we know that $(Y, \mathcal{L}_Y)$ is the reduction of $(S, \mathcal{L}_S)$. Assume that s is not an embedding. Then by the results of Chapter 10 we know that $(S, \mathcal{L}_S)$ belongs to one of the four special cases (10.3.1.1)—(10.3.1.4). Therefore Theorem (11.2.3) gives a contradiction. □

11.3 Very ampleness of the adjoint bundle

Let $\mathcal{M}$ be a smooth, projective variety of dimension $n \geq 2$, and let $\mathcal{L}$ be a very ample line bundle on $\mathcal{M}$. From the results of §9.2 and §11.2 we can assume that the adjoint bundle $K_{\mathcal{M}} + (n-1)\mathcal{L}$ is spanned and big. Let (M, L), $r : \mathcal{M} \to M$, be the first reduction of $(\mathcal{M}, \mathcal{L})$. In this section we show that if $n \geq 3$ then $K_M + (n-1)L$ is always very ample.

The following result from Sommese and Van de Ven [586] is the main result of the chapter.

Theorem 11.3.1. *Let $\mathcal{L}$ be a very ample line bundle on a connected projective manifold $\mathcal{M}$ of dimension $n \geq 2$. Assume that $K_{\mathcal{M}} + (n-1)\mathcal{L}$ is spanned and big, and let (M, L), $r : \mathcal{M} \to M$, be the first reduction of $(\mathcal{M}, \mathcal{L})$. Then $K_M + (n-1)L$ is very ample, unless $n = 2$ and $(\mathcal{M}, \mathcal{L})$ is as in* (10.3.1.1)—(10.3.1.4).

Proof. If $n = 2$, the result is proved in Chapter 10. Thus we can assume that $n \geq 3$. Assuming the result is true in dimension $n-1$, let us show that it is true in dimension n.

Note that $K_M + (n-1)L$ is spanned since $K_{\mathcal{M}} + (n-1)\mathcal{L}$ is spanned and $K_{\mathcal{M}} + (n-1)\mathcal{L} \cong r^*(K_M + (n-1)L)$. Take two distinct points $x_1', x_2' \in M$ and let $x_1, x_2 \in \mathcal{M}$ be two distinct points such that $r(x_i) = x_i'$ for $i = 1, 2$. From (1.7.9) we know that there exists a smooth divisor $A \in |\mathcal{L}|$ passing through x_1, x_2. Then the image $A' := r(A)$ is a smooth divisor in $|L|$ passing through x_1', x_2'. Consider the exact sequence

$$0 \to K_M + (n-2)L \to K_M + (n-1)L \to K_{A'} + (n-2)L_{A'} \to 0.$$

Since $h^1(K_M + (n-2)L) = 0$, we see that $\Gamma(K_M + (n-1)L)$ surjects onto $\Gamma(K_{A'} + (n-2)L_{A'})$. Therefore the restriction to A' of the morphism, s, given by $\Gamma(K_M + (n-1)L)$ coincides with the morphism given by $\Gamma(K_{A'} + (n-2)L_{A'})$. Since $K_{A'} + (n-2)L_{A'}$ is very ample by the induction hypothesis, it follows that $s(x_1') \neq s(x_2')$, i.e., $\Gamma(K_M + (n-1)L)$ separates x_1', x_2'.

To conclude that $K_M + (n-1)L$ is very ample we have to show that $\Gamma(K_M + (n-1)L)$ separates infinitely near points (x', v'), where v' is a nonzero element of $\mathcal{T}_{M,x'}$ for $x' \in M$. Let x be a point on $\mathcal{M}$ such that $r(x) = x'$ and let v be a nonzero element of $\mathcal{T}_{\mathcal{M},x}$ such that $dr_x(v) = v'$, where $dr_x : \mathcal{T}_{\mathcal{M},x} \to \mathcal{T}_{M,x'}$ denotes the differential map. By (1.7.9), there exists a smooth $A \in |\mathcal{L}|$ with $x \in A$ and $v \in \mathcal{T}_{A,x}$. Then the image $A' := r(A)$ is a smooth divisor in $|L|$ passing through x' and with $v' \in \mathcal{T}_{A',x'}$. Now an argument exactly like the above shows that s is an embedding, i.e., $K_M + (n-1)L$ is very ample.

By combining the inductive argument above together with Theorem (11.2.3) we are done. □

11.4 Applications to hyperelliptic curve sections

Recall that a smooth connected projective curve, C, of genus at least 2 is said to be hyperelliptic if there exists a two-to-one branched covering onto $\mathbb{P}^1$. In this section we answer the following question.

Question 11.4.1. *Let $\mathcal{M}$ be a connected n-dimensional submanifold of $\mathbb{P}^N$, $n \geq 2$. When does there exist an $(N-n+1)$-dimensional linear subspace of $\mathbb{P}^N$, intersecting $\mathcal{M}$ along a hyperelliptic curve?*

This question, which goes back to Castelnuovo, has motivated much of the work in the papers [145, 216, 571, 604, 554] of Castelnuovo, Enriques, Sommese, Van de Ven, and Serrano. The following result is more complete than earlier answers to that question. It gives in particular the classification of pairs, $(\mathcal{M}, \mathcal{L})$, with $g(\mathcal{L}) = 2$ since genus 2 curves are all hyperelliptic.

Theorem 11.4.2. (Sommese-Van de Ven [586, (2.1)]) *Let $\mathcal{M}$ be a connected projective manifold of dimension $n \geq 2$, and $\mathcal{L}$ a very ample line bundle on $\mathcal{M}$. Assume that there is a smooth hyperelliptic curve, C, obtained as a transversal intersection of $n-1$ smooth members of $|\mathcal{L}|$. Then one of the following cases occurs:*

1. *$(\mathcal{M}, \mathcal{L})$ is a scroll over a smooth hyperelliptic curve, and therefore a $\mathbb{P}^{n-1}$-bundle over the curve with $\mathcal{L}_F \approx \mathcal{O}_{\mathbb{P}^{n-1}}(1)$ for any fiber F;*
2. *$(\mathcal{M}, \mathcal{L})$ is a quadric fibration over $\mathbb{P}^1$;*
3. *$n = 2$, $\mathcal{M}$ is a $\mathbb{P}^1$-bundle over an elliptic curve, Y, of invariant $e = -1$ and $\mathcal{L} \sim 2E + f$, where E, f denote a section of minimal self-intersection and a fiber of $\mathcal{M} \to Y$;*
4. *$n = 2$ and $(\mathcal{M}, \mathcal{L})$ is a Del Pezzo surface, $\mathcal{L} \approx -3K_{\mathcal{M}}$, $K_{\mathcal{M}}^2 = 1$ as described in* (10.4.3)*;*
5. *$n = 2$ and $(\mathcal{M}, \mathcal{L})$ is a Del Pezzo surface, $\mathcal{L} \approx -2K_{\mathcal{M}}$, $K_{\mathcal{M}}^2 = 2$ as described in* (10.2.4)*;*

6. $(\mathcal{M}, \mathcal{L})$ *has the pair as in* 5) *as its first reduction,* (M, L), *and the reduction map* $r : \mathcal{M} \to M$ *is the blowing up at a single point as described in* (10.2.5).

Proof. By Corollary (9.2.2) we see that either $K_{\mathcal{M}} + (n-1)\mathcal{L}$ is spanned or we are in case (11.4.2.1) above.

Therefore we can assume that $K_{\mathcal{M}} + (n-1)\mathcal{L}$ is spanned. Let $\phi : \mathcal{M} \to \mathbb{P}_{\mathbb{C}}$ be the morphism given by $\Gamma(K_{\mathcal{M}} + (n-1)\mathcal{L})$, and let $\phi = s \circ r$ be the Remmert-Stein factorization of ϕ. Let $Y := r(\mathcal{M})$, and let r_C be the restriction of r to C.

If $\dim \phi(\mathcal{M}) = 0$, then $K_{\mathcal{M}} \approx -(n-1)\mathcal{L}$, so that $g(\mathcal{L}) = 1$, which contradicts the assumptions on C. Therefore we have to consider the cases when $\dim \phi(\mathcal{M}) =$ 1, 2, or n.

Let ϕ_C be the restriction of ϕ to C, and let $\varphi : C \to \varphi(C) \subset \mathbb{P}^{g(\mathcal{L})-1}$ be the canonical map associated to $\Gamma(K_C)$. Let S be the smooth surface obtained as transversal intersection of $n-2$ elements of $|\mathcal{L} \otimes \mathcal{J}_C|$ where $\mathcal{J}_C$ is the ideal sheaf of C in $\mathcal{M}$. Then $C \in |\mathcal{L}_S|$, where $\mathcal{L}_S$ denotes the restriction of $\mathcal{L}$ to S. From the exact sequence

$$0 \to K_S \to K_S + \mathcal{L}_S \to K_C \to 0$$

we see that the restriction ϕ_C factors through φ, $\phi_C = p \circ \varphi$, $p : \varphi(C) \to \phi(C)$. Then in particular $\deg \phi_C \geq \deg \varphi = 2$, with equality if $q(S) = 0$. Furthermore $\phi(C) = \mathbb{P}^1$ since $\varphi(C) = \mathbb{P}^1$.

Let $\dim \phi(\mathcal{M}) = 1$. Then $(\mathcal{M}, \mathcal{L})$ is a quadric fibration over a smooth curve, Y, under $r : \mathcal{M} \to Y$. Since the curve C is the transversal intersection of $n-1$ (very) ample divisors on $\mathcal{M}$, we see that $\phi(\mathcal{M}) = \phi(C) = \mathbb{P}^1$. If $Y \cong \mathbb{P}^1$ we are in the case (11.4.2.2). Thus we can assume $g(Y) > 0$.

Then clearly $s : Y \to \phi(\mathcal{M}) = \mathbb{P}^1$ cannot be an embedding. Therefore from Theorem (10.1.3) we know that S is a $\mathbb{P}^1$-bundle over an elliptic curve of invariant $e = -1$ under the restriction r_S of r to S, and $\mathcal{L} \sim 2E + f$ where E, f denote a section of minimal self-intersection, $E^2 = 1$, and a fiber of r_S. If $n = 2$, we are in the case (11.4.2.3). The case when $n > 2$ does not occur by Theorem (11.2.4).

Let $\dim \phi(\mathcal{M}) = 2 < n$. Then $(\mathcal{M}, \mathcal{L})$ is a scroll over Y under r. Let $\mathcal{L}_Y :=$ $(r_{S*}\mathcal{L}_S)^{**}$. By (11.1.2.1) we know that $(Y, \mathcal{L}_Y)$ is the reduction of $(S, \mathcal{L}_S)$. Hence in particular $K_S + \mathcal{L}_S$ is nef and big, and $C \in |\mathcal{L}_S|$ goes down isomorphically under r. Since $\deg \phi_C \geq 2$, the morphism $s : Y \to \mathbb{P}_{\mathbb{C}}$ is not an embedding. Since $K_Y + \mathcal{L}_Y \approx s^*\mathcal{O}_{\mathbb{P}_{\mathbb{C}}}(1)$, this is equivalent to saying that $K_Y + \mathcal{L}_Y$ is not very ample. This contradicts Theorem (11.2.4).

Let $\dim \phi(\mathcal{M}) = n$. Let $\mathcal{L}_Y := (r_*\mathcal{L})^{**}$. Then $(Y, \mathcal{L}_Y)$ is the first reduction of $(\mathcal{M}, \mathcal{L})$. Hence in particular $K_{\mathcal{M}} + (n-1)\mathcal{L}$ is nef and big, and C goes down isomorphically under r. Since $\deg \phi_C \geq 2$, $s : Y \to \mathbb{P}_{\mathbb{C}}$ is not an embedding, i.e., $K_Y + (n-1)\mathcal{L}_Y$ is not very ample. By Theorem (11.3.1) we can assume that we are in cases (11.4.2.4), (11.4.2.5), (11.4.2.6) or $(\mathcal{M}, \mathcal{L})$ is a $\mathbb{P}^1$-bundle of invariant $e = -1$, $\pi : \mathcal{M} \to B$ over a smooth elliptic curve B, and $\mathcal{L} \sim 3E$, E a section of minimal self-intersection $E^2 = 1$. The following argument rules out this last case (see also Serrano [554]).

Let f be a fiber of $\pi : \mathcal{M} \to B$. Then $K_{\mathcal{M}} \sim -2E + f$ and the genus formula yields $g(\mathcal{L}) = 4$. Look at the exact sequence

$$0 \to K_{\mathcal{M}} \to K_{\mathcal{M}} + \mathcal{L} \to K_C \to 0.$$

Since $h^0(K_{\mathcal{M}}) = 0$, $h^1(K_{\mathcal{M}}) = 1$, and $h^0(K_C) = 4$ we find that $h^0(K_{\mathcal{M}} + \mathcal{L}) = 3$. Furthermore $(K_{\mathcal{M}} + \mathcal{L})^2 = (E + f)^2 = 3$, so we conclude that the adjoint morphism ϕ is a degree 3 cover, $\phi : \mathcal{M} \to \mathbb{P}^2$, of $\mathbb{P}^2$.

Note that since $h^1(\mathcal{O}_{\mathcal{M}}) = 1$ it follows from the sequence

$$0 \to \mathcal{O}_{\mathcal{M}} \to \mathcal{L} \to \mathcal{L}_C \to 0$$

that $h^0(\mathcal{L}) \geq 6$. Using Theorem (8.4.2) with $k = 1$ we see that there is a positive dimensional family of smooth hyperelliptic curves in $|\mathcal{L}|$. Thus it can be assumed without loss of generality that C does not belong to the branch locus of ϕ. Using this, and the fact that ϕ_C is at most three-to-one, and since $\phi_C = p \circ \varphi$ for some morphism p with φ a two-to-one map, we see that p is generically one-to-one.

Hence in particular $\deg \phi_C = 2$. Since $\deg \phi_C = 2$ and ϕ has degree 3 on $\phi^{-1}(\ell)$ where ℓ is the reduced curve associated to $\phi(C)$, we conclude that ϕ has degree 1 on R, where R is an effective curve such that $\phi^{-1}(\ell) = C + R$. Therefore R is irreducible and reduced since it is birational to an image of $\varphi(C) = \mathbb{P}^1$. Since the only irreducible curves on $\mathcal{M}$ that are birational to $\mathbb{P}^1$ are fibers of the map $\mathcal{M} \to B$, we conclude that $R \sim f$ and $\phi^*\ell \sim C + f \sim 3E + f$. This gives $3\ell^2 = (\phi^*\ell)^2 = 15$. This gives the absurdity that ℓ is a curve on $\mathbb{P}^2$ with $\ell^2 = 5$. □

Remark 11.4.3. In the above theorem all curve sections of $(\mathcal{M}, \mathcal{L})$ are hyperelliptic except in cases (11.4.2.4), (11.4.2.5), and (11.4.2.6). In case (11.4.2.4) (see (10.4.3)), the hyperelliptic hyperplane sections are precisely the pullbacks of the smooth, twisted cubics on the singular quadric Q that pass through the vertex of Q and are transverse to the branch curve obtained by the intersection of Q with the cubic surface. In case (11.4.2.5) (see (10.2.4)), the hyperelliptic hyperplane sections are precisely the pullbacks of the smooth conics on $\mathbb{P}^2$ transverse to the branch quartic.

See Serrano [554] for a nice discussion of case (11.4.2.3).

Remark 11.4.4. In Biancofiore, Fania, and Lanteri [123] the result above has been generalized to the case when the line bundle $\mathcal{L}$ is merely ample and spanned. We also refer to [21, §3] by Andreatta and the authors of this book for a result on normal surfaces carrying an ample and spanned line bundle, $\mathcal{L}$, such that $|\mathcal{L}|$ contains a smooth hyperelliptic curve.

We refer to Del Centina and Gimigliano [183, 184] for the classification of manifolds having bielliptic curves, C, as hyperplane curve sections. Here bielliptic means that C is a two-to-one cover of an elliptic curve.

The classification of ample and spanned vector bundles, $\mathscr{E}$, with $c_2(\mathscr{E}) = 2$ on a smooth projective surface, S, is due to Lanteri and Russo [388] and Noma [478]. We give here the proof of the result in the case when $\mathscr{E}$ is very ample, as one more useful application of Theorem (11.4.2). This is an analogue of Theorem (11.1.3). We call attention also to the articles of Ballico and Lanteri [62] and [63]. See also Noma [479] for results in the case $c_2(\mathscr{E}) > 2$.

Theorem 11.4.5. *Let $\mathscr{E}$ be a very ample vector bundle on a smooth projective surface, X. If $c_2(\mathscr{E}) = 2$, then either*

1. $(X, \mathscr{E}) \cong (\mathbb{P}^1 \times \mathbb{P}^1, \mathcal{O}_{\mathbb{P}^1 \times \mathbb{P}^1}(1, 1) \oplus \mathcal{O}_{\mathbb{P}^1 \times \mathbb{P}^1}(1, 1))$*; or*

2. $(X, \mathscr{E}) \cong (\mathbb{P}^2, \mathcal{O}_{\mathbb{P}^2}(2) \oplus \mathcal{O}_{\mathbb{P}^2}(1))$.

Proof. Since $c_2(\mathscr{E}) = 2$, we see that the rank of $\mathscr{E}$ is ≥ 2. If we prove the theorem when the rank is 2, then the same argument used to rule out rank ≥ 3 in Theorem (11.1.3) will apply to the theorem we are now proving. Therefore we can assume without loss of generality that $\operatorname{rank}(\mathscr{E}) = 2$.

Let $\mathscr{L} := \det \mathscr{E}$. By Theorem (11.1.2) we see that there is a smooth $C \in |\mathscr{L}|$ which is a two sheeted branched cover of $\mathbb{P}^1$. Then either $g(\mathscr{L}) = 0, 1$ or C is hyperelliptic.

Assume first that $g(\mathscr{L}) = 0$. The same argument as in Theorem (11.1.3) shows that $(X, \mathscr{L}) \cong (\mathbb{P}^2, \mathcal{O}_{\mathbb{P}^2}(2))$. The assumption that $c_2(\mathscr{E}) = 2$ affects that proof for this case only in implying that $\chi(\mathscr{E} \otimes \mathcal{O}_{\mathbb{P}^2}(-1)) \geq 3$. It leads to the same conclusion, which in this case is contradictory, that $\mathscr{E} \cong \mathcal{O}_{\mathbb{P}^2}(1) \oplus \mathcal{O}_{\mathbb{P}^2}(1)$.

Assume now that $g(\mathscr{L}) = 1$. We use Theorem (8.9.3). Since $\mathscr{L} \cdot D \geq 2$ for any effective curve $D \subset X$, we conclude that scrolls are not possible. Thus $\mathscr{L} \cong -K_X$. Since there are no lines on X, we see that $(X, \mathscr{L})$ is either $(\mathbb{P}^1 \times \mathbb{P}^1, \mathcal{O}_{\mathbb{P}^1 \times \mathbb{P}^1}(2, 2))$ or $(\mathbb{P}^2, \mathcal{O}_{\mathbb{P}^2}(3))$.

In the case $X \cong \mathbb{P}^1 \times \mathbb{P}^1$ we restrict to vertical and horizontal curves to see that in both cases the bundle $\mathscr{E}$ restricts to $\mathcal{O}_{\mathbb{P}^1}(1) \oplus \mathcal{O}_{\mathbb{P}^1}(1)$. Thus we see that $\mathscr{F} := \mathscr{E} \otimes \mathcal{O}_{\mathbb{P}^1 \times \mathbb{P}^1}(-1, -1)$ pulls back from a rank 2 bundle on $\mathbb{P}^1$ which splits. This bundle on $\mathbb{P}^1$ must be the trivial bundle since the restriction $\mathscr{F}_\ell$ for a horizontal $\ell = \mathbb{P}^1$ is the trivial bundle. This leads to the first example of the theorem.

In the case when $X \cong \mathbb{P}^2$, we see that $\mathscr{E}_\ell \cong \mathcal{O}_{\mathbb{P}^1}(1) \oplus \mathcal{O}_{\mathbb{P}^1}(2)$ for any line $\ell \subset \mathbb{P}^2$. This implies that we have the second example of the theorem, by the basic results on uniform bundles (see Okonek, Schneider, and Spindler [493]).

Thus we can assume without loss of generality that $g(\mathscr{L}) \geq 2$ and C is hyperelliptic.

If C is hyperelliptic we must only go down the list in Theorem (11.4.2). As before the scroll cases are ruled out since $\mathcal{L} \cdot D \geq 2$ for any curve D on X. This also rules out case 6). It also implies that the conic bundle in case 2) is a $\mathbb{P}^1$-bundle. Thus in case 2), $X \cong \mathbb{F}_r$ with $\mathcal{L} \cdot f = 2$ for any fiber f. Let E be a section of minimal self-intersection. Then in this case $\det(\mathcal{E} \otimes \mathcal{O}_X(-E)) \cdot f = (\mathcal{L} - 2E) \cdot f = 0$ and hence $\mathcal{E} \otimes \mathcal{O}_X(-E)$ is the pullback of a bundle from $\mathbb{P}^1$. Thus since bundles on $\mathbb{P}^1$ split, $\mathcal{E} \cong A \oplus B$ for two ample line bundles, A, B, on X of degree 1 on fibers. A simple computation shows that the only Hirzebruch surface with 2 ample line bundles with $A \cdot B = 2$ is $\mathbb{F}_0$ and in this case $g(\mathcal{L}) = 1$.

In case 3), X is a $\mathbb{P}^1$-bundle over an elliptic curve. We consider the surface, $S := \mathbb{P}(\mathcal{E}_E)$ where E is a section of X. Note that $\xi_{\mathcal{E}_E}$ is very ample of degree $\deg \mathcal{E}_E = \mathcal{L} \cdot E = 3$. Since $h^1(\mathcal{O}_S) = 1$, we see that $h^0(\xi_{\mathcal{E}_E}) \geq 5$. But this is impossible since $\xi_{\mathcal{E}_E}$ is very ample and $g(\xi_{\mathcal{E}_E}) = 1$ which contradicts Castelnuovo's bound for the genus of $g(\xi_{\mathcal{E}_E})$.

This same argument works in case 4) with E taken to be any smooth curve in the system $|-K_X|$.

In case 5) take for E a smooth curve in $|-K_X|$. Arguing as in case 3) we conclude by using Castelnuovo's bound, (1.3), that $S := \mathbb{P}(\mathcal{E}_E)$ embeds as a degree 4 scroll in $\mathbb{P}^4$ of sectional genus $g(\xi_{\mathcal{E}_E}) = 1$. The double point formula (8.2.1) gives an immediate contradiction. □

11.5 Projective normality of adjoint bundles

One can make a very suggestive analogy between adjoint bundles and canonical bundles. Let $\mathcal{L}$ be a very ample line bundle on $\mathcal{M}$, an n-dimensional projective manifold. General results about canonical bundles of k-dimensional manifolds should have analogues for the pairs, $(\mathcal{M}, \mathcal{L})$, with respect to the bundle $K_{\mathcal{M}} + (n-k)\mathcal{L}$. Let us consider this analogy in the case $k = 1$. There the canonical bundle is spanned except for $\mathbb{P}^1$. The analogous result is that $K_{\mathcal{M}} + (n-1)\mathcal{L}$ is spanned except for the pairs, $(\mathcal{M}, \mathcal{L})$, which have $\mathbb{P}^1$ as a curve section, or are scrolls over curves.

There is a famous theorem about smooth connected projective curves, C, which says that if $g(C) \geq 1$, i.e., K_C is spanned, then K_C is very ample except if C is elliptic or hyperelliptic. To see the analogue of this, assume that $K_{\mathcal{M}} + (n-1)\mathcal{L}$ is spanned by global sections. Let $\phi : \mathcal{M} \to \mathbb{P}_{\mathbb{C}}$ denote the morphism associated to $K_{\mathcal{M}} + (n-1)\mathcal{L}$. Let $\phi = s \circ r$ be the Remmert-Stein factorization of ϕ with $r : \mathcal{M} \to Y$ a morphism with connected fibers onto a normal variety, Y, and $s : Y \to \mathbb{P}_{\mathbb{C}}$ a finite-to-one morphism. The theorems in this chapter show that s is an embedding except for a few very special pairs (most of which contain hyperelliptic curves).

Assume now that K_C is very ample. Then it is a famous result of M. Noether that K_C is in fact projectively normal, i.e., $H^0(xK_C) \otimes H^0(yK_C) \to H^0((x+y)K_C)$ is onto for $x \geq 0$, $y \geq 0$. The following question has been around since the middle 80's when the results in this chapter were proved.

Question 11.5.1. *Let $\mathcal{L}$ be a very ample line bundle on $\mathcal{M}$, an n-dimensional projective manifold. Assume now that the first reduction, (M, L), of $(\mathcal{M}, \mathcal{L})$ exists and $K_M + (n-1)L$ is very ample. What can be said about the projective normality of $K_M + (n-1)L$?*

This question remains open. Andreatta, Ballico, and Sommese [27] (see also [14]), and Ein and Lazarsfeld [207] independently found some positive results on the projective normality of some related, but slightly more ample bundles. We would like to state one of the results in this direction, with a proof of the main part of the statement. The paper [207] has further interesting results about syzygy conditions beyond projective normality. Andreatta and Ballico [15] investigate the theorem we state next when some mild singularities are allowed. There has been an extensive study of the projective normality of low degree varieties, see, Alzati, Bertolini, and Besana [6], Fujita [257], Ionescu [333, 334, 337], and Ohbuchi [484].

Theorem 11.5.2. *Let $\mathcal{L}$ be a very ample line bundle on a connected n-dimensional projective manifold, $\mathcal{M}$. Assume that $K_{\mathcal{M}} + (n-1)\mathcal{L}$ is spanned by global sections. The bundle $aK_{\mathcal{M}} + b\mathcal{L}$ is projectively normal if $b \geq a(n-1)+1$ with $a \geq 1$.*

Proof. We follow the proof of [27]. Let $P := \mathcal{M} \times \mathcal{M}$ with $p : P \to \mathcal{M}$ and $q : P \to \mathcal{M}$ being the product projections onto the first and second factors respectively. Let $\overline{A}_{x,y} := xp^*A + yq^*A$ where A is some line bundle on $\mathcal{M}$. Let Δ denote the diagonal of P and let $\mathcal{J}_\Delta$ denote the ideal sheaf of Δ. Using the projection formula for p followed by the projection formula for the map of $\mathcal{M}$ to a point, we see that $H^0(\overline{A}_{x,y}) \cong H^0(xA) \otimes H^0(yA)$. With this isomorphism the restriction map of $H^0(\overline{A}_{x,y})$ to Δ is precisely the map $H^0(xA) \otimes H^0(yA) \to H^0((x+y)A)$ occurring in the definition of projective normality of A. Thus the projective normality of A is implied by the vanishing of $H^1(\overline{A}_{x,y} \otimes \mathcal{J}_\Delta)$ for all $x \geq 1$ and $y \geq 1$. Let $r : \overline{P} \to P$ denote the blowup of P along Δ and let $\overline{\Delta} := r^{-1}(\Delta)$. Let $\overline{p} := p \circ r$ and $\overline{q} := q \circ r$. By using the usual arguments based on the Leray spectral sequence for r and the formal function theorem we see that $H^1(\overline{A}_{x,y} \otimes \mathcal{J}_\Delta) \cong H^1(r^*\overline{A}_{x,y} - \overline{\Delta})$. Note that $K_P \cong p^*K_{\mathcal{M}} \otimes q^*K_{\mathcal{M}}$ and $K_{\overline{P}} \cong r^*K_P + (n-1)\overline{\Delta}$. Thus to prove the projective normality of A it suffices to show that $r^*\overline{A}_{x,y} - \overline{p}^*K_{\mathcal{M}} - \overline{q}^*K_{\mathcal{M}} - n\overline{\Delta}$ is nef and big for all $x \geq 1$ and $y \geq 1$. Thus in our case it suffices by the Kawamata-Viehweg vanishing theorem to show that

$$\overline{p}^*((xa-1)K_{\mathcal{M}} + xb\mathcal{L}) + \overline{q}^*((ya-1)K_{\mathcal{M}} + yb\mathcal{L}) - n\overline{\Delta} \tag{11.1}$$

is nef and big for all $x \geq 1$ and $y \geq 1$.

We claim that given any very ample line bundle H on $\mathcal{M}$, then $r^*\overline{H}_{1,1} - \overline{\Delta}$ is spanned and hence nef. Using the sections of the form $r^*(p^*s - q^*s) - \overline{\Delta}$ for $s \in H^0(H)$, this is easy to see. We can rewrite the bundle in (11.1) as

$$\overline{p}^*((xa-1)K_{\mathcal{M}} + (xb-n)\mathcal{L}) + \overline{q}^*((ya-1)K_{\mathcal{M}} + (yb-n)\mathcal{L}) + n(r^*\overline{\mathcal{L}}_{1,1} - \overline{\Delta}).$$

Note that $(xa-1)K_{\mathcal{M}} + (xb-n)\mathcal{L}$ is spanned if $xb - n \geq (xa-1)(n-1)$. This is equivalent to $xb \geq xan - xa + 1$. This is true for all $x \geq 1$ only if $b \geq a(n-1)+1$. This shows that the bundle in (11.1) is nef. Since $r^*\overline{\mathcal{L}}_{1,1} - \overline{\Delta}$ is nef, bigness and the theorem would follow if $(xa-1)K_{\mathcal{M}} + (xb-n)\mathcal{L}$ is big for all $x \geq 1$. This would be true if $b \geq a(n-1)+2$. In the case when $b = a(n-1)+1$ an analysis of the bigness of the bundle $r^*\overline{\mathcal{L}}_{1,1} - \overline{\Delta}$ is required. For this we refer the reader to the appendix of [27]. □

11.6 Manifolds of sectional genus ≤ 4

In this section we classify the pairs, $(\mathcal{M}, \mathcal{L})$, where $\mathcal{M}$ is a connected n-dimensional projective manifold, and $\mathcal{L}$ is a very ample line bundle satisfying $3 \leq g(\mathcal{L}) \leq 4$. We follow the papers of Livorni [409, 410, 411, 412] and Ionescu [333, 334] to which we refer the reader for further results in the same direction, and in particular the classification when $g(\mathcal{L}) = 5, 6$. We incorporate the simplifications that come from using the full strength of the results on the adjunction mapping.

One fascinating aspect of this work is the extent to which integrality properties of numbers control the existence and properties of projective varieties. The striking quality of many of these results have led to extensive activity in this area, which we discuss in §14.3.

In what follows we use ϕ to denote the adjunction mapping given by $K_{\mathcal{M}} + (\dim \mathcal{M} - 1)\mathcal{L}$ in the cases when $h^0(K_{\mathcal{M}} + (\dim \mathcal{M} - 1)\mathcal{L}) \neq 0$. Moreover in this case we write the Remmert-Stein factorization $\phi = s \circ r$ with r having connected fibers and s having finite fibers.

We start with the classification in the case when $g(\mathcal{L}) = 3$. When $n = 2$ the classification was done in §10.2.

Theorem 11.6.1. *Let $\mathcal{L}$ be a very ample line bundle on a connected n-dimensional projective manifold, $\mathcal{M}$, with $n \geq 2$ and $g(\mathcal{L}) = 3$. Then either:*

1. *$(\mathcal{M}, \mathcal{L})$ is a scroll over a curve of genus 3; or*

2. *$(\mathcal{M}, \mathcal{L})$ is a quadric fibration over $\mathbb{P}^1$; or*

3. $\dim \mathcal{M} = 2$ *and $\mathcal{M}$ is a $\mathbb{P}^1$-bundle over an elliptic curve of invariant $e = -1$, $\mathcal{L} \sim 2E + f$, where E and f denote a section of minimal self-intersection $E^2 = 1$ and a fiber of the ruling respectively; or*

4. $\dim \mathcal{M} = 2$ *and the reduction (Y, L_Y), $r : \mathcal{M} \to Y$, of $(\mathcal{M}, \mathcal{L})$ is a Del Pezzo surface with $K_Y^2 = 2$, $L_Y \approx -2K_Y$; or*

5. *the adjunction map expresses $(\mathcal{M}, \mathcal{L})$ as a scroll over $(\mathbb{P}^2, \mathcal{O}_{\mathbb{P}^2}(4))$ (possibly with* $\dim \mathcal{M} = 2$*); or*

6. *$\mathcal{M} \subset \mathbb{P}^{n+1}$ and $L \cong \mathcal{O}_{\mathbb{P}^{n+1}}(1)_{\mathcal{M}}$, with* $\deg \mathcal{M} = 4$.

Proof. By Theorem (10.2.3), $n \geq 3$. Using the existence theorem for the adjunction mapping we see that $K_{\mathcal{M}} + (n-1)\mathcal{L}$ is spanned unless we are in case 1) of the theorem. By using Theorem (11.4.2) we can assume without loss of generality that none of the curve sections are hyperelliptic.

By the results of §11.2 we can assume that the adjunction map $r : \mathcal{M} \to Y$ expresses $\mathcal{M}$ as either a quadric fibration over a curve, Y, as a $\mathbb{P}^1$-bundle over a smooth surface, Y, or as the blowup of a smooth n-fold, Y.

In the case $\dim \phi(\mathcal{M}) = 1$ we claim that Y is $\mathbb{P}^1$. To see this note that if $g(Y) > 0$, then using Theorem (10.2.7.3b), we see that $(\mathcal{M}, \mathcal{L})$ has a surface section as in case 3) above, which contradicts Theorem (7.9.5). Thus $(\mathcal{M}, \mathcal{L})$ is as in 2).

In the case $\dim \phi(\mathcal{M}) = 2$, we have that $H := \det \phi_* \mathcal{L}$ is very ample with $g(H) = 3$ and $K_Y + H$ is very ample by (11.2.4). Moreover a surface section of $(\mathcal{M}, \mathcal{L})$ has (Y, H) as a nontrivial reduction. Thus using (10.2.7) we see that $(Y, H) \cong (\mathbb{P}^2, \mathcal{O}_{\mathbb{P}^2}(4))$, and we are in case 5).

In the case $\dim \phi(\mathcal{M}) = n$, we can further assume using the classification in degree ≤ 4 given in §8.10 that $\deg \mathcal{M} \geq 5$, since otherwise we are in case 6). Thus, by using the genus formula, we have that surface sections, S, have negative Kodaira dimension. Then by a straightforward induction argument down to S we see that

$$h^0(K_{\mathcal{M}} + (n-1)\mathcal{L}) = h^0(K_S + \mathcal{L}_S) = g(\mathcal{L}) - q(S) \leq 3.$$

Since $K_{\mathcal{M}} + (n-1)\mathcal{L}$ is spanned and big we must have $h^0(K_{\mathcal{M}} + (n-1)\mathcal{L}) \geq 4$ which gives a contradiction. □

We need the following lemma in our discussion.

Lemma 11.6.2. *Let $\mathcal{L}$ be a very ample line bundle on a connected n-dimensional projective manifold, $\mathcal{M}$, with $n \geq 2$. Assume that there exist $n-2$ smooth elements of $|\mathcal{L}|$ meeting transversally in a $K3$ surface, S, i.e., a smooth surface with $h^1(\mathcal{O}_S) =$*

0 and $K_S \cong \mathcal{O}_S$. If $g(\mathcal{L}) = 4$ then $|\mathcal{L}|$ embeds $\mathcal{M}$ in $\mathbb{P}^{n+2}$ as the complete intersection of a quadric hypersurface and a cubic hypersurface.

Proof. Note that $d := \mathcal{L}^n = 2g(\mathcal{L}) - 2$. We have the formulae for $n \geq 2$:

$$\begin{aligned} h^0(\mathcal{L}) &= g(\mathcal{L}) + n - 1 \\ h^0(2\mathcal{L}) &= (n+2)g(\mathcal{L}) + \frac{n^2 - n - 6}{2} \\ h^0(3\mathcal{L}) &= \frac{(n+1)(n+4)}{2} g(\mathcal{L}) + \frac{(n+4)(n^2 - 4n - 3)}{6} \end{aligned} \qquad (11.2)$$

To see relations (11.2), note that if $n = 2$, we can use the vanishing of $h^i(t\mathcal{L})$ for positive i and t, plus the Riemann-Roch theorem to conclude that $h^0(t\mathcal{L}) = (g(\mathcal{L}) - 1)t^2 + 2$, which is just the equation (11.2) when $n = 2$. If $n \geq 3$ then it follows by induction using (5.3.1) that $-K_{\mathcal{M}} \approx (n-2)\mathcal{L}$. Choose a smooth $A \in |\mathcal{L}|$. We can assume as an induction hypothesis that we know the formulae (11.2) for $(A, \mathcal{L}_A)$. We can use the exact sequence

$$0 \to t\mathcal{L} \to (t+1)\mathcal{L} \to (t+1)\mathcal{L}_A \to 0$$

for $t = 0, 1, 2$, and the fact that $h^1(t\mathcal{L}) = 0$ for $t = 1, 2, 3$ to reduce the general case for $(\mathcal{M}, \mathcal{L})$ to the formulae for $(A, \mathcal{L}_A)$. Note that $|\mathcal{L}|$ embeds $\mathcal{M}$ in $\mathbb{P}^{g(\mathcal{L})+n-2}$.

If $g(\mathcal{L}) = 4$ we see from (11.2) that $h^0(2\mathcal{L}) = (n^2 + 7n + 10)/2$ and $h^0(3\mathcal{L}) = (n+4)(n^2 + 8n + 9)/6$. We have also $h^0(\mathcal{O}_{\mathbb{P}^{n+2}}(2)) = (n+3)(n+4)/2$ and $h^0(\mathcal{O}_{\mathbb{P}^{n+2}}(3)) = (n+3)(n+4)(n+5)/6$. From this and the fact that $\mathcal{M}$ is not contained in a linear $\mathbb{P}^{n+1} \subset \mathbb{P}^{n+2}$ we conclude that there is at least one irreducible quadric, Q, and at least one irreducible cubic, C, containing $\mathcal{M}$. Since they are irreducible the scheme-theoretic intersection $I := Q \cap C$ is of pure dimension n and degree 6. Since the degree of $\mathcal{M}$ is $2g(\mathcal{L}) - 2 = 6$, and $\mathcal{M} \subset I$, we see that I is generically reduced. Since I is a complete intersection, it is Cohen-Macaulay, and therefore it is reduced. This shows that $\mathcal{M} = I$. □

Theorem 11.6.3. *Let $\mathcal{L}$ be a very ample line bundle on a connected n-dimensional projective manifold, $\mathcal{M}$, with $n \geq 2$ and $g(\mathcal{L}) = 4$. Then either:*

1. *$(\mathcal{M}, \mathcal{L})$ is a scroll over a curve of genus 4; or*
2. *$(\mathcal{M}, \mathcal{L})$ is a quadric fibration over a curve of genus 0 or 1; or*
3. $\dim \mathcal{M} = 2$ *and $\mathcal{M}$ is a $\mathbb{P}^1$-bundle over an elliptic curve of invariant $e = -1$, $\mathcal{L} \sim 3E$, where E and f denote a section of minimal self-intersection $E^2 = 1$ and a fiber of the ruling respectively; or*
4. $\dim \mathcal{M} = 2$ *and the reduction (Y, L_Y), $r : \mathcal{M} \to Y$, of $(\mathcal{M}, \mathcal{L})$ is a Del Pezzo surface with $K_Y^2 = 1$, $L_Y \approx -3K_Y$; or*

5. *the adjunction map expresses* $(\mathcal{M}, \mathcal{L})$ *as a scroll over* $(\mathbb{P}^1 \times \mathbb{P}^1, \mathcal{O}_{\mathbb{P}^1 \times \mathbb{P}^1}(3, 3))$ *(possibly with* $\dim \mathcal{M} = 2$*); or*

6. *the adjunction map expresses* $(\mathcal{M}, \mathcal{L})$ *as a scroll over* $(Y, -2K_Y)$*, with* Y *a smooth cubic surface in* $\mathbb{P}^3$ *(possibly with* $\dim \mathcal{M} = 2$*); or*

7. $(\mathcal{M}, \mathcal{L})$ *is as in* (11.6.2).

Proof. If $g(\mathcal{L}) = h^1(\mathcal{O}_{\mathcal{M}})$ then by Theorem (8.7.1), $(\mathcal{M}, \mathcal{L})$ is a scroll over a smooth curve of genus $g(\mathcal{L})$ as in 1). Moreover by Theorem (8.7.1) we can assume that we are in the case when

$$g(\mathcal{L}) > h^1(\mathcal{O}_{\mathcal{M}}). \tag{11.3}$$

By using (8.10.1) we can assume without loss of generality that $d := \mathcal{L}^n \geq 5$. Since there are no plane curves of genus 4, we conclude that $h^0(\mathcal{L}) \geq n + 3$.

Let C be a smooth curve section. As noted above we have $C \subset \mathbb{P}^k$ where $k \geq 3$. By Castelnuovo's bound we see that $d \geq 6$. If $d = 6$ then from $h^0(\mathcal{L}_C) \geq 4$ we see by (1.4.6) that $h^0(\mathcal{L}_C) = 4$ and that $K_C \cong \mathcal{L}_C$. Thus $K_{S|C} \cong \mathcal{O}_C$, for any smooth surface S containing C. From (11.3) and (2.3.6) we see that $K_S \cong \mathcal{O}_S$. Since $5 \leq h^0(\mathcal{L}_S) \leq h^0(\mathcal{L}_C) + 1 = 5$ we see that $h^1(\mathcal{O}_S) \leq h^1(\mathcal{L}_S)$. Since $h^1(\mathcal{L}_S) = h^1(\mathcal{L}_S + K_S) = 0$ we conclude that $h^1(\mathcal{O}_S) = 0$, i.e., that S is a $K3$ surface. This case is classified in Lemma (11.6.2) and $(\mathcal{M}, \mathcal{L})$ is as in 7). Thus we can assume that $d \geq 7 > 2g(\mathcal{L}) - 2$. Choose a smooth surface section S of $\mathcal{M}$. Then $K_S \cdot \mathcal{L}_S < 0$, and therefore we conclude that S is birationally ruled and therefore that $p_g(S) = 0$.

By using Theorem (9.2.1) we can assume without loss of generality that $K_{\mathcal{M}} + (n-1)\mathcal{L}$ is spanned by global sections. Let $s \circ r$ be the Remmert-Stein factorization of the morphism associated to $|K_{\mathcal{M}} + (n-1)\mathcal{L}|$. We can assume $\dim r(\mathcal{M}) = 1, 2, n$ by using Theorem (11.2.4) and the fact that $g(\mathcal{L}) \neq 1$, which excludes the case $\dim r(\mathcal{M}) = 0$. Since a surface section is birationally ruled we see by a straightforward induction argument down to a surface section that

$$h^0(K_{\mathcal{M}} + (n-1)\mathcal{L}) = h^0(K_S + \mathcal{L}_S) = g(\mathcal{L}) - q(S) = 4 - q(S).$$

Assume first that $\dim r(\mathcal{M}) = n$. Let $Y := r(\mathcal{M})$. Then we see that $n = 2$, since otherwise $Y \cong \mathbb{P}^3$ and we would have the numerical contradiction $K_Y + 2(r_*\mathcal{L})^{**} \cong \mathcal{O}_{\mathbb{P}^3}(1)$, and Y is embedded by s as a smooth quadric or cubic surface in $\mathbb{P}^3$. Note that $K_Y + (r_*\mathcal{L})^{**} \cong s^*\mathcal{O}_{\mathbb{P}^3}(1)$, and thus we have $(r_*\mathcal{L})^{**} \cong s^*\mathcal{O}_{\mathbb{P}^3}(3)$ in the quadric case and $(r_*\mathcal{L})^{**} \cong -2K_Y$ in the cubic case. This gives the cases 5), 6) of the theorem with $n = 2$.

Assume next that $\dim r(\mathcal{M}) = 1$. By using (11.2.4), (10.1.3), and the fact that $g(\mathcal{L}) = 4$ we see that s is an embedding. We have $K_{\mathcal{M}} + (n-1)\mathcal{L} \cong r^*H$ for an ample line bundle, H, on $r(\mathcal{M})$. By an induction down to the surface case we see that $h^0(K_{\mathcal{M}} + (n-1)\mathcal{L}) = 4 - q(S)$. Since $(\mathcal{M}, \mathcal{L})$ is a quadric fibration we have $r^*H \cdot \mathcal{L}^{n-1} = 2\deg H$ (see the proof of (10.1.3)), and hence

$$2g(\mathcal{L}) - 2 = (K_{\mathcal{M}} + (n-1)\mathcal{L}) \cdot \mathcal{L}^{n-1} = 2\deg H.$$

Then we see that since $g(\mathcal{L}) = 4$, we have $\deg H = 3$. If $q(S) \geq 2$ then $h^0(K_{\mathcal{M}} + (n-1)\mathcal{L}) \leq 2$ which contradicts the fact that s is an embedding. Thus $q(S) = 0$ or 1 as in case 2).

Assume now that $\dim r(\mathcal{M}) = 2 \neq n$. Then by using Theorem (11.2.4) we can assume without loss of generality that s is an embedding. Thus $(\mathcal{M}, \mathcal{L})$ is a scroll over a smooth quadric or cubic surface in $\mathbb{P}^3$ as in 5), 6). □

Remark 11.6.4. The study of pairs with sectional genus, $g(\mathcal{L}) = 5$, and $n \geq 2$ proceeds along similar lines. Using low degree arguments we reduce to the case when $K_{\mathcal{M}} + (n-1)\mathcal{L}$ is very ample and $\mathcal{L}^n \geq 9$. In this case it is straightforward to show that $h^0(K_{\mathcal{M}} + (n-1)\mathcal{L}) = 5$. Remmert-Stein factorize the adjunction mapping associated to $|K_{\mathcal{M}} + (n-1)\mathcal{L}|$ as $s \circ r$. From $h^0(K_{\mathcal{M}} + (n-1)\mathcal{L}) = 5$ it follows $n = 2$ and, s, embeds the first reduction, $r(\mathcal{M})$, of $\mathcal{M}$ in $\mathbb{P}^4$. As pointed out in [586] the double point formula in this case gives an extra relation which allows an enumeration of all invariants.

11.7 The Fano-Morin adjunction process

Let $\mathcal{M}$ be a smooth connected projective 3-fold and let $\mathcal{L}$ be a very ample line bundle on $\mathcal{M}$. Roth's book [527] (see also the valuable sourcebook put together by Brigaglia, Ciliberto, and Sernesi [140]) discusses two different generalizations to 3-folds of the adjunction process for surfaces. The original one due to Fano was to replace $\mathcal{L}$ by $K_{\mathcal{M}} + \mathcal{L}$. Though very natural, this approach runs into serious technical problems. Morin's approach, which also runs into difficulties, was to replace $\mathcal{L}$ by $K_{\mathcal{M}} + 2\mathcal{L}$. On smooth $S \in |\mathcal{L}|$ this bundle restricts to $K_S + \mathcal{L}_S$, and one can hope that understanding $K_{\mathcal{M}} + 2\mathcal{L}$ will reduce to understanding $K_S + \mathcal{L}_S$. If all worked out well one could hope to understand $K_{\mathcal{M}} + \mathcal{L}$ in the limit. It is very easy using the results of this chapter and Chapter 9 to carry out the Morin approach and as a consequence obtain very complete information about $K_{\mathcal{M}} + \mathcal{L}$. Though the results in this section are obtained in more generality in Chapter 7, we work out the Fano-Morin approach for a number of reasons beyond simple historical interest. For the reader interested in classical adjunction theory and the material in the next few chapters in the generality of very ample divisors, this very quick approach lets the reader skip Chapter 7 and the more involved technique needed there. This approach has proven useful repeatedly in adjunction theory. The Fano-Morin process and its variants have many times yielded results well ahead of other approaches, e.g., [576, 227, 583, 229, 86]. Though it requires assumptions on $\mathcal{L}$ stronger than ampleness this sort of adjunction process typically is less sensitive to singularities than other methods.

For an n-fold, X, and a line bundle, L, the Morin adjunction process considers the sequence, $L_0, L_1, L_2 \dots$ where $L_0 := L$ and for $t \geq 1$, $L_t := K_X + (n-1)L_{t-1}$. Note that for $t \geq 1$,

$$L_t = \frac{(n-1)^t - 1}{n-2} K_X + (n-1)^t L. \tag{11.4}$$

In particular if all the L_t were nef for $t \gg 0$, then $K_X + (n-2)L$ is nef. To see this choose an irreducible curve $C \subset X$ and note that

$$(K_X + (n-2)L) \cdot C = \lim_{t\to\infty} \frac{n-2}{(n-1)^t} L_t \cdot C.$$

The main goal of this section is to prove the following basic result, which follows more generally for ample bundles by the results of Chapter 7.

Theorem 11.7.1. *Let $\mathcal{L}$ be a very ample line bundle on $\mathcal{M}$, a connected n-dimensional projective manifold with $n \geq 3$. Assume that $\kappa(K_{\mathcal{M}} + (n-1)\mathcal{L}) = n$, and let (X, L), $\pi : \mathcal{M} \to X$ be the first reduction mapping. Either:*

1. *all the bundles L_t of the Morin sequence are very ample for $t \geq 1$ and $K_X + (n-2)L$ is semiample, i.e., some positive multiple of $K_X + (n-2)L$ is spanned by global sections; or*

2. *$\kappa(K_X + (n-2)L) = -\infty$, i.e., $h^0(N(K_X + (n-2)L) = 0$ for all $N > 0$ and either:*

   (a) *$(X, L) \cong (\mathbb{P}^n, \mathcal{O}_{\mathbb{P}^n}(d))$ with $(n, d) = (3, 3), (4, 2)$; or*

   (b) *$X \subset \mathbb{P}^4$ is a quadric hypersurface and $L \cong \mathcal{O}_{\mathbb{P}^4}(2)_X$; or*

   (c) *$n = 3$, $(X, K_X + 2L)$ is a scroll, and in particular a $\mathbb{P}^2$-bundle, over a smooth curve with $L_F \cong \mathcal{O}_{\mathbb{P}^2}(2)$ for any fiber F of the scroll projection.*

Proof. First assume that the bundles, L_t, are very ample for $t \geq 1$. Then by the discussion before the theorem it follows that $K_X + (n-2)L$ is nef. Since L is ample, it follows from the Kawamata-Shokurov basepoint free theorem, (1.5.1), that $K_X + (n-2)L$ is semiample. Thus without loss of generality we can assume that L_t is not very ample for all $t \geq 1$. By Theorem (11.3.1), L_1 is very ample.

Let J denote the smallest $t \geq 1$ such that L_J is very ample, but L_{J+1} is not very ample. Since $L_{J+1} := K_X + (n-1)L_J$, we can use the theory developed in this chapter to analyze the deviation of this bundle from being a very ample bundle.

First we note that if L_{J+1} is not spanned by global sections, that either

1. $(X, L_J) \cong (\mathbb{P}^n, \mathcal{O}_{\mathbb{P}^n}(1))$; or

2. $X \subset \mathbb{P}^{n+1}$ is a quadric hypersurface and $L_J \cong \mathcal{O}_{\mathbb{P}^{n+1}}(1)_X$; or

3. (X, L_J) is a scroll, and in particular a $\mathbb{P}^{n-1}$-bundle over a smooth curve with $L_{J|F} \cong \mathcal{O}_{\mathbb{P}^{n-1}}(1)$ for any fiber F of the scroll projection.

Consider the first case. By the equality (11.4) we see that with $L \cong \mathcal{O}_{\mathbb{P}^n}(d)$,

$$-(n+1)((n-1)^J - 1) + d(n-2)(n-1)^J = n-2.$$

This implies that $(n+1-d(n-2))(n-1)^J = 3$. Since $J \geq 1$, this implies that $(n-1)^J$ divides 3. Since $n \geq 3$ we see that $(n, J) = (4, 1)$. Thus we have that $5 - 3d = 1$, and we have case 2a) of the theorem. Doing the same argument for the quadric yields case 2b) of the theorem. Doing the same argument applied to a fiber of the scroll gives case 2c) of the theorem.

Since we can assume that L_{J+1} is spanned we know that either $-K_X \approx (n-1)L_J$, (X, L_J) is a quadric fibration over a curve, (X, L_J) is a scroll over a surface, or L_{J+1} is nef and big. The first gives $nK_X + (n-1)^2 L_{J-1} \cong \mathcal{O}_X$. Since $(n-1)^2$ is relatively prime to n we can find a line bundle H on X such that $-K_X \cong (n-1)^2 H$ and $L_{J-1} \cong nH$. Since $J \geq 1$ and $L_0 := L$ is ample we see that H is ample and therefore we conclude from $-K_X \cong (n-1)^2 H$ and (3.3.2) that $(n-1)^2 \leq n+1$. Since $n \geq 3$ we see that $n = 3$, and by the Kobayashi-Ochiai theorem we conclude that $(X, H) \cong (\mathbb{P}^3, \mathcal{O}_{\mathbb{P}^3}(1))$. An argument as in the last paragraph shows that $J = 1$ and we have the case $(X, L) \cong (\mathbb{P}^3, \mathcal{O}_{\mathbb{P}^3}(3))$. The same argument applied to the fibers of the quadric fibration and the scroll show that they do not exist. Thus we can assume that $K_X + (n-1)L_J$ is spanned and big. Since it is not very ample by assumption, we know from Theorem (11.3.1) that there must exist a fiber $F \cong \mathbb{P}^{n-1}$ of the first reduction map of (X, L_J), with $L_{J|F} \cong \mathcal{O}_{\mathbb{P}^{n-1}}(1)$. Since $K_{X|F} \cong \mathcal{O}_{\mathbb{P}^{n-1}}(1-n)$ we conclude from $(K_X + (n-1)L_J)_F \cong \mathcal{O}_F$ and (11.4) with $L_F \cong \mathcal{O}_{\mathbb{P}^{n-1}}(d)$ that

$$-(n-1)((n-1)^J - 1) + d(n-2)(n-1)^J = n-2.$$

This implies the absurdity that $n-1$ is an integer factor of $n-2$. □

Remark 11.7.2. Note that by (7.6.1) it follows that

$$h^0(N(K_X + (n-2)L)) = h^0(N(K_{\mathcal{M}} + (n-2)\mathcal{L}))$$

for all $N \geq 0$. It is a theorem of the second author (see (13.2.4) and [584]) that $K_X + (n-2)L$ is nef if and only if $h^0(K_{\mathcal{M}} + (n-2)\mathcal{L}) \geq 1$.

Note that nefness of $K_X + (n-2)L$ implies that the canonical bundle of any smooth surface S obtained as a transversal intersection of $n-2$ smooth elements of $|L|$ is nef, and therefore such an S is a minimal model of a surface of nonnegative Kodaira dimension. To emphasize the strength of this result assume that $n = 3$ and that there is a smooth $A \in |\mathcal{L}|$ of nonnegative Kodaira dimension. Then using (11.7.1) we see that either $(\mathcal{M}, \mathcal{L})$ is a scroll over a smooth surface with A a meromorphic section, or $K_{\mathcal{M}} + 2\mathcal{L}$ is nef and big and the restriction to A of the first reduction mapping, $(\mathcal{M}, \mathcal{L}) \to (X, L)$, maps A onto its minimal model with $K_X + L$ nef.

Chapter 12

The second reduction in dimension three

In this chapter we work out the structure of the 2nd reduction in the 3-dimensional case. The structure of the first reduction, (M, L), of an n-dimensional polarized pair, $(\mathcal{M}, \mathcal{L})$, $\mathcal{L}$ ample line bundle, is studied in full generality in Chapter 7 in dimension $n \geq 4$. To study the structure of the second reduction we needed the assumption $n \geq 4$ in the proof of Theorem (7.5.3.5) to insure that the exceptional loci of all the extremal rays subordinate to $K_M + (n-2)L$ are disjoint and hence to conclude that the second reduction morphism is the simultaneous contraction of a finite number of divisorial extremal rays. For $n = 3$, the argument in the proof of Theorem (7.5.3) fails. Indeed the structure of the second reduction map, $\varphi : M \to X$, is much more complicated in dimension $n = 3$. In this chapter we work it out in the case when $\mathcal{M}$ is smooth and the polarization, $\mathcal{L}$, on $\mathcal{M}$ is ample and spanned. Different configurations of exceptional divisors meeting each other can occur, as it is shown in Theorem (12.2.1). The proof of this theorem is rather involved. In §12.1 we give a number of preliminary results, describing all possible configurations of the exceptional divisors for φ. In §12.2 we conclude the proof of the theorem. The argument in the proof of Theorem (12.2.1) needs in an essential way the facts that the original variety, $\mathcal{M}$, is smooth and the polarization, $\mathcal{L}$, of $\mathcal{M}$ is ample and spanned, which are the assumptions we make throughout this chapter.

In §12.3 we recall a result from our paper with Schneider [96] which describes the structure of the first and the second reductions in the special case of 3-folds in $\mathbb{P}^5$. We refer to §14.3 of Chapter 14 for more on 3-folds in $\mathbb{P}^5$.

We follow the second author's original argument of [579] with the refinements added by Fania [219] and the authors of this book and Fania [86].

12.0.1. Let $\mathcal{M}$ be a connected 3-dimensional projective manifold with $\mathcal{L}$ an ample and spanned line bundle on $\mathcal{M}$. Assume that $K_{\mathcal{M}}+2\mathcal{L}$ is nef and big and let (M, L), with $L := (\pi_*\mathcal{L})^{**}$ and $\pi : \mathcal{M} \to M$, be the first reduction as in (7.3.3). Further assume that $K_M + L$ is nef and big. Thus there exists the second reduction $(X, \mathcal{D})$ with $\mathcal{D} := (\varphi_*L)^{**}$ and $\varphi : M \to X$ the morphism with connected fibers and normal image defined by $|m(K_M + L)|$ for $m \gg 0$. Recall that $K_M + L \cong \varphi^*\mathcal{K}$ for an ample line bundle $\mathcal{K}$ on X. As in (7.5.7) and (7.5.8) of Chapter 7, $\mathcal{K} \approx K_X + \mathcal{D}$.

In what follows, E, f denote as usual a section of minimal self-intersection, $E^2 = -r$, and a fiber of a Hirzebruch surface $\mathbb{F}_r$, $r \geq 0$.

12.1 Exceptional divisors of the second reduction morphism

The proof of the main result of this chapter, Theorem (12.2.1), consists in a sequence of several lemmas which make up this section.

Lemma 12.1.1. *Let $(\mathcal{M}, \mathcal{L})$, (M, L), $\pi : \mathcal{M} \to M$ be as in* (12.0.1). *Let B be the set of points blown up under π. Let D be an irreducible and reduced divisor on M. Then a general $S \in |L|$ which contains B is smooth and meets D in an irreducible reduced curve.*

Proof. Recall that there is a one-to-one correspondence between smooth divisors of $|\mathcal{L}|$ and smooth divisors of $|L|$ which contain the set of points blown up under the first reduction map π. By Corollary (2.2.6), the proper transform of S on $\mathcal{M}$ meets the proper transform D' of D on $\mathcal{M}$ in an irreducible reduced curve. □

Lemma 12.1.2. *Let $(\mathcal{M}, \mathcal{L})$, (M, L), $\pi : \mathcal{M} \to M$ be as in* (12.0.1). *Let B be the set of points blown up under π. Let C be an effective curve on M such that $L \cdot C = 1$. Then C is a smooth rational curve and $C \cap B = \emptyset$.*

Proof. Since L is ample, $L \cdot C = 1$ implies that C is irreducible and reduced. Let C' be the proper transform of C under π. If C meets B then C' would meet $\pi^{-1}(B)$. This would imply that

$$\mathcal{L} \cdot C' = (\pi^* L - \pi^{-1}(B)) \cdot C' = L \cdot C - \pi^{-1}(B) \cdot C' \leq 1 - 1 = 0.$$

Since $\mathcal{L}$ is ample and spanned we conclude that $C \cap B = \emptyset$. Therefore $C' \cong C$ and $\mathcal{L} \cdot C' = 1$. Since $\mathcal{L}$ is ample and spanned, this implies that C' and hence C is a smooth rational curve. □

Lemma 12.1.3. *Let $\alpha : V \to W$ be a holomorphic proper map from a 3-dimensional complex manifold, V, to a complex analytic space W. Assume that there exists a curve C on V with C biholomorphic to $\mathbb{P}^1$, $\alpha(C)$ a point, and* $\dim \alpha(V) = 3$. *If $\mathcal{N}_{C/V} \cong \mathcal{O}_{\mathbb{P}^1}(a) \oplus \mathcal{O}_{\mathbb{P}^1}(b)$ then either $a < 0$ or $b < 0$.*

If $a \geq 1$ and $b = -1$, then an irreducible divisor, $D \subset V$, with $\dim \alpha(D) = 0$, *can be chosen such that there is an open neighborhood U of C in V with some component D' of $D \cap U$ such that $C \subset D'$, and D' is is smooth. Furthermore $\mathcal{N}_{C/D'} \cong \mathcal{O}_{\mathbb{P}^1}(a)$.*

Proof. We prove the result for V projective, which suffices for this book. If $a \geq 0$ and $b \geq 0$ then $\mathcal{N}_{C/V}$ is spanned and $h^1(\mathcal{N}_{C/V}) = 0$. Since $\alpha(C)$ is a point, by the rigidity lemma, (4.1.13), there exists a tubular neighborhood U_C of C in V such that $\alpha(U_C)$ is contained in an affine neighborhood $\mathcal{A}_C$ of $\alpha(C)$. Now, since $\mathcal{N}_{C/V}$ is spanned and $h^1(\mathcal{N}_{C/V}) = 0$, Kodaira-Spencer deformation theory, see, e.g., (4.1.9), implies that deformations of C exist to fill up a neighborhood $\mathcal{U}_C$ of C. Therefore, since $\mathcal{A}_C$ contains no compact subvarieties, $\dim \alpha(C') = 0$ for all $C' \in U_C \cap \mathcal{U}_C$. Thus we have a family of 1-dimensional fibers of α of dimension $h^0(\mathcal{N}_{C/V}) \geq 2$. This gives the contradiction that $\dim \alpha(V) \leq 2$. We conclude that either $a < 0$ or $b < 0$.

If $a \geq 1$ and $b = -1$, then we have $h^1(\mathcal{N}_{C/V}) = 0$. It thus follows by using (4.1.9) that there is an open neighborhood U of C and a smooth connected divisor, $D' \subset U$, which contains C and such that $\mathcal{N}_{C/D'} \cong \mathcal{O}_{\mathbb{P}^1}(a)$. The divisor D' contains all small deformations C' of C. Since $a \geq 1$, we have $\deg \mathcal{N}_{C/V} \geq 0$ and hence the smooth deformations C' of C meet C. Therefore $\alpha(C') = \alpha(C)$ and since $\dim \alpha(V) = 3$, it follows that there must be an irreducible divisor $D \subset V$ contained in $\alpha^{-1}(\alpha(C))$ which contains D'. □

Lemma 12.1.4. *Let L be an ample line bundle on a connected projective 3-dimensional manifold, X. Let D be an irreducible reduced divisor on X. Assume that there is a smooth $S \in |L|$ that meets D transversally in a smooth rational curve, C. Then (D, L_D) is biholomorphic to either*

1. $(\mathbb{P}^2, \mathcal{O}_{\mathbb{P}^2}(a))$, $a = 1, 2$; *or*

2. $(\mathbb{F}_r, E + (r+1+k)f)$, $r \geq 0$, $k \geq 0$; *or*

3. $(\widetilde{\mathbb{F}}_2, G)$ *with* $K_D \approx -2G$.

Proof. Since $S \cap D \subset D_{\mathrm{reg}}$, $\mathrm{Sing}(D)$ is finite. Therefore since D is a divisor on X and X is smooth, it follows that D is Cohen-Macaulay and normal. Then by Corollary (3.2.10) we conclude that either we are in cases 1), 2) or $D \cong \widetilde{\mathbb{F}}_r$ with $r \geq 1$. In this case, $D \cong \mathbb{P}^2$ if $r = 1$. Therefore we can assume that $D \cong \widetilde{\mathbb{F}}_r$ for $r \geq 2$. By Lemma (1.1.9) we have $r = 2$. □

Proposition 12.1.5. *Let $(\mathcal{M}, \mathcal{L})$, (M, L) be as in* (12.0.1) *with $\varphi : M \to X$ the second reduction map of $(\mathcal{M}, \mathcal{L})$. Let D be an irreducible and reduced divisor on M such that* $\dim \varphi(D) = 0$. *Then either*

1. $D \cong \mathbb{P}^2$, $\mathcal{N}_{D/M} \cong \mathcal{O}_{\mathbb{P}^2}(-2)$, $L_D \cong \mathcal{O}_{\mathbb{P}^2}(1)$; *or*

2. $D \cong \mathbb{P}^2$, $\mathcal{N}_{D/M} \cong \mathcal{O}_{\mathbb{P}^2}(-1)$, $L_D \cong \mathcal{O}_{\mathbb{P}^2}(2)$; *or*

3. $D \cong \widetilde{\mathbb{F}}_2$, $\mathcal{N}_{D/M} \cong -G$ *where* $-2G \cong K_D$, $L_D \cong G$; *or*

4. $D \cong \mathbb{F}_r$ *for some* $r \geq 0$, $\mathcal{N}_{D/M} \cong -E - f$, $L_D \cong E + (r+1)f$.

Proof. Let B be the set of points blown up under the first reduction map $\pi : \mathcal{M} \to M$ and let $\mathcal{J}_B$ be the ideal sheaf of B in M. By Lemma (12.1.1), a general $S \in |L \otimes \mathcal{J}_B|$ is smooth and meets D in an irreducible reduced curve, C. Since $\dim \varphi(C) = 0$, we have $(K_M + L) \cdot C = 0$. Therefore $K_S \cdot C = 0$. Since $K_M + L$ is nef and big, it follows that some multiple of K_S is the pullback of some ample divisor under the restriction of φ to S, and thus that K_S is nef and big (see also (5.1.1)). Thus, by the Hodge index theorem on S, $K_S \cdot C = 0$ implies that either C is numerically trivial or $C^2 < 0$. The first case is not possible since C is effective. Therefore $C^2 < 0$, so that the genus formula implies that C is a smooth rational curve which satisfies $C^2 = -2$ on S. Since C is smooth and it is a Cartier divisor on D, it follows that D is smooth in a neighborhood of C and that D meets S transversally along C. Since D has a smooth rational curve as an ample Cartier divisor, and since D is a divisor on a smooth 3-fold, it follows that (D, L_D) is as described in Lemma (12.1.4). Note that

$$(K_M + L)_D \approx \mathcal{O}_D \tag{12.1}$$

since $\dim \varphi(D) = 0$. Also, since $(C^2)_S = D_S^2 = L \cdot D^2 = -2$,

$$L_D \cdot \mathcal{N}_{D/M} = -2. \tag{12.2}$$

Let $(D, L_D) \cong (\mathbb{P}^2, \mathcal{O}_{\mathbb{P}^2}(a))$, $a = 1, 2$. Let $\mathcal{N}_{D/M} \cong \mathcal{O}_{\mathbb{P}^2}(b)$ for an integer b. Then by (12.2), the pair (a, b) is either $(1, -2)$ or $(2, -1)$. So we find the cases 1), 2).

Let $D \cong \widetilde{\mathbb{F}}_2$. Then by the adjunction formula, $K_D \approx \mathcal{O}_D(-2)$. From (12.2) we get $L_D \cong -\mathcal{N}_{D/M} \cong \mathcal{O}_D(1)$. This gives the conclusion 3) of the proposition with $G = \mathcal{O}_D(1)$.

Let $D \cong \mathbb{F}_r$ with $r \geq 0$ and $L_D \approx E + (k+1+r)f$ for some integer $k \geq 0$. The adjunction formula gives

$$K_D \approx (K_M + D)_D \approx -2E - (2+r)f. \tag{12.3}$$

If $k = 0$ we are done. Indeed in this case, by combining (12.1) and (12.3) we have

$$\mathcal{N}_{D/M} \approx -2E - (2+r)f + L_D \approx -2E - (2+r)f + E + (r+1)f = -E - f.$$

Therefore we can assume to the contrary that $k \geq 1$. Then $L_D - f \approx E + (r+k)f$ is very ample by Lemma (3.2.4). Note that $L_D - 2f \approx E + (r+k-1)f$ is spanned since $r + k - 1 \geq r$. Hence a general $\Gamma \in |L_D - 2f|$ is a smooth curve of genus $g(\Gamma) = 0$ by the genus formula.

From (12.2) we have $-2 = \Gamma \cdot \mathcal{N}_{D/M} + 2f \cdot \mathcal{N}_{D/M}$ and hence either $\Gamma \cdot \mathcal{N}_{D/M} \geq 0$ or $f \cdot \mathcal{N}_{D/M} \geq 0$. Note that $\Gamma^2 = \deg \mathcal{N}_{\Gamma/D} \geq r \geq 0$ on D. Then if $\Gamma \cdot \mathcal{N}_{D/M} \geq 0$ we would have $\mathcal{N}_{\Gamma/M} \cong \mathcal{O}_{\mathbb{P}^1}(a) \oplus \mathcal{O}_{\mathbb{P}^1}(b)$ with a, b both nonnegative. Since $\dim \varphi(\Gamma) = 0$ and $\dim \varphi(M) = 3$, this is not possible by Lemma (12.1.3). Therefore it must be $f \cdot \mathcal{N}_{D/M} \geq 0$. Since $L_D \cdot f = (E + (r+1+k)f) \cdot f = 1$, (12.1) yields $K_M \cdot f = -1$ and hence (12.3) gives the contradiction $f \cdot \mathcal{N}_{D/M} = -1$. Thus we conclude that $k = 0$, so we have case 4). □

Lemma 12.1.6. *Let $(\mathcal{M}, \mathcal{L})$, (M, L) be as in (1.6) and let $\varphi : M \to X$ be the second reduction map of $(\mathcal{M}, \mathcal{L})$. Let D_1 and D_2 be two irreducible reduced surfaces in M such that $D_1 \cap D_2$ is nonempty and $\dim \varphi(D_i) = 0$ for $i = 1, 2$. Then neither D_1 nor D_2 is isomorphic to $\widetilde{\mathbb{F}}_2$. Also D_1 and D_2 meet transversally in a smooth rational curve C satisfying $L \cdot C = 1$.*

Proof. Let C be the curve intersection of D_1 and D_2. By (12.1.1) we choose a general $S \in |L|$ which is smooth and meets D_1 and D_2 in irreducible reduced curves C_1, C_2 respectively. The same argument as in the proof of (12.1.5) shows that C_1, C_2 are smooth rational curves of self-intersection -2 meeting transversally at a single point. Therefore $L \cdot D_1 \cdot D_2 = L \cdot C = 1$. Since intersections of Cartier divisors on $\widetilde{\mathbb{F}}_2$ are even we see that neither D_1 nor D_2 is isomorphic to $\widetilde{\mathbb{F}}_2$. Hence by (12.1.5), each D_i is isomorphic to either $\mathbb{F}_r$ with $r \geq 0$ or $\mathbb{P}^2$. Since $L \cdot D_1 \cdot D_2 = L \cdot C = 1$, (12.1.2) implies that C is a smooth rational curve. Since $D_1 \cap D_2$ is a smooth curve, D_1, D_2 meet transversally along C. □

Lemma 12.1.7. *Let (M, L), D_1, D_2 be as in (12.1.6). If D_1 is biholomorphic to $\mathbb{P}^2$, then D_2 is biholomorphic to $\mathbb{F}_2$. The curve $C = D_1 \cap D_2$ is a line on D_1 and it is the unique curve E with self-intersection -2 on D_2.*

Proof. By Lemma (12.1.6) we have $L \cdot C = L_{D_1} \cdot C = 1$ and hence C is a line in $D_1 \cong \mathbb{P}^2$. Note that

$$\mathcal{N}_{C/M} \cong \mathcal{N}_1 \oplus \mathcal{N}_2 \tag{12.4}$$

where $\mathcal{N}_i$ is the normal bundle of C in D_i, $i = 1, 2$. Since $(K_M + L)_C \approx \mathcal{O}_C$ and since $L \cdot C = 1$ one has $K_M \cdot C = -1$. Therefore, from the adjunction formula $K_C \approx K_{M|C} + \det \mathcal{N}_{C/M}$ and (12.4), it follows that

$$\deg \mathcal{N}_{C/M} = \deg \mathcal{N}_1 + \deg \mathcal{N}_2 = -1. \tag{12.5}$$

Since $\mathcal{N}_1 \cong \mathcal{O}_{\mathbb{P}^1}(1)$, it follows that $\mathcal{N}_2 \cong \mathcal{O}_{\mathbb{P}^1}(-2)$. Thus $D_2 \not\cong \mathbb{P}^2$. Therefore, recalling Proposition (12.1.5) and Lemma (12.1.6) we conclude that $D_2 \cong \mathbb{F}_r$ for some $r \geq 0$. As shown in Lemma (3.2.13) the only curve with negative self-intersection on $\mathbb{F}_r$ can be E. Thus from the fact that the self-intersection of C on $\mathbb{F}_r$ is -2, we conclude that $r = 2$ and $E = C$. □

Lemma 12.1.8. *Let (M, L), D_1, D_2 be as in* (12.1.6). *Then it is not possible that D_1 is biholomorphic to $\mathbb{F}_a$ and D_2 is biholomorphic to $\mathbb{F}_b$ with $a \geq 0$ and $b \geq 0$.*

Proof. Assume that $D_1 \cong \mathbb{F}_a$ and $D_2 \cong \mathbb{F}_b$, $a \geq 0$, $b \geq 0$. Let E, f denote a basis for $H_2(\mathbb{F}_a, \mathbb{Z})$ and let E', f' denote a basis for $H_2(\mathbb{F}_b, \mathbb{Z})$. Let $C = D_1 \cap D_2$ be the transversal intersection of D_1, D_2. Exactly the same argument as in the proof of Lemma (12.1.7), by using (12.1.5.4) and $L \cdot C = 1$, shows that either

1. $C \sim f$ on D_1 and $C \sim f'$ on D_2; or
2. $C \sim f$ on D_1 and $C \sim E'$ on D_2, or $C \sim E$ on D_1 and $C \sim f'$ on D_2; or
3. $C \sim E$ on D_1 and $C \sim E'$ on D_2.

Note that the condition (12.5) in the proof of the previous lemma is still true. Let C be as in 1). Then $\deg \mathcal{N}_{C/M} = 0$, in contradiction with that condition. So case 1) cannot happen. By noting that the role of E and f can be switched on $\mathbb{F}_0$, it follows that not both a and b can be zero. Therefore, after possibly renumbering, it can be assumed without loss of generality that $a > 0$.

Let C be as in 3). Then $\deg \mathcal{N}_{C/M} = -a - b$, so that $a + b = 1$ by (12.5). Since $a > 0$ we conclude that $a = 1$, $b = 0$. By switching the roles of E', f' in $\mathbb{F}_b = \mathbb{F}_0$ we fall in the second case of 2).

Therefore it remains to consider the case when $C \sim E$ on $\mathbb{F}_a$ and $C \sim f'$ on $\mathbb{F}_b$ as in 2). Since by (12.1.5.4), $\mathcal{N}_{D_1/M} \approx -E-f$ we have $\mathcal{N}_{D_1/M|f} \cong \mathcal{O}_{\mathbb{P}^1}(-1)$. It thus follows by the Nakano criterion, (3.2.8), that we can smoothly contract $D_1 \cong \mathbb{F}_a$ along f, to get an analytic manifold M' and a proper holomorphic modification $q : M \to M'$. Since $\dim \varphi(f) = 0$, there exists a morphism $\sigma : M' \to X$ such that $\varphi = \sigma \circ q$. Denote $D_2' := q(D_2) \cong \mathbb{F}_b$ on M'. From $\mathcal{N}_{D_2/M} \cong -E' - f'$ and since f meets $D_2 \cong \mathbb{F}_b$ on M in a point we infer that $\mathcal{N}_{D_2'/M'} \cong -E'$ on M'. Note that $E' + bf'$ is spanned on $D_2' \cong \mathbb{F}_b$ in M' and take a smooth curve $\Gamma \in |E' + bf'|$. Note also that Γ is a rational curve. One has $\mathcal{N}_{D_2'/M'} \cdot \Gamma = \mathcal{N}_{D_2'/M'} \cdot (E' + bf') = 0$. Therefore $\deg \mathcal{N}_{D_2'/M'|\Gamma} = 0$ and hence $\mathcal{N}_{D_2'/M'|\Gamma} \approx \mathcal{O}_{\mathbb{P}^1}$ since $\Gamma \cong \mathbb{P}^1$. Thus we have an exact sequence

$$0 \to \mathcal{N}_{\Gamma/D_2'} \cong \mathcal{O}_{\mathbb{P}^1}(b) \to \mathcal{N}_{\Gamma/M'} \to \mathcal{N}_{D_2'/M'|\Gamma} \cong \mathcal{O}_{\mathbb{P}^1} \to 0,$$

so that $\mathcal{N}_{\Gamma/M'}$ is spanned and $h^1(\mathcal{N}_{\Gamma/M'}) = 0$. Since $\dim \varphi(\Gamma) = 0$, Lemma (12.1.3) applies to give a contradiction. □

Lemma 12.1.9. *Let $(\mathcal{M}, \mathcal{L})$, (M, L) be as in* (12.0.1) *and let $\varphi : M \to X$ be the second reduction map of $(\mathcal{M}, \mathcal{L})$. Let D be an irreducible reduced divisor on M such that* $\dim \varphi(D) = 0$. *If D is biholomorphic to $\mathbb{F}_r$ and $r \geq 2$, then $r = 2$ and there is a divisor Δ on M biholomorphic to $\mathbb{P}^2$ such that the pair (Δ, D) has the same properties of the pair (D_1, D_2) as in* (12.1.7).

Proof. Since, by (12.1.5.4), $\mathcal{N}_{D/M} \approx -E - f$, we have $\mathcal{N}_{D/M|f} \cong \mathcal{O}_{\mathbb{P}^1}(-1)$. As in (12.1.8), by using (3.2.8), we can smoothly contract $D \cong \mathbb{F}_r$ along f, to get an analytic manifold M' and a proper modification $q : M \to M'$ together with a morphism $\sigma : M' \to X$ such that $\varphi = \sigma \circ q$. Let $q_{\mathbb{F}_r} : \mathbb{F}_r \to E'$ be the restriction of q to $\mathbb{F}_r$, where E' is a smooth rational curve on M', biholomorphic to E under q. Since $E + rf$ is spanned we can take a smooth curve $B \in |E + rf|$. Note that B is rational and $B \cdot E = 0$. Now $\deg \mathcal{N}_{D/M|B} = (-E - f) \cdot (E + rf) = -1$ and $\deg \mathcal{N}_{D/M|E} = (-E - f) \cdot E = r - 1$ so that $\mathcal{N}_{D/M|B} \cong \mathcal{O}_{\mathbb{P}^1}(-1)$ and $\mathcal{N}_{D/M|E} \cong \mathcal{O}_{\mathbb{P}^1}(r-1)$. Therefore

$$\mathcal{N}_{E'/M'} \cong \mathcal{N}_{\mathbb{F}_r/M|E} \oplus \mathcal{N}_{\mathbb{F}_r/M|B} \cong \mathcal{O}_{\mathbb{P}^1}(-1) \oplus \mathcal{O}_{\mathbb{P}^1}(r-1).$$

If $r \geq 2$, then we conclude from this and (12.1.3), by noting that $\dim \sigma(B') = 0$ where $B' := q(B)$, that there is an irreducible divisor, D', on M' containing B' such that $\dim \sigma(D') = 0$. Let Δ be the proper transform of D' under $q : M \to M'$. Clearly Δ maps to a point under φ and Δ meets D in at least E. By (12.1.5), Δ is either $\mathbb{P}^2$, $\mathbb{F}_r$ with $r \geq 0$, or $\widetilde{\mathbb{F}}_2$. By (12.1.6), Δ cannot be biholomorphic to $\widetilde{\mathbb{F}}_2$. By (12.1.7), Δ cannot be biholomorphic to $\mathbb{F}_r$. Thus $\Delta \cong \mathbb{P}^2$. By (12.1.7) we conclude that $D \cong \mathbb{F}_2$. □

Lemma 12.1.10. *Let* $(\mathcal{M}, \mathcal{L})$, (M, L) *be as in* (12.0.1) *and let* $\varphi : M \to X$ *be the second reduction map of* $(\mathcal{M}, \mathcal{L})$. *There are no distinct irreducible divisors* D_1, D_2 *and* D_3 *such that* $D_1 \cap D_2 \neq \emptyset$, $D_1 \cap D_3 \neq \emptyset$ *and* $\dim \varphi(D_i) = 0$, $i = 1, 2, 3$.

Proof. Assume on the contrary that three such divisors existed. Then by combining Lemmas (12.1.6), (12.1.7) and (12.1.8) we conclude that either

1. $D_1 \cong \mathbb{P}^2$ and $D_2 \cong D_3 \cong \mathbb{F}_2$, or
2. $D_1 \cong \mathbb{F}_2$ and $D_2 \cong D_3 \cong \mathbb{P}^2$.

In case 1), (12.1.7) implies that $D_2 \cap D_1 = \ell_2$, $D_3 \cap D_1 = \ell_3$, ℓ_2, ℓ_3 lines in $D_1 \cong \mathbb{P}^2$. Since ℓ_2, ℓ_3 meet in (at least) a point on $D_1 \cong \mathbb{P}^2$ it follows that $D_2 \cap D_3 \neq \emptyset$. This is not possible by Lemma (12.1.8).

In case 2), (12.1.7) implies that D_2 and D_3 both meet $D_1 \cong \mathbb{F}_2$ in the unique section, E, of $\mathbb{F}_2$, with $E^2 = -2$. Hence $D_2 \cap D_3 \neq \emptyset$. Therefore (12.1.7) applies again to give the contradiction that either D_2 or D_3 is biholomorphic to $\mathbb{F}_2$. □

12.2 The structure of the second reduction

We can prove now the main result of the chapter, due to the second author [579, (1.0.1)] and Fania [219, (1.0)].

Theorem 12.2.1. *Let* $(\mathcal{M}, \mathcal{L})$, (M, L) *be as in* (12.0.1) *and let* $(X, \mathcal{D})$, $\varphi : M \to X$ *be the second reduction of* $(\mathcal{M}, \mathcal{L})$. *Then* φ *is an isomorphism outside of an algebraic subset* Z *of* X *such that* $\dim Z \leq 1$. *We have:*

a) *Any 1-dimensional component of* Z *is a smooth curve,* C, *such that* $\operatorname{reg}(X) \supset C$, $\varphi^{-1}(C) = D \cup \{\cup D_i\}_{i \in I}$ *where* D *and the* D_i*'s are smooth irreducible divisors. The restriction* φ_D *of* φ *to* D *is a* $\mathbb{P}^1$*-bundle* $\varphi_D : D \to C$ *and* $\mathcal{N}_{D/M|F} \cong \mathcal{O}_{\mathbb{P}^1}(-1)$ *for a fiber* F *of* φ_D. *Furthermore if* I *is not empty,* $\varphi(D_i)$ *is a point for each* $i \in I$. *Let* $x_i = \varphi(D_i)$, $i \in I$. *Then* $\varphi^{-1}(x_i) = \mathbb{F}_1$, *and for each* $i \in I$, D *meets* $\mathbb{F}_1$ *transversally along the irreducible curve,* E, *of* $\mathbb{F}_1$, *with* $(E^2)_{\mathbb{F}_1} = -1$.

b) *If* x *is a* 0*-dimensional component of* Z *then* $\varphi^{-1}(x)$ *is a reduced divisor,* D, *and either*

   b1) $D \cong \mathbb{P}^2$, $\mathcal{N}_{D/M} \cong \mathcal{O}_{\mathbb{P}^2}(-2)$; *or*

   b2) $D \cong \mathbb{P}^2$, $\mathcal{N}_{D/M} \cong \mathcal{O}_{\mathbb{P}^2}(-1)$; *or*

   b3) $D \cong \widetilde{\mathbb{F}}_2$, $\mathcal{N}_{D/M} \cong -G$ *where* $-2G \cong K_D$; *or*

   b4) $D \cong \mathbb{F}_0$, $\mathcal{N}_{D/M} \cong \mathcal{O}_{\mathbb{F}_0}(-1,-1)$; *or*

   b5) $D = D_1 \cup D_2$ *where* $D_1 \cong \mathbb{F}_2$, $D_2 \cong \mathbb{P}^2$, $\mathcal{N}_{\mathbb{F}_2/M} \approx -E - f$, $\mathcal{N}_{\mathbb{P}^2/M} \cong \mathcal{O}_{\mathbb{P}^2}(-2)$ *and* D_1, D_2 *meet transversally in a smooth rational curve which is the irreducible curve,* E, *of* $\mathbb{F}_2$ *with* $(E^2)_{D_1} = -2$ *and is a line on* $\mathbb{P}^2$.

Proof. Let S be a smooth surface in $|L|$ corresponding to a general element of $|\mathcal{L}|$. Let φ_S be the restriction of φ to S. From the Kawamata-Viehweg vanishing theorem we have $h^1(mK_M + (m-1)L) = 0$ for $m \geq 2$. By using the exact sequence

$$0 \to mK_M + (m-1)L \to m(K_M + L) \to mK_S \to 0$$

we get a surjection $\Gamma(m(K_M + L)) \to \Gamma(mK_S)$, $m \geq 2$. This implies that φ_S is the map associated to $|mK_S|$ for $m \gg 0$ and thus φ_S has connected fibers and normal image.

Claim 12.2.2. $\varphi^{-1}(\varphi(y))$ is pure dimensional for each $y \in M$.

Proof. Assume otherwise. Then for some $y \in M$, $\varphi^{-1}\varphi(y)$ contains irreducible components C and D which are respectively one and two dimensional. Hence $S \cap \varphi^{-1}\varphi(y)$ is disconnected, for a general $S \in |L|$. But, by the above, $S \cap \varphi^{-1}\varphi(y)$ is a fiber of the restriction map φ_S and φ_S has connected fibers. □

We prove part b) of the theorem first. By looking over the above lemmas we see that the only remaining assertion to be proved is that $\varphi^{-1}\varphi(y)$ is reduced when

$\dim \varphi^{-1}\varphi(y) = 2$, $y \in M$. Choose a general $S \in |L|$. If $S \cap \varphi^{-1}\varphi(y) = \varphi_S^{-1}\varphi(y)$ is reduced we are done. Note that $\varphi_S^{-1}\varphi(y)$ is a curve, C, such that $K_S \cdot C = 0$. Since φ_S has normal image and C is the fiber $\varphi_S^{-1}\varphi(y)$ we see that C is a -2 rational curve on S. Since $(\varphi_S^{-1}\varphi(y))_{\text{red}}$ has at most two irreducible components by Lemma (12.1.10), we conclude that the only possible graphs to classify $\varphi_S^{-1}\varphi(y)$ are the Dynkin diagrams A_1 or A_2 (see Table (1.1)). The multiplicities of the irreducible curves of the cycles represented by A_1 and A_2 are both one.

To prove part a), we need the following

Claim 12.2.3. Let f be a 1-dimensional fiber of φ. Then f is a smooth rational curve satisfying $L \cdot f = 1$. Furthermore there exists an irreducible divisor $D \subset M$ such that $f \subset D$, $\dim \varphi(D) = 1$ and $D \cdot f = -1$.

Proof. Choose a general $S \in |L|$. By the above the fibers of the restriction φ_S are connected. By the generality of S, f is not contained in S. Therefore since $\varphi_S^{-1}\varphi(f) = f \cap S$, we conclude that $f \cap S$ is a single, possible nonreduced point. Note that $f \cap S$ is a zero dimensional fiber of the restriction map φ_S and φ_S is an embedding at this point. It thus follows that $f \cap S$ is a reduced point and therefore $L \cdot f = 1$. From $L \cdot f = 1$ and Lemma (12.1.2) we conclude that f is a smooth rational curve.

Since $(K_M + L) \cdot f = 0$ we get $K_M \cdot f = -1$ or $K_{M|f} \cong \mathcal{O}_{\mathbb{P}^1}(-1)$ and therefore the adjunction formula yields $\det \mathcal{N}_{f/M} \cong K_f - K_{M|f} \cong \mathcal{O}_{\mathbb{P}^1}(-1)$. By the Riemann-Roch formula, $\chi(\mathcal{N}_{f/M}) = \deg(\det \mathcal{N}_{f/M}) + (\text{rank}\mathcal{N}_{f/M})\chi(\mathcal{O}_f) = -1 + 2 = 1$. Therefore by (4.1.9), there are nontrivial smooth deformations of f in M. Let $\mathcal{D}$ denote the closure of the union of the smooth deformations of f. By a standard Hilbert scheme argument using (4.1.9), (cf., Sommese [575, (0.7)]), $\mathcal{D}$ is either a divisor or $\mathcal{D} = M$, according to whether $h^0(f, \mathcal{N}_{f/M}) = 1$ or ≥ 2. By the usual rigidity argument (as in the proof of (12.1.3)) we have that $\dim \varphi(f') = 0$ for any smooth deformation f' of f. Therefore, since $\dim \varphi(M) = 3$, the case $\mathcal{D} = M$ is excluded, so we conclude that $\mathcal{D}$ is a divisor. Choose an irreducible component D of $\mathcal{D}$ that contains small deformations of f. Note that $\dim \varphi(D) \geq 1$. Indeed, if $\dim \varphi(D) = 0$, f would be strictly contained in $\varphi^{-1}\varphi(D)$, contradicting the fact that f is a fiber of φ. Since D is the union of the closure of curves, f', satisfying $\dim \varphi(f') = 0$, it follows by dimension reasons that $\dim \varphi(D) \leq 1$. Thus $\dim \varphi(D) = 1$. Since $K_M \cdot f = -L \cdot f = -1$, the adjunction formula gives $D \cdot f = \mathcal{N}_{D/M} \cdot f = K_D \cdot f - K_M \cdot f = \deg K_f - K_M \cdot f = -1$. □

Let D be as in the claim above and let $\varphi_D : D \to \varphi(D)$ be the restriction of φ to D. Let f be a general fiber of φ_D. Then by the above, f is a smooth rational curve satisfying $L \cdot f = 1$. Since $(K_M + L) \cdot f = 0$, Lemma (4.2.14) implies that in the cone of effective curves $NE(M)$, f can be written as a finite sum $f = \sum_i \lambda_i \ell_i$ where $\lambda_i \in \mathbb{R}_+$, ℓ_i are extremal rational curves and $(K_M + L) \cdot \ell_i = 0$ for all i. Since $D \cdot f = -1$, we conclude that $D \cdot \ell_i < 0$ for some index i. Let $R := \mathbb{R}_+[\ell_i]$ and let

$\rho := \mathrm{cont}_R : M \to V$ be the contraction of R. Then φ factors through ρ, $\varphi = \alpha \circ \rho$. Hence ρ is birational. Since $D \cdot R < 0$, we conclude from the contraction theorem, (4.3.1), than D coincides with the exceptional locus of ρ. Since $\dim \varphi(D) = 1$, we have $\dim \rho(D) = 1$. Hence $\rho(D)$ is a smooth curve and D is a smooth $\mathbb{P}^1$-bundle onto $\rho(D)$ under ρ.

Let $\varphi^{-1}\varphi(D) = D \cup \{\cup D_i\}_{i \in I}$ where the D_i's are irreducible divisors.

We claim that $\dim \varphi(D_i) = 0$ for each index i. Assume otherwise that $\varphi(D_i) = \varphi(D)$ for some index i. Then by (12.2.3), we get the contradiction that $(D_i \cup D) \cap \varphi^{-1}(x)$ is a smooth irreducible curve for a general point $x \in \varphi(D)$.

Since the intersection $D \cap D_i$ is of dimension 1 and smooth by Claim (12.2.3), D meets D_i transversally along the fiber, f_i, of φ_D over $x_i = \varphi(D_i)$.

Fix an index i, say $i = 1$. Since $\dim \varphi(D_1) = 0$, we know from Proposition (12.1.5) that D_1 is biholomorphic to either $\mathbb{P}^2$, $\widetilde{\mathbb{F}}_2$ or $\mathbb{F}_r$ with $r \geq 0$. We go on by a case by case analysis.

Let $D_1 \cong \mathbb{P}^2$. Since D meets D_i transversally along the fiber, f_i, of φ_D over $x_i = \varphi(D_i)$, the restriction $\rho_{\mathbb{P}^2} : \mathbb{P}^2 \to \rho(\mathbb{P}^2)$ contracts f_i. Since $\mathrm{Pic}(\mathbb{P}^2) \cong \mathbb{Z}$ this contradicts Lemma (3.1.8).

Similarly, since $\mathrm{Pic}(\widetilde{\mathbb{F}}_2) \cong \mathbb{Z}$, we rule out the case $D_1 \cong \widetilde{\mathbb{F}}_2$.

Let $D_1 \cong \mathbb{F}_0$. Then by Proposition (12.1.5), $\mathcal{N}_{D_1/M} \cong -E - f'$ and $E + f'$ is ample on $\mathbb{F}_0$ (here and in the sequel E, f' denote the unique irreducible curve with self-intersection $E^2 = -r$ and a fiber of $\mathbb{F}_r$, $r \geq 0$). Therefore, by the Grauert contractibility criterion, (3.2.7), $\mathbb{F}_0$ can be contracted to a point. Let $\sigma : M \to M'$ be such a contraction. Since σ is an isomorphism on $M - \mathbb{F}_0$, clearly we have $\dim \sigma(D) = 2$. Since D is a $\mathbb{P}^1$-bundle and a fiber (given by the intersection $D \cap D_1$) gets contracted to a point, then all nearby fibers should be contracted to a point and hence $\dim \sigma(D) = 1$. This contradiction implies that $D_1 \ncong \mathbb{F}_0$.

Let $D_1 \cong \mathbb{F}_r$ with $r \geq 1$. Recall that D, D_1 meet transversally along the fiber, f_1, over $\varphi(D_1)$. Since f_1 is contracted under φ, the curve f_1 on $D_1 \cong \mathbb{F}_r$ coincides with the unique section, E, such that $E^2 = -r$. Therefore we have $-1 = D \cdot f = \deg \mathcal{N}_{D/M|f} = \deg \mathcal{N}_{f_1/D_1} = \deg \mathcal{N}_{E/\mathbb{F}_r} = -r$ and hence $r = 1$.

Thus we have shown that $\varphi^{-1}\varphi(D) = D \cup \{\cup D_i\}_{i \in I}$ where $D_i \cong \mathbb{F}_1$, $\dim \varphi(D_i) = 0$ and D, D_i meet transversally along a fiber f_i of $\varphi_D : D \to \varphi(D)$, such that $f_i^2 = -1$ on $D_i \cong \mathbb{F}_1$ for each index i. Recall that $\varphi : M \to X$ factors through the Mori contraction $\rho : M \to V$, $\varphi = \alpha \circ \rho$. Fix an index i, say $i = 1$, and let $\Delta := \rho(D_1) = \rho(\mathbb{F}_1)$. Note that $\rho(f_1)$, $f_1 = D \cap D_1$, is a point, v_1, on $\rho(D)$, that is f_1 is a fiber of the $\mathbb{P}^1$-bundle $\rho : D \to \rho(D)$. Indeed otherwise, $\rho(f_1) = \rho(D)$ would map onto $\varphi(D)$ under α in contradiction to the fact that $\dim \varphi(f_1) = 0$. Recall that V is smooth and let $\delta = 0$ be a local equation of Δ in a neighborhood of v_1. Let $\tilde{\delta}$ be the pullback of δ on M under ρ. Since $\tilde{\delta} = 0$ locally agrees with the Cartier divisor $\mathbb{F}_1$ outside of a codimension two subset, then $\tilde{\delta} = 0$ defines $\mathbb{F}_1$ in a neighborhood of f_1. Since D, $\mathbb{F}_1$ meet transversally along f_1 it thus follows that Δ is smooth at v_1 and meets $\rho(D)$ transversally at v_1. Hence $\Delta \cong \mathbb{P}^2$ and the restriction $\rho_{\mathbb{F}_1} : \mathbb{F}_1 \to \mathbb{P}^2$ is simply the contraction of the section, $f_1 = E$, of $\mathbb{F}_1$. The map α contracts Δ and the curves $\rho(D)$, $\varphi(D)$ are isomorphic under α. Therefore $\varphi(D)$ is a smooth curve and D is a smooth $\mathbb{P}^1$-bundle over $\varphi(D)$ under the restriction φ_D of φ to D. □

To study the singularities of the second reduction $(X, \mathscr{D})$ we need the following lemma of our paper with Fania [86, (0.2.3)].

Lemma 12.2.4. *Let $f : \mathscr{X} \to Y$ be a birational proper morphism from a manifold $\mathscr{X}$ to a germ of a normal variety, Y, of dimension at least 2. Assume that* $\mathrm{Sing}(Y)$ *is a single point, y, $f : \mathscr{X} - f^{-1}(y) \to Y - \{y\}$ is a biholomorphism and $(f^{-1}(y))_{\mathrm{red}}$ is a union of distinct irreducible reduced divisors, $D_1, \dots, D_s$, with D_i meeting D_j transversally in a smooth curve for $1 \le i < j \le s$.*

1. *Assume that there are integers $a_1, \dots, a_s, r$ with $r > 0$ such that $rK_{\mathscr{X}} + \sum_i a_i D_i \cong \mathcal{O}_{\mathscr{X}}$. Then K_Y is r-Cartier and $f^*(rK_Y) \cong rK_{\mathscr{X}} + \sum_i a_i D_i$.*
2. *If $h^1(\mathcal{O}_{\mathscr{X}}) = 0$, then there is an injection of* $\mathrm{Pic}(\mathscr{X})$ *into $H^2(D_1 \cup \dots \cup D_s, \mathbb{Z})$.*
3. *If $(f^{-1}(y))_{\mathrm{red}} = D_1 \cup D_2$ with $h^1(D_1 \cap D_2, \mathbb{Z}) = 0$, then there is an injection of $H^2(D_1 \cup D_2, \mathbb{Z})$ into $H^2(D_1, \mathbb{Z}) \oplus H^2(D_2, \mathbb{Z})$.*

Proof. To see 1), let $V = \mathscr{X} - f^{-1}(y)$. Then

$$(rK_{\mathscr{X}} + \sum_i a_i D_i)_V \cong \mathcal{O}_{\mathscr{X}|V} \cong \mathcal{O}_{Y-\{y\}}$$

and also

$$(rK_{\mathscr{X}} + \sum_i a_i D_i)_V \cong (rK_{\mathscr{X}})_V \cong (rK_Y)_{Y-\{y\}}.$$

So $(rK_Y)_{Y-\{y\}} \cong \mathcal{O}_{Y-\{y\}}$ and thus $rK_Y \cong \mathcal{O}_Y$. Hence

$$f^*(rK_Y) \cong \mathcal{O}_{\mathscr{X}} \cong rK_{\mathscr{X}} + \sum_i a_i D_i.$$

Part 2) follows immediately from the exponential exact sequence, by noting that $H^2(\mathscr{X}, \mathbb{Z}) \cong H^2(\cup D_i, \mathbb{Z})$.

Part 3) follows from the Mayer-Vietoris exact sequence, e.g., Spanier [589, Chap. 4, §6]. □

Remark 12.2.5. Using Table (1.1) and (12.2.1) it is straightforward to recover the result of Sommese [579] describing the possible configurations of smooth rational curves of self-intersection -2 on the minimal models of smooth $S \in |\mathscr{L}|$ for a pair $(\mathcal{M}, \mathscr{L})$ as in (12.2.1). For a general $S \in |\mathscr{L}|$ only A_1, A_2 are possible. For a smooth but nongeneral $S \in |\mathscr{L}|$ the configurations A_1, A_2, A_3, D_4 are possible.

It is worth noting that only (12.2.1.bi) with $i = 1, 3, 4$ lead to singular points. (12.2.1.b1) leads to a 2-factorial singularity and (12.2.1.b3) leads to a factorial singularity. (12.2.1.b4) leads to a singularity which is analytically not even $\mathbb{Q}$-factorial. Algebraically it leads to a singularity which is not $\mathbb{Q}$-factorial if the restriction map $\mathrm{Pic}(X) \to \mathrm{Pic}(D)$ has a rank 2 image, but is factorial if the image of $\mathrm{Pic}(X)$ in $\mathrm{Pic}(D)$ has a rank 1 image.

We fix the following notation.
$D_a :=$ the union of all divisors ($\cong \mathbb{P}^1$-bundle) occurring in (12.2.1.a) and which contract to a curve under φ;
$D_{ai} :=$ the union of all divisors ($\cong \mathbb{F}_1$) occurring in (12.2.1.a) and which contract to a point under φ;
$D_{bi} :=$ the union of all divisors occurring in (12.2.1.bi), $i = 1, 2, 3, 4$;
$D_{b5,\mathbb{P}^2} :=$ the union of all divisors biholomorphic to $\mathbb{P}^2$ occurring in (12.2.1.b5);
$D_{b5,\mathbb{F}_2} :=$ the union of all divisors biholomorphic to $\mathbb{F}_2$ occurring in (12.2.1.b5).
The following is the 3-dimensional analogue of Proposition (7.5.6).

Proposition 12.2.6. *Let* $(\mathcal{M}, \mathcal{L})$, (M, L) *be as in* (12.0.1) *and let* $(X, \mathcal{D})$, $\varphi : M \to X$ *be the second reduction of* $(\mathcal{M}, \mathcal{L})$. *Then*

$$2K_M \approx \varphi^*(2K_X) + D_{b1} + 2(D_a + 2D_{ai} + 2D_{b2} + D_{b3} + D_{b4} + 3D_{b5,\mathbb{F}_2} + 2D_{b5,\mathbb{P}^2}).$$

In particular X is 2-Gorenstein and has isolated terminal rational singularities.

Proof. First note that for a given point $x \in X$ there exists an affine neighborhood $\mathcal{U}$ of x such that $K_{\varphi^{-1}(\mathcal{U})} + L_{\varphi^{-1}(\mathcal{U})} \cong \mathcal{O}_{\varphi^{-1}(\mathcal{U})}$ since $K_M + L \cong \varphi^*\mathcal{K}$. Therefore, denoting by $\varphi_{(i)}(\mathcal{F})$ the i-th direct image sheaf of a coherent sheaf, $\mathcal{F}$, $\varphi_{(i)}(\mathcal{O}_{\varphi^{-1}(\mathcal{U})}) \cong \varphi_{(i)}((K_M + L)_{\varphi^{-1}(\mathcal{U})})$ and the right term vanishes for $i > 0$ by the relative Kawamata-Viehweg vanishing theorem, (2.2.1). This implies that $\varphi_{(i)}(\mathcal{O}_M) = 0$ for $i > 0$ and hence X has rational singularities.

We claim that

1. $(2K_M - D_{b1})_D = 0$ in $H^2(D, \mathbb{Z})$ for any divisor $D \cong \mathbb{P}^2$ in D_{b1};
2. $(K_M - 2D_{b2})_D = 0$ in $H^2(D, \mathbb{Z})$ for any divisor $D \cong \mathbb{P}^2$ in D_{b2};
3. $(K_M - D_{bi})_D = 0$ in $H^2(D, \mathbb{Z})$ for any divisor D in D_{bi}, $i = 3, 4$;
4. $(K_M - 3D_{b5,\mathbb{F}_2} - 2D_{b5,\mathbb{P}^2})_D = 0$ in $H^2(D, \mathbb{Z})$ for any divisor D either in $D_{b5,\mathbb{F}_2}$ or in $D_{b5,\mathbb{P}^2}$.

Let us explicitly compute 1) and 4); 2) and 3) follow in the same way.

For 1) we can assume $D_{b1} = D \cong \mathbb{P}^2$. Then

$$K_{M|D} \cong K_{\mathbb{P}^2} - \mathcal{N}_{\mathbb{P}^2/M} \cong \mathcal{O}_{\mathbb{P}^2}(-1)$$

and $\mathcal{N}_{D/M} \cong \mathcal{O}_{\mathbb{P}^2}(-2)$ give $(2K_M - D)_D = 0$ in $H^2(\mathbb{P}^2, \mathbb{Z})$.

For 4) let $D \cong \mathbb{P}^2$. We can assume $D_{b5,\mathbb{P}^2} = D$. Then

$$K_{M|D} \cong K_{\mathbb{P}^2} - \mathcal{N}_{\mathbb{P}^2/M} \cong \mathcal{O}_{\mathbb{P}^2}(-1),$$

$\mathcal{N}_{D/M} \cong \mathcal{O}_{\mathbb{P}^2}(-2)$, and $\mathcal{O}_D(D_{b5,\mathbb{F}_2}) \cong \mathcal{O}_{\mathbb{P}^2}(1)$ give

$$(K_M - 3D_{b5,\mathbb{F}_2} - 2D)_D = 0 \text{ in } H^2(\mathbb{P}^2, \mathbb{Z}).$$

Let $D \cong \mathbb{F}_2$. We can assume $D_{b5,\mathbb{F}_2} = D$. Then

$$K_{M|D} \cong K_{\mathbb{F}_2} - \mathcal{N}_{\mathbb{F}_2/M} \cong \mathcal{O}_{\mathbb{F}_2}(-E - 3f),$$

and

$$\mathcal{N}_{D/M} \cong \mathcal{O}_{\mathbb{F}_2}(-E - f), \quad \mathcal{O}_D(D_{b5,\mathbb{P}^2}) \cong \mathcal{O}_{\mathbb{F}_2}(E)$$

give $(K_M - 3D - 2D_{b5,\mathbb{P}^2})_D = 0$ in $H^2(\mathbb{F}_2, \mathbb{Z})$.

Recalling that X has rational singularities, thus the above lemma applies to conclude that K_X is 2-Cartier, i.e., X is 2-Gorenstein and

$$2K_M \approx \varphi^*(2K_X) + D_{b1} + 2(D_a + 2D_{ai} + 2D_{b2} + D_{b3} + D_{b4} + 3D_{b5,\mathbb{F}_2} + 2D_{b5,\mathbb{P}^2}).$$

Thus X has only terminal singularities. □

Lemma 12.2.7. *Let* $(\mathcal{M}, \mathcal{L})$, (M, L), $\mathcal{K}$, *be as in* (12.0.1) *and let* $(X, \mathcal{D})$, $\varphi : M \to X$ *be the second reduction of* $(\mathcal{M}, \mathcal{L})$ *with* $\mathcal{D} = (\varphi_* L)^{**}$. *Then* $\mathcal{D}$ *is a 2-Cartier divisor,* $\mathcal{K} \cong K_X + \mathcal{D}$, $\mathcal{D}$ *is* φ*-ample, and* $2\mathcal{K}$ *is spanned.*

Proof. The fact that $2\mathcal{K}$ is spanned follows from the fact that $|2(K_M + L)|$ is basepoint free. This is proved in Proposition (13.2.5). The remaining assertions are proved in Lemma (7.5.8). □

12.3 The second reduction for threefolds in $\mathbb{P}^5$

We recall in this section a result of the authors of this book and Schneider [96] where the structure of the second reduction is studied in the special case of 3-folds in $\mathbb{P}^5$. We refer to [96] and the papers [138, 139] of Braun, Ottaviani, Schneider, and Schreyer for further related results for 3-folds in $\mathbb{P}^5$.

Theorem 12.3.1. ([96, (1.3), (4.3), (5,3)]) *Let* $\mathcal{M}$ *be an irreducible 3-dimensional projective manifold, and let* $\mathcal{L}$ *be a very ample line bundle on* $\mathcal{M}$ *such that* $\Gamma(\mathcal{L})$ *embeds* $\mathcal{M}$ *in* $\mathbb{P}^5$. *Let* (M, L) *be the first reduction of* $(\mathcal{M}, \mathcal{L})$ *and let* $(X, \mathcal{D})$, $\varphi : M \to X$ *be the second reduction of* $(\mathcal{M}, \mathcal{L})$ *as in* (12.0.1). *Let* $\mathcal{K} \approx K_X + \mathcal{D}$ *and* $d := \mathcal{L}^3$ *be the degree of* $(\mathcal{M}, \mathcal{L})$.

1. *If* $d \neq 7$, *then* $(\mathcal{M}, \mathcal{L}) \cong (M, L)$;

2. *If* $(d \neq 7$ *and*$)$ $d \neq 10, 13$, *then* X *is smooth,* φ *is an isomorphism outside of an algebraic subset* $\mathcal{C}$ *which is a disjoint union of smooth curves. Let* C *be an irreducible component of* $\mathcal{C}$ *and let* $D := \varphi^{-1}(C)$. *Then the restriction* φ_D *of* φ *to* D *is a* $\mathbb{P}^1$*-bundle* $\varphi_D : D \to C$ *and* $\mathcal{N}_{D/\mathcal{M}|f} \cong \mathcal{O}_{\mathbb{P}^1}(-1)$ *for any fiber* f *of* φ_D. *Furthermore* φ *is simply the blowing up along* $\mathcal{C}$.

Assume now that $d \geq 15$. *Then* $K_X + \mathcal{K}$ *is nef and big unless either:*

a) X *is a* $\mathbb{P}^1$*-bundle,* $\psi : X \to Y$, *over a smooth surface* Y, *the restriction of* $\mathcal{K}$ *to each fiber,* f, *is* $\mathcal{O}_{\mathbb{P}^1}(1)$ *and* $\mathcal{D}_f \cong \mathcal{O}_{\mathbb{P}^1}(3)$; *or*

b) $(X, \mathcal{K})$ *is a quadric fibration over a normal surface.*

Chapter 13

Varieties $(\mathcal{M}, \mathcal{L})$ with $\kappa(K_{\mathcal{M}} + (\dim \mathcal{M} - 2)\mathcal{L}) \geq 0$

In this chapter we discuss some numerical results for n-dimensional smooth polarized pairs $(\mathcal{M}, \mathcal{L})$. Most of these results hold true under the assumption that $\kappa(\mathcal{M}) \geq 0$, or, more generally, $\kappa(K_{\mathcal{M}} + (n-2)\mathcal{L}) \geq 0$.

In §13.1 we define the pluridegrees of a polarized pair and we prove some general numerical relations that they satisfy. These invariants turn out to be very useful in the classification of polarized pairs. The double point formula, (13.1.5), and its consequence, (13.1.6), as well as consequences of Tsuji's inequality, (13.1.7), (13.1.8), give useful numerical bounds expressed in terms of pluridegrees.

In §13.2 westate one of the main open problems in adjunction theory, Conjecture (13.2.1). We prove some partial results in this direction which give strong evidence for the conjecture. We also discuss some related results, Theorems (13.2.6) and (13.2.7), which give a lower bound for the dimension of the space of the global sections of $K_{\mathcal{M}}+(n-2)\mathcal{L}$ and conditions for global spannedness of $K_{\mathcal{M}}+(n-2)\mathcal{L}$ respectively.

In §13.3 we show a result of Livorni and the second author [415], Theorem (13.3.1), which describes the asymptotic behavior of the Chern classes of a surface obtained by intersecting a projective manifold of nonnegative Kodaira dimension with general hyperplane sections. We also conjecture a stronger version of the Miyaoka-Yau inequality for a smooth surface, S, with $\kappa(S) = 2$, which is an ample divisor on a 3-fold, $\mathcal{M}$, such that $\kappa(K_{\mathcal{M}} + S) \geq 0$.

The main references for the results of this chapter are Sommese [575, 584], Livorni and Sommese [415], Beltrametti, Biancofiore, and Sommese [85], and Beltrametti and Sommese [112].

13.1 The double point formula for threefolds

In this section we give some definitions and recall some general useful results, mainly the double point formula and a consequence of it, which play a key role in the proof of the results of §13.2.

Pluridegrees. Let L be a line bundle on an irreducible normal n-dimensional projective variety X. Assume that X is 1-Gorenstein, i.e., that K_X is invertible. For $j = 0, \cdots, n$ define the j-th *pluridegree of the pair*, (X, L), as

$$d_j(L) := (K_X + (n-2)L)^j \cdot L^{n-j}.$$

If L and $K_X + (n-2)L$ are nef then one has $d_j(L)^2 \geq d_{j+1}(L)d_{j-1}(L)$ for $j = 1, \ldots, n-1$ by the generalized Hodge index theorem, (2.5.1).

Lemma 13.1.1. ([85, (1.4)]) *Let (X, L) be as above. For $j = 0, \ldots, n-1$ and j even, $d_j(L) \equiv d_{j+1}(L)$ mod 2.*

Proof. Write $d_j := d_j(L)$. Note that

$$d_{j+1} = (K_X + (n-2)L)^j \cdot K_X \cdot L^{n-j-1} + (n-2)d_j.$$

If n is even, then we have

$$d_{j+1} \equiv (K_X + (n-2)L)^j \cdot K_X \cdot L^{n-j-1} \text{ mod } 2.$$

Noting that

$$(K_X + (n-2)L)^j \cdot K_X \cdot L^{n-j-1} = K_X \cdot L \cdot ((K_X + (n-2)L)^2)^{j/2} \cdot (L^2)^{(n-j-2)/2}$$

and that j is even, we use Lemma (1.1.11) to conclude that $d_{j+1} \equiv d_j$ mod 2. If n is odd, then

$$d_{j+1} \equiv (K_X + (n-2)L)^j \cdot K_X \cdot L^{n-j-1} + d_j \text{ mod } 2,$$

and since j is even it follows from Lemma (1.1.11) that $d_{j+1} \equiv d_j$ mod 2. □

Let $(\mathcal{M}, \mathcal{L})$ be a smooth n-fold polarized with an ample line bundle $\mathcal{L}$. Let (M, L) be the first reduction of $(\mathcal{M}, \mathcal{L})$ as in (7.3.3). Let $\pi : \mathcal{M} \to M$ be the first reduction morphism and let γ be the number of points blown up under π. For $j = 0, \ldots, n$ we denote the pluridegrees of $(\mathcal{M}, \mathcal{L})$ by $\widehat{d_j}$, and the pluridegrees of (M, L) by d_j. The invariants $\widehat{d_j}$, d_j, $j = 0, \ldots, n$, are related by

$$\widehat{d_j} = d_j - (-1)^n\gamma. \tag{13.1}$$

We put $\widehat{d} = \widehat{d_0}$, $d = d_0$.

Moreover if $K_M + (n-2)L$ is nef and big the numbers d_j are positive.

If the second reduction, $(X, \mathcal{D})$, $\varphi : M \to X$ with $\mathcal{D} := (\varphi_* L)^{**}$, of $(\mathcal{M}, \mathcal{L})$ exists, then letting $\mathcal{K} := K_X + (n-2)\mathcal{D}$, we define

$$d'_j := \mathcal{K}^j \cdot \mathcal{D}^{n-j}; \quad j = 0, \ldots, n, \quad d' := d'_0.$$

Note that $d_{n-1} = d'_{n-1}$ and $d_n = d'_n$. To see this, recall that $2L \approx \varphi^*(2\mathcal{D}) - \Delta$ for some effective Cartier divisor Δ which is φ-exceptional (see (12.2.7)) and compute

$$d_n = (K_M + (n-2)L)^n = (\varphi^* \mathcal{K})^n = \mathcal{K}^n = d'_n;$$

$$2d_{n-1} = 2(K_M + (n-2)L)^{n-1} \cdot L = (\varphi^* \mathcal{K})^{n-1} \cdot (\varphi^*(2\mathcal{D}) - \Delta) = 2\mathcal{K}^{n-1} \cdot \mathcal{D} = 2d'_{n-1}.$$

For more on the pluridegrees we refer to [85, §1], [96, §0] and [86, §5] by the authors and Biancofiore, Schneider, and Fania respectively.

The following example shows that the numbers d_j's can vary quite a lot (see [85, (2.4)]).

Example 13.1.2. Fix the notation as in (7.6.8). Let $X = \mathbb{P}^3$, $\mathcal{A} = \mathcal{O}_{\mathbb{P}^3}(4)$, $\mathcal{H} = \mathcal{O}_{\mathbb{P}^3}(1)$. Let $\varphi : M \to X$ be the blowing up of $\mathbb{P}^3$ along the smooth curve, Z, obtained as transversal intersection of 2 general elements of $|\mathcal{A}|$. Let $F := \varphi^* \mathcal{A} - \varphi^{-1}(Z)$ and $L := F + \varphi^* \mathcal{H}$. Then F is spanned and gives a map $q : M \to \mathbb{P}^1$ with connected fibers and with $F \cong q^* \mathcal{O}_{\mathbb{P}^1}(1)$. Let $\mathcal{K}_{\mathbb{P}^3} := K_{\mathbb{P}^3} + \mathcal{A} + \mathcal{H} = \mathcal{O}_{\mathbb{P}^3}(1)$. Then as shown in (7.6.8), $(\mathbb{P}^3, \mathcal{O}_{\mathbb{P}^3}(1))$, $\varphi : M \to \mathbb{P}^3$, is the second reduction of (M, L).

Let $\xi := \varphi^* \mathcal{O}_{\mathbb{P}^3}(1)$. Then $L := F + \xi = 5\xi - \varphi^{-1}(Z)$, $K_M + L = \xi$. We have $\xi^3 = 1$, $F^3 = \xi \cdot F^2 = 0$, $\xi^2 \cdot F = 4$, and therefore we see that:

$$d = (F + \xi)^3 = 13, \quad d_1 = (F + \xi)^2 \cdot \xi = 9, \quad d_2 = (F + \xi) \cdot \xi^2 = 5, \quad d_3 = \xi^3 = 1.$$

By taking $\mathcal{A} = \mathcal{O}_{\mathbb{P}^3}(1)$ and $\mathcal{H} = \mathcal{O}_{\mathbb{P}^3}(4)$, the same construction as above gives rise to a pair (M, L) with pluridegrees $d_3 = 1$, $d_2 = 5$, $d_1 = 24$ and $d = 112$.

We prove here some numerical bounds we will use in the sequel (see Sommese [575, §1] and Livorni and Sommese [415, §0]).

Lemma 13.1.3. *Let $\mathcal{M}$ be a smooth connected variety of dimension $n \geq 3$, polarized by an ample line bundle $\mathcal{L}$. Assume that $\kappa(\mathcal{M}) \geq 0$. Let (M, L) be the first reduction of $(\mathcal{M}, \mathcal{L})$. Then for $j = 1, \ldots, n$, $d_j \geq (n-2)d_{j-1}$. Furthermore if $\kappa(\mathcal{M}) \geq 1$, then the inequalities are strict.*

Proof. Since $\kappa(\mathcal{M}) \geq 0$ the first reduction, (M, L), of $(\mathcal{M}, \mathcal{L})$ exists. Since $\kappa(\mathcal{M}) = \kappa(M) \geq 0$, there exists some positive integer t such that tK_M is effective.

Furthermore $\kappa(K_M + (n-2)L) = n$ and hence from (7.6.9) we know that $K_M + (n-2)L$ is nef and big. Thus we have

$$\begin{aligned} d_j &= (K_M + (n-2)L)^j \cdot L^{n-j} \\ &= K_M \cdot (K_M + (n-2)L)^{j-1} \cdot L^{n-j} + (n-2)(K_M + (n-2)L)^{j-1} \cdot L^{n-j+1} \\ &= K_M \cdot (K_M + (n-2)L)^{j-1} \cdot L^{n-j} + (n-2)d_{j-1}. \end{aligned}$$

Since in our present assumption $K_M \cdot (K_M + (n-2)L)^{j-1} \cdot L^{n-j} \geq 0$ we have the first part of the statement.

If $\kappa(\mathcal{M}) \geq 1$, then the result follows from Lemma (2.5.9). □

Lemma 13.1.4. *Let $\mathcal{M}$ be a smooth 3-fold polarized by a very ample line bundle $\mathcal{L}$. Let (M, L) be the first reduction of $(\mathcal{M}, \mathcal{L})$. Let $\widehat{S}$ be a smooth element of $|\mathcal{L}|$ and let S be the corresponding smooth surface in $|L|$. Then $d_2 + 4d_1 + 6d \geq e(S)$.*

Proof. Let $J_1(\mathcal{L})$ be the first jet bundle of $\mathcal{L}$ on $\mathcal{M}$. A standard Chern classes computation, by using the defining exact sequence (1.6) of §1.6, shows that

$$c_1(J_1(\mathcal{L})) = K_{\mathcal{M}} + 4\mathcal{L}, \quad c_2(J_1(\mathcal{L})) = c_2(\mathcal{M}) + 3K_{\mathcal{M}} \cdot \mathcal{L} + 6\mathcal{L}^2.$$

Since $\mathcal{L}$ is very ample, we know from §1.6 that $J_1(\mathcal{L})$ is spanned by global sections. Therefore we conclude that (see Fulton [267, p. 216]):

$$(c_1^2(J_1(\mathcal{L})) - c_2(J_1(\mathcal{L}))) \cdot \mathcal{L} \geq 0,$$

or

$$(K_{\mathcal{M}}^2 + 5K_{\mathcal{M}} \cdot \mathcal{L} + 10\mathcal{L}^2 - c_2(\mathcal{M})) \cdot \mathcal{L} \geq 0. \tag{13.2}$$

Let $\mathcal{L}_{\widehat{S}}$ be the restriction of $\mathcal{L}$ to $\widehat{S}$. Since

$$K_{\mathcal{M}}^2 \cdot \mathcal{L} = (K_{\widehat{S}} - \mathcal{L}_{\widehat{S}})^2 = \widehat{d}_2 - 2\widehat{d}_1 + \widehat{d},$$

$$K_{\mathcal{M}} \cdot \mathcal{L}^2 = (K_{\widehat{S}} - \mathcal{L}_{\widehat{S}}) \cdot \mathcal{L}_{\widehat{S}} = \widehat{d}_1 - \widehat{d},$$

$$c_2(\mathcal{M}) \cdot \mathcal{L} = e(\widehat{S}) - \widehat{d}_1,$$

by substituting in (13.2) we find $6\widehat{d} + 4\widehat{d}_1 + \widehat{d}_2 \geq e(\widehat{S})$. Since $e(\widehat{S}) \geq e(S)$, and $6d + 4d_1 + d_2 \geq 6\widehat{d} + 4\widehat{d}_1 + \widehat{d}_2$ from (13.1), we have the desired equality. □

Theorem 13.1.5. (Double point formula) *Let $(\mathcal{M}, \mathcal{L})$ be a smooth 3-fold, polarized with a very ample line bundle $\mathcal{L}$. Let $N = h^0(\mathcal{L}) - 1$. Let $\widehat{d_j}$, $j = 0, 1, 2, 3$, be the pluridegrees of $(\mathcal{M}, \mathcal{L})$. Let S be a smooth element of $|\mathcal{L}|$. Then*

$$e(\mathcal{M}) - 48\chi(\mathcal{O}_{\mathcal{M}}) + 84\chi(\mathcal{O}_S) - 11\widehat{d_2} - 17\widehat{d_1} - \widehat{d_3} + \widehat{d}(\widehat{d} - 20) \geq 0,$$

with equality if $N \leq 6$.

Proof. We can assume that $\mathcal{M} \subset \mathbb{P}^N$ with $N \geq 6$ by using the natural inclusion of $\mathbb{P}^a \subset \mathbb{P}^6$ of a linear $\mathbb{P}^a$ when $a \leq 5$. The formula is simply a particular case of the general formula of Kleiman of [350, (I, 37), Section D, p. 313]. It should be noted that the virtual normal bundle, $\mathcal{V}$, in that formula is defined in our situation by the exact sequence

$$0 \to \mathcal{T}_{\mathcal{M}} \to p^*\mathcal{T}_{\mathbb{P}^6} \to \mathcal{V} \to 0,$$

where $p : \mathcal{M} \to \mathbb{P}^6$ is the restriction to $\mathcal{M}$ of the projection from a general $\mathbb{P}^{N-7}$ if $N > 6$ and $\mathcal{V} = \mathcal{N}_{\mathcal{M}/\mathbb{P}^6}$, the usual normal bundle, if $N = 6$. □

The following is a consequence of the double point formula above.

Proposition 13.1.6. *Let $(\mathcal{M}, \mathcal{L})$ be a smooth projective 3-fold, polarized with a very ample line bundle $\mathcal{L}$. Assume that the first reduction, (M, L), of $(\mathcal{M}, \mathcal{L})$ exists. Let $\widehat{d}_j$, d_j, $0 \leq j \leq 3$, be the pluridegrees of $(\mathcal{M}, \mathcal{L})$, (M, L) respectively. Let γ, be the number of points blown up under the first reduction map. Let S be a smooth element in $|L|$. Then*

$$44h^0(K_{\mathcal{M}} + \mathcal{L}) + 58\chi(\mathcal{O}_S) + 2h^0(K_{\mathcal{M}}) + 4 \geq 12d_2 + 17d_1 + d_3 + (20 - \widehat{d})\widehat{d} + 5\gamma.$$

Proof. See our result [112, (0.5.2)]. □

The following is the log version of the usual Yau inequality. For the definition of log cotangent bundle in the theorem below see Iitaka's book [330]. For a very helpful survey of Chern inequalities and their relation to stability of bundles, see Miyaoka's very nice survey [443].

Theorem 13.1.7. (Tsuji inequality [598, (5.2)]) *Let X be a connected projective manifold of dimension $n \geq 2$ and let D be a smooth divisor on X. Let $\mathcal{T}_X^*(\log D)$ be the log cotangent bundle. Assume that $K_X + D$ is nef. Then*

$$c_1^n(\mathcal{T}_X^*(\log D)) \leq \frac{2(n+1)}{n} c_1^{n-2}(\mathcal{T}_X^*(\log D)) c_2(\mathcal{T}_X^*(\log D)).$$

As noted in [584, §1] the above result gives the following useful bounds.

Corollary 13.1.8. *Let $(\mathcal{M}, \mathcal{L})$, (M, L), S be as in* (13.1.6). *Assume that $K_M + L$ is nef. Then we have*

$$(K_M + L)^3 + \frac{8}{3} K_S \cdot L_S \leq 32(2h^0(K_M + L) - \chi(\mathcal{O}_S)),$$

or $h^0(K_M + L) \geq d_3/64 + d_1/24 + \chi(\mathcal{O}_S)/2$.

Proof. Let $\mathcal{T}^*_M(\log S)$ be the log cotangent bundle, defined by the residue exact sequence

$$0 \to \mathcal{T}^*_M \to \mathcal{T}^*_M(\log S) \to \mathcal{O}_S \to 0. \tag{13.3}$$

Tsuji's inequality (13.1.7) yields

$$c_1^3(\mathcal{T}^*_M(\log S)) \leq \frac{8}{3} c_1(\mathcal{T}^*_M(\log S)) c_2(\mathcal{T}^*_M(\log S)). \tag{13.4}$$

From the exact sequence (13.3) we get the equality of total Chern classes

$$c(\mathcal{T}^*_M(\log S)) = c(\mathcal{T}^*_M) c(\mathcal{O}_S). \tag{13.5}$$

From the exact sequence

$$0 \to -S \to \mathcal{O}_M \to \mathcal{O}_S \to 0$$

we find $c(-S)c(\mathcal{O}_S) = c(\mathcal{O}_M) = 1$ and hence $c(\mathcal{O}_S) = 1/(1-S)$. Therefore (13.5) gives

$$c(\mathcal{T}^*_M(\log S)) = (1 + K_M + c_2(M) - c_3(M))(1 + S + S^2 + S^3) =$$

$$1 + (K_M + S) + (c_2(M) + K_M \cdot S + S^2) + (S^3 + K_M \cdot S^2 + c_2(M) \cdot S - c_3(M)).$$

This implies $c_2(\mathcal{T}^*_M(\log S)) = c_2(M) + K_M \cdot S + S^2$. Note that $K_M + S = c_1(\mathcal{T}^*_M(\log S))$. Therefore from (13.4), (13.5) we get

$$\begin{aligned} (K_M + S)^3 &\leq \frac{8}{3}(K_M + S)(c_2(M) + K_M \cdot S + S^2) \\ &= \frac{8}{3}(c_2(M) \cdot K_M + c_2(M) \cdot S + (K_M + S)^2 \cdot S). \end{aligned} \tag{13.6}$$

Since by Riemann-Roch

$$c_2(M) \cdot K_M = -24\chi(\mathcal{O}_M), \quad c_2(M) \cdot S = e(S) - (K_M + S) \cdot S^2,$$

we find from (13.6)

$$(K_M + S)^3 \leq \frac{8}{3}(-24\chi(\mathcal{O}_M) + e(S) - (K_M + S) \cdot S^2 + (K_M + S)^2 \cdot S),$$

or

$$d_3 + \frac{8}{3} d_1 \leq -64\chi(\mathcal{O}_M) + \frac{8}{3}(e(S) + d_2).$$

Since $e(S) + d_2 = 12\chi(\mathcal{O}_S)$, the above inequality yields

$$\frac{d_1}{24} + \frac{d_3}{64} \leq -\chi(\mathcal{O}_M) + \frac{\chi(\mathcal{O}_S)}{2} = \chi(K_M) + \frac{\chi(\mathcal{O}_S)}{2}. \tag{13.7}$$

Since

$$h^0(K_M + L) = \chi(K_M + S) = \chi(K_M) + \chi(K_S) = \chi(K_M) + \chi(\mathcal{O}_S),$$

the inequality (13.7) gives the result. □

13.2 The linear system $|K_M + (n-2)L|$ on the first reduction (M, L)

Let $\mathcal{M}$ be a smooth connected n-fold, $n \geq 3$, polarized with a very ample line bundle $\mathcal{L}$. Recall that by (7.6.9) the first reduction (M, L) of $(\mathcal{M}, \mathcal{L})$ as in (7.3.3) exists with $K_M + (n-2)L$ nef (respectively nef and big) if $\kappa(K_{\mathcal{M}} + (n-2)\mathcal{L}) \geq 0$ (respectively $\kappa(K_{\mathcal{M}} + (n-2)\mathcal{L}) = n$). One of the main open problems in adjunction theory is the following conjecture of the second author [584, (2.3)].

Conjecture 13.2.1. *Let $(\mathcal{M}, \mathcal{L})$ be a smooth projective n-fold polarized by a very ample line bundle $\mathcal{L}$. Assume that $\kappa(\mathcal{M}) \geq 0$. Let (M, L) be the first reduction of $(\mathcal{M}, \mathcal{L})$. Then $K_M + (n-2)L$ is spanned.*

Note that the assumption $\kappa(\mathcal{M}) \geq 0$ is needed as the Example (11.2.1) of §11.2 shows. Nevertheless we expect that even if $K_M + (n-2)L$ is merely nef, $K_M + (n-2)L$ will be spanned except for a few exceptions. The following partial results of the second author from [584] give strong evidence for the conjecture above.

Proposition 13.2.2. *Let $\mathcal{M}$ be a smooth connected n-fold polarized with an ample and spanned line bundle $\mathcal{L}$, $n \geq 3$. Assume that $0 \leq \kappa(K_{\mathcal{M}} + (n-2)\mathcal{L}) \leq 1$. Let (M, L) be the first reduction of $(\mathcal{M}, \mathcal{L})$. Let $\widehat{S}$ be a smooth surface obtained as transversal intersection of $n-2$ general members of $|\mathcal{L}|$ and let S be the corresponding smooth surface in $|L|$. Then $K_M + (n-2)L$ is nef, $h^0(K_M + (n-2)L) > 0$, and $K_M + (n-2)L$ is spanned unless $\chi(\mathcal{O}_S) \leq 1$ and $q(S) \geq 1$.*

Proof. Let φ be the morphism given by $\Gamma(m(K_M + (n-2)L))$, for $m \gg 0$. If $\kappa(K_M + (n-2)L) = 0$, then $K_M \sim -(n-2)L$. Hence in particular $-K_M$ is ample. It thus follows that $\mathrm{Pic}(M)$ is torsion free and therefore $K_M \approx -(n-2)L$. The assertions are clear in this case.

Thus we can assume that $\kappa(K_M + (n-2)L) = 1$, i.e., (M, L), $\varphi : M \to B$, is a Del Pezzo fibration over a smooth curve, B, of genus $g(B)$. We have $K_M + (n-2)L \approx \varphi^* H$ for some ample line bundle H on B. Note that $g(B) = h^1(\mathcal{O}_M) = h^1(\mathcal{O}_S)$. Since $-K_M$ is ample relative to φ, the first equality follows from Theorem (2.2.4) and the Leray spectral sequence for φ and $\mathcal{O}_M$. The second equality $h^1(\mathcal{O}_M) = h^1(\mathcal{O}_S)$ follows from the Kodaira vanishing theorem. If $h^1(\mathcal{O}_M) = 0$ we have $B \cong \mathbb{P}^1$. Then H is spanned and therefore $K_M + (n-2)L$ is also spanned.

Thus we can assume that $h^1(\mathcal{O}_M) > 0$. Let $\varphi_S : S \to B$ be the restriction of φ to S. Note that S is an elliptic surface, i.e., the general fiber of φ_S is an elliptic curve. From the canonical bundle formula we have (see, e.g., Griffiths and Harris [293, p. 572])

$$K_S \approx \sum_{i \in I} (m_i - 1) F_i + \varphi_S^* D,$$

where D is a divisor of degree $\deg D = 2g(B) - 2 + \chi(\mathcal{O}_S)$, I is a possibly empty set indexing the finite set of multiple fibers, $m_i F_i$, with multiplicities m_i, $i \in I$. On the other hand, we have by adjunction $K_S \approx \varphi_S^* H$. From [584, (0.5.1)] we know that φ_S has no multiple fibers, i.e., $m_i = 1$ for each index i. Thus we conclude that $\varphi_S^* H \approx \varphi_S^* D$. It follows that $\deg H = \deg D = 2g(B) - 2 + \chi(\mathcal{O}_S)$. Assume $\chi(\mathcal{O}_S) \geq 2$. Then $\deg H \geq 2g(B)$ and hence H is spanned (see Hartshorne [318, p. 308]). Therefore we can assume $\chi(\mathcal{O}_S) \leq 1$. The Riemann-Roch theorem on B gives

$$h^0(H) \geq \deg H + 1 - g(B) = h^1(\mathcal{O}_M) - 1 + \chi(\mathcal{O}_S).$$

If $\chi(\mathcal{O}_S) = 1$, then $h^0(K_M + (n-2)L) = h^0(H) \geq h^1(\mathcal{O}_M) \geq 1$. Assume $\chi(\mathcal{O}_S) = 0$. In this case $\deg H > 0$ implies $h^1(\mathcal{O}_S) = h^1(\mathcal{O}_M) \geq 2$ and hence we have again $h^0(K_M + (n-2)L) = h^0(H) \geq h^1(\mathcal{O}_M) - 1 \geq 1$. Note that $\chi(\mathcal{O}_S) \geq 0$ since $\kappa(S) = 1$. □

Remark 13.2.3. (The case of a quadric fibration over a surface) Let $(\mathcal{M}, \mathcal{L})$, (M, L) be as in Proposition (13.2.2). Assume that $K_M + (n-2)L$ is nef and $\kappa(K_M + (n-2)L) = 2$. This is equivalent to say that (M, L) is a quadric fibration, $\varphi : M \to Y$, over a normal surface Y. In this case $K_M + (n-2)L$ is spanned unless a few possible exceptions of "low degree." For a precise statement and the proofs we refer to Besana [118, (1.3), (1.4) and §8] (see also (14.2.3)).

Proposition 13.2.4. (Sommese) *Let $\mathcal{M}$ be a smooth connected n-fold polarized with an ample and spanned line bundle $\mathcal{L}$, $n \geq 3$. Then the first reduction, (M, L), of $(\mathcal{M}, \mathcal{L})$ exists with $K_M + (n-2)L$ nef if and only if $h^0(K_{\mathcal{M}} + (n-2)\mathcal{L}) > 0$.*

Proof. The "if" part is an immediate consequence of Theorem (7.6.9).

To show the converse assume first $n = 3$. Let S be a smooth surface in $|L|$ corresponding to a smooth general member of $|\mathcal{L}|$. By Proposition (13.2.2) we may assume $\kappa(K_M + L) \geq 2$. Hence in particular K_S is nef and big (see Remark (7.6.10)), so that S is a minimal surface of general type. Let L_S be the restriction of L to S. From Tsuji's inequality (13.1.8) we get

$$h^0(K_M + L) \geq (K_M + L)^3/64 + (K_S \cdot L_S)/24 + \chi(\mathcal{O}_S)/2.$$

Since $(K_M + L)^3 \geq 0$, $K_S \cdot L_S > 0$, $\chi(\mathcal{O}_S) \geq 1$, we have $h^0(K_M + L) \geq 1$.

If $n \geq 4$, take a smooth $A \in |L|$. Then $K_A + (\dim A - 2)L_A \approx (K_M + (n-2)L)_A$ is nef. Since $h^1(K_M + (n-3)L) = 0$, we have $h^0(K_M + (n-2)L) \geq h^0(K_A + (\dim A - 2)L_A)$. Thus, if $K_M + (n-2)L$ is nef, we conclude by induction on n that $h^0(K_M + (n-2)L) > 0$. □

Proposition 13.2.5. *Let $\mathcal{M}$ be a smooth connected n-fold polarized with an ample and spanned line bundle $\mathcal{L}$, $n \geq 3$. Assume that the first reduction, (M, L), of*

$(\mathcal{M}, \mathcal{L})$ exists with $K_M + (n-2)L$ nef and big. Then $2(K_M + (n-2)L)$ is spanned by its global sections.

Proof. If $n \geq 4$, take a smooth $A \in |L|$ and consider the exact sequence

$$0 \to K_M^{\otimes 2} \otimes L^{\otimes 2(n-2)-1} \to K_M^{\otimes 2} \otimes L^{\otimes 2(n-2)} \to (K_A^{\otimes 2} \otimes L_A^{\otimes 2(\dim A-2)}) \to 0.$$

Note that $n \geq 4$ gives $2n-5 > n-2$ and therefore $K_M + (2(n-2)-1)L$ is ample. Then $h^1(2K_M + (2(n-2)-1)L) = 0$ by Kodaira vanishing.

Thus without loss of generality we assume that $n = 3$. Let S be a smooth surface in $|L|$ corresponding to a smooth general member of $|\mathcal{L}|$. Since $K_M + L$ is nef and big, S is a minimal surface of general type (see Remark (7.6.10)). From Proposition (13.2.4) we know that $h^0(K_M + L) > 0$. Let $\sigma \in H^0(K_M + L)$ be a not identically zero section. By the generality of S, the restriction, σ_S, of σ to S is not identically zero on S. By the adjunction formula, σ_S gives rise to a nonzero element in $H^0(K_S)$. Therefore $p_g(S) > 0$. By results on the bicanonical map of a surface of general type (see Catanese and Ciliberto [151] and Francia [237]), we conclude that $2K_S$ is spanned. Since $h^1(2K_M + L) = 0$ by the Kawamata-Viehweg vanishing theorem, we conclude that $2(K_M + L)$ is spanned from the exact sequence

$$0 \to 2K_M + L \to 2(K_M + L) \to 2K_S \to 0.$$

□

In the cases when the hyperplane threefold section, V, of $(\mathcal{M}, \mathcal{L})$ is of log-general type or when $\kappa(V) \geq 0$, we have the following lower bounds for $h^0(K_{\mathcal{M}} + (n-2)\mathcal{L}) (= h^0(K_M + (n-2)L))$. This improves a result of Sommese [584].

Theorem 13.2.6. *Let $(\mathcal{M}, \mathcal{L})$ be a smooth connected n-fold, $n \geq 3$, polarized with a very ample line bundle $\mathcal{L}$. Let V be the smooth 3-fold obtained as transversal intersection of $n-3$ general members of $|\mathcal{L}|$. Let $\mathcal{L}_V$ be the restriction of $\mathcal{L}$ to V. Let $\widehat{S}$ be the smooth surface obtained as transversal intersection of $n-2$ general members of $|\mathcal{L}|$. We have:*

1. *If $(V, \mathcal{L}_V)$ is of log-general type, then $h^0(K_{\mathcal{M}} + (n-2)\mathcal{L}) \geq 2$.*
2. *If $\kappa(V) \geq 0$, e.g., if $\kappa(K_{\mathcal{M}} + (n-3)\mathcal{L}) \geq 0$, then $h^0(K_{\mathcal{M}} + (n-2)\mathcal{L}) \geq 5$ with equality only if $n = 3$ and $(\mathcal{M}, \mathcal{L})$ is a smooth quintic hypersurface in $\mathbb{P}^4$. Furthermore either $p_g(\widehat{S}) \geq 6$ or $\widehat{S}$ is a degree 5 surface in $\mathbb{P}^3$ with $p_g(\widehat{S}) = 4$.*

Proof. We refer to our paper [112] for a complete proof. To give the flavor of it, let us recall the proof from [584] of the weaker bounds $h^0(K_{\mathcal{M}} + \mathcal{L}) \geq 3$ and $p_g(\widehat{S}) \geq 3$, when $\kappa(\mathcal{M}) \geq 0$ and $n = 3$.

First note that we can assume

$$h^0(\mathcal{L}) \geq 6; \; \widehat{d} \geq 8. \tag{13.8}$$

Indeed, let $\mathcal{M} \hookrightarrow \mathbb{P}^N$, $N \geq 4$, be the embedding given by $\Gamma(\mathcal{L})$. Let $N = 4$. Then the assumption $\kappa(\mathcal{M}) \geq 0$ implies $\widehat{d} \geq 5$, so that $h^0(K_{\mathcal{M}} + \mathcal{L}) \geq 5$. Therefore we can assume $N \geq 5$, or $h^0(\mathcal{L}) \geq 6$. Hence from (8.1.1) we have $\widehat{d} := \mathcal{L}^3 \geq 3(N-3)+2 \geq 8$.

Let (M, L) be the first reduction of $(\mathcal{M}, \mathcal{L})$ and let $S \in |L|$ be a smooth surface corresponding to a smooth element $\widehat{S} \in |\mathcal{L}|$. Recall that, since $\kappa(\mathcal{M}) \geq 0$, $(\mathcal{M}, \mathcal{L})$ is of log-general type and hence S is a minimal surface of general type (cf., (7.6.10)).

Assume that $\chi(\mathcal{O}_S) = \chi(\mathcal{O}_{\widehat{S}}) \leq 2$. Letting $d = L^3$, we have from the double point formula for surfaces as in Lemma (8.2.1)

$$\widehat{d}(\widehat{d}-5) - 10(g(\mathcal{L})-1) + 12\chi(\mathcal{O}_{\widehat{S}}) \geq 2K_{\widehat{S}}^2. \tag{13.9}$$

Passing to the reduction (M, L) we get

$$d(d-5) - 10(g(L)-1) + 12\chi(\mathcal{O}_S) \geq 2K_S^2. \tag{13.10}$$

Indeed from (13.1) we know that $d = \widehat{d} + x$ and $K_S^2 = K_{\widehat{S}}^2 + x$ for some integer $x \geq 0$. Therefore, since $g(\mathcal{L}) = g(L)$, (13.10) follows from (13.9) as soon as we show that $\widehat{d}(\widehat{d}-5) - 2K_{\widehat{S}}^2 \leq d(d-5) - 2K_S^2$. This is equivalent to $\widehat{d} \geq (7-x)/2$. So we are done by (13.8). From (13.1.3) we know that $K_S^2 \geq K_S \cdot L_S \geq d$ with $K_S^2 = d$ if and only if $tK_M \approx \mathcal{O}_M$ for some $t > 0$. In this case $L - K_M$ is ample, so that $\chi(L) = h^0(L)$. Since $K_M \sim \mathcal{O}_M$ we also have $\chi(L) = \chi(K_M + L) = h^0(K_M + L)$. Thus $h^0(K_M + L)(= h^0(K_{\mathcal{M}} + \mathcal{L})) = h^0(L) \geq h^0(\mathcal{L}) \geq 6$. Since $d_1 \neq d+1$ by the parity condition as in Lemma (13.1.1) we can therefore assume that

$$K_S^2 \geq K_S \cdot L_S \geq d + 2. \tag{13.11}$$

Thus, recalling the genus formula $2g(L) - 2 = d + d_1$, by (13.10) we get $d(d-17) + 12\chi(\mathcal{O}_S) \geq 14$. If $\chi(\mathcal{O}_S) \leq 2$, then we get $d(d-17)+10 \geq 0$, or $d \geq 17$. By (13.11) we find $K_S^2 \geq 19$, contradicting the Miyaoka inequality $K_S^2 \leq 9\chi(\mathcal{O}_S) \leq 18$.

Thus we can assume $\chi(\mathcal{O}_S) \geq 3$. Since $d_3 := (K_M + L)^3 \geq d_1$, by combining (13.8) with (13.11) we have $d_3 \geq 10$. Therefore Tsuji's inequality (13.1.7) gives $h^0(K_M+L) \geq 2$. Then we can assume $h^0(K_M+L) = 2$, $\chi(\mathcal{O}_S) \geq 3$. By using again Tsuji's inequality, combined with $d_3 \geq d_1$, we find $2 \geq d_1/64 + d_1/24 + 3/2$, which gives $d_1 \leq 9$. Therefore (13.11) leads to $\widehat{d} \leq d \leq 7$. This contradicts (13.8). So we conclude that $h^0(K_M + L) \geq 3$.

To show that $p_g(\widehat{S})(= p_g(S)) \geq 3$, let $\mathcal{L}_{\widehat{S}}$ be the restriction of $\mathcal{L}$ to $\widehat{S} \in |\mathcal{L}|$ and note that $p_g(S) \geq h^0(K_{\mathcal{M}} + \mathcal{L}) - h^0(K_{\mathcal{M}}) \geq 3 - h^0(K_{\mathcal{M}})$. If $h^0(K_{\mathcal{M}}) > 0$ then for a general $\widehat{S}$

$$p_g(S) = h^0(K_{\mathcal{M}|\widehat{S}} + \mathcal{L}_{\widehat{S}}) \geq h^0(K_{\mathcal{M}|\widehat{S}}) + h^0(\mathcal{L}_{\widehat{S}}) - 1 \geq h^0(\mathcal{L}_{\widehat{S}}).$$

By (13.8) we may assume that $h^0(\mathcal{L}_{\widehat{S}}) \geq 5$. □

In the direction of the Conjecture (13.2.1), there is also the following result from our paper with Schneider [95]. It is proved by using the Ein-Lazarsfeld criterion [208].

Theorem 13.2.7. ([95, Theorem 15]) *Let $\mathcal{M}$ be a smooth n-fold, $n \geq 3$, with no rational curves. Let $\mathcal{L}$ be a very ample line bundle on $\mathcal{M}$ such that $\mathcal{L}^n \geq 850$. Then $K_{\mathcal{M}} + (n-2)\mathcal{L}$ is spanned by global sections.*

As usual, by a rational curve we mean an irreducible reduced curve whose desingularization is of genus zero. Note that if $\mathcal{M}$ is hyperbolic in the sense of Kobayashi, or if $\mathcal{M}$ has a nef cotangent bundle, then the condition on no rational curves is satisfied. Note also that from the results of [95] one concludes that given a very ample line bundle, $\mathcal{L}$, on a smooth projective n-fold $\mathcal{M}$ with $n \geq 3$ and $\mathcal{L}^n \geq 850$, it follows that $\Gamma(K_{\mathcal{M}} + (n-2)\mathcal{L})$ is spanned by global sections on $\mathcal{M} - B$, where B is the union of all elliptic and rational curves on $\mathcal{M}$.

Let us conclude this section by proposing the following conjecture.

Conjecture 13.2.8. *Let $\mathcal{M}$ be a smooth n-fold, $n \geq 3$. Let $\mathcal{L}$ be a very ample line bundle on $\mathcal{M}$. Assume that $K_{\mathcal{M}} + t\mathcal{L}$ is nef for some integer $t \geq n/2 + 1$. Then $K_{\mathcal{M}} + t\mathcal{L}$ is spanned by global sections.*

To give some more evidence for the conjecture above note that from §6.6 the result is true if either $(\mathcal{M}, \mathcal{L})$ is a scroll or $(\mathcal{M}, \mathcal{L})$ is a quadric fibration with $t \geq 2n/3 + 1$.

13.3 Some Chern inequalities for ample divisors

The following theorem of Livorni and Sommese [415, (1.2)] describes the asymptotic behavior of the Chern classes of a surface gotten by intersecting a projective manifold of nonnegative Kodaira dimension with general hyperplane sections.

Theorem 13.3.1. *Let $\mathcal{M}$ be a smooth connected n-dimensional variety polarized by a very ample line bundle, $\mathcal{L}$. Assume that $\kappa(\mathcal{M}) \geq 0$. Let (M, L), $\pi : \mathcal{M} \to M$ be the first reduction of $(\mathcal{M}, \mathcal{L})$. Let $\widehat{S}$ be the smooth surface obtained by transversal intersection of $n-2$ general members of $|\mathcal{L}|$. Let $S := \pi(\widehat{S})$ be the corresponding smooth surface in M. Then*

$$\frac{c_1^2(S)}{c_2(S)} \geq \frac{(n-2)^2}{n^2+2}.$$

In particular $\lim_{n\to\infty} \frac{c_1^2(S)}{c_2(S)} \geq 1.$

Proof. Since $\kappa(\mathcal{M}) \geq 0$ we have $\kappa(K_{\mathcal{M}} + (n-2)\mathcal{L}) = n$ and hence by combining Proposition (7.6.9) and Remark (7.6.10) we know that the first reduction, (M, L), of $(\mathcal{M}, \mathcal{L})$ exists, the surface $\widehat{S}$ is of general type and the restriction $\pi_{\widehat{S}} : \widehat{S} \to S$ maps $\widehat{S}$ onto its minimal model. Let L_S be the restriction of L to S and let $d = L_S^2$, $d_1 = K_S \cdot L_S$, $d_2 = K_S^2 = c_1^2(S)$ be the pluridegrees of (S, L_S). Note that Lemma (13.1.3) gives $d_2 \geq (n-2)d_1$, $d_1 \geq (n-2)d$. Then, by using Lemma (13.1.4), we have

$$d_2 + 4d_2/(n-2) + 6d_2/(n-2)^2 \geq d_2 + 4d_1 + 6d \geq c_2(S).$$

This leads to $d_2(n^2+2)/(n-2)^2 \geq c_2(S)$. □

Conjecture 13.3.2. (Sommese) *Let $\mathcal{M}$ be a smooth connected 3-fold, $\mathcal{L}$ an ample line bundle on $\mathcal{M}$. Assume that there exists a smooth surface $S \in |\mathcal{L}|$ such that $\kappa(S) = 2$. Further assume that $\kappa(K_{\mathcal{M}} + \mathcal{L}) \geq 0$. Then $c_1^2(S) \leq 2c_2(S)$.*

Note that in the conjecture above we may assume that S is minimal. Indeed since $\kappa(K_{\mathcal{M}} + \mathcal{L}) \geq 2$ we know from Proposition (7.6.9) and Remark (7.6.10) that the first reduction, (M, L), $\pi : \mathcal{M} \to M$, exists and the restriction, $\pi_S : S \to \pi(S) := S_0$, of π to S maps S onto its minimal model. Since $c_1^2(S) \leq c_1^2(S_0)$ and $c_2(S_0) \leq c_2(S)$, the conjectured inequality for S_0 yields the inequality for S.

Note that $c_1^2(S) \leq 3c_2(S)$ by the well-known Miyaoka-Yau inequality (see Miyaoka [440]). Note that the equality cannot happen. Otherwise S is a ball quotient and hence standard arguments show that S would be a $K(\Pi, 1)$, contradicting Corollary (5.1.6).

Under the assumption $\kappa(S) = 2$, the condition $\kappa(K_{\mathcal{M}} + \mathcal{L}) \geq 0$ in the conjecture is equivalent to saying that $(\mathcal{M}, \mathcal{L})$ is not a scroll over a surface, and that in fact $\kappa(K_{\mathcal{M}} + \mathcal{L}) \geq 2$. In the scroll case it is easy to give examples where the conclusion of the conjecture is false.

Chapter 14

Special varieties

For definitions of special varieties arising in adjunction theory we refer to (3.3). In this chapter we discuss a number of results, mainly about scrolls and quadric fibrations. Most of these results are consequences of results of Chapter 6. We also keep the same notation as in that chapter.

In §14.1 we prove some structure results for scrolls. In Theorem (14.1.1) we show that the structural morphism, $\varphi : X \to Y$, of an n-dimensional scroll (X, L) onto a m-dimensional variety Y is a fiber type contraction of an extremal ray if $n \geq 2m-1$. In (14.1.10) we make the conjecture that φ is always a $\mathbb{P}^{n-m}$-bundle if $n \geq 2m-1$. A consequence of (14.1.1), Proposition (14.1.3), proves the conjecture if $m \leq 3$ and L is very ample.

In §14.2 we prove the analogous result of Theorem (14.1.1) for quadric fibrations and we discuss the relation between quadric fibrations in the adjunction theoretic sense and in the classical sense. In (14.2.10) we propose the general (and very optimistic) conjecture that the nefvalue morphism, $\phi : X \to W$, of a smooth polarized pair (X, L) is flat, assuming that ϕ is not birational and that the nefvalue, τ, of (X, L) satisfies the condition $\tau \geq \dim W$. We note that the scroll conjecture, (14.1.10), is a special case of the above conjecture. Furthermore, the structure Theorem (14.2.3) shows that the Conjecture (14.2.10) is true for quadric fibrations. The main reference for the results of §14.1 and §14.2 is our paper with Wiśniewski [114]. Further references are our papers [105, 104] and the paper [118] of Besana.

In §14.3 we give a guide to the main papers on classification of projective varieties with small invariants, such as sectional genus, Δ-genus, degree, codimension.

In §14.4 we prove some results about manifolds with positive defect. We prove a structure result of ours and Fania reducing the classification of such manifolds to the study of a very special class of high index Fano manifolds. We are led to the conjecture that there are no such manifolds beyond the few known examples.

In §14.5 we give a survey of some of the recent work on Hilbert functions of hyperplane sections of curves.

14.1 Structure results for scrolls

Let X be a smooth connected variety of dimension n and let L be an ample line bundle on X. As usual we denote by $(\mathcal{F}, \mathcal{P}, p, q, \nu)$ a family of rational curves where $\mathcal{P}$ is the parameter space, $p : \mathcal{F} \to \mathcal{P}$ and $q : \mathcal{F} \to X$ are induced by the product projections and $\nu = \nu(\mathcal{F})$ is the normal degree of the family. In this section we will assume that (X, L) is a scroll, $\varphi : X \to Y$, onto a normal variety Y. We have the following structure theorem for $\varphi : X \to Y$.

Theorem 14.1.1. ([114, (3.2.1)]) *Let X be a smooth connected variety of dimension n and let L be an ample line bundle on X. Assume that (X, L) is a scroll, $\varphi : X \to Y$, onto a normal variety Y of dimension m. Let $\mathbb{P}^d$, $d = n - m$, be a general fiber of φ and let ℓ be a line in $\mathbb{P}^d$. Assume $n \geq 2m - 1$. Then ℓ is an extremal rational curve and $\varphi = \mathrm{cont}_R$ is the fiber type contraction of $R = \mathbb{R}_+[\ell]$. There are no effective divisors which are components of fibers of φ unless $m = 1$.*

Proof. We have $K_X + (d+1)L \approx \varphi^* H$ for some ample line bundle H on Y. Then $K_X + (d+1)L$ is nef but not ample and therefore $d+1 = \tau$, the nefvalue of (X, L). Take a line ℓ in a smooth general fiber of φ. Hence $L \cdot \ell = 1$ so that we can construct a nonbreaking family $(\mathcal{F}, \mathcal{P}, p, q, \nu)$ of rational curves which are deformations of ℓ and which fill up X. Now compute $\nu := -K_X \cdot \ell - 2 = -K_{\mathbb{P}^d} \cdot \ell - 2 = d - 1$. Since $(K_X + \tau L) \cdot \ell = 0$ we know from Lemma (4.2.14) that there exists an extremal ray, R, subordinate to φ, i.e., such that $(K_X + \tau L) \cdot R = 0$. Let μ be an extremal rational curve generating R.

If $\varphi = \mathrm{cont}_R$, then since $\dim \varphi(\ell) = 0$, it follows that $\ell \in R$. From $(K_X + \tau L) \cdot \ell = 0$, $L \cdot \ell = 1$ and $\tau = n - m + 1$ we see that $1 \leq -K_X \cdot \ell \leq n + 1$. Hence ℓ would be an extremal rational curve and the first part of the statement would be proved.

Thus we can assume that $\varphi \neq \mathrm{cont}_R$. Let $\rho : X \to Z$ be the contraction associated to the ray R. Then φ factors through ρ, $\varphi = \alpha \circ \rho$.

We can assume that $\rho(\ell)$ is not a point since otherwise $\ell \in R$, which would imply that all the extremal rays subordinate to φ are of the form $R = \mathbb{R}_+[\ell]$, i.e., that ρ was the unique contraction subordinate to φ. Since $\varphi \neq \mathrm{cont}_{\mathbb{R}_+[\ell]}$, this is not possible. To see this note that if all curves C such that $\varphi(C)$ is a point satisfy $C \in \mathbb{R}_+[\ell]$, then clearly $\varphi = \mathrm{cont}_{\mathbb{R}_+[\ell]}$. If there exists a curve, C, contracted by φ and with $C \notin \mathbb{R}_+[\ell]$, by Lemma (4.2.14) we can write C in $NE(X)$ as finite sum $C = \sum_i \lambda_i \ell_i$ where $\lambda_i \in \mathbb{R}_+$ and $\mathbb{R}_+[\ell_i]$ are extremal rays such that $(K_X + \tau L) \cdot \ell_i = 0$. This contradicts the fact that ρ is the unique contraction subordinate to φ.

Thus we conclude that ρ does not contract curves from the family $\mathcal{F}$ to points. By combining Theorem (6.3.2) and Lemma (6.4.6) we conclude that any fiber of ρ is of dimension $\leq n - \nu - 1$. Moreover if there exists a fiber, F, of ρ such that $\dim F = n - \nu - 1$, then $NE(X) = NE(F) + \mathbb{R}_+[C]$ for a curve C in the family.

In the latter case, since all effective curves in F have images in the same extremal ray R in $\overline{NE}(X)$, we see that X is a Fano manifold with Picard number two. By looking at the decomposition $\varphi = \alpha \circ \rho$, it thus follows that Y is a point and X is a projective space. This contradicts the fact that the Picard number is two.

Therefore we may assume that every fiber, F, of the contraction ρ is of dimension $\dim F \le n - \nu - 2 = n - d - 1 = m - 1$. Since $n \ge 2m - 1$, or $d \ge m - 1$, we thus conclude that $\dim F \le d = \tau - 1$.

From $(K_X + \tau L) \cdot \mu = 0$, $R = \mathbb{R}_+[\mu]$, we find length$(R) \ge \tau$. Therefore by Theorem (6.3.6) we conclude that all positive dimensional fibers of ρ have dimension $\ge \tau - 1$ and that ρ is of fiber type. Then by the above all fibers of ρ have dimension $\tau - 1 = n - m$, the dimension of the general fiber of φ. It thus follows that the general fiber of α is zero dimensional. On the other hand, since ρ does not contract curves from the family to points, we have that the fibers of α are at least d-dimensional. So we get a contradiction.

To see that, if $m > 1$, φ has no divisorial fibers note that since $\varphi = \mathrm{cont}_R$, $R = \mathbb{R}_+[\ell]$ (and $L \cdot \ell = 1$) we have the exact sequence of contraction theorem, (4.3.1). Any divisorial fiber F of φ satisfies $F \cdot \ell = 0$ and hence $F \in \varphi^*\mathrm{Pic}(Y)$, which is clearly not possible. □

Remark-Example 14.1.2. (Existence of divisorial fibers for scrolls) Note that the result above is sharp. This is shown by the following example, from [105, (4.2)] of a $(2n-2)$-dimensional scroll over an n-fold with a divisorial fiber.

Let $\pi : Y \to W$ express a smooth projective n-fold Y as the blowing up of a smooth projective n-fold W at a point $w \in W$. Let H be a very ample line bundle on Y such that $H_{\pi^{-1}(w)} \cong \mathcal{O}_{\pi^{-1}(w)}(1)$ and $(W, (\pi_*H)^{**})$ is the reduction of (Y, H) (e.g., $H = \pi^*(2\mathcal{L}) - [\pi^{-1}(w)]$ for a very ample line bundle $\mathcal{L}$ on W). Let X be the $(2n-2)$-fold defined by $X := \mathbb{P}(\oplus^{n-1} H)$ with $\varphi : X \to Y$ the projection. Let L be the tautological bundle of X (which is ample and in fact very ample). One has

$$K_X + (n-1)L \approx \varphi^*(K_Y + (n-1)H) \approx \varphi^*\pi^*(K_W + (n-1)H')$$

where $H' := (\pi_*H)^{**}$. Thus (X, L) is a scroll under $\psi = \pi \circ \varphi$ and we have a divisorial fiber $\Delta = \psi^{-1}(w)$ which is a $\mathbb{P}^{n-2}$-bundle over a $\mathbb{P}^{n-1}$.

A consequence of Theorem (14.1.1) is the following result which states that, in a number of cases, a scroll (X, L) is a $\mathbb{P}^d$-bundle.

Proposition 14.1.3. ([114, (3.2.3)]) *Let L be a very ample line bundle on an n-dimensional connected projective manifold, X. Assume that (X, L) is a scroll, $\varphi : X \to Y$, over a normal projective variety Y of dimension $m \le 3$. If $d := n - m \ge m - 1$, then φ is a linear $\mathbb{P}^d$-bundle.*

Proof. By (3.2.1) it suffices to prove that all fibers are equal dimensional. If $m = 1$ this is obvious. If $m = 2$ and not all fibers are equal dimensional, then there is an irreducible divisor, $D \subset X$, which is a component of a fiber of φ. This contradicts Theorem (14.1.1).

So we can assume $m = 3$. Since $n \geq 2m - 1$, Theorem (14.1.1) applies to say that φ is the contraction of a numerically effective extremal ray $R = \mathbb{R}_+[\ell]$, ℓ a line in a general fiber $\mathbb{P}^d$ and φ has no divisorial fibers.

Thus it is enough to show that there are no fibers F of dimension $n - 2$.

Assume otherwise that there is a fiber F of dimension $n - 2$. By slicing with general elements of $|L|$ we reduce to the $n = 5$ case where $\dim F = 3$.

Let V be the smooth 3-fold obtained as transversal intersection of 2 general members $D_1, D_2 \in |L|$. Consider the restriction $\psi : V \to Y$ of φ to V. Since $K_X + 3L \approx \varphi^*\mathcal{L}$ for some ample line bundle $\mathcal{L}$ on Y, we have $K_V + L_V \approx \psi^*\mathcal{L}$ and, since Y is normal and the fibers of ψ are connected, ψ is the morphism associated to $|N(K_V + L_V)|$ for $N \gg 0$. This is the second reduction morphism studied in Chapter 12 (see, e.g., Theorem (12.2.1)). Let C be the curve obtained as transversal intersection of the fiber F of φ with V, $C = F \cap V$. Then C is a 1-dimensional fiber of ψ. Thus by the results of Chapter 12 we know that C is the fiber of a $\mathbb{P}^1$-bundle S such that $\mathcal{N}_{S/V|C} \cong \mathcal{O}_C(-1)$ and $L_C \cong \mathcal{O}_C(1)$. Therefore $C \cong \mathbb{P}^1$ and $\mathcal{N}_{C/V} \cong \mathcal{O}_{\mathbb{P}^1} \oplus \mathcal{O}_{\mathbb{P}^1}(-1)$.

Let $G := F_{\text{red}}$ denote F with its reduced structure. We claim that $(G, L_G) \cong (\mathbb{P}^3, \mathcal{O}_{\mathbb{P}^3}(1))$. To see this note that $h^0(L_G) \geq 4$ since L is ample and spanned, and that $L_G^3 = L^2 \cdot G = 1$. Thus by (3.1.4) we know that $(G, L_G) \cong (\mathbb{P}^3, \mathcal{O}_{\mathbb{P}^3}(1))$.

Since the curve C is transversal intersection of F and V we have that $\mathcal{N}_{G/X|C} \cong \mathcal{O}_{\mathbb{P}^1} \oplus \mathcal{O}_{\mathbb{P}^1}(-1)$. Since $L \cdot \ell = 1$ for a line ℓ on $\mathbb{P}^3$ and L is very ample any such ℓ is a transverse intersection of F and a V as above. Therefore the normal bundle $\mathcal{N}_{G/X}$ is a uniform rank 2 vector bundle on $G \cong \mathbb{P}^3$ and hence $\mathcal{N}_{G/X} \cong \mathcal{O}_{\mathbb{P}^3} \oplus \mathcal{O}_{\mathbb{P}^3}(-1)$ (see, e.g., [493, I, §3]). Then $h^0(\mathcal{N}_{G/X}) = 1$, $h^1(\mathcal{N}_{G/X}) = 0$. It thus follows that there exists a 1-dimensional family of deformations of G. Furthermore since G is a component of a fiber, (4.1.13) implies that deformations of G are contained in nearby fibers Δ of φ. By the semicontinuity of the dimension of fibers we have $3 = \dim G \geq \dim \Delta$. Therefore $\dim \Delta = 3$ so that one has in X a 1-dimensional family of 3-dimensional fibers of φ. This family fills up a divisor, D, such that $D \cdot \ell = 0$ and hence $D \in \varphi^*\mathrm{Pic}(Y)$ by the exact sequence of the contraction theorem, (4.3.1). This is not possible since D maps down to a curve. □

The following shows that the condition $n \geq 2m - 1$ in Proposition (14.1.3) above is sharp.

Remark-Example 14.1.4. (The case $n = 2m - 2$, cf., [114, (3.2.4)]) Let (X, L) be a $(2n - 2)$-dimensional scroll, $\varphi : X \to Y$, over an n-fold Y with L very ample. Then the map φ does not have to be a projective bundle even if it is a Mori contraction (so there are no divisorial fibers) and the statement of Proposition (14.1.3) is false in this case (compare with (14.1.2)).

However, for $n = 3$, it turns out that the 2-dimensional fibers of φ are isomorphic to $\mathbb{P}^2$ with normal bundle $\mathcal{T}^*_{\mathbb{P}^2}(1)$. To see this, let F be a 2-dimensional fiber of

φ. Then by looking over the proof of Proposition (14.1.3) we see that $F \cong \mathbb{P}^2$ and $\mathcal{N}_{F/X}$ is a uniform rank 2 vector bundle over $\mathbb{P}^2$ such that $\mathcal{N}_{F/X|\mathbb{P}^1} \cong \mathcal{O}_{\mathbb{P}^1} \oplus \mathcal{O}_{\mathbb{P}^1}(-1)$. Then either $\mathcal{N}_{F/X} \cong \mathcal{O}_{\mathbb{P}^2} \oplus \mathcal{O}_{\mathbb{P}^2}(-1)$ or $\mathcal{N}_{F/X} \cong \mathcal{T}_{\mathbb{P}^2}(a)$ for some integer a (see, e.g., [493], p. 59). The first case leads to the same contradiction as in the proof of Proposition (14.1.3), while in the second case we see that $a = -2$ since $(\det \mathcal{T}_{\mathbb{P}^2}(a))_{\mathbb{P}^1} (\cong \mathcal{O}_{\mathbb{P}^1}(2a+3)) \cong \mathcal{O}_{\mathbb{P}^1}(-1)$. Note that $\mathcal{T}_{\mathbb{P}^2}(-2) \cong \mathcal{T}^*_{\mathbb{P}^2}(1)$.

Let us give explicit examples from [114, (3.2.4)] of two types of special fibers which can occur for $n \geq 3$. For $n = 3$ both have codimension 2.

Example 14.1.5. (Fibers of dimension $n-1$ and codimension $n-1$) Let Y be a smooth projective n-fold and $W := \mathbb{P}^{n-1} \times Y$. Let $q : W \to \mathbb{P}^{n-1}$, $p : W \to Y$ be the projections. Take $\mathcal{L} := q^*\mathcal{O}_{\mathbb{P}^{n-1}}(1) + p^*\mathcal{H}$ for a very ample line bundle $\mathcal{H}$ on Y. Hence $\mathcal{L}$ is very ample and let X be a smooth divisor in W corresponding to a general member of $|\mathcal{L}|$. Let π be the restriction of p to X. An easy check shows that $(X, \mathcal{O}_X(1))$ is a scroll, $\pi : X \to Y$, over Y with $\mathcal{H}^n$ $(n-1)$-dimensional fibers F such that $\mathcal{N}_{F/X} \cong \mathcal{T}^*_{\mathbb{P}^{n-1}}(1)$. Such $(n-1)$-dimensional fibers correspond to the points of Y which are zeroes of a general section of $p_*\mathcal{L} \cong \mathcal{H} \oplus \cdots \oplus \mathcal{H}$, n times.

Example 14.1.6. (Fibers of dimension $2n-4$ and codimension 2) Let $Y = \mathbb{P}(V^*)$ where V is an $(n+1)$-dimensional complex vector space. Let us choose a point y_0 in Y which represents a 1-dimensional linear subspace, V_0, of V. Now let X be the incidence variety of projective 2-dimensional planes in Y meeting y_0 :

$$X = \{(y, \Pi) \text{ such that the projective plane } \Pi \text{ in } Y \text{ contains the points } y \text{ and } y_0\}.$$

We have a projection $q : X \to \mathrm{Grass}(2, V/V_0)$, the Grassmannian of two dimensional vector subspaces of the quotient V/V_0. The map, q, makes X into a $\mathbb{P}^2$-bundle over $\mathrm{Grass}(2, V/V_0)$ so that X is smooth. On the other hand we have a projection $p : X \to Y$ which makes $(X, \mathcal{O}_X(1))$ a scroll over Y. The special fiber over the point y_0 is then isomorphic to $\mathrm{Grass}(2, V/V_0)$.

The following is an easy consequence of Proposition (14.1.3) (compare with [583, (3.3), (3.3.2)]).

Corollary 14.1.7. ([114, (3.2.5)]) *Let (X, L) be a smooth n-dimensional scroll, $\varphi : X \to Y$, over a normal projective variety Y of dimension $m \geq 3$ and let L be a very ample line bundle on X. Let $\mathcal{Z} = \{y \in Y, \dim \varphi^{-1}(y) > n-m\}$. If the general fiber of φ has dimension bigger or equal to 2, then* $\mathrm{cod}_Y \mathcal{Z} \geq 4$.

Proof. By slicing with general hyperplane sections on Y we can assume $m = 3$. Therefore the assumption $n - m \geq 2$ is equivalent to $n \geq 2m - 1$. Hence $\mathcal{Z} = \emptyset$ by Proposition (14.1.3) and we are done. □

The following example shows that the condition $\mathrm{cod}_Y \mathcal{Z} \geq 4$ in the above corollary cannot be improved.

Example 14.1.8. Let $\mathcal{P} = \mathbb{P}(\mathcal{O}_{\mathbb{P}^4}(1) \oplus \mathcal{O}_{\mathbb{P}^4}(1) \oplus \mathcal{O}_{\mathbb{P}^4}(1) \oplus \mathcal{O}_{\mathbb{P}^4}(1))$. Let s be the section of the tautological bundle ξ on $\mathcal{P}$ that corresponds to the section $z_0 \oplus z_1 \oplus z_2 \oplus z_3$, where z_0, z_1, z_2, z_3 are homogeneous coordinates considered as sections of $\mathcal{O}_{\mathbb{P}^4}(1)$. Let $X := s^{-1}(0) \subset \mathcal{P}$. Let ξ_X be the restriction of ξ to X. Then $\dim X = 6$, (X, ξ_X) is a scroll over $\mathbb{P}^4$ with precisely one 2-dimensional fiber, the fiber over $(0, 0, 0, 1) \in \mathbb{P}^4$.

Remark 14.1.9. Notation as in (14.1.7). Note that the corollary above improves [583, (3.3)] under the assumption that $m \geq 3$. Note also that the result is false without the assumption on the dimension of the general fiber of φ. This is shown in [583, (3.3.2)].

The following conjecture of ours was the motivation for Proposition (14.1.3).

Conjecture 14.1.10. *Let (X, L) be a smooth n-dimensional scroll, $\varphi : X \to Y$, with L an ample line bundle, over a normal projective variety Y of dimension m. If $n \geq 2m - 1$ then φ is a linear $\mathbb{P}^d$-bundle, $d = n - m$.*

To give some more evidence for the conjecture above, note that if $n \geq 2m + 1$ and L is very ample the result follows from a result of Ein [203, I, (1.7)].

The conjecture is also true in the special case when L is merely ample and for any fiber, F, of φ, $\dim F \leq n - m + 1$, i.e., when the fibers of wrong dimension are of dimension $n - m + 1$ at most. This follows from the results of [28, §4].

In [66] the above conjecture is proved in some special cases.

Assume first that X is the projectivization of a coherent sheaf, $\mathcal{E}$, of rank $r \geq 2$. Following [66], $\mathcal{E}$ is called a *Bănică sheaf* in this case. Then if $r \geq m$, i.e., $n \geq 2m - 1$, φ is a $\mathbb{P}^{n-m}$-bundle (see [66, (2.6)]). Note that this is not the general case, i.e., there are scrolls, (X, L), which are not projectivizations of sheaves (see [66, (3.2)]). However, recall that if all fibers of φ are $(n - m)$-dimensional, then $X \cong \mathbb{P}(p_*L)$ and φ is a linear $\mathbb{P}^{n-m}$-bundle by Proposition (3.2.1).

A second special case is the following. Let L be merely ample and let F be a fiber of φ with $k := \dim F > n - m$. Assume that $(F, L_F) \cong (\mathbb{P}^k, \mathcal{O}_{\mathbb{P}^k}(1))$. Under this assumption on F one has $k \leq m$ (see [66, (2.4)]). Thus, if we further assume $n \geq 2m$, we have the numerical contradiction $m \leq n - m < k \leq m$.

The following example shows that the smoothness hypothesis of Proposition (14.1.3) and Conjecture (14.1.10) cannot be weakened to the assumption that the variety X has at worst rational Gorenstein singularities.

Example 14.1.11. (Sommese [583, (3.3.3)]) Choose a smooth projective surface S. Let $\pi : S' \to S$ be the blowup of S at one point y and let $E = \pi^{-1}(y)$. Let $\mathcal{L}$ be a very ample line bundle on S. Then $\mathcal{L}' := \pi^*(2\mathcal{L}) - E$ is very ample on S'

by Lemma (1.7.7). Let $X' := \mathbb{P}(\mathcal{L}' \oplus \pi^*\mathcal{L})$. We have the commutative diagram

$$\begin{array}{ccccc} & X' & \stackrel{\overline{\pi}}{\rightarrow} & S' & \\ \varphi & \downarrow & & \downarrow & \pi \\ & X & \stackrel{p}{\rightarrow} & S & \end{array}$$

Here $\varphi : X' \to \mathbb{P}_{\mathbb{C}}$ with $X := \varphi(X')$ is the map associated to $\Gamma(\xi)$, where ξ is the tautological line bundle on X' and X is precisely X' with the curve C ($\cong \mathbb{P}^1$) over E corresponding to the quotient $(\mathcal{L}' \oplus \pi^*\mathcal{L})_E \to \mathcal{O}_E \to 0$ contracted. Furthermore $\overline{\pi}$ is the canonical projection and p is the holomorphic map that sends $x \in X$ to $\pi(\overline{\pi}(\varphi^{-1}(x)))$. Since $\mathcal{L}'_E \cong \mathcal{O}_E(1)$ and $(\pi^*\mathcal{L})_E \cong \mathcal{O}_E$ we see that $\overline{\pi}^{-1}(E) \cong \mathbb{F}_1$ and the restriction of $\overline{\pi}$ to $\overline{\pi}^{-1}(E)$ is the projection onto $E \cong \mathbb{P}^1$. Furthermore $\varphi(\mathbb{F}_1) \cong \mathbb{P}^2$ and the restriction $\varphi_{\mathbb{F}_1}$ is the contraction of C to a point. Since $\mathcal{N}_{C/\mathbb{F}_1} \cong \mathcal{O}_{\mathbb{P}^1}(-1)$ and $\mathcal{N}_{\mathbb{F}_1/X'|C} \cong \mathcal{N}_{E/S'} \cong \mathcal{O}_{\mathbb{P}^1}(-1)$, we conclude, by the standard exact sequence of normal bundles, that

$$\mathcal{N}_{C/X'} \cong \mathcal{O}_{\mathbb{P}^1}(-1) \oplus \mathcal{O}_{\mathbb{P}^1}(-1). \tag{14.1}$$

We claim that X is Gorenstein. To prove this, let us first show that

$$K_{X'|U} \cong \mathcal{O}_U \tag{14.2}$$

for some Zariski neighborhood, U, of C in X'. Let $\mathcal{J}_C$ be the ideal sheaf defining C in X' and consider the exact sequence

$$0 \to K_{X'} \otimes \mathcal{J}_C \to K_{X'} \to K_{X'|C} \to 0. \tag{14.3}$$

From (14.1) it follows by the adjunction formula that $K_{X'|C} \cong \mathcal{O}_C$. Therefore, since $\varphi(C)$ is a point, $\varphi_*(K_{X'|C}) \cong \varphi_*(\mathcal{O}_C) \cong \mathbb{C}$. Thus, by considering the long exact sequence of higher derived functors

$$0 \to \varphi_*(K_{X'} \otimes \mathcal{J}_C) \to \varphi_* K_{X'} \to \varphi_*(K_{X'|C}) \cong \mathbb{C} \to \varphi_{(1)}(K_{X'} \otimes \mathcal{J}_C) \to \cdots$$

associated to (14.3), we conclude that (14.2) follows as soon as we show that $\varphi_{(1)}(K_{X'} \otimes \mathcal{J}_C) = 0$. To prove this it is enough by the formal function theorem (see Hartshorne [318, p. 277]) to show that

$$h^1(K_{X'} \otimes (\mathcal{J}_C/\mathcal{J}_C^k)) = 0 \tag{14.4}$$

for all integers $k \geq 1$. Write $\mathcal{F}_j := K_{X'} \otimes \mathcal{J}_C^j$, $j \geq 1$, and look at the exact sequence

$$0 \to \mathcal{F}_{i+1}/\mathcal{F}_k \to \mathcal{F}_i/\mathcal{F}_k \to \mathcal{F}_i/\mathcal{F}_{i+1} \to 0,$$

for $i = 1, \ldots, k-1$. Note that

$$\mathcal{F}_i/\mathcal{F}_{i+1} \cong K_{X'} \otimes (\mathcal{N}^*_{C/X'})^{(i)} \cong (\mathcal{O}_{\mathbb{P}^1}(1) \oplus \mathcal{O}_{\mathbb{P}^1}(1))^{(i)} \cong \oplus_{\alpha=1}^{i+1} \mathcal{O}_{\mathbb{P}^1}(i).$$

Then $h^1(\mathcal{F}_i/\mathcal{F}_{i+1}) = 0$ for each $i = 1, \ldots, k-1$. Therefore we find $h^1(K_{X'} \otimes (\mathcal{J}_C/\mathcal{J}_C^k)) = 0$ This proves (14.4) and hence (14.2).

From (14.2) it follows that $K_{X'} \cong \varphi^*\mathcal{H}$ for some line bundle $\mathcal{H}$ on X. Therefore $\varphi_*K_{X'} \cong \mathcal{H}$ is a line bundle. Since $\varphi_*K_{X'}$ and K_X agree outside of the point $\varphi(C)$ and K_X is a reflexive sheaf, we have $(\varphi_*K_{X'})^{**} \cong K_X$. Since, by the above, $\varphi_*K_{X'}$ is a line bundle, we conclude that K_X is a line bundle too. This means that X is 1-Gorenstein. To prove that X is Gorenstein we need to show that X is Cohen-Macaulay. Note that the same argument as above, starting from the exact sequence

$$0 \to \mathcal{J}_C \to \mathcal{O}_{X'} \to \mathcal{O}_C \to 0,$$

instead of (14.3), shows that X has rational singularities. Hence X is Cohen-Macaulay. Thus we conclude that X is Gorenstein.

Let $L := \mathcal{O}_{\mathbb{P}_\mathbb{C}}(1)_{|X}$. Note that $\xi = \varphi^*L$,

$$K_{S'} \approx \pi^*K_S + E = \pi^*K_S + \pi^*(2\mathcal{L}) - \mathcal{L}'.$$

Therefore

$$\varphi^*(K_X + 2L) \cong K_{X'} + 2\xi \cong \overline{\pi}^*(K_{S'} + \mathcal{L}' + \pi^*\mathcal{L}) \cong \overline{\pi}^*(\pi^*(K_S + 3\mathcal{L})).$$

Hence $K_X + 2L \approx p^*(K_S + 3\mathcal{L})$. Using the fact that $K_S + 3\mathcal{L}$ is ample for most choices of $(S, \mathcal{L})$, e.g., Proposition (9.1.3), we have an example of a scroll (X, L) with projection $p : X \to S$, $\mathbb{P}^2 = p^{-1}(y)$ as exceptional fiber and $\mathbb{P}^1$ as general fiber.

Another interesting question arising here is the following. Let (X, L) be a *classical scroll*, that is a $\mathbb{P}^d$-bundle over a variety Y such that L is ample and the restriction L_F is $\mathcal{O}_{\mathbb{P}^d}(1)$ for each fiber $F \cong \mathbb{P}^d$. Note that the adjunction theoretic definition of scrolls behaves much better than the classical definition, e.g., a hyperplane section of a scroll is a scroll. What are the differences between the two definitions?

In [105] we classify scrolls that are not scrolls in the classical sense if the dimension of the base, Y, is ≤ 3. So, by using Proposition (14.1.3), for scrolls of dimension $n \geq 5$ over Y with $\dim Y \leq 3$ and L very ample we have complete results. It is interesting to note that if $\dim Y = 1$, a classical scroll is an adjunction theoretic scroll and vice versa except for $\mathbb{P}^1 \times \mathbb{P}^1$ with $L = \mathcal{O}_{\mathbb{P}^1 \times \mathbb{P}^1}(1, 1)$. For the classical examples the two definitions agree except for the 2-dimensional quadric (which fails to have a well-defined scroll structure even classically).

We refer to Andreatta, Ballico, and Wiśniewski [18] for the generalization to any dimension of [105, Theorem (1.3)] where ample vector bundles, $\mathcal{E}$, of rank 2 on 3-folds Y, such that $K_Y + \det\mathcal{E}$ is nef and big but not ample were described. The general result was conjectured in [105, (4.1)].

14.2 Structure results for quadric fibrations

Let X be a smooth connected variety and L an ample line bundle on X. In this section we consider the case when (X, L) is a quadric fibration. We have the following structure theorem of the authors and of Wiśniewski.

Theorem 14.2.1. ([114, (3.2.6)]) *Let X be a smooth connected n-dimensional variety and let L be an ample line bundle on X. Assume that (X, L) is a quadric fibration, $\varphi : X \to Y$, over a normal projective variety, Y, of dimension m. Furthermore assume $n - m \geq 3$ and $n \geq 2m + 1$. Let ℓ be a line in a smooth general fiber of φ. Then ℓ is an extremal rational curve and φ is the fiber type contraction of $R = \mathbb{R}_+[\ell]$.*

Proof. We have $K_X + (n-m)L \approx \varphi^*\mathcal{L}$ for some ample line bundle $\mathcal{L}$ on Y. Then $K_X+(n-m)L$ is nef but not ample and therefore $n-m = \tau$, the nefvalue of (X, L). Take a line ℓ in a smooth fiber, Q, of φ. We have $L\cdot\ell = 1$ so that we can construct a nonbreaking family $(\mathcal{F}, \mathcal{P}, p, q, \nu)$ of rational curves which are deformations of ℓ and which fill up X. Now compute $\nu := -K_X\cdot\ell-2 = -K_Q\cdot\ell-2 = n-m-2$.

Exactly the same argument as in the proof of Theorem (14.1.1) shows that either we are done or φ factors through the contraction, $\rho : X \to Z$, of an extremal ray $R = \mathbb{R}_+[\mu]$, $\varphi = \alpha \circ \rho$ for some nontrivial morphism α. Furthermore ρ does not contract curves from the family $\mathcal{F}$ to points.

Again, the same argument as in the proof of (14.1.1), by using (6.3.2) and (6.4.6), and the fact that, if $m = 0$, the Picard number of X is 1, shows that every fiber, F, of the contraction ρ is of dimension $\dim F \leq n - \nu - 2 = m$. Since $n \geq 2m + 1$, we have $\dim F \leq n - m - 1 = \tau - 1$. As in the proof of (14.1.1), by using (6.3.6), we thus conclude that all fibers of ρ have dimension $\tau-1 = n-m-1$. Let F be a general fiber of ρ. Then $(K_X + \tau L)_F \approx K_F + \tau L_F \sim \mathcal{O}_F$, so by the Kobayashi-Ochiai characterization of projective spaces (3.1.6) we conclude that $(F, L_F) \cong (\mathbb{P}^{\tau-1}, \mathcal{O}_{\mathbb{P}^{\tau-1}}(1))$. Thus by using Proposition (3.2.1), we see that $\rho : X \to Z$ is a $\mathbb{P}^{\tau-1}$-bundle. Therefore φ is the composition of a $\mathbb{P}^{\tau-1}$-bundle projection, ρ, with a nontrivial morphism α. Since the general fiber of φ is a hyperquadric of dimension $n - m \geq 3$, and hence of Picard number 1, this is clearly not possible. □

Remark 14.2.2. Note that the condition $n-m \geq 3$ in the theorem above is needed if $n = 2m$. To see this, let $X := \mathbb{P}^1 \times \mathbb{P}^1 \times \mathbb{P}^2$ and let $p : X \to \mathbb{P}^2$ be the product projection. Let $L := \mathcal{O}_X(1, 1, 2)$. Then $K_X + 2L = p^*\mathcal{O}_{\mathbb{P}^2}(1)$, so that (X, L) is a quadric fibration under p. In this case p is not the contraction of an extremal ray. Here $p = \sigma \circ q_1 = \sigma \circ q_2$, where $q_1 : X \to \mathbb{P}^1 \times \mathbb{P}^2$ is the projection onto the product of the second two factors of X, and $q_2 : X \to \mathbb{P}^1 \times \mathbb{P}^2$ is the projection onto the product of the first and third factor of X. Here $\sigma : \mathbb{P}^1 \times \mathbb{P}^2 \to \mathbb{P}^2$.

In the case of a quadric fibration over a surface we have the following structure results. In what follows E, f denote, as usual, a section of self-intersection $E^2 = -r$ and a fiber of a Hirzebruch surface $\mathbb{F}_r$, $r \geq 0$.

Theorem 14.2.3. *Let (X, L) be the first reduction of an n-dimensional pair $(\widehat{X}, \widehat{L})$ with $\widehat{X}$ smooth and $\widehat{L}$ very ample. Assume that (X, L) is a quadric fibration, $\varphi : X \to Y$, over a normal surface Y.*

1. *Either φ has equal dimensional fibers or $n = 3$ and the only divisorial fibers are*

   a) *isomorphic to $\mathbb{F}_0$ with $L_{\mathbb{F}_0} \sim \mathcal{O}_{\mathbb{F}_0}(1, 2)$, or*

   b) *isomorphic to $\mathbb{F}_0 \cup \mathbb{F}_1$ with $L_{\mathbb{F}_0} \sim \mathcal{O}_{\mathbb{F}_0}(1, 1)$, $L_{\mathbb{F}_1} \sim E + 2f$. In this case φ is described as in (14.2.4) below.*

2. *The surface Y is smooth.*

Lemma 14.2.4. *Notation as in (14.2.3). Let F' be a divisorial fiber of $\varphi : X \to Y$ and let $F = F'_{\text{red}}$. Let D_1, D_2 be irreducible components of F such that the intersection $D_1 \cap D_2$ is not empty. Then after renaming, D_1 is isomorphic to $\mathbb{F}_1$ and D_2 is isomorphic to $\mathbb{F}_0$, D_1, D_2 meet along a section E of $\mathbb{F}_1$, and $L_{\mathbb{F}_0} \sim \mathcal{O}_{\mathbb{F}_0}(1, 1)$, $L_{\mathbb{F}_1} \sim E + 2f$. The fiber F' is reduced, i.e., $F' = F'_{\text{red}}$. Furthermore φ factors, $\varphi = p \circ \sigma'' \circ \sigma'$, where $\sigma' : X \to W$ is the smooth contraction of $\mathbb{F}_0$ along the fibers $f' = E$, and $\sigma'' : W \to Z$ is the blowing up at a smooth point of Z. The map p is locally a product projection in an inverse image $p^{-1}(U)$ of a neighborhood U of $\varphi(F)$.*

We proved the statement 1) of Theorem (14.2.3) and Lemma (14.2.4) above in our paper [104, (2.3), (2.6)] where we also show that the base surface Y has at worst Gorenstein, rational singular points, y, of type A_1, such that $\varphi^{-1}(y)$ is a 1-dimensional nonreduced fiber. The fact that Y is smooth has been proved in Besana [118, (1.3), (8.2)].

We also refer to [118, (1.4), (8.2)] for the proof of the fact that $K_X + (n-2)L$ is spanned by its global sections unless certain invariants are small (compare with our results [104, (3.2), (3.3)]).

Let us give the following explicit example of a quadric fibration over a surface with $\mathbb{P}^1 \times \mathbb{P}^1$ as a divisorial fiber.

Remark-Example 14.2.5. ([104, (2.8)]) Let (X, L) and $\varphi : X \to Y$ be as in Theorem (14.2.3) with $\dim X = 3$. The case of a divisorial fiber $F \cong \mathbb{F}_0$ of $\varphi : X \to Y$ really occurs. Indeed, let $X := \mathbb{P}^2 \times \mathbb{P}^1$. Let $p : X \to \mathbb{P}^2$ be the projection and $\pi : \widehat{\mathbb{P}^2} \to \mathbb{P}^2$ be the blowing up at a point y. Look at the base change diagram

$$\begin{array}{ccc} \widehat{X} & \to & X = \mathbb{P}^2 \times \mathbb{P}^1 \\ \widehat{p} \downarrow & & \downarrow p \\ \widehat{\mathbb{P}^2} & \to & \mathbb{P}^2 \end{array}$$

Let $q : \widehat{X} \to \mathbb{P}^1$ be the canonical projection over $\mathbb{P}^1$. Then $K_{\widehat{X}} \approx \widehat{p}^* K_{\widehat{\mathbb{P}^2}} + q^* K_{\mathbb{P}^1}$. Let $E = \pi^{-1}(y)$ and define $\mathcal{L} := \widehat{p}^*(\pi^*\mathcal{O}_{\mathbb{P}^2}(4) - E) + q^*\mathcal{O}_{\mathbb{P}^1}(2)$. Thus $\mathcal{L}$ is very ample since $\pi^*\mathcal{O}_{\mathbb{P}^2}(4) - E$ is very ample by (1.7.7) and, since $K_{\widehat{\mathbb{P}^2}} \approx \pi^* K_{\mathbb{P}^2} + E$,

$$K_{\widehat{X}} + \mathcal{L} \approx \widehat{p}^*(\pi^*(K_{\mathbb{P}^2} + \mathcal{O}_{\mathbb{P}^2}(4)) \approx \widehat{p}^*\pi^*\mathcal{O}_{\mathbb{P}^2}(1).$$

This shows that $(\widehat{X}, \mathcal{L})$ is a quadric fibration over $\mathbb{P}^2$, with the only divisorial fiber $\mathbb{F}_0 = (\pi \circ \widehat{p})^{-1}(y)$ over y.

Let (X, L), $\varphi : X \to Y$, be as in (14.2.3) with $\dim X = 3$ and assume that there exists a unique divisorial fiber $F = \mathbb{F}_0$ of φ. Let $\sigma : Y' \to Y$ be the blowup of Y at $\varphi(F)$. Then there is a morphism $\varphi' : X \to Y'$ such that $\varphi = \sigma \circ \varphi'$ and all the fibers of φ' are 1-dimensional since $\varphi'^{-1}(s) \cong \mathbb{P}^1$ for any $s \in \sigma^{-1}(\varphi(F))$. Furthermore $K_X + L \approx \varphi'^*\sigma^*\mathcal{L}$ where $\mathcal{L}$ is an ample line bundle on Y such that $K_X + L \approx \varphi^*\mathcal{L}$. Therefore φ' expresses (X, L) as a quadric fibration over Y' except for the fact that $\sigma^*\mathcal{L}$ is merely nef.

Note that also the case of a divisorial fiber $F \cong \mathbb{F}_0 \cup \mathbb{F}_1$ really occurs. We refer for this to an explicit example given in Besana [118, (3.6)].

For a quadric fibration (X, L) such that the structural morphism is a contraction of an extremal ray (note that this is always the case under suitable numerical conditions as in Theorem (14.2.1)) and is equidimensional we have the following result of Andreatta, Ballico, and Wiśniewski [20, Theorem B].

Theorem 14.2.6. *Let (X, L) be a quadric fibration, $\varphi : X \to Y$, with L an ample line bundle on X. Let $n = \dim X$, $m = \dim Y$. Assume that φ is a contraction of an extremal ray (or that $n \geq 2m + 1$ and $n - m \geq 3$ by* (14.2.1)) *and that φ is equidimensional. Then Y is smooth, $\mathcal{E} := \varphi_* L$ is a locally free sheaf of rank $n - m + 2$ and L embeds X into $\mathbb{P}(\mathcal{E})$ as a divisor of relative degree* 2.

Following Besana [118, §3], we briefly discuss the relations between adjunction theoretic conic fibrations over surfaces and classical conic bundles in the sense of Beauville [76].

Definition 14.2.7. A *quadric bundle* (a "fibré en quadriques" in the sense of [76]) is a connected projective manifold, X, with a morphism $\varphi : X \to S$, where S is a smooth surface, such that the fibers are hyperquadrics of the same dimension, m. Thus φ is flat by (4.1.2). The *discriminant curve* of the quadric bundle is the possibly empty curve, $C \subset S$, such that $s \in C$ if and only if $\varphi^{-1}(s)$ is singular.

If $m = 1$, we refer to $\varphi : X \to S$ as a *conic bundle*, or a *classical conic bundle*.

Example 14.2.8. (The cubic in $\mathbb{P}^4$, cf., [118, (3.2)]) Let V be a smooth cubic hypersurface in $\mathbb{P}^4$. It is well-known (see, e.g., Beauville [76]) that the blowup, $\pi : X \to V$, of V along a general line $\ell \subset V$ is a classical conic bundle over

$\mathbb{P}^2$ without double lines as fibers. Let $\varphi : X \to \mathbb{P}^2$ be the classical conic bundle morphism. Let $E = \pi^{-1}(\ell)$ be the exceptional divisor of π. Let

$$L := \varphi^*\mathcal{O}_{\mathbb{P}^2}(1) \otimes \pi^*\mathcal{O}_{\mathbb{P}^4}(2)_{|V} \otimes \mathcal{O}_X(-E) \cong \varphi^*\mathcal{O}_{\mathbb{P}^2}(1) \otimes \pi^*(-K_V) \otimes \mathcal{O}_X(-E).$$

We claim that (X, L), $\varphi : X \to \mathbb{P}^2$, is a conic fibration in the adjunction theoretic sense. Since $K_X + L \approx \varphi^*\mathcal{O}_{\mathbb{P}^2}(1)$, it is enough to show that L is ample. Note that $\varphi^*\mathcal{O}_{\mathbb{P}^2}(1)$ is spanned by global sections since $\mathcal{O}_{\mathbb{P}^2}(1)$ is very ample. Let $M := \mathcal{O}_{\mathbb{P}^4}(1)_{|V}$. Then M is very ample and $L = \varphi^*\mathcal{O}_{\mathbb{P}^2}(1) \otimes \pi^*(2M) \otimes \mathcal{O}_X(-E)$. By Lemma (1.7.7) we know that $\pi^*(2M) \otimes \mathcal{O}_X(-E)$ is very ample. This combined with the spannedness of $\varphi^*\mathcal{O}_{\mathbb{P}^2}(1)$ implies that L is very ample.

The next result of Besana [118, (3.4)] shows that, as the example above suggests, any conic bundle in the classical sense is an adjunction theoretic conic fibration.

Proposition 14.2.9. (Besana) *Let $\varphi : X \to Y$ be a classical conic bundle. Then there exist ample line bundles L and H on X and Y respectively such that $K_X + L \approx \varphi^*H$, i.e., (X, L) is a conic fibration in the adjunction theoretic sense.*

Proof. The anticanonical bundle $-K_X$ is φ-very ample (see Beauville [76] or Sarkisov [536]). This implies by Kleiman's ampleness criterion (4.2.1) that $N\varphi^*\mathcal{H} - K_X$ is ample for any ample line bundle $\mathcal{H}$ on Y and $N \gg 0$. Let $H := N\mathcal{H}$ and $L := \varphi^*H - K_X$. Then by the above L is ample and $K_X + L \approx \varphi^*H$, so we are done. □

We refer to Besana [118], §3 for further examples of conic fibrations over a Del Pezzo surface and a Godeaux surface. We refer to Wiśniewski [624, 625] for results on Fano manifolds of index r and dimension $2r$, and some of their relations to quadric fibrations.

Let us propose the following conjecture, as conclusion of this section.

Conjecture 14.2.10. *Let X be a smooth connected projective variety of dimension n and let L be an ample line bundle on X. Let τ be the nefvalue of (X, L) and let $\phi : X \to W$ be the nefvalue morphism of (X, L). Assume that ϕ is not birational. If $\tau \geq \dim W$, then ϕ is flat.*

Let us give some evidence for the conjecture above. Let (X, L) be a n-dimensional scroll over a variety Y of dimension m. Then the nefvalue of (X, L) is $\tau = n - m + 1$. Therefore $\tau \geq m$ is equivalent to $n \geq 2m - 1$. Thus the Conjecture (14.1.10) is a special case of the Conjecture (14.2.10). Let (X, L) be an n-dimensional quadric fibration over a surface. Then the nefvalue of (X, L) is $\tau = n - 2$. Therefore $\tau \geq 2$ gives $n \geq 4$ and the conjecture above is true in this range by Theorem (14.2.3).

14.3 Varieties with small invariants

From its beginning a motivating application of adjunction theory has been the classification of projective manifolds with small invariants. In this book we have only done classification as far as need for the general theory, though all the tools needed to do classification have been developed. In this section we give a guide to the main papers concerned with classification.

Let L be a line bundle on an irreducible projective variety X of dimension n. In classification problems some degree of ampleness of L (e.g., nef and big, ample, very ample, k-very ample, etc.) and some degree of smoothness of X (e.g., smooth, Gorenstein, normal, etc.) are assumed. The general philosophy is that for each given invariant (e.g., the degree $d = L^n$, the sectional genus, $g(L)$, the Δ-genus, or whatever) there is some number up to which the classification leads to a complete or almost complete answer and beyond which the results quickly become very incomplete, as the number of possible cases explodes. The number is typically very small if L is merely ample and increases significantly as stronger conditions on L (such as very ampleness or k-very ampleness) are assumed.

Some of the results are classical. We refer to Roth [526] for an overview of these results. A good general reference for classification results on d, $g(L)$, $\Delta(X, L)$ for L merely ample is Fujita's book [257].

Varieties with small sectional genus. Let us first consider the case of surfaces, X, and assume that L is very ample. There is a complete classification in the smooth case up to $g(L) \leq 7$. See Castelnuovo and Enriques [148] for classical work on genus up to 3 and Roth [526] for classical work on genus up to 6. For $g(L) \leq 3$, the second author [571] gave biregular results based on his results on the adjunction mapping. This classification was fully worked out mainly due to Livorni, who completed it in several papers [409, 410, 411, 412, 414] and Ionescu [333].

The classical approach to many classification problems was by iterating adjunction and knowing all the ways that the process could terminate. Due to the incomplete knowledge of the spannedness of the adjunction bundles this approach classically led to birational, and not to biregular results. A good reference to get a feel for the classical work is Roth [527, Chapter V]. To carry out the biregular version of the classical adjunction process is at the root of the modern theory [571], e.g., the modern revival of interest Roth's important paper [526] goes back to Sommese [571]. The iterated adjunction procedure was extensively used by Livorni to carry out the classification up to genus 7. Following Biancofiore and Livorni [129], let us give some definitions and recall this procedure (see also Fujita [257, II, §13]).

Let X be a smooth surface polarized with a very ample line bundle L. Assume $\kappa(X) < 0$. Except for the exceptions listed in (9.2.2) we have that $K_X + L$ is spanned and the connected part of the morphism associated to $\Gamma(K_X + L)$ is the reduction map, $\pi : X \to X'$. Let $L' := (\pi_* L)^{**}$. To the pair (X, L) we can associate the new pair $(X_1, L_1) := (X', K_{X'} + L')$. If (X, L) is not one of the four pairs listed in (10.3.1), then $K_{X'} + L'$ is very ample and we can consider the new pair $(X_2, L_2) := (X'_1, K_{X'_1} + L'_1)$, where (X'_1, L'_1) is the reduction of (X_1, L_1). Thus we can iterate this procedure until we reach a new pair (X_t, L_t) such that $X_t = X'_{t-1}$ is a smooth surface, $L_t = K_{X'_{t-1}} + L'_{t-1}$ is very ample and either $K_{X_t} + L_t$ is not spanned and big or (X_t, L_t) is as in one of the four cases listed in (10.3.1). Following Biancofiore and Livorni [129], we call this pair, (X_t, L_t), *terminal pair*. Then a terminal pair is either one of the surfaces listed in (10.3.1), or one of the surfaces listed in (9.2.2), or a conic fibration over a smooth curve. We call (X_m, L_m), $i = 1, \ldots, t$, the *m-iterated reduction* of (X, L). Denote $(X_0, L_0) := (X, L)$, $g_0 = g(L)$, $g_m := g(L_m)$, $i = 1, \ldots, t$.

Note that if $\kappa(X) \geq 0$, there are no terminal pairs and by the procedure described above, after the contraction of finitely many rational curves, we reach the minimal model, S, of X, i.e., K_S is nef.

The following striking theorem of Biancofiore and Livorni plays an important role in the classification of polarized surfaces, (X, L), of negative Kodaira dimension and of given sectional genus.

Theorem 14.3.1. (Biancofiore and Livorni ([129, §1])) *Let* (X, L), (X_m, L_m), g_m, $m = 0, \ldots, t$, $g_0 = g(L)$ *be as above. Assume that* $\kappa(X) < 0$. *Then we have:*

1. *If* $K_{X_1} \cdot L_1 \leq 0$, *then* $g_m \leq g_0$*;*

2. *If* $K_{X_1} \cdot L_1 > 0$, *then there exists a positive integer,* α, *such that* $K_{X_{\alpha+1}} \cdot L_{\alpha+1} \geq 0$ *and* $K_{X_{\alpha+2}} \cdot L_{\alpha+2} < 0$*;*

3. *If* $K_{X_1} \cdot L_1 > 0$, *the curve described by the* g_m*'s reaches the maximum at* $m = \alpha + 1$ *and after it decreases up to* $g_m \leq g_0$.

In Biancofiore, Livorni, and Sommese [131] there are some results on the asymptotic number of possible polarized surfaces to be classified as the sectional genus or degree become very large.

If L is merely ample and $g(L) = 0$ the polarized surface (X, L) can be easily described by using, e.g., Reider's criterion (see (8.9.1) and Fujita [257]). For $g(L) = 1$, we refer to the n-dimensional case below. The classification in the case $g(L) = 2$ was done by Lanteri, Palleschi, and the first author of this book [90] and Fujita [254], and in the case $g(L) = 3$ by Maeda [421]. If we relax the assumption on L to be nef and big, the complete classification for $g(L) \leq 2$ is given in our paper [99]. The description in the case $g(L) \leq 1$ with L simply *numerically positive*, i.e., $L \cdot C > 0$ for each irreducible curve, C, on X, has been recently done in Lanteri and Rondena [386]. See also Lanteri and Palleschi [379] for the cases $g(L) = 3, 4$ with L ample and spanned. If X is singular there are results up to $g(L) \leq 3$ (see Andreatta and Sommese [26]).

Assume now $\dim X \geq 2$ and L ample. We refer to §3.1 for the classification in the case $g(L) = 0$. The case $g(L) = 1$ is an easy consequence of the results of §7.2. We include here the discussion of this case for the reader's convenience (see also Fujita [257, II, §12]).

Theorem 14.3.2. *Let (X, L) be a smooth n-fold, $n \geq 2$, polarized with an ample line bundle L. Assume $g(L) = 1$. Then (X, L) is either a Del Pezzo manifold or a $\mathbb{P}^1$-bundle over an elliptic curve.*

Proof. By combining (7.2.2), (7.2.3) and (7.3.2) we conclude that either (X, L) is a $\mathbb{P}^1$-bundle over a smooth curve, C, or $K_X + (n-1)L$ is nef. In the first case, since C is isomorphic to the curve obtained as transversal intersection of $n-1$ general members of $|L|$, $g(L) = g(C)$. So C is an elliptic curve. In the latter case, we conclude that there exists a morphism $\varphi : X \to W$, given by $\Gamma(m(K_X + (n-1)L))$ for $m \gg 0$, and $K_X + (n-1)L \approx \varphi^* H$ for some ample line bundle H on W. Then the genus formula yields

$$(K_X + (n-1)L) \cdot L^{n-1} = \varphi^* H \cdot L^{n-1} = 2g(L) - 2 = 0.$$

Since H is ample this is impossible unless $\dim W = 0$. Thus $K_X \approx -(n-1)L$ and (X, L) is a Del Pezzo manifold. □

Currently there are complete results up to $g(L) \leq 2$. The classification in the case $g(L) = 2$ is due to Fujita [257, II, §15] and covers also the case of surfaces studied in [90]. In the case when $\dim X \geq 3$ and L is very ample, Ionescu [333] did the classification in the cases $g(L) = 3, 4, 5$. Allowing normal Gorenstein singularities on X, there are results of ours [100] for $g(L) = 3, 4$.

Varieties of small codimension. Here we assume that L is very ample and we denote by $X \hookrightarrow \mathbb{P}^N$ the embedding given by $\Gamma(L)$. Assume $n = \dim X \geq 3$. Recall that $\operatorname{Pic}(X) \cong \mathbb{Z}$ as soon as $2n - N \geq 2$ by the generalized Lefschetz theorem, (2.3.11). There is also a famous conjecture due to Hartshorne which says that if $N - n = 2$ and $N \geq 6$, then X is a complete intersection in the scheme theoretic sense. For these reasons the case of 3-folds in $\mathbb{P}^5$ holds a special interest. A classification, including almost complete existence and uniqueness results, is done up to degree $d \leq 12$. The classification up to degree 8 was completed by Okonek and Ionescu (see the following paragraph on varieties of small degree). See Beltrametti, Schneider and Sommese [93, 94, 96] for $d = 9, 10, 11$ and Edelmann [198] for $d = 12$. We also refer to [96, §6] for a detailed list of numerical invariants up to $d \leq 12$. For these results adjunction theory is a major tool to get maximal lists of all possible cases. Liaison is the major tool used to get existence results.

The complete lists of such 3-folds in $\mathbb{P}^5$ are too long to be reported. We limit ourselves to a description of two examples from [93] which seem of particular interest, since they are the only ones up to degree 10 of nonnegative Kodaira dimension (up to complete intersections). Note that the first case of not complete intersection 3-fold in $\mathbb{P}^5$ of general type occurs for $d = 11$ (see [94]). We refer to the papers quoted above for more details.

Let X be a smooth 3-fold embedded by a very ample line bundle, L, in $\mathbb{P}^5$. In the following $X \overset{(a,b)}{\sim} Y$ means that X is *linked* or in *liaison* to Y inside the complete intersection of two hypersurfaces H_a, H_b of degrees a and b, i.e., $X \cup Y = H_a \cap H_b$. Let $\mathcal{J}$ be the ideal sheaf defining X in $\mathbb{P}^5$. Let S be a smooth surface in $|L|$. For more on liaison, which is a major tool in the construction of varieties with given projective invariants, we refer to Peskine and Szpiro [501] and Migliore [436].

Example 14.3.3. $d = 9$, $X \overset{(2,5)}{\sim} \mathbb{P}^3$ and there are resolutions

$$0 \to \mathcal{O}_{\mathbb{P}^5}(-6)^{\oplus 2} \to \mathcal{O}_{\mathbb{P}^5}(-5)^{\oplus 2} \oplus \mathcal{O}_{\mathbb{P}^5}(-2) \to \mathcal{J} \to 0,$$

$$0 \to \mathcal{O}_{\mathbb{P}^5}(-6) \to \mathcal{O}_{\mathbb{P}^5}(-1)^{\oplus 2} \oplus \mathcal{O}_{\mathbb{P}^5}(-4) \to \mathcal{O}_{\mathbb{P}^5}^{\oplus 2} \to K_X \to 0.$$

Furthermore the invariants are $\kappa(X) = 1$, $\chi(\mathcal{O}_X) = -1$, $\chi(\mathcal{O}_S) = 9$, $d_1 = 13$, $d_2 = 17$, $d_3 = 21$, $g(L) = 12$. In this case $g(L)$ reaches the maximum with respect to the Castelnuovo bound, so that X is a Castelnuovo variety.

Example 14.3.4. $d = 10$, $X \overset{(3,4)}{\sim} Q$, Q hyperquadric in $\mathbb{P}^5$. There are resolutions

$$0 \to \mathcal{O}_{\mathbb{P}^5}(-6) \oplus \mathcal{O}_{\mathbb{P}^5}(-5) \to \mathcal{O}_{\mathbb{P}^5}(-4)^{\oplus 2} \oplus \mathcal{O}_{\mathbb{P}^5}(-3) \to \mathcal{J} \to 0,$$

$$0 \to \mathcal{O}_{\mathbb{P}^5}(-6) \to \mathcal{O}_{\mathbb{P}^5}(-2)^{\oplus 2} \oplus \mathcal{O}_{\mathbb{P}^5}(-3) \to \mathcal{O}_{\mathbb{P}^5} \oplus \mathcal{O}_{\mathbb{P}^5}(-1) \to K_X \to 0.$$

Furthermore the invariants are $\kappa(X) = 0$, $\chi(\mathcal{O}_X) = 0$, $\chi(\mathcal{O}_S) = 7$, $d_1 = d_2 = d_3 = 12$, $g(L) = 12$.

We refer to Livorni and Sommese [415], Beltrametti, Biancofiore, and Sommese [85], Chang [153] and Miro-Roig [438] for further results on 3-folds in $\mathbb{P}^5$. We refer to Braun, Ottaviani, Schneider, and Schreyer [139], Beltrametti and Sommese [104, §4] and Ottaviani [496] for the study of special varieties arising in adjunction theory and lying in $\mathbb{P}^5$. We also refer to Braun, Ottaviani, Schneider, and Schreyer [138] for the proof of the fact that there are only finitely many families of nongeneral type 3-folds in $\mathbb{P}^5$. The analogous results for surfaces in $\mathbb{P}^4$ were proved by Ellingsrud and Peskine [215].

It is interesting to study surfaces in $\mathbb{P}^4$. There is a complete classification up to degree 10 due to the work of several people. We refer for this subject to the papers of Alexander [1, 2], Decker, Ein and Schreyer [181], Mezzetti and Ranestad [433], Popescu [506], Popescu and Ranestad [507], and Ranestad [513].

As soon as the codimension of X in $\mathbb{P}^N$ increases not much is known. For results in this direction we refer to Toma [597] for the study of 3-dimensional scrolls in $\mathbb{P}^6$ and especially to Okonek's paper [489]. We also refer to Tendian [594] for results on surfaces of degree d and sectional genus g in $\mathbb{P}^{d-g+1}$.

We conclude this paragraph by recalling the following result which shows that if $(\mathcal{M}, \mathcal{L})$ has a Del Pezzo fibration, (M, L), as first reduction, then $\mathcal{M}$ cannot lie in $\mathbb{P}^5$ except for two cases.

Proposition 14.3.5. *Let $\mathcal{M}$ be a smooth connected threefold, $\mathcal{L}$ a very ample line bundle on $\mathcal{M}$, such that $\Gamma(\mathcal{L})$ embeds $\mathcal{M}$ in $\mathbb{P}^N$. Let (M, L), $\pi : \mathcal{M} \to M$, be the first reduction of $(\mathcal{M}, \mathcal{L})$. Assume that (M, L) is a Del Pezzo fibration $\varphi : M \to B$ over a smooth curve B.*

1. *Let F be a general fiber of φ and let $\deg F = K_F^2 = f$. Then there are no fibers of φ containing more than $f-3$ distinct points blown up under $\pi : \mathcal{M} \to M$. (Hence $\mathcal{M} \cong M$ if $f = 3$.)*
2. *One has $N \geq 6$ unless $d := \mathcal{L}^3 = 7, 8$.*

Proof. Assume that there exists a fiber, F, of φ containing $x_1, \ldots, x_j$ distinct points of M blown up under π with $j \geq f-2$. We consider only the case $j = f-2$ since the other cases are simpler and easier. Write $b = \varphi(F)$, ${\mathbb{P}^2}_i = \pi^{-1}(x_i)$, $i = 1, \ldots, f-2$, and let $p = \varphi \circ \pi$. Then

$$p^{-1}(b) = R \cup {\mathbb{P}^2}_1 \cup \ldots \cup {\mathbb{P}^2}_{f-2}$$

where R is a surface in $\mathcal{M}$. Since $\deg F = f$ and π is an isomorphism outside of a finite number of points, R is a (possibly singular, reducible or not reduced) quadric. If R is a smooth quadric, then the intersection ${\mathbb{P}^2}_i \cap R$ is a curve on R which contracts to a point x_i, $i = 1, \ldots, f-2$, and this is not possible. In all the other cases a component R' of R_{red} meets one of the ${\mathbb{P}^2}_i$'s in a curve which again contracts to a point x_i. Since $\operatorname{Pic}(R') \cong \mathbb{Z}$, the restriction of π to R' is finite-to-one by Lemma (3.1.8), which leads to a contradiction.

To prove 2) first recall that $3 \leq f \leq 9$ by general results on classification of Del Pezzo surfaces. Furthermore note that $N \geq 5$, since, if $N = 4$, the Lefschetz hyperplane section theorem, (2.3.1), implies $H^2(\mathcal{M}, \mathbb{Q}) \cong \mathbb{Q}$, but $\dim H^2(\mathcal{M}, \mathbb{Q}) \geq 2$ since φ has positive dimensional fibers and positive dimensional image.

For any smooth 3-fold $\mathcal{M}$ embedded by a very ample line bundle $\mathcal{L}$ in $\mathbb{P}^5$, and of degree $d = \mathcal{L}^3$, the exact sequence

$$0 \to \mathcal{T}_{\mathcal{M}} \to \mathcal{T}_{\mathbb{P}^5|\mathcal{M}} \to \mathcal{N}_{\mathcal{M}/\mathbb{P}^5} \to 0$$

easily gives the following useful form of the double point formula:

$$15\mathcal{L}^2 = c_2(\mathcal{M}) - K_{\mathcal{M}} \cdot (6\mathcal{L} + K_{\mathcal{M}}) + d\mathcal{L}^2.$$

Applying this formula to our situation we have

$$15\mathscr{L}_F^2 = e(F) - \mathbb{C}_F(F) \cdot K_F - K_F \cdot (6\mathscr{L}_F + K_F) + d\mathscr{L}_F^2,$$

where $\mathscr{L}_F$ denotes the restriction of $\mathscr{L}$ to the general fiber F of φ. Since $K_F \approx -\mathscr{L}_F$ we find

$$(11-d)f = 12.$$

Since $3 \leq f \leq 9$ the only possible cases are $(f, d) = (3, 7), (4, 8), (6, 9)$. Of these three cases the degree 9 is shown not to exist in [93]. Ionescu and Okonek (see the following paragraph on varieties of small degree) have shown that the degree 7 and 8 cases do exist. For a complete discussion of such "log-special" threefolds in $\mathbb{P}^5$ we refer to the paper [139] of Braun, Ottaviani, Schneider and Schreyer. □

Varieties with small degree. The complete classification of all varieties, including the singular ones, of degree $d \leq 3$ goes back in modern times to Weil [615], and for $d = 4$ to Swinnerton-Dyer [592]. In the smooth case the classification up to degree 8 has been completed in several papers by Okonek [485, 486, 487, 488], and Ionescu [333, 337]. Recently, by using the classification of surfaces in $\mathbb{P}^4$ and 3-folds in $\mathbb{P}^5$, Fania and Livorni classified all varieties of degree 9, 10 [224, 225]. For $d = 9$ they have almost complete existence results. For degree $d = 10$ they give a maximal list of all possible cases.

Let us also mention Halanay's results [302, 303, 304] for singular surfaces up to degree 6.

We explicitly find all smooth surfaces of degree $d \leq 4$ in §8.10 since we use such a classification in the following Chapter 10. Note also that the results of Chapter 10 are in fact one step away to give a complete list of surfaces up to degree 8.

If L is k-very ample on a smooth surface, X, then an almost complete classification of pairs (X, L) when $d \leq 4k + 4$ is carried out in Ballico and Sommese [65].

Varieties with small Δ-genus. The classification of smooth polarized n-folds, (X, L), $n \geq 2$, with small Δ-genus is due to Fujita and Ionescu. If L is merely ample, there is a complete classification for $\Delta(X, L) \leq 1$ and almost complete for $\Delta(X, L) = 2$. Recall that $\Delta(X, L) = 0$ if and only if $g(L) = 0$ (see (3.1.3) and Fujita [257, II, (12.1)]). We refer to [257, I, §§5, 6, 9, 10] for details and references. If L is very ample Ionescu [333] classified the cases $\Delta(X, L) = 2, 3, 4$.

Remark 14.3.6. We refer to Zak [631, 632] for a few results on the classification of varieties whose dual varieties have small degree.

We refer to Lanteri and Turrini [395] and Turrini and Verderio [599] for classification of threefolds and surfaces of small class. We recall that if $X \subset \mathbb{P}^N$ is a complex smooth n-fold, its *class* is the number of hyperplanes in a general pencil which are tangent to X.

To conclude, let us point out the following general problem.

Question 14.3.7. *Find asymptotic estimates of the number of objects to be classified for the given invariant as the invariant becomes large. For example for smooth threefolds in $\mathbb{P}^5$. Is the number of different families of smooth 3-folds of degree d bounded by $p(d)$ or $p(d \log d)$ for some polynomial $p(x)$ for large d? If yes, is the lowest degree of such a polynomial equal to 2, the codimension?*

14.4 Projective manifolds with positive defect

A *positive defect pair* (X, L) is a pair consisting of a very ample line bundle, L, on a connected n-dimensional projective manifold, X, such that $k(X, L)$, the defect of (X, L), is positive. Recall that the defect of (X, L) equals $\operatorname{cod}_{\mathbb{P}(\Gamma(L)^*)}\mathcal{D} - 1$, where $\mathcal{D} \subset |L|$ is the set of singular hyperplane sections of X with the convention that $k(X, L) = \dim |L|$ if $\mathcal{D}$ is empty: by (1.6.12) the discriminant locus $\mathcal{D}$ is empty only if $(X, L) \cong (\mathbb{P}^n, \mathcal{O}_{\mathbb{P}^n}(1))$. As we saw earlier $\mathcal{D}$ is irreducible. In this section we will discuss the structure theory of pairs (X, L) with $k(X, L) > 0$. This structure theory reduces the study of positive defect pairs to a very restricted class of high index Fano manifolds, called *maximal positive defect pairs.*

We need some preliminary results. In §1.6 we noted that if the defect, $k(X, L)$, is positive, then an $A \in |L|$ corresponding to a general point of $\mathcal{D} \subset |L|$ has a $\mathbb{P}^{k(X,L)}$ of nondegenerate quadratic singularities. The following fundamental observation is due to Ein [203, 204].

Theorem 14.4.1. (Ein) *Let X be a smooth variety of dimension $n \geq 2$, let L be a very ample line bundle on X, and let $\mathcal{D} \subset |L|$ denote the discriminant variety of (X, L). Assume that $k := \operatorname{def}(X, L) > 0$, and let $\mathbb{P}^k$ be the singular set of a general $A \in \mathcal{D}$. Then one has $\mathcal{N}_{\mathbb{P}^k/X} \cong \mathcal{N}^*_{\mathbb{P}^k/X}(1)$.*

Proof. A general $A \in \mathcal{D} \subset |L|$ has a $\mathbb{P}^k$ of nondegenerate quadratic singularities and $L_{\mathbb{P}^k} \cong \mathcal{O}_{\mathbb{P}^k}(1)$. Let $s \in \Gamma(L)$ be the section defining A. Consider the map

$$j_1 : X \times \Gamma(L) \to J_1(L) \to 0$$

induced by the natural map $j_1 : L \to J_1(L)$ and the exact sequence

$$0 \to \mathcal{T}_X^{*(2)} \otimes L \to J_2(L) \xrightarrow{\alpha} J_1(L) \to 0.$$

Here $J_2(L)$ denotes the second jet bundle of L defined analogously to $J_1(L)$ as in (1.6.3), with $\mathcal{J}_\Delta^3$ in place of $\mathcal{J}_\Delta^2$. There is a natural map $j_2 : L \to J_2(L)$, which is a sheaf but not a bundle map, defined analogously to j_1. The map $\alpha : J_2(L) \to J_1(L)$ is the canonical surjection with kernel $\mathcal{T}_X^{*(2)} \otimes L$ obtained by sending a 2-jet of a section, s, to the 1-jet of s (see [363] for more details). Since s vanishes on $\mathbb{P}^k$ and it is singular on it, the 1-jet, $j_1(s)$, is zero in $J_1(L)_{\mathbb{P}^k}$. Since the image under α of the 2-jet $j_2(s)$ is $j_1(s)$, we conclude that $j_2(s) \in \Gamma((\mathcal{T}_X^{*(2)} \otimes L)_{\mathbb{P}^k})$. Since $s(\mathbb{P}^k) = 0$, all partial derivatives in the $\mathbb{P}^k$ direction are zero. This implies that

$$j_2(s) \in \Gamma({\mathcal{N}^*_{\mathbb{P}^k/X}}^{(2)} \otimes L_{\mathbb{P}^k}) \subseteq \Gamma((\mathcal{T}_X^{*(2)} \otimes L)_{\mathbb{P}^k}).$$

From the isomorphism

$$\begin{aligned}({\mathcal{N}^*_{\mathbb{P}^k/X}}^{(2)} \otimes L_{\mathbb{P}^k}) \oplus (\wedge^2 \mathcal{N}^*_{\mathbb{P}^k/X} \otimes L_{\mathbb{P}^k}) &\cong \mathcal{N}^*_{\mathbb{P}^k/X} \otimes \mathcal{N}^*_{\mathbb{P}^k/X} \otimes L_{\mathbb{P}^k} \\ &\cong \mathcal{H}om_{\mathcal{O}_{\mathbb{P}^k}}(\mathcal{N}_{\mathbb{P}^k/X}, \mathcal{N}^*_{\mathbb{P}^k/X} \otimes L_{\mathbb{P}^k})\end{aligned}$$

we conclude that $j_2(s)$ gives rise to a morphism $h(s) : \mathcal{N}_{\mathbb{P}^k/X} \to \mathcal{N}^*_{\mathbb{P}^k/X} \otimes L_{\mathbb{P}^k}$. The fact that A has nondegenerate quadratic singularities on $\mathbb{P}^k$ implies that $h(s)$ is an isomorphism. Since $L_{\mathbb{P}^k} \cong \mathcal{O}_{\mathbb{P}^k}(1)$ we are done. $\square$

As shown in (1.6.9) the $\mathbb{P}^k$'s arising as singular sets of $A \in \mathcal{D}$ are dense in X. From this it follows that $\mathcal{N}_{\mathbb{P}^k/X}$ is spanned by global sections at a general point. Thus if ℓ is a general line in $\mathbb{P}^k$, then $\mathcal{N}_{\mathbb{P}^k/X|\ell}$ is spanned by global sections. Thus $\mathcal{N}_{\mathbb{P}^k/X|\ell} \cong \oplus_{i=1}^{n-k}\mathcal{O}_{\mathbb{P}^1}(a_i)$ with $a_i \geq 0$ for all i. Thus by (14.4.1) we conclude that the set of integers $1 - a_i$ after reordering coincides with the set of integers a_i. From this we see that

$$\mathcal{N}_{\mathbb{P}^k/X|\ell} \cong \mathcal{O}_{\mathbb{P}^1}^{\oplus \frac{n-k}{2}} \oplus \mathcal{O}_{\mathbb{P}^1}(1)^{\oplus \frac{n-k}{2}}. \tag{14.5}$$

Since $\mathcal{N}_{\ell/\mathbb{P}^k} \cong \mathcal{O}_{\mathbb{P}^1}(1)^{\oplus k-1}$ we see that

$$\mathcal{N}_{\ell/X} \cong \mathcal{O}_{\mathbb{P}^1}^{\oplus \frac{n-k}{2}} \oplus \mathcal{O}_{\mathbb{P}^1}(1)^{\oplus \frac{n+k}{2}-1}. \tag{14.6}$$

In particular we have Landman's parity result that $n \equiv k \bmod 2$ and thus using (1.6.12) that if $(X, L) \not\cong (\mathbb{P}^n, \mathcal{O}_{\mathbb{P}^n}(1))$, then $k \leq n-2$. We also obtain the important count

$$K_{X|\mathbb{P}^k} \cong \mathcal{O}_{\mathbb{P}^k}((-n-k-2)/2). \tag{14.7}$$

We can prove now the two main results, Theorems (1.1) and (1.2), from our paper with Fania [87]. These results give a structure theorem for (X, L) with positive defect in terms of the nefvalue morphism, ϕ, of (X, L) and complement Corollary (6.6.8), which says that ϕ contracts the linear $\mathbb{P}^{\operatorname{def}(X,L)}$'s which cover X and the nefvalue, τ, is

$$\tau = \frac{n + \operatorname{def}(X, L)}{2} + 1. \tag{14.8}$$

Theorem 14.4.2. *Let X be a smooth variety of dimension $n \geq 2$. Let L be a very ample line bundle on X. Assume that $k := \operatorname{def}(X, L) > 0$. Let $\phi : X \to W$ be the nefvalue morphism of (X, L). Let F be a general fiber of ϕ. Then* $\operatorname{def}(F, L_F) > 0$.

Proof. Let $m := \dim W$, $\theta := \operatorname{def}(F, L_F)$. Note that $\dim F \geq n - m - k + 1$ since $k \geq 1$. Let $J_1(L)_F$ be the restriction of the first jet bundle $J_1(X, L)$ to F. Then an argument similar to that used in the proof of Lemma (1.6.10) shows that $c_{n-m}(J_1(L)_F) = 0$ (we refer to [87, (0.16)] for more details). By using the sequence (1.6) we see that there is an exact sequence

$$0 \to \mathcal{N}^*_{F/X} \otimes L_F \to J_1(L)_F \to J_1(L_F) \to 0.$$

Furthermore $\mathcal{N}^*_{F/X} \otimes L_F \cong L_F \oplus \cdots \oplus L_F$ (m copies). The total Chern classes satisfy the relation $c(J_1(L)_F) = (1 + L_F)^m \cdot c(J_1(L_F))$. Hence

$$c_{n-m}(J_1(L)_F) = c_{n-m}(J_1(L_F)) + \sum_{i=1}^{m} \frac{m!}{(m-i)!i!} c_{n-m-i}(J_1(L_F)) \cdot L_F^i.$$

Note that all summands on the right are nonnegative since L_F is very ample and $J_1(L_F)$ is spanned. Since $c_{n-m}(J_1(L)_F) = 0$, we conclude that all the summands on the right are zero, and $c_{n-m}(J_1(L_F)) = 0$. This implies that $\theta > 0$ (see also (1.6.11)). □

The following key theorem builds on (14.4.2) and on a result of Lanteri and Struppa [393, (3.5)]. The result below reduces the classification of pairs (X, L) with positive defect to the study of Fano varieties of positive defect.

Theorem 14.4.3. *Let X be a smooth variety of dimension $n \geq 2$. Let L be a very ample line bundle on X. Let τ be the nefvalue of (X, L) and let $\phi : X \to W$ be the nefvalue morphism of (X, L). Let F be a general fiber of ϕ and let $\dim W := m$. Then the following conditions are equivalent:*

1. $\operatorname{def}(F, L_F) > m$;
2. $\operatorname{def}(X, L) > 0$;
3. $\operatorname{def}(X, L) = \operatorname{def}(F, L_F) - m > 0$.

Moreover if one of the above conditions holds true, then $\tau = \dfrac{n + \operatorname{def}(X, L)}{2} + 1$ *and* $\operatorname{Pic}(F) \cong \mathbb{Z}$.

Proof. Let $k = \operatorname{def}(X, L)$, $\theta = \operatorname{def}(F, L_F)$. From Lanteri and Struppa [393, (3.5)] we know that $k \geq \theta - n + \dim F = \theta - m$. Thus (14.4.3.1) implies (14.4.3.2). Clearly (14.4.3.3) implies (14.4.3.1). Then it is enough to show that (14.4.3.2) implies (14.4.3.3). Since $k > 0$ one has $\tau = (n+k)/2 + 1$ by Corollary (6.6.8). Then for a general fiber, F, of ϕ we have $K_F + ((n+k)/2+1)L_F \approx \mathcal{O}_F$. Let τ_F be the nefvalue of (F, L_F). Hence $\tau_F = (n+k)/2 + 1$. We know that $\theta > 0$ by (14.4.2). Thus by Corollary (6.6.8) we get $\tau_F = (n - m + \theta)/2 + 1$. Therefore $(n+k)/2 = (n - m + \theta)/2$, or equivalently, $k = \theta - m$. The isomorphism $\mathrm{Pic}(F) \cong \mathbb{Z}$ follows from (6.6.4) and the fact that $\mathrm{index}(F) \geq \tau_F > \dim F/2 + 1$. □

Given a pair (X, L) with defect $k(X, L) \geq 2$ we see by equation (1.7) that we can get a new positive defect pair, (A, L_A), with $k(A, L_A) = k(X, L) - 1$, by passing to smooth $A \in |L|$.

If (X, L) is a positive defect pair then let M be a projective manifold of dimension $y < k := k(X, L)$. Let $\phi : X \to W$ denote the nefvalue morphism of (X, L). Let H denote a very ample line bundle on M such that $K_M + H$ is very ample. Let $p : X \times M \to X$ and $q : X \times M \to M$ denote the product projections. Let $\mathcal{L} = p^*L + q^*H$ and let $\mathcal{M} = X \times M$. Note that the nefvalue of $(\mathcal{M}, \mathcal{L})$ is $(n+k)/2+1$ and the nefvalue morphism, $\rho : \mathcal{M} \to W \times M$, is simply the product morphism, (ϕ, id_M). By Theorem (14.4.3) we see that $k(\mathcal{M}, \mathcal{L}) = k - y > 0$.

The above two paragraphs show that given a positive defect pair, (X, L), with $k := k(X, L)$ and $n := \dim X$, then by taking hyperplane sections and making fibrations we can construct new pairs, $(A, \mathcal{H})$, with $(\dim A, k(A, \mathcal{H})) = (a, \kappa)$ for any $k \geq \kappa > 0$ satisfying $\kappa \equiv a \bmod 2$, $a + \kappa \leq n + k$. Up to construction of certain fibrations the study of positive defect pairs is reduced to the study of certain high index Fano manifolds. A pair (X, L) consisting of a very ample line bundle, L, on a connected projective manifold, X, is said to be a *maximal positive defect pair* if

1. there is no pair $(M, \mathcal{L})$ consisting of a very ample line bundle, $\mathcal{L}$, on a projective manifold, M, such that $X \in |\mathcal{L}|$ and $\mathcal{L}_X \cong L$;

2. $\mathrm{Pic}(X) = \mathbb{Z}$;

3. $k(X, L) > 0$.

It is a striking fact that all known examples of positive defect pairs, (X, L), are derived from the following very short and well-known list of homogenous maximal positive defect pairs; see Knop and Menzel [354], and also Lascoux [398] and Snow [566].

Example 14.4.4. It is a result of Lascoux [398] (see also Kleiman [353]) that if L is a very ample line bundle on a Grassmannian $X := \mathrm{Grass}(a, b)$, of $\mathbb{C}^a$'s in $\mathbb{C}^b$, and $\mathrm{def}(X, L) = k > 0$, then L is a generator of $\mathrm{Pic}(\mathrm{Grass}(a, b))$, and either

1. $a = 1$ or $a = b - 1$, i.e., $(\mathrm{Grass}(a, b), L) \cong (\mathbb{P}^{b-1}, \mathcal{O}_{\mathbb{P}^{b-1}}(1))$; or

2. $b = a + 2$, a is odd, and $k = 2$.

In the second case, the Plücker embedding of the Grassmannian $\mathrm{Grass}(a, a + 2)$ is into $\mathbb{P}^{a(a+3)/2}$. Note that in this case $K_X \approx \mathcal{O}_X(-a - 2)$, so that (X, L) is a Fano manifold of positive defect. By taking a general hyperplane section A of $\mathrm{Gr}(2, a+2)$ we find a manifold with $\mathrm{def}(A, L_A) = k - 1 = 1$ (see equation (1.7)).

Example 14.4.5. Let (X, L) be the pair consisting of the spinor 10-fold embedded in $\mathbb{P}^{15}$ by $|L|$. It is known that $k = \mathrm{def}(X, L) = 4$ (see Mukai [459] and Lanteri and Struppa [393]). *Except for scrolls there is no known example with greater defect.* Note that $K_X \approx -8L$, in other words $K_X + ((n + k)/2 + 1)L \approx \mathcal{O}_X$, so that (X, L) is a Fano 10-fold of positive defect. By taking a general hyperplane section A of X we find a manifold with $\mathrm{def}(A, L_A) = k - 1 = 3$ (see equation (1.7)).

There is the obvious conjecture.

Conjecture 14.4.6. *All maximal positive defect pairs are homogeneous.*

There is a basic result of Ein [203, (4.4)] that applies to the Fano manifolds of positive defect.

Theorem 14.4.7. (Ein) *Let X be a smooth variety of dimension $n \geq 2$ and let L be a very ample line bundle on X. Assume that $k := \mathrm{def}(X, L) > 0$. If $(X, L) \not\cong (\mathbb{P}^n, \mathcal{O}_{\mathbb{P}^n}(1))$ and $K_X \cong \mathcal{O}_X(b)$ for some $b \in \mathbb{Z}$ then $k \leq (n - 2)/2$.*

From Ein [204, (4.1)] we see that if $k \geq n/2$ it follows that the nefvalue morphism is a scroll map. In fact (X, L) is a linear $\mathbb{P}^{(n+k)/2}$-bundle over a $((n - k)/2)$-fold.

For $\dim X \leq 10$ a very complete classification of positive defect pairs (X, L) follows from (14.4.2), (14.4.3), and (14.4.7). For $n \leq 6$ and $n = 7$, $k = 5$ this classification is due to Ein [204]: the role of Corollary (6.6.8) is played by the earlier adjunction mapping results of Sommese [571, 574]. For $n = 7$, $k = 3$ the classification is due to Lanteri and Struppa [393]. If $n = 7$, $k = 1$ the classification is done in [393] under the extra assumption that $K_X + 5L$ is spanned.

We will sketch this classification up to $\dim X = 10$. First note that a general fiber F of the nefvalue morphism of (X, L) satisfies $-K_F \cong (\dim F + 1 - t)L_F$ for $t \leq 4$. To see this let $f = \dim F$, $n = \dim X$, and note that since (X, L) and (F, L_F) have the same nefvalue, $f + 1 - t$, we have by (14.8) that $f + 1 - t = (n + k)/2 + 1$, i.e., $t = f - (n + k)/2$. Assume that $t \geq 5$ so that $2f \geq 10 + n + k$. By (14.4.3) we have $k(F, L_F) = k + n - f$ and thus $f \geq 10 + k(F, L_F)$ contradicting the fact $f \leq n \leq 10$. If $t = 4$ the above reasoning shows that $n = f = 9$ with $k = 1$.

For $t = 0$ we get $(F, L_F) \cong (\mathbb{P}^{\dim F}, \mathcal{O}_{\mathbb{P}^{\dim F}}(1))$. We have that $t \neq 1$ since quadrics have 0 defect. If $t = 2$ then we have that F is a Del Pezzo manifold and

$k(F, L_F) = \dim F - 4$ by using (14.8). Thus by (14.4.7) and (14.4.3) we see that $\dim F - 4 = k(F, L_F) \leq (\dim F - 2)/2$, i.e., $\dim F \leq 6$, with equality implying $k(F, L_F) = 2$. It can be shown that Conjecture (14.4.6) is true for Del Pezzo manifolds. For $t = 3$, the same reasoning shows that $\dim F \leq 10$ with equality implying $k = 4$. Mukai [458, 459, 460] has extended Fano's work on 3-folds to give a classification of projective manifolds, X, with a very ample line bundle, L, such that $-K_X \cong (n-2)L$. For $t = 4$ we have a Fano 9-fold with $\mathrm{Pic}(X) \cong \mathbb{Z}$ and $-K_X \cong 6L$. We refer to [87] for further details on the classification of positive defect pairs up to dimension 10.

14.5 Hyperplane sections of curves

The overwhelming majority of results in this book are on the hyperplane sections of n-folds with $n \geq 2$. In this section we would like to discuss some of the current work on hyperplane sections of curves. The second author would like to thank Juan Migliore for helpful discussions without which this section would not have been written. Migliore's notes [434] were particularly helpful.

As noted in (1.4.9), Castelnuovo's bound and the Gruson and Peskine bound give sharp bounds for the genus of a nondegenerate irreducible curve, $C \subset \mathbb{P}^{N+1}$, in terms of $N+1$ and the cardinality of a hyperplane section of C, i.e., the degree of C.

What sort of fine structure can a 0-dimensional subscheme, Z, in $\mathbb{P}^N$ have beyond its cardinality? One obvious invariant is the *Hilbert function*, $H_Z(t)$, of Z, i.e., the integer valued function defined for integers, $t \geq 0$, by

$$H_Z(t) := \dim \operatorname{image}\{H^0(\mathbb{P}^N, \mathcal{O}_{\mathbb{P}^N}(t)) \to H^0(Z, \mathcal{O}_{\mathbb{P}^N}(t)_Z)\}.$$

Note that for large t the Hilbert function is equal to the Hilbert polynomial of Z, i.e., $H_Z(t) = \chi(\mathcal{O}_{\mathbb{P}^N}(t)_Z) = \deg Z$, for $t \gg 0$. It is a basic fact that the Hilbert function contains a lot of information about the configuration of the points, Z.

A finite set of distinct points, $Z \subset \mathbb{P}^N$, is said to have the *Uniform Position Property* (or UPP for short), if given any subset, $Z' \subset Z$, all subsets of Z of the same cardinality as Z' have the same Hilbert function. Geometrically this is equivalent to the condition that the dimension of the vector space of homogeneous functions of degree d vanishing on a subset $Z' \subset Z$ of cardinality m is the same as the dimension of the vector space of homogeneous functions of degree d vanishing on any other subset $Z'' \subset Z$ of cardinality m.

There is a result of Harris [307] in this direction which showed the importance of the UPP.

Theorem 14.5.1. *Let C be an irreducible nondegenerate curve in $\mathbb{P}^N$. Then the general hyperplane section of C has the uniform position property.*

Eisenbud and Harris [310] ask the following questions:

1. What may be the Hilbert function of finite set of points with the UPP?
2. What may be the Hilbert function of the general hyperplane sections of an irreducible curve?
3. Are the answers to 1) and 2) the same?

For sets of points in $\mathbb{P}^2$ and curves in $\mathbb{P}^3$, Gruson and Peskine [299], Maggioni and Ragusa [424, 425], and Sauer [543] proved results in rather different forms, which answered question 2). For a comparison of the different techniques used, see the paper of Geramita and Migliore [276]. Some results about UPP, including the fact that the answer to question 1) coincides with that of question 2) for points in $\mathbb{P}^2$, can be seen for instance in Maggioni and Ragusa, Davis, Ellia and Peskine, and Geramita and Migliore [424, 177, 212, 274].

Passing to finite sets of points, $Z \subset \mathbb{P}^3$, much less is known. In [132], Bigatti, Geramita, and Migliore show how—in a fashion analogous to the case of $Z \subset \mathbb{P}^2$—it is possible to conclude that if Z has the UPP and if the growth of the Hilbert function of Z, from $t = d$ to $t = d + 1$, is the maximum allowed by a classical result of Macaulay [420], then $Z \subset V$ where V is an irreducible curve or surface in $\mathbb{P}^3$ (you can decide which from the Hilbert function), and the elements of I_Z are the same as the elements of I_V for all degrees $\leq d + 1$. Here I_Z, I_V, denote the homogeneous ideals of Z and V respectively.

A related line of research is on when appropriate conditions of an algebraic nature on a hyperplane section of a curve lift to the curve.

Here is one result in this direction conjectured by Geramita and Migliore [275], and shown by Strano [590]. Note that in what follows, "complete intersection" is understood, as usual, in the scheme-theoretic sense.

Theorem 14.5.2. *Let Z be a hyperplane section of an irreducible curve, C, in $\mathbb{P}^3$. If C is not contained in a quadric, then C is a complete intersection of multidegrees (a, b) in $\mathbb{P}^3$ if Z is a complete intersection of multidegrees (a, b) in $\mathbb{P}^2$.*

It should be noted that the analogous higher dimensional case of Theorem (14.5.2) is a well-known fact. I.e., if X is an n-dimensional nondegenerate projective variety in $\mathbb{P}^N$ and a hyperplane section, $X \cap H$, of X is a complete intersection of multidegrees $(a_1, \dots, a_{N-n})$ in $H = \mathbb{P}^{N-1}$, then X is a complete intersection of multidegrees $(a_1, \dots, a_{N-n})$ in $\mathbb{P}^N$.

Further work on this relaxing the conditions on $\mathbb{P}^3$ and allowing hypersurface sections is due to Re [514], Huneke and Ulrich [327], Migliore [435], and Migliore and Nagel [437].

One can also ask, following Eisenbud, Ein, and Katz [211], which sets of points, Z, can be hyperplane sections of smooth curves, and one can ambitiously try to find special curves, C, with this property, i.e., C contains a given set of points as hyperplane section. Positive results have been obtained by Chiantini and Orecchia [155], who found smooth projectively normal curves, for $Z \subset \mathbb{P}^2$, and by Ballico and Migliore [64] and Walter [613] for $Z \subset \mathbb{P}^N$ with $N \geq 3$. Negative results for general sets of points and certain classes of curves have been found by Ballico [56] and Walter [614].

Finally we would like to call attention to a related result of Laudal [399] (see also Strano [591]) that complements a classical result of Roth [526]. Roth showed that given a general hyperplane section, C, of a nondegenerate smooth surface, S, of degree d in $\mathbb{P}^4$, then if $d \geq \sigma^2 + 1$, and C is contained in a hypersurface of degree σ in $\mathbb{P}^3$, then S is contained in a hypersurface of degree σ in $\mathbb{P}^4$. As noted by Braun, Ottaviani, Schneider, and Schreyer in [138, Theorem 1.2] this statement is true with S replaced by a smooth 3-fold in $\mathbb{P}^5$, and C replaced by a general hyperplane section of X. Laudal showed that if Z is a general hyperplane section of an irreducible curve, C, of degree d in $\mathbb{P}^3$, and if $d \geq \sigma^2 + 2$ with Z lying on a curve in $\mathbb{P}^2$ of degree σ, then C lies on a surface of degree σ in $\mathbb{P}^3$. For some further developments in the same direction, see Mezzetti [432].

Bibliography

[1] J. Alexander, "Surfaces rationnelles non spéciales dans $\mathbb{P}^4$," Math. Z. 200 (1988), 87–110.

[2] J. Alexander, "Surfaces rationnelles spéciales de degré 9 dans $\mathbb{P}^4$," preprint.

[3] V.A. Alexeev, "Ample Weil divisors on $K3$ surfaces with Du Val singularities," Duke Math. J. 64 (1991), 617–624.

[4] A. Altman and S. Kleiman, *Introduction to Grothendieck Duality Theory*, Lecture Notes in Math. 146 (1970), Springer-Verlag, New York.

[5] A. Alzati and M. Bertolini, "Sulla razionalità delle 3-varietà di Fano con $B_2 \geq 2$," Matematiche 47 (1992), 63–74.

[6] A. Alzati, M. Bertolini, and G.M. Besana, "Projective normality of varieties of small degree," preprint, 1994.

[7] T. Ando, "On extremal rays of the higher dimensional varieties," Invent. Math. 81 (1985), 347–357.

[8] M. Andreatta, "The stable adjunction mapping," Math. Ann. 275 (1986), 305–315.

[9] M. Andreatta, "Surfaces of sectional genus $g \leq 8$ with no trisecant lines," Arch. Math. (Basel) 60 (1993), 85–95.

[10] M. Andreatta, "Contraction of Gorenstein polarized varieties with high nef value," to appear in Math. Ann.

[11] M. Andreatta and E. Ballico, "On the adjunction process over a surface in char. p," Manuscripta Math. 62 (1988), 227–244.

[12] M. Andreatta and E. Ballico, "Classification of projective surfaces with small sectional genus: char. $p \geq 0$," Rend. Sem. Mat. Univ. Padova 84 (1990), 175–193.

[13] M. Andreatta and E. Ballico, "On the adjunction process over a surface in char. p II: the singular case," J. Reine Angew. Math. 417 (1991), 77–85.

[14] M. Andreatta and E. Ballico, "Projectively normal adjunction bundles," Proc. Amer. Math. Soc. 112 (1991), 919–924.

[15] M. Andreatta and E. Ballico, "On the projective normality of adjunction bundles: the singular case," in *Geometry of Complex Projective Varieties, Cetraro, Italy, 1990*, ed. by A. Lanteri, M. Palleschi, and D. Struppa, Seminars and Conferences 9 (1993), 1–7, Mediterranean Press.

[16] M. Andreatta and E. Ballico, “Hyperplane sections of projective surfaces with $K_S = \mathcal{O}_S$,” Comm. Algebra 18 (1990), 3639–3646.

[17] M. Andreatta and E. Ballico, “Curves and 0-cycles on projective surfaces,” Geom. Dedicata 51 (1994), 29–45.

[18] M. Andreatta, E. Ballico, and J.A. Wiśniewski, “Projective manifolds containing large linear subspaces,” in *Classification of Irregular Varieties: Proceedings Trento, 1990*, ed. by E. Ballico, F. Catanese, and C. Ciliberto, Lecture Notes in Math. 1515 (1991), 1–11, Springer-Verlag, New York.

[19] M. Andreatta, E. Ballico, and J.A. Wiśniewski, “Vector bundles and adjunction,” Internat. J. Math. 3 (1992), 331–340.

[20] M. Andreatta, E. Ballico, and J.A. Wiśniewski, “Two theorems on elementary contractions,” Math. Ann. 297 (1993), 191–198.

[21] M. Andreatta, M. Beltrametti, and A.J. Sommese, “Generic properties of the adjunction mapping for singular surfaces and applications,” Pacific J. Math. 142 (1990), 1–15.

[22] M. Andreatta and A. Lanteri, ”On projective surfaces arising from an adjunction process,” Japan. J. Math. (N.S.) 17 (1991), 285–298.

[23] M. Andreatta and M. Palleschi, “On the 2-very ampleness of the adjoint bundle, (with an appendix by E. Ballico)”, Manuscripta Math. 73 (1991), 45–62.

[24] M. Andreatta and A.J. Sommese, “On the adjunction mapping for singular projective varieties,” Forum Math. 1 (1989), 143–152.

[25] M. Andreatta and A.J. Sommese, “Generically ample divisors on normal Gorenstein surfaces,” in *Singularities*, Contemp. Math. 90 (1989), 1–20.

[26] M. Andreatta and A.J. Sommese, “Classification of irreducible projective surfaces of smooth sectional genus ≤ 3,” Math. Scand. 67 (1990), 197–214.

[27] M. Andreatta and A.J. Sommese, “On the projective normality of the adjunction bundles (with an appendix by M. Andreatta, E. Ballico, and A.J. Sommese),” Comment. Math. Helv. 66 (1991), 362–367.

[28] M. Andreatta and J.A. Wiśniewski, “A note on non-vanishing and its applications,” Duke Math. J. 72, (1993), 739–755.

[29] A. Andreotti, “The Lefschetz theorem on hyperplane sections,” Ann. of Math. 69 (1959), 713–717.

[30] A. Andreotti, “On a theorem of Torelli,” Amer. J. Math. 80 (1958), 801–828.

[31] A. Andreotti and T. Frankel, “The second Lefschetz theorem on hyperplane sections,” in *Global Analysis: Papers in honor of K. Kodaira*, ed. by D.C. Spencer and S. Iyanaga, 1–20, (1969), Princeton Univ. Press.

[32] A. Andreotti and H. Grauert, “Théorèmes de finitude pour la cohomologie des espaces complexes,” Bull. Soc. Math. France, 90 (1962), 193–259.

[33] E. Arbarello, M. Cornalba, P.A. Griffiths, and J. Harris, *Geometry of Algebraic Curves*, Volume I, Grundlehren Math. Wiss. 267, Springer-Verlag, New York, (1985).

[34] M. Artin, "Some numerical criteria for contractibility of curves on algebraic surfaces," Amer. J. Math. 84 (1962), 485–496.

[35] M. Artin, "On isolated rational singularities of surfaces," Amer. J. Math. 88 (1966), 129–136.

[36] M. Artin, "Théorèmes de répresentabilité pour les espaces algébriques," S.M.S. 1970, Université de Montréal, 44 (1973).

[37] A. Aure, W. Decker, K. Hulek, S. Popescu, K. Ranestad, "The geometry of bielliptic surfaces in $\mathbb{P}^4$," Internat. J. Math. 4 (1993), 873–902.

[38] L. Bădescu, "A remark on the Grothendieck-Lefschetz theorem about the Picard group," Nagoya Math. J. 71 (1978), 169–179.

[39] L. Bădescu, "On ample divisors," Nagoya Math. J. 86 (1982), 155–171.

[40] L. Bădescu, "On ample divisors: II," in *Proceedings of the Week of Algebraic Geometry, Bucharest, 1980*, Teubner-Texte Math. 40 (1981), 12–32.

[41] L. Bădescu, "The projective plane blown-up at a point as an ample divisor," Atti Accad. Ligure Sci. Lett. 38 (1981), 3–7.

[42] L. Bădescu, "Hyperplane sections and deformations," in *Proceedings of the Week of Algebraic Geometry, Bucharest, 1982*, Lecture Notes in Math. 1056 (1984), 1–33, Springer-Verlag, New York.

[43] L. Bădescu, "On a criterion for hyperplane sections," Math. Proc. Cambridge Philos. Soc. 103 (1988), 59–67.

[44] L. Bădescu, "Infinitesimal deformations of negative weights and hyperplane sections," in *Algebraic Geometry, Proceedings of Conference on Hyperplane Sections, L'Aquila, Italy, 1988*, ed. by A. Sommese, A. Biancofiore, and E.L. Livorni, Lecture Notes in Math. 1417 (1990), 1–22, Springer-Verlag, New York.

[45] L. Bădescu, "Polarized varieties with no deformations of negative weights," in *Geometry of Complex Projective Varieties, Cetraro, Italy, 1990*, ed. by A. Lanteri, M. Palleschi, and D. Struppa, Seminars and Conferences 9 (1993), 9–33, Mediterranean Press.

[46] L. Bădescu, "On a result of Zak-L'vovsky," in *Projective Geometry with Applications*, ed. by E. Ballico, Lecture Notes in Pure and Applied Math. 166 (1994), 57–73, Marcel Dekker, New York.

[47] L. Bădescu, "Algebraic Barth-Lefschetz theorems", preprint.

[48] E. Ballico, "On vector bundles on threefolds with sectional genus 1," Trans. Amer. Math. Soc. 324 (1991), 135–147.

[49] E. Ballico, "On ample and spanned rank-3 bundles with low Chern numbers," Pacific J. Math. 140 (1989), 209–216.

[50] E. Ballico, "Spanned and ample vector bundles with low Chern numbers," Manuscripta Math. 68 (1990), 9–16.

[51] E. Ballico, "On ample and spanned vector bundles with zero Δ-genera," Manuscripta Math. 70 (1991), 153–155.

[52] E. Ballico, "On k-spanned projective surfaces," in *Algebraic Geometry, Proceedings of Conference on Hyperplane Sections, L'Aquila, Italy, 1988*, ed. by A.J. Sommese, A. Biancofiore, and E.L. Livorni, Lecture Notes in Math. 1417 (1990), p. 23, Springer-Verlag, New York.

[53] E. Ballico, "A characterization of the Veronese surface," Proc. Amer. Math. Soc. 105 (1989), 531–534.

[54] E. Ballico, "On k-spanned embeddings of projective manifolds," Manuscripta Math. 75 (1992), 103–107.

[55] E. Ballico, "On projective varieties with projectively equivalent zero-dimensional linear sections," Canad. Math. Bull. 35 (1992), 3–13.

[56] E. Ballico, "Points not as hyperplane sections of linearly normal curves," to appear in Proc. Amer. Math. Soc.

[57] E. Ballico and M.C. Beltrametti, "On 2-spannedness for the adjunction mapping," Manuscripta Math. 61 (1988), 447–458.

[58] E. Ballico and M.C. Beltrametti,"On the k-spannedness of the adjoint line bundle," Manuscripta Math. 76 (1992), 407–420.

[59] E. Ballico and M.C. Beltrametti, "On the k-spannedness of the adjoint line bundle for degree $\leq 4k+4$," to appear in Abh. Math. Sem. Univ. Hamburg.

[60] E. Ballico, L. Chiantini, and V. Monti, "On the adjunction mapping for surfaces of Kodaira dimension ≤ 0 in char. p," Manuscripta Math. 73 (1991), 313–318.

[61] E. Ballico and C. Ciliberto, "On Gaussian maps for projective varieties," in *Geometry of Complex Projective Varieties, Cetraro, Italy, 1990*, ed. by A. Lanteri, M. Palleschi, and D. Struppa, Seminars and Conferences 9 (1993), 35–53, Mediterranean Press.

[62] E. Ballico and A. Lanteri, "An indecomposable rank-2 vector bundle the complete linear system of whose determinant consists of hyperelliptic curves," Boll. Un. Mat. Ital. 7 (1989), 225–230.

[63] E. Ballico and A. Lanteri, "Ample and spanned rank-2 vector bundles with $c_2 = 2$ on complex surfaces," Arch. Math. (Basel) 56 (1991), 611–615.

[64] E. Ballico and J. Migliore, "Smooth curves whose hyperplane section is a given set of points," Comm. Algebra 18 (1990), 3015–3040.

[65] E. Ballico and A.J. Sommese, "Projective surfaces with k-very ample line bundles of degree $\leq 4k+4$," to appear in Nagoya Math. J.

[66] E. Ballico and J.A. Wiśniewski, "On Bănică sheaves and Fano manifolds," Max Planck Institut für Math. preprint, 94-3.

[67] D. Barlet, "Espace analytique réduit des cycles analytiques complexes compacts d'un espace analytique complexe de dimension finie," in *Fonctions de Plusiers Variables Complexes II*, ed. by F. Norguet, Lecture Notes in Math. 482 (1975), 1–158, Springer-Verlag, New York.

[68] W. Barth, "Transplanting cohomology classes in complex-projective space," Amer. J. Math. 92 (1970), 951–967.

[69] W. Barth, "Submanifolds of low codimension in projective space," in *Proc. Intern. Cong. Math.*, Vancouver (1974), 409–413.

[70] W. Barth, "Larsen's theorem on the homotopy groups of projective manifolds of small embedding codimension," in *Algebraic Geometry Arcata 1974*, Proc. Sympos. Pure Math. 29 (1975), 307–313.

[71] W. Barth and M.E. Larsen, "On the homotopy groups of complex manifolds," Math. Scand. 30 (1972), 88–94.

[72] W. Barth, C. Peters, and A. Van de Ven, *Compact Complex Surfaces*, Ergeb. Math. Grenzgeb. (3) 4, Springer-Verlag, Berlin, (1984).

[73] W. Barth and A. Van de Ven, "Fano-varieties of lines on hypersurfaces," Arch. Math. (Basel) 31 (1978), 96–104.

[74] V.V. Batyrev, "The cone of effective divisors of threefolds," Contemp. Math. 131 (1992) Part 3, 337–352.

[75] L. Bayle, "Classification des variétés complexes projective de dimension trois dont une section hyperplane générale est une surface d'Enriques, " J. Reine Angew. Math. 449 (1994), 9–63.

[76] A. Beauville, "Variétés de Prym et Jacobiennes intermédiaires," Ann. Sci. École Norm. Sup. (4) 10 (1977), 309–391.

[77] A. Beauville, "Surfaces algébriques complexes," Astérisque 54 (1978).

[78] A. Beauville, "L'application canonique pour les surfaces de type general," Invent. Math. 55 (1979), 121–140.

[79] A. Beauville, "Letter continuing Reid's letter on Reider's method," dated March 1986.

[80] A. Beauville, "Sur les variétés dont les sections hyperplanes sont à module constant," preprint.

[81] A. Beauville and J.Y. Merindol, "Sections hyperplanes des surfaces $K3$," Duke Math. J. 55 (1987), 873–878.

[82] M.C. Beltrametti, "On d-folds whose canonical bundle is not numerically effective, according to Mori and Kawamata," Ann. Mat. Pura Appl. 147 (1987), 151–172.

[83] M.C. Beltrametti, "Contractions of non-numerically effective extremal rays in dimension 4," in *Proceedings of Conference on Algebraic Geometry, Berlin, 1985*, Teubner-Texte Math. 92 (1987), 24–37.

[84] M.C. Beltrametti, G.M. Besana, and A.J. Sommese, "On the dimension of the adjoint linear system for quadric fibrations," in *Algebraic Geometry and its Applications, Proceedings of the 8-th Algebraic Geometry Conference, Yaroslavl', 1992*, ed. by A. Tikhomirov and A. Tyurin, Aspects of Math. E 25 (1994), 9–20, Vieweg, Braunschweig.

[85] M.C. Beltrametti, A. Biancofiore, and A.J. Sommese, "Projective n-folds of log general type. I," Trans. Amer. Math. Soc. 314 (1989), 825–849.

[86] M.C. Beltrametti, M.L. Fania, and A.J. Sommese, "On the adjunction theoretic classification of projective varieties," Math. Ann. 290 (1991), 31–62.

[87] M.C. Beltrametti, M.L. Fania, and A.J. Sommese, "On the discriminant variety of a projective manifold," Forum Math. 4 (1992), 529–547.

[88] M.C. Beltrametti, P. Francia, and A.J. Sommese, "On Reider's method and higher order embeddings," Duke Math. J. 58 (1989), 425–439.

[89] M.C. Beltrametti and A. Lanteri, "On the 2 and the 3-connectedness of ample divisors on a surface," Manuscripta Math. 58 (1987), 109–128.

[90] M.C. Beltrametti, A. Lanteri, and M. Palleschi, "Algebraic surfaces containing an ample divisor of arithmetic genus two," Ark. Mat. 25 (1987), 189–210.

[91] M.C. Beltrametti and M. Palleschi, "On threefolds with low sectional genus," Nagoya Math. J. 101 (1986), 27–36.

[92] M.C. Beltrametti and L. Robbiano, "Introduction to the theory of weighted projective spaces," Exposition. Math. 4 (1986), 111–162.

[93] M.C. Beltrametti, M. Schneider, and A.J. Sommese, "Threefolds of degree 9 and 10 in $\mathbb{P}^5$," Math. Ann. 288 (1990), 413–444.

[94] M.C. Beltrametti, M. Schneider, and A.J. Sommese, "Threefolds of degree 11 in $\mathbb{P}^5$," in *Complex Projective Geometry*, ed. by G. Ellingsrud, C. Peskine, G. Sacchiero, and S.A. Stromme, London Math. Soc. Lecture Note Ser. 179 (1992), 59–80.

[95] M.C. Beltrametti, M. Schneider, and A.J. Sommese, "Applications of the Ein-Lazarsfeld criterion for spannedness of adjoint bundles," Math. Z. 214 (1994), 593–599.

[96] M.C. Beltrametti, M. Schneider, and A.J. Sommese, "Special properties of the adjunction theory for threefolds in $\mathbb{P}^5$," to appear in Mem. of the Amer. Math. Soc.

[97] M.C. Beltrametti, M. Schneider, and A.J. Sommese, "Chern inequalities and spannedness of adjoint bundles," in *Proceedings of the Hirzebruch 65 Conference, Bar-Ilan, 1993*, to appear.

[98] M.C. Beltrametti and A.J. Sommese, "A criterion for a variety to be a cone," Comment. Math. Helv. 62 (1987), 417–422.

[99] M.C. Beltrametti and A.J. Sommese, "On generically polarized Gorenstein surfaces of sectional genus 2," J. Reine Angew. Math. 386 (1988), 172–186.

[100] M.C. Beltrametti and A.J. Sommese, "On normal Gorenstein polarized varieties of sectional genus 3 and 4," Indiana Univ. Math. J. 37 (1988), 667–686.

[101] M.C. Beltrametti and A.J. Sommese, "On the relative adjunction mapping," Math. Scand. 65 (1989), 189–205.

[102] M.C. Beltrametti and A.J. Sommese, "On k-spannedness for projective surfaces," in *Algebraic Geometry, Proceedings of Conference on Hyperplane Sections, L'Aquila, Italy, 1988*, ed. by A.J. Sommese, A. Biancofiore, and E.L. Livorni, Lecture Notes in Math. 1417 (1990), 24–51, Springer-Verlag, New York.

[103] M.C. Beltrametti and A.J. Sommese, "Zero cycles and k-th order embeddings of smooth projective surfaces (with an appendix by L. Göttsche)," in *Problems in the Theory of Surfaces and their Classification, Cortona, Italy, 1988*, ed. by F. Catanese and C. Ciliberto, Sympos. Math. 32 (1992), 33–48, INDAM, Academic Press, London.

[104] M.C. Beltrametti and A.J. Sommese, "New properties of special varieties arising from adjunction theory," J. Math. Soc. Japan 43 (1991), 381–412.

[105] M.C. Beltrametti and A.J. Sommese, "Comparing the classical and the adjunction theoretic definition of scrolls," in *Geometry of Complex Projective Varieties, Cetraro, Italy 1990*, ed. by A. Lanteri, M. Palleschi, and D. Struppa, Seminars and Conferences 9 (1993), 55–74, Mediterranean Press.

[106] M.C. Beltrametti and A.J. Sommese, "On the adjunction theoretic classification of polarized varieties," J. Reine Angew. Math. 427 (1992), 157–192.

[107] M.C. Beltrametti and A.J. Sommese, "On k-jet ampleness," in *Complex Analysis and Geometry*, ed. by V. Ancona and A. Silva, 355–376 (1993), Plenum Press, New York.

[108] M.C. Beltrametti and A.J. Sommese, "On the preservation of k-very ampleness under adjunction," Math. Z. 212 (1993), 257–284.

[109] M.C. Beltrametti and A.J. Sommese, "Special results in adjunction theory in dimension four and five," Ark. Mat. 31 (1993), 197–208.

[110] M.C. Beltrametti and A.J. Sommese, "A remark on the Kawamata rationality theorem," J. Math. Soc. Japan 45 (1993), 557–568.

[111] M.C. Beltrametti and A.J. Sommese, "Some effects of the spectral values on reductions," in *Classification of Algebraic Varieties, L'Aquila, Italy, 1992*, ed. by C. Ciliberto, E.L. Livorni, and A.J. Sommese, Contemp. Math. 162 (1994), 31–48.

[112] M.C. Beltrametti and A.J. Sommese, "On the dimension of the adjoint linear system for threefolds,", to appear in Ann. Scuola Norm. Sup. Pisa.

[113] M.C. Beltrametti and A.J. Sommese, "Remarks on numerically positive and big line bundles," in *Projective Geometry with Applications*, ed. by E. Ballico, Lecture Notes in Pure and Applied Math. 166 (1994), 9–18, Marcel Dekker, New York.

[114] M.C. Beltrametti, A.J. Sommese, and J.A. Wiśniewski, "Results on varieties with many lines and their applications to adjunction theory (with an appendix by M.C. Beltrametti and A.J. Sommese)," in *Complex Algebraic Varieties, Bayreuth 1990*, ed. by K. Hulek, T. Peternell, M. Schneider, and F.-O. Schreyer, Lecture Notes in Math. 1507 (1992), 16–38, Springer-Verlag, New York.

[115] X. Benveniste, "Sur le cone des 1-cycles effectifs en dimension 3," Math. Ann. 272 (1985), 257–265.

[116] E. Bertini, *Introduzione alla geometria proiettiva degli iperspazi*, 2^a edizione riveduta ed ampliata, Messina, casa editrice G. Principato, 1923.

[117] A. Bertram, L. Ein, and R. Lazarsfeld, "Vanishing theorems, a theorem of Severi, and the equations defining projective varieties," J. Amer. Math. Soc. 4 (1991), 587–602.

[118] G.M. Besana, "On the geometry of conic bundles arising in adjunction theory," Math. Nachr. 160 (1993), 223–251.

[119] E. Bese, "On the spannedness and very ampleness of certain line bundles on the blow-ups of $\mathbb{P}^2_{\mathbb{C}}$ and $\mathbb{F}_r$," Math. Ann. 262 (1983), 225–238.

[120] A. Biancofiore, "On the degree of the discriminant locus of a smooth hyperplane section of a threefold with nonnegative Kodaira dimension," Arch. Math. (Basel) 48 (1987), 538–542.

[121] A. Biancofiore, "On the hyperplane sections of ruled surfaces," in *Algebraic Geometry, Proceedings of Conference on Hyperplane Sections, L'Aquila, Italy, 1988*, ed. by A. Sommese, A. Biancofiore, and E.L. Livorni, Lecture Notes in Math. 1417 (1990), 52–66, Springer-Verlag, New York.

[122] A. Biancofiore and G. Ceresa, "Remarks on k-spannedness for algebraic surfaces," Rend. Sem. Mat. Univ. Politec. Torino, 46 (1988), 343–352.

[123] A. Biancofiore, M.L. Fania, and A. Lanteri, "Polarized surfaces with hyperelliptic sections," Pacific J. Math. 143 (1990), 9–24.

[124] A. Biancofiore, A. Lanteri, and E.L. Livorni, "Ample and spanned vector bundles of sectional genera three," Math. Ann. 291 (1991), 87–101.

[125] A. Biancofiore and E.L. Livorni, "Algebraic non ruled surfaces with sectional genus equal to seven," Ann. Univ. Ferrara Sez. VII (N.S.) 32 (1986), 1–14.

[126] A. Biancofiore and E.L. Livorni, "Algebraic ruled surfaces with low sectional genus," Ricerche Mat. 36 (1987), 17–32.

[127] A. Biancofiore and E.L. Livorni, "On the iteration of the adjunction process in the study of rational surfaces," Indiana Univ. Math. J. 36 (1987), 167–188.

[128] A. Biancofiore and E.L. Livorni, "On the genus of a hyperplane section of a geometrically ruled surface," Ann. Mat. Pura Appl. 147 (1987), 173–185.

[129] A. Biancofiore and E.L. Livorni, "On the iteration of the adjunction process for surfaces of negative Kodaira dimension," Manuscripta Math. 64 (1989), 35–54.

[130] A. Biancofiore and E.L. Livorni, "The asymptotic behavior of the set of polarized, negative Kodaira dimension, surfaces as the sectional genus or the degree goes to infinity," preprint, 1992.

[131] A. Biancofiore, E.L. Livorni, and A.J. Sommese "On the complexity of the projective classification of surfaces," to appear in Monatsh. Math.

[132] A. Bigatti, A.V. Geramita, and J. Migliore, "Geometric consequences of extremal behavior in a theorem of Macaulay," preprint.

[133] P. Blass and J. Lipman, "Remarks on adjoints and arithmetic genera of algebraic varieties," Amer. J. Math. 101 (1979), 331–336.

[134] S. Bloch, "Semi-regularity and de Rham cohomology," Invent. Math. 17 (1972), 51–66.

[135] E. Bombieri, "Canonical models of surfaces of general type," Inst. Hautes Études Sci. Publ. Math. 42 (1973), 171–219.

[136] R. Bott, "On a theorem of Lefschetz," Michigan Math. J. (1959), 211–216.

[137] R. Braun, "On a geometric property of the normal bundle of surfaces in $\mathbb{P}_4$," Math. Z. 206 (1991), 535–550.

[138] R. Braun, G. Ottaviani, M. Schneider, and F.-O. Schreyer, "Boundedness of non-general type 3-folds in $\mathbb{P}^5$," in *Complex Analysis and Geometry*, ed. by V. Ancona and A. Silva, 311–338 (1993), Plenum Press, New York.

[139] R. Braun, G. Ottaviani, M. Schneider, F.-O. Schreyer, "Classification of log-special 3-folds in $\mathbb{P}^5$," preprint Bayreuth, 1992.

[140] A. Brigaglia, C. Ciliberto, and E. Sernesi, "Italian algebraic geometers from 1850 to 1970: a bibliography and a few biographical notes," 305 pages, (1992), Rome.

[141] A. Buium, "Espaces projectifs anisotrophes et diviseurs amples," C.R. Acad. Sci. Paris 292 (1981), 585–586.

[142] A. Buium, "On surfaces of degree at most $2n + 1$ in $\mathbb{P}^n$," in *Algebraic Geometry, Bucharest, 1982*, Lecture Notes in Math. 1056 (1984), 47–67, Springer-Verlag, New York.

[143] F. Campana and H. Flenner, "Projective threefolds containing a smooth rational surface with ample normal bundle," J. Reine Angew. Math. 440 (1993), 77–98.

[144] J.B. Carrell and A.J. Sommese, "Some topological aspects of $\mathbb{C}^*$ actions on compact Kaehler manifolds," Comment. Math. Helv. 54 (1979), 567–582.

[145] G. Castelnuovo, "Sulle superficie algebriche le cui sezioni piane sono curve iperellittiche," Rend. Circ. Mat. Palermo 4 (1890), 73–88.

[146] G. Castelnuovo, "Sui multipli di una serie lineare di gruppi di punti appartenenti ad una curva algebrica," Rend. Circ. Mat. Palermo 7 (1893), 89–110.

[147] G. Castelnuovo, "Sulle serie algebriche di punti appartenenti ad una curva algebrica," Atti Reale Accad. Naz. Lincei Rend. Cl. Sci. Fis. Mat. Natur. (5) 15 (1906), 337–359.

[148] G. Castelnuovo and F. Enriques, "Sur quelques résultats nouveaux dans la théorie des surfaces algébriques," in [504] E. Picard and G. Simart, *Théorie des Fonctions Algébriques, I, II.*

[149] G. Castelnuovo and F. Enriques, "Sopra alcune questioni fondamentali nella teoria delle superficie algebriche," Ann. Mat. Pura Appl. (3) 6 (1901), 165–225.

[150] F. Catanese, "Footnotes to a theorem of I. Reider," in *Algebraic Geometry, Proceedings of Conference on Hyperplane Sections, L'Aquila, Italy, 1988*, ed. by A.J. Sommese, A. Biancofiore, and E.L. Livorni, Lecture Notes in Math. 1417 (1990), 67–74, Springer-Verlag, New York.

[151] F. Catanese and C. Ciliberto, "Surfaces with $p_g = q = 1$," in *Problems in the Theory of Surfaces and their Classification, Cortona, Italy, 1988*, ed. by F. Catanese and C. Ciliberto, Sympos. Math. 32 (1992), INDAM, 49–79, Academic Press, London.

[152] F. Catanese and L. Göttsche, "d-very-ample line bundles and embeddings of Hilbert schemes of 0-cycles," Manuscripta Math. 68 (1990), 337–341.

[153] M.-C. Chang, "On the hyperplane sections of certain codimension 2 subvarieties of $\mathbb{P}^n$," Ark. Mat. 58 (1992), 547–550.

[154] L. Chiantini, C. Ciliberto, V. Di Gennaro, "The genus of projective curves," Duke Math. J. 70 (1993), 229–245.

[155] L. Chiantini and F. Orecchia, "Plane sections of arithmetically normal curves in $\mathbb{P}^3$," in *Algebraic Curves and Projective Geometry, Proceedings Trento, 1988*, ed. by E. Ballico and C. Ciliberto, Lecture Notes in Math. 1389 (1989), 32–42, Springer-Verlag, New York.

[156] K. Cho and Y. Miyaoka, "Characterizations of projective spaces and hyperquadrics in terms of the minimum degree of rational curves," preprint, 1993.

[157] C. Ciliberto, "Hilbert functions of finite sets of points and the genus of a curve in projective space," in *Space Curves, Rocca di Papa, 1985*, Lecture Notes in Math. 1266 (1987), 24–73, Springer-Verlag, New York.

[158] C. Ciliberto, "On a property of Castelnuovo varieties," Trans. Amer. Math. Soc. 303 (1987), 201–210.

[159] C. Ciliberto, "On the Hilbert scheme of curves of maximal genus in a projective space," Math. Z. 194 (1987), 351–363.

[160] C. Ciliberto, "On the degree and genus of smooth curves in a projective space," Adv. Math. 81, (1990), 198–248.

[161] C. Ciliberto, J. Harris, and R. Miranda, "General components of the Noether-Lefschetz locus and their density in the space of all surfaces," Math. Ann. 282 (1988), 667–680.

[162] C. Ciliberto and E. Sernesi, "Families of varieties and the Hilbert scheme," in *Proceedings College on Riemann Surfaces, ICTP Trieste, 1988.*

[163] C. Ciliberto and G. Van der Geer, "On the Jacobian of a hyperplane section of a surface," in *Classification of Irregular Varieties: Proceedings Trento 1990*, ed. by E. Ballico, F. Catanese, and C. Ciliberto, Lecture Notes in Math. 1515 (1991), 31–40, Springer-Verlag, New York.

[164] H. Clemens, J. Kollár, and S. Mori, *Higher dimensional complex geometry*, Astérisque 166 (1988).

[165] H. Clemens and P.A. Griffiths, "The intermediate Jacobian of the cubic threefold," Ann. of Math. 95 (1972), 281–356.

[166] A. Conte and J.P. Murre, "On threefolds whose hyperplane sections are Enriques surfaces," in *Algebraic Threefolds, Proceedings Varenna, 1981*, ed. by A. Conte, 947 (1982), 221–228.

[167] A. Conte and J.P. Murre, "Algebraic varieties of dimension three whose hyperplane sections are Enriques surfaces," Ann. Scuola Norm. Sup. Pisa Cl. Sci. Ser. (4) 12 (1985), 43–80.

[168] A. Conte and J.P. Murre, "On the definition and on the nature of the singularities of Fano threefolds," Math. Inst. Univ. of Leiden, Report n. 25 (1985).

[169] M. Coppens and G. Martens, "Secant spaces and Clifford's theorem," Compositio Math. 78 (1991), 193–212.

[170] M. Cornalba, "Two theorems on modifications of analytic spaces," Invent. Math. 20 (1973), 227–247.

[171] A. Corti, "Families of Del Pezzo surfaces," preprint, 1992.

[172] F.R. Cossec, "Projective models of Enriques surfaces," Math. Ann. 256 (1983), 283–334.

[173] F.R. Cossec and I. Dolgachev, *Enriques surfaces, I*, Progr. Math. 76 (1989), Birkhäuser, Boston.

[174] D. Cox, "The Noether-Lefschetz locus of regular elliptic surfaces with section and $p_g \geq 2$," Amer. J. Math. 112 (1990), 289–329.

[175] S.D. Cutkowsky, "Elementary contractions of Gorenstein threefolds," Math. Ann. 280 (1988), 521–525.

[176] E.D. Davis, "Complete intersections of codimension 2 in $\mathbb{P}^r$: the Bezout-Segre theorem revisited," Rend. Sem. Mat. Univ. Politec. Torino 43 (1985), 333–353.

[177] E.D. Davis, "0-dimensional subschemes of $\mathbb{P}^2$: new application of Castelnuovo's function," Ann. Univ. Ferrara Sez. VII (N.S.) 32 (1986), 93–107.

[178] E.D. Davis and A.V. Geramita, "Birational morphisms to $\mathbb{P}^2$: an ideal-theoretic perspective," Math. Ann. 279 (1988), 435–448.

[179] M.A. De Cataldo, " The genus of curves on the three dimensional quadric," preprint, 1994.

[180] M.A. De Cataldo and M. Palleschi, "Polarized surfaces of positive Kodaira dimension with canonical bundle of small degree," Forum Math. 4 (1992), 217–229.

[181] W. Decker, L. Ein and F.-O. Schreyer, "Construction of surfaces in $\mathbb{P}_4$," J. Algebraic Geom. 2 (1993), 185–237.

[182] W. Decker and S. Popescu, "On surfaces in $\mathbb{P}^4$ and 3-folds in $\mathbb{P}^5$," preprint, 1994.

[183] A. Del Centina and A. Gimigliano, "Projective surfaces with bielliptic hyperplane sections," Manuscripta Math. 71 (1991), 253–282.

[184] A. Del Centina and A. Gimigliano, "On projective varieties admitting a bielliptic or trigonal curve-section," Matematiche XLVIII (1993), 101–107.

[185] J.-P. Demailly, "Singular hermitian metrics on positive line bundles," in *Complex Algebraic Varieties, Proceedings Bayreuth, 1990*, ed. by K. Hulek, T. Peternell, M. Schneider, and F.-O. Schreyer, Lecture Notes in Math. 1507 (1992), 87–104, Springer-Verlag, New York.

[186] J.-P. Demailly, "A numerical criterion for very ample line bundles," J. Differential Geometry 37 (1993), 323–374.

[187] J.-P. Demailly, T. Peternell, M. Schneider, "Compact complex manifolds with numerically effective tangent bundles," J. Algebraic Geom. 3 (1994), 295–345.

[188] M. Demazure, "Surfaces de Del Pezzo", in *Seminaire sur les Singularités des Surfaces*, ed. by M. Demazure, H. Pinkham, and B. Teissier, Lectures Notes in Math. 777 (1980), 23–69, Springer-Verlag, New York.

[189] J. Dieudonné and A. Grothendieck, *Éléments de Géometrie Algébrique*, Inst. Hautes Études Sci. Publ. Math. 4, 8, 11, 17, 20, 24, 28, 32.

[190] V. Di Gennaro, "Generalized Castelnuovo varieties," to appear in Manuscripta Math.

[191] S. Di Rocco, "k-very ample line bundles on Del Pezzo surfaces," preprint, 1994.

[192] I. Dolgachev, "Weighted projective varieties," in *Proceedings Vancouver, 1981*, ed. by J.B. Carrell, Lectures Notes in Math. 956 (1982), 34–71, Springer-Verlag, New York.

[193] H. D'Souza, "Threefolds whose hyperplane sections are elliptic surfaces," Pacific J. Math. 134 (1988), 57–78.

[194] H. D'Souza and M.L. Fania, "Varieties whose surface sections are elliptic," Tôhoku Math. J. 42 (1990), 457–474.

[195] A. Durfee, "Fifteen characterizations of rational double points and simple critical points," Enseign. Math. 25 (1979), 131–163.

[196] M. Ebihara, "On unirationality of threefolds which contain toric surfaces with ample normal bundles," preprint Gakushuin University, 1991.

[197] G. Edelmann, "3-Mannigfaltigkeiten im $\mathbb{P}^5$ vom Grad 12," Ph.D. Thesis, University of Bayreuth, Germany (1993).

[198] G. Edelmann, "3-folds in $\mathbb{P}^5$ of degree 12," Manuscripta Math. 82 (1994), 393–406.

[199] L. Ein, "The ramification divisors for branched coverings of $\mathbb{P}^n$," Math. Ann. (1982), 483–485.

[200] L. Ein, "Surfaces with a hyperelliptic hyperplane section," Duke Math. J. 50 (1984), 1–11.

[201] L. Ein, "Nondegenerate surfaces of degree $n+3$ in $\mathbb{P}_{\mathbb{C}}$," J. Reine Angew. Math. 351 (1984), 1–11.

[202] L. Ein, " An analogue of Max Noether's theorem," Duke Math. J. 52 (1985), 689–706.

[203] L. Ein, "Varieties with small dual varieties, I," Invent. Math. 96 (1986), 63–74.

[204] L. Ein, "Varieties with small dual varieties, II," Duke Math. J. 52 (1985), 895–907.

[205] L. Ein, O. Küchle, and R. Lazarsfeld, "Local positivity of ample line bundles," preprint, 1994.

[206] L. Ein and R. Lazarsfeld, "Stability of restrictions of Picard bundles, with an application to the normal bundles of elliptic curves," in *Complex Projective Geometry*, ed. by G. Ellingsrud, C. Peskine, G. Sacchiero, and S.A. Stromme, London Math. Soc. Lecture Note Ser. 179 (1992),149–156.

[207] L. Ein and R. Lazarsfeld, "Syzygies and Koszul cohomology of smooth projective varieties of arbitrary dimension," Invent. Math. 111 (1993), 51–67.

[208] L. Ein and R. Lazarsfeld, "Global generation of pluricanonical and adjoint linear series on smooth projective threefolds," J. Amer. Math. Soc. 6 (1993), 875–903.

[209] L. Ein and R. Lazarsfeld, "Seshadri constants on smooth surfaces," in *Journées de Géométrie Algébrique d'Orsay, July 1992,* Astérisque 282 (1993), 177–186.

[210] L. Ein, R. Lazarsfeld and V. Masek, "Global generation of linear series on terminal threefolds," preprint.

[211] D. Eisenbud, L. Ein and S. Katz, "Varieties cut out by quadrics: scheme-theoretic versus homogeneous generation of ideals," in *Algebraic Geometry, Sundance 1986*, Lecture Notes in Math. 1311 (1988), 51–70, Springer-Verlag, New York.

[212] P. Ellia and C. Peskine, "Groupes de points de $\mathbb{P}^2$: caractère et position uniforme," preprint.

[213] R. Elkik, "Singularités rationelles et déformations," Invent. Math. 47 (1978), 139–147.

[214] R. Elkik, "Rationalité des singularités canoniques," Invent. Math. 64 (1981), 1–6.

[215] G. Ellingsrud and C. Peskine, "Sur les surfaces lisses de $\mathbb{P}^4$," Invent. Math. 95 (1989), 1–11.

[216] F. Enriques, "Sui sistemi lineari di superficie algebriche ad intersezioni variabili iperellittiche," Math. Ann. 46 (1885), 179–199.

[217] H. Esnault and E. Viehweg, *Lectures on Vanishing Theorems*, DMV-Sem. 20 (1992), Birkhäuser, Boston.

[218] M.L. Fania, "Extension of modifications of ample divisors on fourfolds," J. Math. Soc. Japan 36 (1984), 107–120; "II," ibid. 38 (1986), 285–294.

[219] M.L. Fania, "Configurations of -2 rational curves on sectional surfaces of n-folds," Math. Ann. 275 (1986), 317–325.

[220] M.L. Fania, "Trigonal hyperplane sections of projective surfaces," Manuscripta Math. 68 (1990), 17–34.

[221] M.L. Fania, "When $K + (n - 4)L$ fails to be nef," Manuscripta Math. 79 (1993), 209–223.

[222] M.L. Fania and E.L. Livorni, "Polarized surfaces of Δ-genus 3," Trans. Amer. Math. Soc. 328 (1991), 445–463.

[223] M.L. Fania and E.L. Livorni, "Polarized manifolds (X, L) of dimension $\geq$ three, Δ-genus three, $\dim \mathrm{Bs}|L| \leq 0$ and degree $\geq 2\Delta(X, L) - 1$," Saitama Math. J. 11 (1993), 41–58.

[224] M.L. Fania and E.L. Livorni, "Degree nine manifolds of dimension ≥ 3," Math. Nachr. 169 (1994), 117–134.

[225] M.L. Fania and E.L. Livorni, "Degree ten manifolds of dimension ≥ 3," preprint, 1992.

[226] M.L. Fania, E. Sato, and A.J. Sommese, "On the structure of fourfolds with a hyperplane section which is a $\mathbb{P}^1$ bundle over a surface that fibres over a curve," Nagoya Math. J. 108 (1987), 1–14.

[227] M.L. Fania and A.J. Sommese, "On the minimality of hyperplane sections of Gorenstein threefolds," in *Contributions to Several Complex Variables*, ed. by A. Howard and P-M. Wong, Aspects of Math. E 9 (1986), 89–114, Vieweg, Braunschweig.

[228] M.L. Fania and A.J. Sommese, "Varieties whose hyperplane sections are ${\mathbb{P}^k}_{\mathbb{C}}$ bundles," Ann. Scuola Norm. Sup. Pisa Cl. Sci. Ser. (4) 15 (1988), 193–218.

[229] M.L. Fania and A.J. Sommese, "On the projective classification of smooth n-folds with n even," Ark. Mat. 27 (1989), 245–256.

[230] G. Fano, "Sulle varietà algebriche a tre dimensioni a superficie-sezioni razionali," Ann. Mat. Pura Appl. (3) 24 (1915), 49–88.

[231] G. Fano, "Sulle superficie di uno spazio qualunque a sezioni iperpiane collineari," Atti Reale Accad. Naz. Lincei Rend. Cl. Sci. Fis. Mat. Natur. (6) (1925), 115–129.

[232] G. Fano, "Sulle superfici dello spazio S_3 a sezioni piane collineari," Atti Reale Accad. Naz. Lincei Rend. Cl. Sci. Fis. Mat. Natur. (6) (1925), 473–477.

[233] G. Fano, "Nuove ricerche sulle varietá algebriche a tre dimensioni a curve sezioni canoniche," Comment. Pont. Acad. Sci. 11 (1947).

[234] G. Fischer, *Complex Analytic Geometry*, Lecture Notes in Math. 538 (1976), Springer-Verlag, New York.

[235] A. Franchetta, "Sulle curve eccezionali di prima specie appartenenti a una superficie algebrica," Boll. Un. Mat. Ital. Ser. 2, 3, (1940-41), 28–29.

[236] P. Francia, "Some remarks on minimal models I," Compositio Math. 40 (1980), 301–313.

[237] P. Francia, "On the base points of the bicanonical system," in *Problems in the Theory of Surfaces and their Classification, Cortona, Italy, 1988* ed. by F. Catanese and C. Ciliberto, Sympos. Math. 32 (1992), 141–150, INDAM, Academic Press, London.

[238] K. Fritzsche, "q-konvex Restmengen in kompakten komplexen Mannigfaltigkeiten," Math. Ann. 221 (1976), 251–273.

[239] T. Fujita, "On the structure of polarized varieties with Δ-genera zero," J. Fac. Sci. Univ. Tokyo Sect. IA Math. 22 (1975), 103–115.

[240] T. Fujita, "An extendability criterion for vector bundles on ample divisors," Proc. Japan Acad. Ser. A Math. Sci. 54 (1978), 298–299.

[241] T. Fujita, "On the hyperplane section principle of Lefschetz," J. Math. Soc. Japan 32 (1980), 153–169.

[242] T. Fujita, "On the structure of polarized manifolds of total deficiency one," J. Math. Soc. Japan 32 (1980), 709–725; "II," J. Math. Soc. Japan 33 (1981), 415–434; "III," J. Math. Soc. Japan 36 (1984), 75–89.

[243] T. Fujita, "Vector bundles on ample divisors," J. Math. Soc. Japan 33 (1981), 405–414.

[244] T. Fujita, "On polarized varieties of small Δ-genera," Tôhoku Math. J. 34 (1982), 319–347.

[245] T. Fujita, "Impossibility criterion of being an ample divisor," J. Math. Soc. Japan 34 (1982), 355–363.

[246] T. Fujita, "Theorems of Bertini type for certain types of polarized manifolds," J. Math. Soc. Japan 34 (1982), 709–718.

[247] T. Fujita, "Rational retractions onto ample divisors," Sci. Papers College Arts Sci. Univ. Tokyo 33 (1983), 33–39.

[248] T. Fujita, "On hyperelliptic polarized varieties," Tôhoku Math. J. 35 (1983), 1–44.

[249] T. Fujita, "Semipositive line bundles," J. Fac. Sci. Univ. Tokyo Sect. IA Math. 30 (1983), 353–378.

[250] T. Fujita, "On polarized manifolds of Δ-genus two," J. Math. Soc. Japan 36 (1984), 709–730.

[251] T. Fujita, "Projective varieties of Δ-genus 1," in *Algebraic and Topological Theories—to the memory of Dr. Takehiko MIYATA*, (1985), 149–175, Kinokuniya, Tokyo.

[252] T. Fujita, "On polarized manifolds whose adjoint bundles are not semipositive," in *Algebraic Geometry, Sendai 1985*, Adv. Stud. Pure Math. 10 (1987), 167–178.

[253] T. Fujita, contribution to "Birational geometry of algebraic varieties. Open problems," The 23rd Int. Symp. of the Division of Math. of the Taniguchi Foundation, Katata, August 1988.

[254] T. Fujita, "Classification of polarized manifolds of sectional genus two," in *Algebraic Geometry and Commutative Algebra*, I (1988), 73–98, Kinokuniya, Tokyo.

[255] T. Fujita, "Remarks on quasi-polarized varieties," Nagoya Math. J. 115 (1989), 105–123.

[256] T. Fujita, "Ample vector bundles of small c_1-sectional genera," J. Math. Kyoto Univ. 29 (1989), 1–16.

[257] T. Fujita, *Classification Theories of Polarized Varieties*, London Math. Soc. Lecture Note Ser. 155, Cambridge University Press, (1990).

[258] T. Fujita, "On singular Del Pezzo varieties," in *Algebraic Geometry, Proceedings of Conference on Hyperplane Sections, L'Aquila, Italy, 1988*, ed. by A.J. Sommese, A. Biancofiore, and E.L. Livorni, Lecture Notes in Math. 1417 (1990), 117–128, Springer-Verlag, New York.

[259] T. Fujita, "On adjoint bundles of ample vector bundles," in *Complex Algebraic Varieties, Bayreuth 1990*, ed. by K. Hulek, T. Peternell, M. Schneider, and F.-O. Schreyer, Lecture Notes in Math. 1507 (1992), 105–112, Springer-Verlag, New York.

[260] T. Fujita, "On Del Pezzo fibrations over curves," Osaka J. Math. 27 (1990), 229–245.

[261] T. Fujita, "On Kodaira energy and adjoint reduction of polarized manifolds," Manuscripta Math. 76 (1992), 59–84.

[262] T. Fujita, "On certain polarized elliptic surfaces," in *Geometry of Complex Projective Varieties, Cetraro, Italy, 1990*, ed. by A. Lanteri, M. Palleschi, and D. Struppa, Seminars and Conferences 9 (1993), 153–164, Mediterranean Press.

[263] T. Fujita, "Notes on Kodaira energies of polarized varieties," preprint, 1992.

[264] T. Fujita, "On Kodaira energy and classification of polarized varieties," to appear in Sugaku Expositions.

[265] T. Fujita, "On Kodaira energy of polarized log varieties," preprint, 1993.

[266] T. Fujita, "Remarks on Ein-Lazarsfeld criterion of adjoint bundles of polarized threefolds," preprint, 1993.

[267] W. Fulton, *Intersection Theory*, Ergeb. Math. Grenzgeb. (3) 2, Springer-Verlag, Berlin, (1984).

[268] W. Fulton and J. Hansen, "A connectedness theorem for projective varieties, with applications to singularities of mappings," Ann. of Math. 110 (1979), 159–166.

[269] W. Fulton and R. Lazarsfeld, "On the connectedness of the degeneracy loci and special divisors," Acta Math. 146 (1981), 271–283.

[270] W. Fulton and R. Lazarsfeld, "Connectivity and its applications in algebraic geometry," in *Algebraic Geometry, Chicago, 1981*, ed. by A. Libgober and P. Wagreich, Lecture Notes in Math. 862 (1981), 26–92, Springer-Verlag, New York.

[271] W. Fulton and R. Lazarsfeld, "The numerical positivity of ample vector bundles," Ann. of Math. 118 (1983), 35–60.

[272] A.V. Geramita and A. Gimigliano, "Generators for the defining ideal of certain rational surfaces," Duke Math. J. 62 (1991), 61–83.

[273] A.V. Geramita, A. Gimigliano, and B. Harbourne, "Projectively normal but superabundant embeddings of rational surfaces in projective space," to appear in J. Algebra.

[274] A.V. Geramita and P. Maroscia, "The ideal of forms vanishing at a finite set of points in $\mathbb{P}^n$," J. Algebra 90 (1984), 528–555.

[275] A.V. Geramita and J. Migliore, "On the ideal of an arithmetically Buchsbaum curve," J. Pure Appl. Algebra 54 (1988), 215–247.

[276] A.V. Geramita and J. Migliore, "Hyperplane sections of a smooth curve in $\mathbb{P}^3$," Comm. Algebra 17 (1989), 3129–3164.

[277] N. Goldstein, "A second Lefschetz theorem for general manifold sections in complex projective space," Math. Ann. 246 (1979), 41–68.

[278] N. Goldstein, "Ampleness and connectedness in complex G/P," Trans. Amer. Math. Soc. 274 (1982), 361–373.

[279] N. Goldstein, "Examples of non-ample normal bundles," Compositio Math. 51 (1984), 189–192.

[280] N. Goldstein, "A special surface in the 4-quadric," Duke Math. J. 50 (1983), 745–761.

[281] R. Goren, "Characterization and algebraic deformations of projective space," J. Math. Kyoto Univ. 8 (1968), 41–47.

[282] M. Goresky and R. McPherson, *Stratified Morse Theory*, Ergeb. Math. Grenzgeb. (3) 14, Springer-Verlag, Berlin, (1988).

[283] L. Göttsche, *Hilbert Schemes of Zero-Dimensional Subschemes of Smooth Varieties,* Lecture Notes in Math. 1572 (1994), Springer-Verlag, New York.

[284] G. Gotzmann, "Eine Bedingung für die Flachheit und das Hilbertpolynom eines graduierten Ringes," Math. Z. 158 (1978), 61–70.

[285] H. Grauert, "Über Modifikationen und exzeptionelle analytische Mengen," Math. Ann. 146 (1962), 331–368.

[286] H. Grauert and O. Riemenschneider, "Verschwindungssatze fur analytische Kohomologiegruppen auf komplexen Raumen," Invent. Math. 11 (1970), 263–292.

[287] M. Green, "A new proof of the explicit Noether-Lefschetz theorem," J. Differential Geometry 27 (1988), 155–159.

[288] M. Green, "Components of maximal dimension in the Noether-Lefschetz locus," J. Differential Geometry 29 (1989), 295–302.

[289] M. Green, "Restrictions of linear series to hyperplanes, and some results of Macaulay and Gotzmann," in *Algebraic Curves and Projective Geometry, Proceedings Trento, 1988*, ed. by E. Ballico and C. Ciliberto, Lecture Notes in Math. 1389 (1989), 76–86, Springer-Verlag, New York.

[290] M. Green and P.A. Griffiths, "Two applications of algebraic geometry to entire holomorphic mappings," in *Proceedings of the Int. Symp. on Differential Geometry in honor of S.S. Chern, Berkeley 1979*, ed. by W.Y. Hsiang et al., 41–74, Springer-Verlag, New York (1980).

[291] P.A. Griffiths, "The extension problem in complex analysis – II (Embeddings with positive normal bundle)," Amer. J. Math. 88 (1966), 366–446.

[292] P.A. Griffiths and J. Harris, "Residues and zero cycles on algebraic varieties," Ann. of. Math. 108 (1978), 461–505.

[293] P.A. Griffiths and J. Harris, *Principles of Algebraic Geometry*, Wiley-Interscience, New York, (1978).

[294] P.A. Griffiths and J. Harris, "Algebraic geometry and local differential geometry," Ann. Sci. École Norm. Sup. 12 (1979), 355–452.

[295] P.A. Griffiths and J. Harris, "On the Noether-Lefschetz theorem and two remarks on codimension-two cycles," Math. Ann. 271 (1985), 31–51.

[296] A. Grothendieck, "Techniques de construction et théorèmes d'existence en géométrie algébrique, IV: les schémas de Hilbert," Séminaire Bourbaki 221 (1960/61).

[297] A. Grothendieck, *Cohomolgie locale des faisceaux cohérents et théorèmes de Lefschetz locaux et globaux* (SGA 2), North Holland Press (1968).

[298] A. Grothendieck, "Le groupe de Brauer I," 46–66; "II," 67–87; "III: Exemples et compléments," 88–188, North Holland Press, (1968).

[299] L. Gruson and C. Peskine, "Genre des courbes de l'éspace projectif," in *Proceedings, Algebraic Geometry, Tromsø, 1977*, ed. by L.D. Olson, Lecture Notes in Math. 687 (1978), 31–59, Springer-Verlag, New York.

[300] L. Gruson, R. Lazarsfeld, and C. Peskine, "On a theorem of Castelnuovo and the equations defining space curves," Invent. Math. 72 (1983), 491–506.

[301] R.C. Gunning and H. Rossi, *Analytic functions of several complex variables*, (1965), Prentice-Hall, Englewood Cliffs, N.J.

[302] E. Halanay, "Normal surfaces of degree 5 in $\mathbb{P}^n$," Rev. Roumaine Math. Pures Appl. 33 (1988), 297–303.

[303] E. Halanay, "Normal surfaces of degree 6 in $\mathbb{P}^n$," Rev. Roumaine Math. Pures Appl. 34 (1989), 111–116.

[304] E. Halanay, "Non-normal surfaces of degree 5 and 6 in $\mathbb{P}^n$," Geom. Dedicata 32 (1989), 265–279.

[305] H.A. Hamm, "Lefschetz theorems for singular varieties," Proc. Symp. Pure Math. 40-I (1983), 547–557.

[306] B. Harbourne, "Very ample divisors on rational surfaces," Math. Ann. 272 (1985), 139–153.

[307] J. Harris, "The genus of space curves," Math. Ann. 249 (1980), 191–204.

[308] J. Harris, "A bound on the geometric genus of projective varieties," Ann. Scuola Norm. Sup. Pisa Cl. Sci. Ser. (4), 8 (1981), 35–68.

[309] J. Harris, "The Kodaira dimension of the moduli space of curves II: The even genus case," Invent. Math. 75 (1984), 437–466.

[310] J. Harris, *Curves in projective space*, with the collaboration of D. Eisenbud, Université de Montreal, Montreal (Québec), Canada, (1982).

[311] J. Harris, *Algebraic Geometry*, Graduate Texts in Math. 133, Springer-Verlag, New York (1992).

[312] J. Harris and D. Mumford, "On the Kodaira dimension of the moduli space of curves," Invent. Math. 67 (1982), 23–86.

[313] R. Hartshorne, "Ample Vector Bundles," Publ. Math. Inst. Hautes Études Sci. 29 (1966), 63–94.

[314] R. Hartshorne, "Curves with high self-intersection on algebraic surfaces," Inst. Hautes Études Sci. Publ. Math. 36 (1969), 111–126.

[315] R. Hartshorne *Ample Subvarieties of Algebraic Varieties*, Lecture Notes in Math. 156 (1970), Springer-Verlag, New York.

[316] R. Hartshorne, "Ample vector bundles on curves," Nagoya Math. J. 43 (1971), 73–89.

[317] R. Hartshorne, "Varieties of small codimension in projective space," Bull. Amer. Math. Soc. 80 (1974), 1017–1031.

[318] R. Hartshorne, *Algebraic Geometry*, Graduate Texts in Math. 52, Springer-Verlag, New York, (1978).

[319] F. Hidaka and K. Watanabe, "Normal Gorenstein surfaces with ample anti-canonical divisor," Tokyo J. Math. 4 (1981), 319–330.

[320] H. Hironaka, "Resolution of singularities of an algebraic variety over a field of characteristic zero," Ann. of Math. 79 (1964), 109–326.

[321] H. Hironaka, "Smoothing of algebraic cycles of small dimensions," Amer. J. Math. 90 (1968), 1–54.

[322] H. Hironaka, "On resolution of singularities (characteristic zero)," *Proc. Intern. Cong. Math.*, Stockolm (1962), 507–521.

[323] F. Hirzebruch, *Topological Methods in Algebraic Geometry*, Grundlehren Math. Wiss. 131 (3rd enlarged edition), Springer-Verlag, New York, (1966).

[324] F. Hirzebruch and K. Kodaira, "On the complex projective spaces," J. Math. Pures Appl. 36 (1957), 201–216.

[325] W.V.D. Hodge and D. Pedoe, *Methods of algebraic geometry*, Vol. 1 (1953), Vol. 2 (1952), Vol. 3 (1954), Cambridge University Press.

[326] K. Hulek, C. Okonek, and A. Van de Ven, "Multiplicity-2 structures on Castelnuovo surfaces," Ann. Scuola Norm. Sup. Pisa Cl. Sci. Ser. (4) 13 (1986), 427–448.

[327] C. Huneke, B. Ulrich, "General hyperplane sections of algebraic varieties," J. Algebraic Geom. 2 (1993), 487–505.

[328] M. Idá and E. Mezzetti, "Smooth non-special surfaces of $\mathbb{P}^4$," Manuscripta Math. 68 (1990), 57–67.

[329] S. Iitaka, "On *D*-dimensions of algebraic varieties," J. Math. Soc. Japan 23 (1971), 356–373.

[330] S. Iitaka, *Algebraic Geometry*, Graduate Texts in Math. 76, Springer-Verlag, New York, (1982).

[331] P. Ionescu, "Variétés projective lisses de degrés 5 et 6," C.R. Acad. Sci. Paris 293 (1981), 685–687.

[332] P. Ionescu, "On varieties whose degree is small with respect to codimension," Math. Ann. 271 (1985), 339–348.

[333] P. Ionescu, "Embedded projective varieties of small invariants," in *Proceedings of the 1982 Week of Algebraic Geometry, Bucharest*, Lecture Notes in Math. 1056 (1984), 142–187, Springer-Verlag, New York.

[334] P. Ionescu, "Embedded projective varieties of small invariants, II," Rev. Roumaine Math. Pures Appl. 31 (1986), 539–544.

[335] P. Ionescu, "Generalized adjunction and applications," Math. Proc. Cambridge Philos. Soc. 99 (1986), 457–472.

[336] P. Ionescu, "Ample and very ample divisors on surfaces," Rev. Roumaine Math. Pures Appl. 33 (1988), 349–358.

[337] P. Ionescu, "Embedded projective varieties of small invariants, III," in *Algebraic Geometry, Proceedings of Conference on Hyperplane Sections, L'Aquila, Italy, 1988*, ed. by A.J. Sommese, A. Biancofiore, and E.L. Livorni, Lecture Notes in Math. 1417 (1990), 138–154, Springer-Verlag, New York.

[338] V.A. Iskovskih, "Fano 3-folds I," Math. USSR-Izv. 11 (1977), 485–527; "II," Math. USSR-Izv. 12 (1978), 469–506.

[339] V.A. Iskovskih and V.V. Shokurov, "Biregular theory of Fano 3-folds," in *Proceedings of the Algebraic Geometry Conference, Copenhagen 1978*, Lecture Notes in Math. 732 (1979), 171–182, Springer-Verlag, New York.

[340] N. Katz, "Pinceaux de Lefschetz: théorème d'existence," in *Groupes de Monodromie en Géométrie Algébrique*, Lecture Notes in Math. 340 (1973), 212–253, Springer-Verlag, New York.

[341] Y. Kawamata, "A generalization of Kodaira-Ramanujam's vanishing theorem," Math. Ann. 261 (1982), 43–46.

[342] Y. Kawamata, "Elementary contractions of algebraic 3-folds," Ann. of Math. 119 (1984), 95–110.

[343] Y. Kawamata, "The cone of curves of algebraic varieties," Ann. of Math. 119 (1984), 603–633.

[344] Y. Kawamata, "Small contractions of four dimensional algebraic manifolds," Math. Ann. 284 (1989), 595–600.

[345] Y. Kawamata, "On the length of an extremal rational curve," Invent. Math. 105 (1991), 609–611.

[346] Y. Kawamata, K. Matsuda, and K. Matsuki, "Introduction to the Minimal Model Problem," in *Algebraic Geometry, Sendai 1985*, Adv. Stud. Pure Math. 10 (1987), 283–360.

[347] G.R. Kempf et al. *Toroidal embeddings I*, Lecture Notes in Math. 339 (1973), Springer-Verlag, New York.

[348] G.R. Kempf, *Algebraic Varieties*, (1993), Cambridge University Press.

[349] S.L. Kleiman, "Towards a numerical theory of ampleness," Ann. of Math. 84 (1966), 293–344.

[350] S.L. Kleiman, "The enumerative theory of singularities," in *Real and Complex Singularities, Oslo 1976*, ed. by P. Holme, 297–396, Alphen aan den Rijn, Sijthoff and Noordhoff, Rockville, Maryland, 1977.

[351] S.L. Kleiman, "Relative duality for quasi-coherent sheaves," Compositio Math. 41 (1980), 39–60.

[352] S.L. Kleiman, "About the conormal scheme," in *Complete Intersections, Proceedings Acireale, 1983*, ed. by S. Greco and R. Strano, Lecture Notes in Math. 1092 (1984), 161–197, Springer-Verlag, New York.

[353] S.L. Kleiman, "Tangency and duality," in *Proceedings of the 1984 Vancouver Conference in Algebraic Geometry, 1984*, ed. by J.B. Carrell, A.V. Geramita, and P. Russell, Canadian Math. Soc. Conference Proceedings, 6 (1986), 163–225.

[354] F. Knop and G. Menzel, "Duale Varietäten von Fahnenvarietäten," Comment. Math. Helv. 62 (1987), 38–61.

[355] S. Kobayashi and T. Ochiai, "Characterization of the complex projective space and hyperquadrics," J. Math Kyoto Univ. 13 (1972), 31–47.

[356] J. Kollár, "The cone theorem," Ann. of Math. 120 (1984), 1–5.

[357] J. Kollár, "Higher direct images of dualizing sheaves I," Ann. of Math. 123 (1986), 11–42; "II," Ann. of Math. 124 (1986), 171–202.

[358] J. Kollár, "The structure of algebraic threefolds: an introduction to Mori's program," Bull. Amer. Math. Soc. 17 (1987), 221–273.

[359] J. Kollár, "Effective base point freeness," Math. Ann. 296 (1993), 595–605.

[360] J. Kollár, "Rational curves on algebraic varieties," preprint, 1994.

[361] D. Kotschick, "Stable and unstable vector bundles on algebraic surfaces," in *Problems in the Theory of Surfaces and their Classification, Cortona, Italy, 1988*, ed. by F. Catanese and C. Ciliberto, Sympos. Math. 32 (1992), 151–166, INDAM, Academic Press, London.

[362] S. Kosarew, "Ein verschwindungssatz für gewisse Kohomologiegruppen in Umgebung kompakter komplexer Unterräume mit konvex/konkavem Normalenbündel," Math. Ann. 261 (1982), 315–326.

[363] A. Kumpera and D. Spencer, *Lie Equations; Vol. I: General Theory*, Ann. of Math. Stud. 73 (1972), Princeton University Press, Princeton, N.J.

[364] A. Lanteri, "Algebraic surfaces containing a smooth curve of genus $q(S)$ as an ample divisor," Geom. Dedicata 17 (1984), 189–197.

[365] A. Lanteri, "Pluriadjoint bundles of polarized surfaces," Monatsh. Math. 109 (1990), 69–77.

[366] A. Lanteri, "2-spanned surfaces of sectional genus six," Ann. Mat. Pura Appl. (4) 165 (1993), 197–216.

[367] A. Lanteri, "Double covers of smooth quadrics as ample and very ample divisors," Abh. Math. Sem. Univ. Hamburg 64 (1994), 97–103.

[368] A. Lanteri, "Small degree covers of $\mathbb{P}^n$ and hyperplane sections," in *Scritti in Onore di G. Melzi,* ed. by C.F. Manara et al., Vita e Pensiero, Milano (1994), 231–248.

[369] A. Lanteri and E.L. Livorni, "Complex projective surfaces polarized by an ample and spanned line bundle of genus 3," Geom. Dedicata 31 (1989), 267–289.

[370] A. Lanteri and E.L. Livorni, "Triple solids with sectional genus 3," Forum Math. 2 (1990), 297–306.

[371] A. Lanteri and H. Maeda, "Adjoint bundles of ample and spanned vector bundles on algebraic surfaces," J. Reine Angew. Math. 433 (1992), 181–199.

[372] A. Lanteri and M. Palleschi, "On the adjoint system to a very ample divisor on a surface and connected inequalities, I, II," Atti Accad. Naz. Lincei Rend. Cl. Sci. Mat. Natur. (8) 71 (1981), 66–76 and 166–175.

[373] A. Lanteri and M. Palleschi, "Some characterizations of projective bundles and projective spaces," Geom. Dedicata 14 (1983), 203–208.

[374] A. Lanteri and M. Palleschi, "Characterizing projective bundles by means of ample divisors," Manuscripta Math. 45 (1984), 207–218.

[375] A. Lanteri and M. Palleschi, "About the adjunction process for polarized algebraic surfaces," J. Reine Angew. Math. 352 (1984), 15–23.

[376] A. Lanteri and M. Palleschi, "On the ampleness of $K_X \otimes \mathcal{L}^n$ for a polarized threefold $(X, \mathcal{L})$," Atti Accad. Naz. Lincei Rend. Cl. Sci. Fis. Mat. Natur. (8) 78 (1985), 213–217.

[377] A. Lanteri and M. Palleschi, "Projective manifolds containing many rational curves," Indiana Univ. Math. J. 36 (1987), 857–865.

[378] A. Lanteri and M. Palleschi, "Adjunction properties of polarized surfaces via Reider's method," Math. Scand. 65 (1989), 175–188.

[379] A. Lanteri and M. Palleschi, "Complex projective surfaces of nonnegative Kodaira dimension polarized by an ample and spanned line bundle of genus 4," Indiana Univ. Math. J. 39 (1990), 85–104.

[380] A. Lanteri and M. Palleschi, "Remarks on recent work by Sommese," in *Proceedings of Conference on Algebraic Geometry, Berlin, 1985*, Teubner-Texte Math. 92 (1987), 228–235.

[381] A. Lanteri, M. Palleschi, and F. Serrano, "Elliptic surfaces with low genus polarizations," in preparation.

[382] A. Lanteri, M. Palleschi, and A.J. Sommese, "Very ampleness of $K_X \otimes \mathcal{L}^{\dim X}$ for an ample and spanned line bundle $\mathcal{L}$," Osaka J. Math. 26 (1989), 647–664.

[383] A. Lanteri, M. Palleschi, and A.J. Sommese, "Double covers of $\mathbb{P}^n$ as very ample divisors," to appear in Nagoya Math. J.

[384] A. Lanteri, M. Palleschi, and A.J. Sommese, "On triple covers of $\mathbb{P}^n$ as very ample divisors," in *Classification of Algebraic Varieties, L'Aquila, Italy, 1992,* ed. by C. Ciliberto, E.L. Livorni, and A.J. Sommese, Contemp. Math. 162 (1994), 277–292.

[385] A. Lanteri, M. Palleschi, and A.J. Sommese, "Del Pezzo surfaces as hyperplane sections," preprint, 1994.

[386] A. Lanteri and B. Rondena, "Numerically positive divisors on algebraic surfaces," to appear in Geom. Dedicata.

[387] A. Lanteri and F. Russo, "A lower bound for sectional genera of ample and spanned vector bundles on algebraic surfaces," Proc. Amer. Math. Soc. 119 (1993), 1053–1059.

[388] A. Lanteri and F. Russo, "A footnote of a paper by Noma," Atti Accad. Naz. Lincei Rend. Cl. Sci. Fis. Mat. Natur. (9) 4 (1993), 131–132.

[389] A. Lanteri and A.J. Sommese, "A vector bundle characterization of $\mathbb{P}^n$," Abh. Math. Sem. Univ. Hamburg 58 (1988), 89–96.

[390] A. Lanteri and D. Struppa, "Some topological conditions for projective manifolds with degenerate dual varieties: connections with $\mathbb{P}$-bundles," Atti Accad. Naz. Lincei Rend. Cl. Sci. Fis. Mat. Natur. (8) 77 (1984), 155–158.

[391] A. Lanteri and D. Struppa, "Projective manifolds with the same homology as $\mathbb{P}^k$," Monatsh. Math. 101 (1986), 53–58.

[392] A. Lanteri and D. Struppa, "Projective manifolds whose topology is strongly reflected in their hyperplane sections," Geom. Dedicata 21 (1986), 357–374.

[393] A. Lanteri and D. Struppa, "Projective 7-folds with positive defect," Compositio Math. 61 (1987), 329–337.

[394] A. Lanteri and D. Struppa, "Topological properties of cyclic coverings branched along an ample divisor," Canad. J. Math. 41 (1989), 462–479.

[395] A. Lanteri and C. Turrini, "Projective threefolds of small class," Abh. Math. Sem. Univ. Hamburg 57 (1987), 103–117.

[396] A. Lanteri and C. Turrini, "Projective surfaces with class less than or equal to twice the degree," to appear in Math. Nachr.

[397] M.E. Larsen, "On the topology of complex projective manifolds," Invent. Math. 19 (1973), 251–260.

[398] A. Lascoux, "Degree of the dual of a Grassmann variety," Comm. Algebra 9 (1981), 1215–1225.

[399] O.A. Laudal, "A generalized trisecant lemma," in *Proceedings, Algebraic Geometry, Tromsø, 1977*, ed. by L.D. Olson, Lecture Notes in Math. 687 (1978), 112–149, Springer-Verlag, New York.

[400] H.B. Laufer, *Normal two-dimensional singularities*, Ann. of Math. Stud. 71 (1971), Princeton Univ. Press, Princeton, New Jersey.

[401] R. Lazarsfeld, "A Barth-type theorem for branched coverings of projective space," Math. Ann. 249 (1980), 153–162.

[402] R. Lazarsfeld, "Some applications to the theory of positive vector bundles," in *Complete Intersections, Proceedings Acireale, 1983*, ed. by S. Greco and R. Strano, Lecture Notes in Math. 1092 (1984), 29–61, Springer-Verlag, New York.

[403] R. Lazarsfeld, "A sharp Castelnuovo bound for smooth surfaces," Duke Math. J. 55 (1987), 423–429.

[404] R. Lazarsfeld, "A sampling of vector bundle techniques in the study of linear series," in *Lectures on Riemann Surfaces*, ed. by Cornalba et al. World Scientific Press, (1989).

[405] R. Lazarsfeld and A. Van de Ven, *Topics in the Geometry of Projective Space: Recent Work of F. L. Zak*, DMV Sem. 4 (1984), Birkhäuser, Boston.

[406] D. Liebermann and D. Mumford, "Matsusaka's big theorem," in *Algebraic Geometry, Arcata 1974*, Proc. Amer. Math. Soc. Symp. in Pure Math. 29 (1975), 513–530.

[407] J. Lipman, "Rational singularities," Inst. Hautes Études Sci. Publ. Math. 36 (1969), 195–280.

[408] J. Lipman and A.J. Sommese, "On blowing down projective spaces in singular varieties," J. Reine Angew. Math. 362 (1985), 52–62.

[409] E.L. Livorni, "Classification of algebraic surfaces with sectional genus less than or equal to six I: Rational surfaces," Pacific J. Math. 113 (1984), 93–114.

[410] E.L. Livorni, "Classification of algebraic nonruled surfaces with sectional genus less than or equal to six," Nagoya Math. J. 100 (1985), 1–9.

[411] E.L. Livorni, "Classification of algebraic surfaces with sectional genus less than or equal six. II: Ruled surfaces with dim $\phi_{K_X \otimes L}(X) = 1$," Canad. J. Math. 37 (1986), 1110–1121.

[412] E.L. Livorni, "Classification of algebraic surfaces with sectional genus less than or equal six. III: Ruled surfaces with dim $\phi_{K_X \otimes L}(X) = 2$," Math. Scand. 59 (1986), 9–29.

[413] E.L. Livorni, "Some results about a smooth hyperplane section of a 3-fold of non-negative Kodaira dimension," Arch. Math. (Basel) 48 (1987), 530–532.

[414] E.L. Livorni, "On the existence of some surfaces," in *Algebraic Geometry, Proceedings of Conference on Hyperplane Sections, L'Aquila, Italy, 1988*, ed. by A.J. Sommese, A. Biancofiore, and E.L. Livorni, Lecture Notes in Math. 1417 (1990), 155–179, Springer-Verlag, New York.

[415] E.L. Livorni and A.J. Sommese, "Threefolds of non-negative Kodaira dimension with sectional genus less than or equal to 15," Ann. Scuola Norm. Sup. Pisa Cl. Sci. Ser. (4) 13 (1986), 537–558.

[416] T. Luo, "A note on the Hodge index theorem," Manuscripta Math. 67 (1990), 17–20.

[417] S. L'vovsky, "Criterion for a variety to be nonrepresentable as a hyperplane section," Moscow Univ. Math. Bull. 40 (1985), 37–42.

[418] S. L'vovsky, "Extensions of projective varieties and deformations I," Michigan Math. J. 39 (1992), 41–52; "II," Michigan Math. J. 39 (1992), 65–70.

[419] S. L'vovsky, "On curves and surfaces with projectively equivalent hyperplane sections," preprint, 1993.

[420] F.S. Macaulay, "Some properties of enumeration in the theory of modular systems," Proc. London Math. Soc. 26 (1927), 531–555.

[421] H. Maeda, "On polarized surfaces of sectional genus three," Sci. Papers College Arts Sci. Univ. Tokyo 37 (1987), 103–112.

[422] H. Maeda, "Ramification divisors for branched coverings of $\mathbb{P}^n$," Math. Ann. 288 (1990), 195–199.

[423] H. Maeda, "A criterion for a smooth surface to be Del Pezzo," Math. Proc. Cambridge Philos. Soc. 113 (1993), 1–3.

[424] R. Maggioni and A. Ragusa, "Connections between Hilbert function and geometric properties for a finite set of points in $\mathbb{P}^2$," Matematiche 34 (1984), 153–170.

[425] R. Maggioni and A. Ragusa, "The Hilbert function of generic plane sections of curves in $\mathbb{P}^3$," Invent. Math. 91 (1988), 253–258.

[426] B.V. Martynov, "Three-dimensional varieties with rational sections," Math. USSR-Sb. 81 (1970), 569–579.

[427] B.V. Martynov, "Smooth projective threefolds whose hyperplane sections are ruled surfaces of irregularity $q > 0$," Math. Zametki 14 (1973), 821–828.

[428] H. Matsumura, "Geometric structure of the cohomology rings in abstract algebraic geometry," Mem. Coll. Sci. Univ. Kyoto Ser. A 32 (1959), 33–84.

[429] T. Matsusaka and D. Mumford, "Two fundamental theorems on deformations of polarized varieties," Amer. J. Math. 86 (1964), 668–684.

[430] J. McKernan, "Varieties with isomorphic or birational hyperplane sections," Internat. J. Math. 4 (1993), 113–125.

[431] M. Mella and M. Palleschi, "The k-very ampleness of an elliptic quasi-bundle," Abh. Math. Sem. Univ. Hamburg 63 (1993), 215–226.

[432] E. Mezzetti, "Differential-geometric methods for the lifting problem and linear systems on plane curves," preprint, 1993.

[433] E. Mezzetti and K. Ranestad, "The non-existence of a smooth sectionally non-special surface of degree 11 and sectional genus 8 in the projective fourspace," Manuscripta Math. 70 (1991), 279–283.

[434] J. Migliore, "Curves and their hyperplane sections," Lecture Notes for the Notre Dame Algebraic Geometry Seminar (1989).

[435] J. Migliore, "Hypersurface sections of curves," in *Proceedings of the International Conference on Zero-Dimensional Schemes, Ravello, 1992*, ed. by F. Orecchia and L. Chiantini, De Gruyter (1994), 269–282.

[436] J. Migliore, *An introduction to deficiency modules and liaison theory for subschemes of projective space*, monograph, in preparation, to be published by the Global Analysis Research Center, Seoul National University.

[437] J. Migliore and U. Nagel, "On the Cohen-Macaulay type of the general hypersurface section of a curve," to appear in Math. Z.

[438] R.M. Miro-Roig, "The degree of smooth non-arithmetically Cohen-Macaulay threefolds in $\mathbb{P}^5$," Proc. Amer. Math. Soc. 110 (1990), 311–313.

[439] M. Miyanishi, "Algebraic threefolds, algebraic varieties and analytic varieties," Adv. Stud. Pure Math. 1 (1983), 69–99.

[440] Y. Miyaoka, "On the Chern numbers of surfaces of general type," Invent. Math. 42 (1977), 225–237.

[441] Y. Miyaoka, "The maximal number of quotient singularities on surfaces with given numerical invariants," Math. Ann. 268 (1984), 159–171.

[442] Y. Miyaoka, "The Chern classes and Kodaira dimension of a minimal variety," in *Algebraic Geometry, Sendai 1985*, Adv. Stud. Pure Math. 10 (1987), 449–476.

[443] Y. Miyaoka, "Theme and variations–Inequalities between Chern numbers," Sugaku Expositions 4 (1991), 157–176.

[444] Y. Miyaoka and S. Mori, "A numerical criterion for uniruledness," Ann. of Math. 124 (1986), 65–70.

[445] S. Mori, "On a generalization of complete intersections," J. Math. Kyoto Univ. 15-3 (1975), 619–646.

[446] S. Mori, "Projective manifolds with ample tangent bundles," Ann. of Math. 110 (1979), 593–606.

[447] S. Mori, "Threefolds whose canonical bundles are not numerically effective," Ann. of Math. 116 (1982), 133–176.

[448] S. Mori, "Threefolds whose canonical bundles are not numerically effective," in *Algebraic Threefolds, Proceedings Varenna, 1981*, ed. by A. Conte, Lecture Notes in Math. 947 (1982), 125–189, Springer-Verlag, New York.

[449] S. Mori, "On 3-dimensional terminal singularities," Nagoya Math. J. 98 (1985), 43–66.

[450] S. Mori, "Classification of higher dimensional varieties," in *Algebraic Geometry, Bowdoin 1985*, Proc. Sympos. Pure Math. 46 – Part I (1987), 269–331.

[451] S. Mori, "Flip theorem and the existence of minimal models for 3-folds," J. Amer. Math. Soc. 1 (1988), 117–253.

[452] S. Mori, "Hartshorne conjecture and extremal ray," Sugaku Expositions 0 (1988), 15–37.

[453] S. Mori and S. Mukai, "Classification of Fano 3-folds with $B_2 \geq 2$," Manuscripta Math. 36 (1981), 147–162.

[454] U. Morin, "Sulle classificazione proiettiva delle varietà a superficie-sezioni razionali," Ann. Mat. Pura Appl. 18 (1939), 147–161.

[455] U. Morin, "Sui tipi di sistemi lineari di superfici algebriche a curva-caratteristica di genere due," Ann. Mat. Pura Appl. 19 (1940), 257–288.

[456] U. Morin, "Sulle varietà algebriche a curve-sezioni di genere tre," Ann. Mat. Pura Appl. 21 (1940), 113–155.

[457] S. Mukai, contribution to "Birational geometry of algebraic varieties. Open problems," The 23rd Int. Symp. of the Division of Math. of the Taniguchi Foundation, Katata, August 1988.

[458] S. Mukai, "Biregular classification of Fano threefolds and Fano manifolds of coindex 3," Proc. Nat. Acad. Sci. U.S.A. 86 (1989), 3000–3002.

[459] S. Mukai, "New classification of Fano threefolds and manifolds of coindex 3," preprint, 1988.

[460] S. Mukai, "Fano 3-folds," in *Complex Projective Geometry*, ed. by G. Ellingsrud, C. Peskine, G. Sacchiero, and S.A. Stromme, London Math. Soc. Lecture Note Ser. 179 (1992), 255–263, Cambridge University Press.

[461] D. Mumford, "The topology of normal singularities of an algebraic surface and a criterion for simplicity," Inst. Hautes Études Sci. Publ. Math. 9 (1961), 5–22.

[462] D. Mumford, *Lectures on Curves on an Algebraic Surface*, (with a section by G.M. Bergman), Ann. of Math. Stud. 59 (1966), Princeton University Press.

[463] D. Mumford, *The Red Book of Varieties and Schemes*, Lecture Notes in Math. 1358 (1988), Springer-Verlag, New York.

[464] D. Mumford, "Pathologies, III," Amer. J. Math. 89 (1967), 94–104.

[465] D. Mumford, "Varieties defined by quadratic equations," in *Questions on algebraic varieties*, CIME course 1969, Rome (1970), 30–100.

[466] D. Mumford, *Algebraic Geometry I*, Grundlehren Math. Wiss. 221, Springer-Verlag, New York, (1976).

[467] J.P. Murre, "Classification of Fano threefolds according to Fano and Iskovskih," in *Algebraic Threefolds, Proceedings Varenna, 1981*, ed. by A. Conte, Lecture Notes in Math. 947 (1982), 35–92, Springer-Verlag, New York.

[468] M. Nagata, "On rational surfaces I," Mem. Coll. Sci. Univ. Kyoto Ser. A 32 (1960), 351–370; "II," Mem. Coll. Sci. Univ. Kyoto Ser. A 33 (1960), 271–293.

[469] M. Nagata, "On self-intersection number of a section on a ruled surface," Nagoya Math. J. 37 (1970), 191–196.

[470] Y. Nakai, "A criterion of an ample sheaf on a projective scheme," Amer. J. Math. 85 (1963), 14–26.

[471] S. Nakamura, "The classification of the third reductions with a spectral value condition," 1994 Ph.D. Thesis, Notre Dame.

[472] S. Nakamura, "On the classification of the third reductions with a spectral value condition," preprint, 1994.

[473] S. Nakamura, "On the birational third adjoint contractions," preprint, 1993.

[474] S. Nakamura, "A quadric criterion for the existence of a non-trivial section of adjoint systems," preprint, 1994.

[475] S. Nakano, "On the inverse of monoidal transformation," Publ. Res. Inst. Math. Sci. 6 (1971), 483–502.

[476] T. Nakashima, "On Reider's method for surfaces in positive characteristic," J. Reine Angew. Math. 438 (1993), 175–185.

[477] M. Namba, *Families of Meromorphic Functions on Compact Riemann Surfaces*, Lecture Notes in Math. 767 (1979), Springer-Verlag, New York.

[478] A. Noma, "Classification of rank-2 ample and spanned vector bundles on surfaces whose zero loci consist of general points," Trans. Amer. Math. Soc. 342 (1994), 867–894.

[479] A. Noma, "Ample and spanned vector bundles with large c_1^2 relative to c_2 on surfaces," Duke Math. J. 69 (1993), 663–669.

[480] T. Oda, *Lectures on Torus Embedding and Applications*, Tata Inst. Lecture Notes, 57, Springer-Verlag, New York, (1978).

[481] T. Oda, *Convex Bodies and Algebraic Geometry: an Introduction to the Theory of Toric Varieties*, Springer-Verlag, New York, (1988).

[482] K. Oguiso, "On polarized Calabi-Yau 3-folds," J. Fac. Sci. Univ. Tokyo Sect. IA Math. 38 (1991), 395–429.

[483] K. Oguiso and T. Peternell, "Semi-positivity and cotangent bundles," preprint, 1993.

[484] A. Ohbuchi, "On the projective normality of some varieties of degree 5," Pacific J. Math. 144 (1990), 313–325.

[485] C. Okonek, "3-Mannigfaltigkeiten im $\mathbb{P}^5$ und ihre zugehörigen stabilen Garben," Manuscripta Math. 38 (1982), 175–199.

[486] C. Okonek, "Moduli reflexiver Garben und Flachem von kleinem Grad im $\mathbb{P}^4$," Math. Z. 184 (1983), 549–572.

[487] C. Okonek, "Über 2-codimensionale Untermannigfaltigkeiten vom Grad 7 in $\mathbb{P}^4$ und $\mathbb{P}^5$," Math. Z. 187 (1984), 209–219.

[488] C. Okonek, "Flachem von Grad 8 im $\mathbb{P}^4$," Math. Z. 191 (1986), 207–223.

[489] C. Okonek, "On codimension-2 submanifolds in $\mathbb{P}^4$ and $\mathbb{P}^5$," Math. Gottingensis 50 (1986).

[490] C. Okonek, "Concavity, convexity, and complements in complex spaces," in *Proceedings of the Complex Analysis and Algebraic Geometry Conference, Göttingen, 1985*, ed. by H. Grauert, Lecture Notes in Math. 1194 (1986), 104–126, Springer-Verlag, New York.

[491] C. Okonek, "Barth-Lefschetz theorems for singular spaces," J. Reine Angew. Math. 374 (1987), 24–38.

[492] C. Okonek, "Notes on varieties of codimension 3 in $\mathbb{P}^N$," Manuscripta Math. 84 (1994), 421–442.

[493] C. Okonek, M. Schneider, and H. Spindler, *Vector Bundles on Complex Projective Spaces*, Progr. Math. 3 (1980), Birkhäuser, Boston.

[494] C. Okonek and A. Van de Ven, "Vector bundles and geometry: some open problems," Math. Gottingensis.

[495] C. Oliva, "On the pluriadjoint maps of polarized normal Gorenstein surfaces," in *Algebraic Geometry, Proceedings of Conference on Hyperplane Sections, L'Aquila, Italy, 1988*, ed. by A.J. Sommese, A. Biancofiore, and E.L. Livorni, Lecture Notes in Math. 1417 (1990), 180–183, Springer-Verlag, New York.

[496] G. Ottaviani, "On 3-folds in $\mathbb{P}^5$ which are scrolls," Ann. Scuola Norm. Sup. Pisa Cl. Sci. Ser. (4) 19 (1992), 451–471.

[497] M. Palleschi, "On the adjoint line bundle to an ample and spanned one," in *Algebraic Geometry, Proceedings of Conference on Hyperplane Sections, L'Aquila, Italy, 1988*, ed. by A.J. Sommese, A. Biancofiore, and E.L. Livorni, Lecture Notes in Math. 1417 (1990), 184–191, Springer-Verlag, New York.

[498] R. Paoletti, "Free pencils on divisors," to appear in Math. Ann.

[499] R. Paoletti, "Seshadri constants, gonality of space curves and restrictions of stable bundles," to appear in J. Differential Geometry.

[500] R. Paoletti, "Positivity conditions for curves in a projective threefold," preprint, 1994.

[501] C. Peskine and L. Szpiro, "Liaison des variétés algébriques. I," Invent. Math. 26 (1974), 271–302.

[502] T. Peternell, J. Le Potier, and M. Schneider, "Vanishing theorems, linear and quadratic normality," Invent. Math. 87 (1987), 573–586.

[503] T. Peternell, M. Szurek, and J.A. Wiśniewski, "Numerically effective vector bundles with small Chern classes," in *Complex Algebraic Varieties, Bayreuth 1990*, ed. by K. Hulek, T. Peternell, M. Schneider, and F.-O. Schreyer, Lecture Notes in Math. 1507 (1992), 145–156, Springer-Verlag, New York.

[504] E. Picard and G. Simart, *Théorie des Fonctions Algébriques*, I (1897), II (1906); reprinted in one volume: Chelsea, (1971), Bronx, New York.

[505] L. Picco Botta and A. Verra, "The nonrationality of the generic Enriques threefold," Compositio Math. 48 (1983), 167–184.

[506] S. Popescu, "On smooth surfaces of degree ≥ 11 in the projective fourspace," Dissertation, Universität des Saarlandes, Saarbrücken, 1993.

[507] S. Popescu and K. Ranestad, "Surfaces of degree 10 in the projective fourspace via linear systems and linkage," 1993 Saarbrücken/Oslo preprint.

[508] S. Ramanam, "A note on C.P. Ramanujam," in *C.P. Ramanujam—A tribute*, 11–13, Springer-Verlag, New York, (1978).

[509] S. Ramanan, "Ample divisors on Abelian surfaces," Proc. London Math. Soc. (3) 51 (1985), 231–245.

[510] C.P. Ramanujam, "Remarks on the Kodaira vanishing theorem," J. Indian Math. Soc. (N.S.) 36 (1972), 41–51; "A supplement," J. Indian Math. Soc. (N.S.) 38 (1974), 121–124.

[511] Z. Ran, "The structure of Gauss-like mappings," Compositio Math. 52 (1984), 171–177.

[512] Z. Ran, "Hodge theory and the Hilbert scheme," J. Differential Geometry, 37 (1993), 191–198.

[513] K. Ranestad, "Surfaces of degree 10 in the projective fourspace," in *Problems in the Theory of Surfaces and their Classification, Cortona, Italy, 1988,* ed. by F. Catanese and C. Ciliberto, Sympos. Math. 32 (1992), 270–307, INDAM, Academic Press, London.

[514] R. Re, "Sulle sezioni iperpiane di una varietá proiettiva," Matematiche 42 (1987) 211–218.

[515] M. Reid, "Canonical 3-folds," in *Algebraic Geometry, Angers, 1979*, ed. by A. Beauville, 273–310, Alphen aan den Rijn, Sijthoff and Noordhoff, Rockville, Maryland, (1980).

[516] M. Reid, "Minimal models of canonical 3-folds," in *Algebraic Varieties and Analytic Varieties*, Adv. Stud. Pure Math. 1 (1983), 131–180.

[517] M. Reid, "Projective morphisms according to Kawamata," 1983 University of Warwick preprint.

[518] M. Reid, "Young person's guide to canonical singularities," in *Algebraic Geometry, Bowdoin 1985*, Proc. Sympos. Pure Math. 46 – Part I (1987), 45–414.

[519] M. Reid, "Infinitesimal view of extending a hyperplane section—deformation theory and computer algebra," in *Algebraic Geometry, Proceedings of Conference on Hyperplane Sections, L'Aquila, Italy, 1988*, ed. by A.J. Sommese, A. Biancofiore, and E.L. Livorni, Lecture Notes in Math. 1417 (1990), 214–286, Springer-Verlag, New York.

[520] M. Reid, "Decomposition of toric morphisms," in *Arithmetic and Geometry II, papers dedicated to I.R. Shafarevich*, Progr. Math. 36 (1983), 395–418, Birkhäuser, Boston.

[521] M. Reid, "Nonnormal del Pezzo surfaces," preprint, 1994.

[522] I. Reider, "Vector bundles of rank 2 and linear systems on algebraic surfaces," Ann. of Math. 127 (1988), 309–316.

[523] E. Rogora, "Varieties with many lines," Manuscripta Math. 82 (1994), 207–226.

[524] H. Rossi, "Attaching analytic spaces to an analytic space along a pseudoconcave boundary," in *Proceedings of the Conference on Complex Analysis, Minneapolis, 1964*, ed. by A. Aeppli, E. Calabi, and H. Röhrl, 242–256, Springer-Verlag, New York, (1965).

[525] H. Rossi, "Continuation of subvarieties of projective varieties," Amer. J. Math. 91 (1969), 565–575.

[526] L. Roth, "On the projective classification of surfaces," Proc. London Math. Soc. (2) 42 (1937), 142–170.

[527] L. Roth, *Algebraic Threefolds with Special Regard to Problems of Rationality*, Springer-Verlag, New York, (1955).

[528] C. Sacchi, "Special Fano manifolds as ample divisors," preprint, 1994.

[529] F. Sakai, "Enriques classification of normal Gorenstein surfaces," Amer. J. Math. 104 (1982), 1233–1241.

[530] F. Sakai, "Semi-stable curves on algebraic surfaces and logarithmic pluricanonical maps," Math. Ann. 254 (1980), 89–120.

[531] F. Sakai, "Weil divisors on normal surfaces," Duke Math. J. 51 (1984), 877–887.

[532] F. Sakai, "Ample Cartier divisors on normal surfaces," J. Reine Angew. Math. 366 (1986), 121–128.

[533] F. Sakai, "On polarized normal surfaces," Manuscripta Math. 59 (1987), 109–127.

[534] F. Sakai, "Reider-Serrano's method on normal surfaces," in *Algebraic Geometry, Proceedings of Conference on Hyperplane Sections, L'Aquila, Italy, 1988*, ed. by A.J. Sommese, A. Biancofiore, and E.L. Livorni, Lecture Notes in Math. 1417 (1990), 301–319, Springer-Verlag, New York.

[535] Y. Sakane, "On nonsingular hyperplane sections of some Hermitian symmetric spaces," Osaka J. Math. 22 (1985), 107–121.

[536] V.G. Sarkisov, "On the structure of conic bundles," Math. USSR-Izv. 20 (1982), 355–390.

[537] E. Sato, "Varieties which have two projective space bundle structures," J. Math. Kyoto Univ. 25 (1985), 445–457.

[538] E. Sato, "Varieties which contain many lines," Max Planck Institut für Math. preprint, 86-2.

[539] E. Sato, "A variety which contains a $\mathbb{P}^1$-fiber space as an ample divisor," in *Algebraic Geometry and Commutative Algebra, in honor of Masayoshi Nagata*, Kinokuniya (1987), 665–691.

[540] E. Sato, "Smooth projective varieties swept out by large dimensional linear spaces," preprint, 1994.

[541] E. Sato and H. Spindler, "On the structure of 4-folds with a hyperplane section which is a $\mathbb{P}^1$ bundle over a ruled surface," Lecture Notes in Math. 1194 (1986), 145–149, Springer-Verlag, New York.

[542] E. Sato and H. Spindler, "The existence of varieties whose hyperplane section is $\mathbb{P}^r$-bundle," J. Math. Kyoto Univ. 30-3 (1990), 543–557.

[543] T. Sauer, "Smoothing projectively Cohen-Macaulay space curves," Math. Ann. 272 (1985), 83–90.

[544] M. Schneider, "Lefschetzsätze und Hyperconvexität," Invent. Math. 31 (1975), 183–192.

[545] M. Schneider, "3-folds in $\mathbb{P}^5$: classification in low degree and finiteness results," in *Geometry of Complex Projective Varieties, Cetraro, Italy, 1990*, ed. by A. Lanteri, M. Palleschi, and D. Struppa, Seminars and Conferences 9, Mediterranean Press (1993), 275–288.

[546] M. Schneider and A. Tancredi, "Positive vector bundles on complex surfaces," Manuscr. Math. 50 (1985), 133–144.

[547] M. Schneider and A. Tancredi, "Almost-positive vector bundles on projective surfaces," Math. Ann. 280 (1988), 537–547.

[548] M. Schneider and J. Zintl, "The theorem of Barth-Lefschetz as a consequence of Le Potier's vanishing theorem," Manuscr. Math. 80 (1993), 259–263.

[549] G. Scorza, "Le varietá a curve sezioni ellittiche," Ann. Mat. Pura Appl. (3) 15 (1908), 217–273.

[550] B. Segre, "Sulle V_n aventi piu' di ∞^{n-k} S^k, Nota I," Atti Accad. Naz. Lincei Rend. Cl. Sci. Fis. Mat. Natur. (8) V (1948), 193–197; "Nota II," ibid. 275–280.

[551] A. Seidenberg, "The hyperplane sections of normal varieties," Trans. Amer. Math. Soc. 69 (1950), 357–386.

[552] J.G. Semple and L. Roth, *Introduction to Algebraic Geometry*, Clarendon Press, Oxford, (1985).

[553] E. Sernesi, "Topics on families of projective spaces," Queen's Papers in Pure and Appl. Math. 73 (1986).

[554] F. Serrano, "The adjunction mapping and hyperelliptic divisors on a surface," J. Reine Angew. Math. 381 (1987), 90–109.

[555] F. Serrano, "Extension of morphisms defined on a divisor," Math. Ann. 277 (1987), 395–413.

[556] F. Serrano, "Surfaces having a hyperplane section with a special pencil," preprint.

[557] F. Serrano, "Divisors of bielliptic surfaces and embeddings in $\mathbb{P}^4$," Math. Z. 203 (1990), 527–533.

[558] F. Serrano, "Elliptic surfaces with an ample divisor of genus two," Pacific J. Math. 152 (1992), 187–199.

[559] J.P. Serre, "Faisceaux algébriques cohèrents," Ann. of Math. 61 (1955), 197–278.

[560] F. Severi, "La géometrie algébrique italienne: sa rigueur, ses méthodes, ses problèmes," in *Colloque de Géométrie Algébrique, Liège, December 1949*, (1950), 9–55.

[561] N.I. Shepherd-Barron, "Unstable vector bundles and linear systems on surfaces in characteristic p," Invent. Math. 106 (1991), 243–262.

[562] B. Shiffman and A.J. Sommese, *Vanishing theorems on complex manifolds*, Progr. Math. 56 (1985), Birkhäuser, Boston.

[563] V.V. Shokurov, "The nonvanishing theorem," Math. USSR-Izv. 26 (1986), 591–604.

[564] B. Smyth and A.J. Sommese, "On the degree of the Gauss mapping of a submanifold of an Abelian variety," Comment. Math. Helv. 59 (1984), 341–346.

[565] D. Snow, "On the ampleness of homogeneous vector bundles," Trans. Amer. Math. Soc. 294 (1986), 585–594.

[566] D. Snow, 'The nef value and defect of homogeneous line bundles," Trans. Amer. Math. Soc. 340 (1993), 227–241.

[567] A.J. Sommese, "On manifolds that cannot be ample divisors," Math. Ann. 221 (1976), 55–72.

[568] A. J. Sommese, "Theorems of Barth-Lefschetz type for homogeneous complex manifolds," Proc. Nat. Acad. Sci. U.S.A. 74 (1977), 1332–1333.

[569] A.J. Sommese, "Submanifolds of Abelian varieties," Math. Ann. 233 (1978), 229–256.

[570] A.J. Sommese, "Concavity theorems," Math. Ann. 235 (1978), 37–53.

[571] A.J. Sommese, "Hyperplane sections of projective surfaces, I: The adjunction mapping," Duke Math. J. 46 (1979), 377–401.

[572] A.J. Sommese, "Complex subspaces of homogeneous complex manifolds I. Transplanting theorems," Duke Math. J. 46 (1979), 527–548.

[573] A.J. Sommese, "Non-smoothable varieties," Comment. Math. Helv. 54 (1979), 140–146.

[574] A.J. Sommese, "Hyperplane sections," in *Algebraic Geometry, Chicago, 1981*, , ed. by A. Libgober and P. Wagreich, Lecture Notes in Math. 862 (1981), 232–271, Springer-Verlag, New York.

[575] A.J. Sommese, "On the minimality of hyperplane sections of projective threefolds," J. Reine Angew. Math. 329 (1981), 16–41.

[576] A.J. Sommese, "The birational theory of hyperplane sections of projective threefolds," preprint, 1981.

[577] A.J. Sommese, "Ample divisors on threefolds" in *Algebraic Threefolds, Proceedings Varenna, 1981*, ed. by A. Conte, Lecture Notes in Math. 947 (1982), 229–240, Springer-Verlag, New York.

[578] A.J. Sommese, "Complex subspaces of homogeneous complex manifolds II - Homotopy results," Nagoya Math. J. 86 (1982), 101–129.

[579] A.J. Sommese, "Configurations of -2 rational curves on hyperplane sections of projective threefolds," in *Classification of Algebraic and Analytic Manifolds*, ed. by K. Ueno, Progr. Math. 39 (1983), 465–497, Birkhäuser, Boston.

[580] A.J. Sommese, "A convexity theorem," in *Algebraic Geometry, Arcata 1974*, Proc. Sympos. Pure Math. 40 (1983), 497–505.

[581] A.J. Sommese, "Ample divisors on normal Gorenstein surfaces," Abh. Math. Sem. Univ. Hamburg 55 (1985), 151–170.

[582] A.J. Sommese, "Ample divisors on Gorenstein varieties," in *Journées Complexes 1985*, Révue de l'Institut E. Cartan, Nancy 10 (1986), 104–125.

[583] A.J. Sommese, "On the adjunction theoretic structure of projective varieties," in *Proceedings of the Complex Analysis and Algebraic Geometry Conference, Göttingen, 1985*, ed. by H. Grauert, Lecture Notes in Math. 1194 (1986), 175–213, Springer-Verlag, New York.

[584] A.J. Sommese, "On the nonemptiness of the adjoint linear system of a hyperplane section of a threefold," J. Reine Angew. Math. 402 (1989), 211–220; "Erratum," J. Reine Angew. Math. 411 (1990), 122–123.

[585] A.J. Sommese and A. Van de Ven, "Homotopy groups of pullbacks of varieties," Nagoya Math. J. 102 (1986), 79–90.

[586] A.J. Sommese and A. Van de Ven, "On the adjunction mapping," Math. Ann. 278 (1987), 593–603.

[587] J.G. Spandaw, "Noether-Lefschetz problems for vector bundles," Ph.D. Thesis, University of Leiden, The Netherlands (1992).

[588] J.G. Spandaw, "A Noether-Lefschetz theorem for linked surfaces in $\mathbb{P}^4$," Indag. Math. N.S. 3 (1) (1992), 91–112.

[589] E.H. Spanier, *Algebraic topology*, (1966), McGraw-Hill, New York.

[590] R. Strano, "A characterization of complete intersection curves in $\mathbb{P}^3$," Proc. Amer. Math. Soc. 104 (1988), 711–715.

[591] R. Strano, "On generalized Laudal's lemma," in *Complex Projective Geometry*, ed. by G. Ellingsrud, C. Peskine, G. Sacchiero, and S.A. Stromme, London Math. Soc. Lecture Note Ser. 179 (1992), 284–293.

[592] H.P.F. Swinnerton Dyer, "An enumeration of all varieties of degree 4," Amer. J. Math. 95 (1973), 403–418.

[593] S. Tendian, "Deformations of cones and the Gaussian-Wahl map," preprint.

[594] S. Tendian, "Surfaces of degree d with sectional genus g in $\mathbb{P}^{d+1-g}$ and deformations of cones," Duke Math. J. 65 (1992), 157–185.

[595] A.S. Tikhomirov, "Standard bundles on a Hilbert scheme of points on a surface," in *Algebraic Geometry and its Applications, Proceedings of the 8-th Algebraic Geometry Conference, Yaroslavl', 1992*, ed. by A. Tikhomirov and A. Tyurin, Aspects of Math. (1994), Vieweg, Braunschweig.

[596] A.S. Tikhomirov and T.L. Troshina, "Top Segre class of a standard vector bundle E_D^4 on the Hilbert scheme $\mathrm{Hilb}^4 S$ of a surface S," in *Algebraic Geometry and its Applications, Proceedings of the 8-th Algebraic Geometry Conference, Yaroslavl', 1992*, ed. by A. Tikhomirov and A. Tyurin, Aspects of Math. (1994), Vieweg, Braunschweig.

[597] M. Toma, "Three dimensional scrolls in $\mathbb{P}^6$," preprint Bayreuth, 1992.

[598] H. Tsuji, "Stability of tangent bundles of minimal algebraic varieties," Topology 27 (1988), 429–442.

[599] C. Turrini and E. Verderio, "Projective surfaces of small classes," Geom. Dedicata 47 (1993), 1–14.

[600] A.N. Tyurin, "Five lectures on three dimensional varieties," Russian Math. Surveys 27 (1972), 1–53.

[601] A.N. Tyurin, "Cycles, curves and vector bundles on an algebraic surface," Duke Math. J. 54 (1987), 1–26.

[602] K. Ueno, *Classification Theory of Algebraic Varieties and Compact Complex Spaces*, Lecture Notes in Math. 639 (1975), Springer-Verlag, New York.

[603] A. Van de Ven, "On the Chern numbers of surfaces of general type," Invent. Math. 36 (1976), 285–293.

[604] A. Van de Ven, "On the 2-connectedness of very ample divisors on a surface," Duke Math. J. 46 (1979), 403–407.

[605] B.L. Van der Waerden, "Zur algebraischen Geometrie II: Die geraden Linien auf den Hyperflächen des $\mathbb{P}^n$," Math. Ann. 108 (1933), 253–259.

[606] A. Verra, "Smooth surfaces of degree 9 in $G(1,3)$," Manuscripta Math. 62 (1988), 417–435.

[607] E. Viehweg, "Vanishing theorems," J. Reine Angew. Math. 335 (1982), 1–8.

[608] C. Voisin, "Une précision concernant le théorèm de Noether," Math. Ann. 280 (1988), 605–611.

[609] C. Voisin, "Composantes de petit codimension du lieu de Noether-Lefschetz," Comment. Math. Helv. 64 (1989), 515–526.

[610] C. Voisin, "Contre-exemple a une conjecture de Harris," preprint, 1991.

[611] J.M. Wahl, "A cohomological characterization of $\mathbb{P}^n$," Invent. Math. 72 (1983), 315–322.

[612] J.M. Wahl, "The Jacobian algebra of a graded Gorenstein singularity," Duke Math. J. 55 (1987), 843–871.

[613] C. Walter, "Hyperplane sections of curves of small genus," preprint.

[614] C. Walter, "Hyperplane sections of arithmetically Cohen-Macaulay curves," preprint.

[615] A. Weil, "Sur les critères d'équivalence en géométrie algébrique," Math. Ann. 128 (1954), 95–127.

[616] ***, "Correspondence," Amer. J. Math. 79 (1957), 951–952; (A. Weil, Oeuvres Sc. Vol. II, 555–556).

[617] G.W. Whitehead, *Homotopy theory*, M.I.T. Press (1966), Cambridge, Massachusetts.

[618] P.M.H. Wilson, "Fano fourfolds of index greater than one," J. Reine Angew. Math. 379 (1987), 172–181.

[619] J.A. Wiśniewski, "Length of extremal rays and generalized adjunction," Math. Z. 200 (1989), 409–427.

[620] J.A. Wiśniewski, "On a conjecture of Mukai," Manuscripta Math. 68 (1990), 135–141.

[621] J.A. Wiśniewski, "On contractions of extremal rays of Fano manifolds," J. Reine Angew. Math. 417 (1991), 141–157.

[622] J.A. Wiśniewski, "On Fano manifolds of large index," Manuscripta Math. 70 (1991), 145–152.

[623] J.A. Wiśniewski, "On deformation of nef values," Duke Math. J. 64 (1991), 325–332.

[624] J.A. Wiśniewski, "Fano manifolds and quadric bundles," Math. Z. 214 (1993), 261–271.

[625] J.A. Wiśniewski, "A report on Fano manifolds of middle index and $b_2 \geq 2$," in *Projective Geometry with Applications,* ed. by E. Ballico, Lecture Notes in Pure and Applied Math. 166 (1994), 19–26, Marcel Dekker, New York.

[626] Y.-G. Ye and Q. Zhang, "On ample vector bundles whose adjunction bundles are not numerically effective," Duke Math. J. 60 (1990), 671–687.

[627] F.L. Zak, "Surfaces with zero Lefschetz cycles," Math. Zametki 13 (1973), 869–880.

[628] F.L. Zak, "Projections of algebraic varieties," Math. USSR-Sb. 44 (1983), 535–544.

[629] F.L. Zak, "Linear systems of hyperplane sections on varieties of low codimension," Functional Anal. Appl. 19 (1985), 165–173.

[630] F.L. Zak, "Some properties of dual varieties and their application in projective geometry," in *Algebraic Geometry, Proceedings Chicago, 1989*, ed. by S. Bloch, I. Dolgachev, and W. Fulton, Lecture Notes in Math. 1479 (1991), 273–280, Springer-Verlag, New York.

[631] F.L. Zak, "Varieties in codegree three in projective space" in *Geometry of Complex Projective Varieties, Cetraro, Italy, 1990*, ed. by A. Lanteri, M. Palleschi, and D. Struppa, Seminars and Conferences 9, Mediterranean Press (1993), 303–320.

[632] F.L. Zak, *Tangents and secants of algebraic varieties*, Transl. of Math. Monographs 127 (1993), Amer. Math. Soc., Providence, R.I.

[633] O. Zariski, "The theorem of Riemann-Roch for high multiples of an effective divisor on an algebraic surface," Ann. of Math. 76 (1962), 560–615.

[634] O. Zariski, *Algebraic surfaces*, with appendices by S.S. Abhyankar, J. Lipman, and D. Mumford, Ergeb. Math. Grenzgeb. (2), 61 Springer-Verlag, Berlin, (1971).

[635] Q. Zhang, "A theorem on the adjoint system for vector bundles," Manuscripta Math. 70 (1991), 189–202.

[636] Q. Zhang, "Extremal rays in higher dimensional projective varieties," Math. Ann. 291 (1991), 497–504.

[637] Q. Zhang, "On a conjecture of Lanteri-Sommese," preprint, 1993.

[638] Q. Zhang, "A theorem on ample vector bundles," preprint, 1993.

References added in proofs

[1] M. Andreatta and M. Mella, "Contractions on a manifold polarized by an ample vector bundle," preprint, 1994.

[2] E. Ballico and M. Coppens, "On the geometry of projective embeddings of blown-up varieties," preprint 1994.

[3] M.C. Beltrametti and A.J. Sommese, "On the second adjunction mapping. The case of a 1-dimensional image," in preparation.

[4] G.M. Besana, "On polarized surfaces of degree three whose adjoint bundles are not spanned," to appear in Ark. Mat.

[5] G.M. Besana and A. Biancofiore, "Degree eleven manifolds of dimension greater or equal than three," in preparation.

[6] G.M. Besana and S. Di Rocco, "Projective normality of surfaces of degree nine," in preparation.

[7] M.A. De Cataldo, "Codimension two submanifolds of quadrics," in preparation.

[8] G.F. Del Busto, "A Matsusaka-type theorem for surfaces," preprint, 1994.

[9] J.P. Demailly, "L^2 vanishing theorems for positive line bundles and adjunction theory", in *Transcendental Methods in Algebraic Geometry,* CIME Session, Cetraro, Italy, July 1994, Prépublication de l'Institute Fourier, Grenoble, n. 288, 1994.

[10] S. Di Rocco, "k-very ampleness on non singular toric surfaces," preprint, 1994.

[11] T. Fujita, "Approximating Zariski decomposition of big line bundles," Kodai Math. J. 17 (1994), 1–3.

[12] T. Fujita, "Towards a separation theorem of points by adjoint linear systems on polarized threefolds," preprint, 1994.

[13] W. Fulton, "Positive polynomials for filtered ample vector bundles," preprint.

[14] L. Göttsche, "Trisecant formulas for smooth projective varieties," in *Projective Geometry with Applications,* ed. by E. Ballico, Lecture Notes in Pure and Applied Math. 166 (1994), 81–95, Marcel Dekker, New York.

[15] M. Kim, "A Barth-Lefschetz type theorem for branched coverings of Grassmannians," preprint, 1994.

[16] A. Lanteri, "On the class of an elliptic projective surface," to appear in Arch. Math.

[17] A. Lanteri and C. Sacchi, "On some special Fano manifolds," preprint, 1994.

[18] S. Nakamura, "On the third adjoint contractions," preprint, 1994.

[19] S. Nakamura, "Tentative report on the fourth adjoint contractions of divisorial and fiber types," preprint, 1994.

[20] Y. Prokhorov, "Remarks on algebraic threefolds whose hyperplane sections are Enriques surfaces," preprint, 1994.

[21] Y. Prokhorov, "On the existence of good divisors on Fano varieties of coindex 3," preprint, 1994.

[22] E. Sato, "Smooth projective varieties swept out by large dimensional linear spaces," preprint.

[23] H. Terakawa, "On k-spannedness for polarized abelian surfaces," preprint, 1994

Index